Heat Transfer

Heat Transfer

Y V C Rao

Universities Press

Universities Press (India) Private Limited

Registered Office
3-5-747/1/A & 3-6-754/1 Himayatnagar,
Hyderabad 500 029 (A.P.), INDIA
e-mail: info@universitiespress.com

Distributed by
Orient Blackswan Private Limited

Registered Office
3-6-752 Himayatnagar, Hyderabad 500 029 (A.P.), India

Other Offices
Bangalore, Bhopal, Bhubaneshwar, Chennai, Ernakulam, Guwahati,
Hyderabad, Jaipur, Kolkata, Lucknow, Mumbai, New Delhi, Noida, Patna

First published 2001
Reprinted 2008, 2009, 2013

ISBN : 978 81 7371 384 2

Typeset by
OSDATA
Hyderabad 500 029

Printed in India at
B B Press
NOIDA

Published by
Universities Press (India) Private Limited
3-5-747/1/A & 3-6-754/1 Himayatnagar,.
Hyderabad 500 029 (A.P.), INDIA

Dedicated to
Lord Sri Krishna

CONTENTS

PREFACE

Heat transfer is the science dealing with the transfer of energy in the form of heat from one body to another as a result of a temperature difference between them. The science of heat transfer provides an explanation for the different modes of heat transfer and also enables one to predict the rate at which energy transfer takes place under specified conditions.

Nowadays, heat transfer is a compulsory core course in the curriculum of almost all branches of engineering in several engineering and technical institutes and universities. Several textbooks are available on the subject of heat transfer. The merit of a book lies in its well balanced coverage of physical concepts and the demonstration of the underlying principles and concepts through practical situations. This text, which is the outcome of the lecture notes prepared by me is meant primarily for an introductory course in Heat and Mass Transfer. Particular emphasis is placed on precise and logical presentation of the concepts and principles.

The traditional, classical approach has been adopted in presenting the subject matter. The relationship between heat transfer and thermodynamics and the principles of heat transfer, at an elementary level, are presented in Chapter 1. Conduction heat transfer is discussed in detail in Chapters 2–6. Conduction is treated from the analytical, numerical and graphical viewpoints which provide a good understanding of the subject. Radiation heat transfer is discussed in Chapter 7.

Convection heat transfer, both natural and forced convection, is presented in Chapters 8 and 9. Exact mathematical as well as integral analysis of boundary layers are discussed in Chapter 8 and several empirical correlations for predicting the convection heat transfer coefficients are presented in Chapter 9. Condensation and boiling heat transfer are discussed in Chapter 10. The log mean temperature difference and effectiveness approaches are discussed in Chapter 11, which deals with heat exchangers. Finally, a brief introduction to mass transfer is presented in Chapter 12.

Every chapter begins with a set of learning objectives — which is more helpful than a one or two paragraph introduction — outlining the chapter's content. At the end of each chapter, the material covered in that chapter is summarized, followed by a set of review questions and sufficient number of unsolved problems. Heat transfer is a quantitative subject and the reader's ability to understand and to apply the principles and concepts to solve real life problems will be enhanced by illustrating the application of the principles to a few selected problems. Keeping this objective in mind, a sufficient number of solved examples are included in each chapter. Further, several unsolved problems are given at the end of each chapter (with answers at the end of the book) for the readers to solve by applying the principles of heat transfer to gain a better insight into the subject. It is impossible for any textbook to provide enough data through tables or charts to satisfy a practicing engineer. As this book is intended for undergraduate students, sufficient data is included in the appendices. The SI system of units has been adopted.

The material presented in this book can be covered in a one-semester course (or in 38–40 lectures). A solutions manual in which all the problems are solved in detail is also being prepared. This solutions manual, in the author's opinion, will benefit the reader as well as the teacher.

Some errors might have inadvertently crept into this book and the author would greatly appreciate if the readers would bring them to his notice.

The financial support provided by the Curriculum Development Cell, Quality Improvement Programme (at IIT-Kanpur) of the Ministry of Human Resource Development for the preparation of the manuscript is gratefully acknowledged. The author is also grateful to the publishers, Universities Press (India) Ltd., for bringing out the book in its present form in a short period of time. Finally, the preparation of this textbook would not have been possible but for the patience and excellent support of my wife and children. My sincere thanks to them.

Kanpur
May 2001

Y V C RAO

Nomenclature

A	Area of cross-section
b	Width
Bi	Biot modulus
C_f	Friction coefficient
$\bar{C}_f$	Average friction coefficient
C_{th}	Thermal capacity
c	Specific heat
c	Molar concentration
D	Diffusion coefficient
d	Diameter
d_H	Hydraulic diameter
E	Emissive power
Eu	Euler number
ΔE	Electric potential difference
F	View factor or shape factor
F	Correction factor for LMTD
$\mathcal{F}_{12}$	Gray body shape factor
Fo	Fourier number
f	Friction factor
f	Fugacity
G	Irradiation
G	Gibbs free energy
$G_{\max}$	Maximum flow rate per unit of minimum flow area
Gr	Grashof number
Gz	Graetz number
g	Acceleration due to gravity
h	Convection heat transfer coefficient
$\bar{h}$	Average convection heat transfer coefficient
h_{fg}	Latent heat of vaporization
I	Intensity of radiation
i	Electric current
J	Radiosity
J^*	Molar flux relative to local molar average velocity
j	Colburn j factor
j	Mass flux relative to local mass average velocity
K	Thermal conductance
k	Thermal conductivity

k	Mass transfer coefficient
L	Phenomenological coefficient
Le	Lewis number
LMTD	Log Mean Temperature Difference
ℓ	Length
ℓ	Thickness
M	Number of heat flow lanes (in Flux plot)
M	Molar mass
$\dot{m}$	Mass flow rate
N	Number of temperature increments (in Flux plot)
Nu	Nusselt modulus
N	Mole number
N	Molar flux relative to stationary coordinate axis
n	Mass flux relative to stationary coordinate axis
P	Perimeter of fin
P	Wetted perimeter
P	Pressure
P	Parameter to estimate correction factor for LMTD
Pe	Peclet number
Pr	Prandtl number
p	Partial pressure
Q	Electric charge
q	Rate of heat transfer
q'	Heat flux for nucleate boiling
q''	Critical flux for nucleate boiling
$\dot{q}$	Rate of heat generation per unit volume
R	Thermal resistance
R	Universal gas constant
R	Parameter to estimate correction factor for LMTD
Ra	Rayleigh number
Re	Reynolds number
R_f	Fouling factor
r	Radius
r	Cylindrical coordinate
S	Conduction shape factor
S_ℓ	Longitudinal pitch
S_t	Transverse pitch
Sc	Schmidt number
Sh	Sherwood number
St	Stanton number

T Temperature
t Time
U Overall heat transfer coefficient
u Velocity
V Volume
V Local mass average velocity
v Velocity
v^* Local molar average velocity
w Mass transfer
x x-coordinate
x Mole fraction
y y-coordinate
z z-coordinate

Greek Letters

α Thermal diffusivity
α Absorptivity
β Temperature coefficient of thermal conductivity
β Coefficient of volume expansion
δ Thickness of fin
δ Hydrodynamic boundary layer thickness
δ_t Thermal boundary layer thickness
ε Emissivity
ε Heat exchanger effectiveness
η_f Fin efficiency
θ Cylindrical coordinate
μ Dynamic viscosity
μ Chemical potential
ν Frequency
ν Kinematic viscosity
ρ Density
ρ Reflectivity
ρ Mass concentration
σ Stefan–Boltzmann constant
τ Time constant
τ Transmissivity
τ Shear stress
λ Wavelength
ω Solid angle

Subscripts

b	Black body
c	Cold fluid
c	Convection
c	Critical
e	Electrical
f	Film
g	Gray body
h	Hot fluid
i	Inner surface
k	Conduction
m	Mean value
max	Maximum
min	Minimum
o	Outer surface
o	Centre
r	Radiation
s	Surface
s	Saturation
v	Vapor
w	Wall
x	Local value

CHAPTER ONE

INTRODUCTION

Learning Objectives

This chapter will enable the readers to

- define heat.
- know the relation between heat transfer and thermodynamics.
- understand the different modes of heat transfer.
- understand the mechanisms of conduction and convection.
- know the laws of heat transfer—Fourier's law of heat conduction, Stefan–Boltzman law of thermal radiation and Newton's law of cooling.
- define thermal resistance.
- know about the electrical analogy of heat transfer.

1.1 Heat and Thermodynamics

Heat transfer is the science which deals with the rates of energy transfer as heat between two bodies which are initially at different temperatures. The study of heat transfer would be greatly enhanced by a sound understanding of equilibrium, mechanisms or modes of transfer and the laws governing them. It is essential to understand the term *Heat* before one undertakes the study of heat transfer. Heat is intimately related to thermodynamics. We know that all matter possess energy. The energy possessed by a body can be classified as kinetic energy, potential energy and internal energy. Whenever a body interacts with another body or the surroundings, its energy may undergo a change. The interactions between a body and its surroundings can be classified as work interaction and heat interaction.

Thermodynamics deals with the study of equilibrium, energy, its transformation from one form to another and the laws governing such transformation. The laws of thermodynamics are of wide applicability and are used in several branches of engineering and science. The principles of thermodynamics can be summarized in the form of four laws, known as the Zeroth, First, Second and Third laws of thermodynamics. In any scientific or engineering analysis it is necessary to clearly specify the material on which attention is focused. Suppose we are interested in analyzing the motion of a body by applying Newton's laws of motion. For this purpose we draw a free body diagram in which our attention is focused on the body under consideration and represent the various forces acting on it. Similarly, in thermodynamic analysis, we identify the body under analysis by enclosing it in a boundary which separates the body from the rest of the universe or surroundings, and we refer to the body under analysis as a thermodynamic system. That is, a thermodynamic system (or system for simplicity) refers to a definite quantity of matter enclosed by a boundary on which we focus our attention for thermodynamic analysis.

The *Zeroth Law of Thermodynamics* states that if each of two given systems are in thermal equilibrium with a third system, then the two given systems are in thermal equilibrium with each other. This law of thermodynamics forms the basis for the measurement of temperature. The *First Law of Thermodynamics* deals with the conservation of energy and relates to the changes in the energy of the matter under consideration with the energy exchanged as heat and work with the surroundings. In thermodynamics, *heat* is defined as the energy transfer between a system and its surroundings if it occurs by virtue of temperature difference across the boundary which separates the system from its surroundings. In other words, heat is a mode of energy transfer and it is energy in transit. Heat is not a property of the system. Heat is not stored in a body, but energy is stored in a body. Heat is a mode of transferring energy from a system to its surroundings. Therefore, from a thermodynamic point of view the terminology *Heat Transfer* is not appropriate and it should be referred as *Energy Transfer as Heat*. However, we continue to use the terminology *Heat Transfer* because it has been in use for the past several years and will not lead to any confusion.

The *Second Law of Thermodynamics* rules out the possibility of a spontaneous process reversing on its own. Thus it dictates the direction of a process and provides information regarding the feasibility of a process. The transfer of energy as heat from a body at high temperature to a body at low temperature is a spontaneous process. The Second Law of Thermodynamics denies the possibility of a spontaneous process reversing on its own. That is, energy transfer as heat cannot take place spontaneously from a body at low temperature to a body at high temperature. A spontaneous process proceeds till the systems reach a state of equilibrium with the surroundings. Equilibrium is a concept associated with the absence of any tendency for spontaneous changes when the system is isolated. Once the system is allowed to reach a state of equilibrium, all macroscopically observable properties remain constant. In thermodynamics, equilibrium refers to a state of equilibrium with respect to all possible changes—thermal, mechanical and chemical. However, in the study of heat transfer, we are concerned with thermal equilibrium only.

Since no process can violate the laws of thermodynamics, all heat transfer processes obey the laws of thermodynamics. It does not imply that the principles of heat transfer can be deduced from the laws of thermodynamics. Thermodynamics deals with equilibrium and the transformation from one equilibrium state to another equilibrium state through a process. If it is desired to change a system from one equilibrium state to another equilibrium state, while the system is interacting with its surroundings, the laws of thermodynamics can be applied to predict whether such a change is feasible and the amount of energy required for the proposed change (if it is feasible). However, the laws of thermodynamics do not provide any information regarding the rate at which the process takes place. Suppose a lump of steel at high temperature is suddenly immersed in an insulated vessel containing a fixed quantity of water at low temperature. The principles of thermodynamics can be used to predict the final equilibrium state or the final temperature of the lump of steel and water, the amount of energy transferred as heat from the lump of steel to water. However, the laws of thermodynamics cannot be applied to predict the time required to reach the state of equilibrium or the temperatures of the lump of steel and the water as a function of time. The science of heat transfer supplements the required information to predict the rate of heat transfer between the lump of steel and the water. The principles of heat transfer can be used to predict the temperatures of the lump of steel and the water as functions of time.

The rates of heat transfer plays an important role in the estimation of the size of equipment necessary to transfer a specified amount of energy in a given time. The dimensions of heat exchangers, condensers, etc., depend not only on the amount of energy to be transferred, but also on the rate of heat transfer under the specified conditions. An engineer encounters heat transfer problems in the fields of thermal power plants, refrigeration and air-conditioning, thermal control of chemical reactors, etc., and hence it is essential for an engineer to gain a thorough knowledge of heat transfer.

1.2 Modes of Heat Transfer

The transfer of energy as heat from a region of high temperature to a region of low temperature is called heat transfer and the cause of heat transfer is the temperature difference across the boundary of the system. The quantity of energy transferred as heat is not governed by a single or unique relationship. There are three different modes of heat transfer. They are conduction, radiation and convection. Strictly speaking, only conduction and radiation can be called modes of heat transfer, because these processes occur solely due to temperature difference between the two interacting systems. On the other hand, convection depends on the transport of mass from one region to another, in addition to the temperature gradient. In most situations heat transfer occurs by a combination of these modes. A qualitative explanation of these modes of heat transfer is presented in what follows.

Conduction

Conduction is a mode of heat transfer from a region of high temperature to a region of low temperature within a solid, liquid or gas medium or between different mediums which are in physical contact with each other. We know that all matter consists of a large number of molecules which are in random motion. The energy possessed by matter can be broadly classified as macroscopic modes of energy (kinetic energy and potential energy) and microscopic modes of energy (translational energy, rotational energy, vibrational energy, etc., of the molecules).

The macroscopic modes of energy can be estimated in terms of macroscopically measurable quantities like velocity and position of the matter under consideration. On the other hand, the microscopic modes of energy, namely the translational energy, rotational energy and vibrational energy, etc., associated with the molecules constituting the matter cannot be estimated in terms of macroscopically measurable quantities. The total energy possessed by all these microscopic modes is called the *internal energy* of the matter. The internal energy of matter increases with increasing temperature. If a temperature gradient exists in the medium, the molecules in successive layers of the medium will have different internal energies. The molecules with higher internal energy will have higher translational kinetic energy. Then transfer of energy from molecules with higher internal energy to molecules with lower internal energy takes place through collisions in fluids or through diffusion of faster moving electrons from higher to lower temperature regions in metals. If a temperature gradient is maintained in the medium by the addition and removal of energy at different regions, a continuous transfer of energy as heat will take place from the region of high temperature to the region of low temperature.

Radiation

The mode of heat transfer from a body at high temperature to a body at low temperature, when the bodies are not in direct physical contact with each other or when the bodies are separated from each other in space, is called radiation. The energy transfer through radiation exists even if perfect vacuum exists between the two bodies which are separated from each other. Radiation is an electromagnetic phenomenon and the radiation which is propagated as a result of temperature difference is called thermal radiation. The energy transferred by this mechanism is called *radiant heat*. All bodies continuously emit radiant heat which travels with the speed of light. The intensity of the emission depends on the temperature and nature of the surface. Heat transfer by radiation requires no medium for propagation and the motion of radiant heat in space can be described by wave theory. Heat transfer by radiation becomes increasingly important as the temperature of the body increases. At ordinary temperatures, the energy transferred by radiation can be neglected.

Convection

The mode of energy transfer as heat by the combined effect of conduction and material transport is called convection heat transfer. Convection heat transfer from a solid surface to the surrounding fluid takes place through the following steps. (a) Conduction heat transfer takes place from the solid surface to the adjacent fluid particles, thus increasing the internal energy as well as the temperature of the fluid particles. (b) The fluid particles with higher internal energy and temperature move to a region of low temperature and mix with the fluid particles possessing lesser internal energy and transfer a part of their internal energy. Depending on the nature of the forces which cause the material motion, convection heat transfer is classified as *natural convection* or *free convection* and *forced convection*.

If the material transport and mixing of matter occurs due to density differences caused by temperature gradient in the medium, then the convection heat transfer is called *natural* or *free convection*. On the other hand, if the material transport and mixing of matter is induced by some external agency like a pump, blower, fan or impeller, then the convection heat transfer is called *forced convection*. We know that a hot metal plate cools faster when placed in front of a running fan than when it is exposed to stagnant air. The velocity of the air which is blown over the hot plate influences the heat transfer rate. In other words, the effectiveness of heat transfer depends largely on the material transport and mixing of matter. A study of convection heat transfer requires a knowledge of the fluid flow characteristics.

In the analysis of heat transfer problems it is essential to identify the mode of heat transfer as well as whether the process is transient or steady-state. *Steady-state* implies that the properties of a system at any specified location are independent of time and there is no accumulation of energy in the system and hence there is no change in the internal energy of the system. That is, the rate of heat transfer into the system or rate of heat influx is equal to the rate of heat transfer out of the system or rate of heat efflux. In a steady-state the rate of heat transfer in a system is constant and the temperature at any specified location of a system also does not change with time. On the other hand, if the process is *transient* or in an unsteady-state, the temperature at any specified location of the system changes with time and the internal energy of the system also changes with time. In a transient process the rate of heat influx is not equal to the rate of heat efflux and hence there will be accumulation or depletion of energy in the system.

1.3 Heat Transfer Laws

To perform a quantitative analysis of heat transfer problems, it is essential to know the physical laws and relations which govern the different mechanisms of heat transfer. In this section an elementary discussion of the governing equations for conduction, radiation and convection heat transfer is presented.

1.3.1 Conduction Heat Transfer

Joseph Fourier, a French mathematical physicist, made significant contribution to the analytical treatment of conduction heat transfer. *Fourier's law of heat conduction* states that the rate of heat transfer, per unit area, by conduction is proportional to the normal temperature gradient. That is,

$$\frac{q_k}{A} \propto \frac{\partial T}{\partial x} \qquad \text{or} \qquad q_k = -kA\frac{\partial T}{\partial x} \tag{1.1}$$

where

q_k = rate of heat transfer by conduction

A = area of cross-section, through which heat flows, measured perpendicular to the direction of heat flow

k = thermal conductivity of the material

$\dfrac{\partial T}{\partial x}$ = temperature (T) gradient in the direction (x) of heat flow.

The Second Law of Thermodynamics dictates that energy flows spontaneously in the form of heat from a high temperature region to a low temperature region. That is, heat flows in the direction of decreasing temperature. A negative sign is introduced in Eq. (1.1) to make it consistent with the Second Law of Thermodynamics. The coordinate system used in writing Eq. (1.1) is shown in Fig. 1.1.

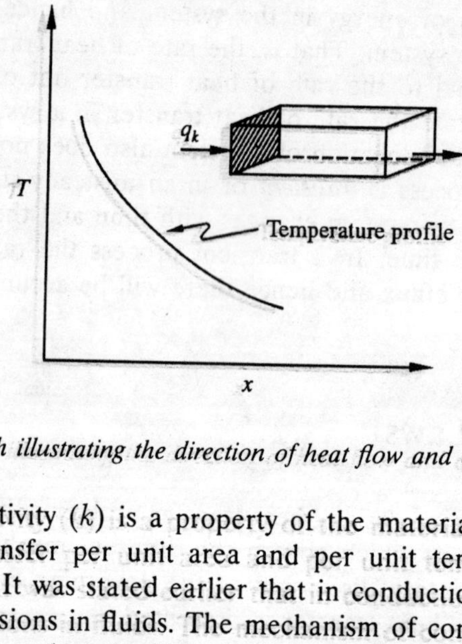

Fig. 1.1 Sketch illustrating the direction of heat flow and coordinate system.

The thermal conductivity (k) is a property of the material and indicates the rate of conduction heat transfer per unit area and per unit temperature gradient. The units of k are W/m K. It was stated earlier that in conduction, energy is transmitted through molecular collisions in fluids. The mechanism of conduction in gases can be qualitatively explained based on the kinetic theory of gases. It may be recalled that the Root Mean Square(RMS) speed of a gas molecule is given by $V_{RMS} = (3kT/m)^{1/2}$, where k is the Boltzmann constant and m is the mass of a molecule. This shows that V_{RMS} depends on the square root of absolute temperature. Hence, on an average the gas molecules move with V_{RMS} and collide with other molecules transferring part of their energy. Therefore, one can expect the thermal conductivity of a gas to depend on the square root of absolute temperature. The thermal conductivity of gases at atmospheric pressure range from 0.02 to 0.2 W/m K. The thermal conductivity of liquids at ordinary temperatures range from 0.2 to 0.5 W/m K. The thermal conductivity of metals vary from 40 to 400 W/m K, whereas that of non-metallic solids vary from 0.04 to 40 W/m K. Materials with high thermal conductivity are called *conductors*, while materials with low thermal conductivity are called *insulators*. In general, the thermal conductivity varies with temperature. The thermal conductivity of metals, non-metals, liquids and gases are presented in Appendix (Table A1).

Now, let us consider steady-state heat transfer through a plane wall as shown in Fig. 1.2.

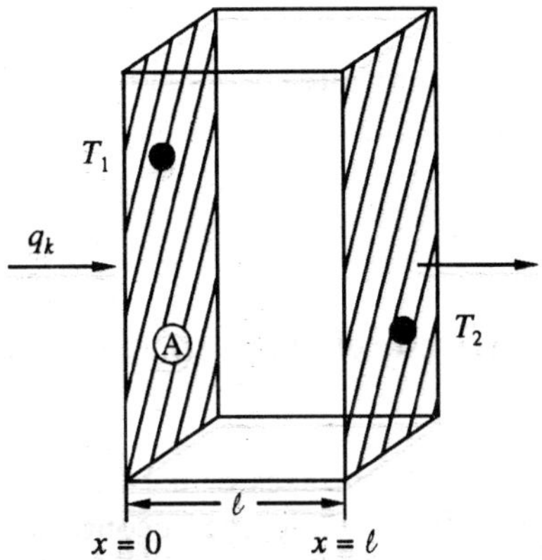

Fig. 1.2 *Temperature profile for steady-state heat conduction through a plane wall.*

Let A = the cross-sectional area of the plane wall, normal to the direction of heat flow

ℓ = thickness of the wall

T_1, T_2 = temperatures of the left-hand and right-hand faces of the wall, respectively.

Then Eq. (1.1) can be rewritten as

$$q_k \int_{x=0}^{x=\ell} dx = -kA \int_{T_1}^{T_2} dT$$

or $q_k \ell = kA(T_1 - T_2)$

$$\text{or} \quad q_k = \frac{kA}{\ell} \Delta T = \frac{\Delta T}{(\ell/kA)} \tag{1.2}$$

In Eq. (1.2) ΔT is the driving force for heat transfer and (ℓ/kA) is the *thermal resistance* (R_k) for conduction. The reciprocal of the thermal resistance is called *thermal conductance* (K_k) for conduction. The thermal resistance and conductance for conduction are given by

$$R_k = \frac{\ell}{kA} \tag{1.3}$$

and

$$K_k = \frac{kA}{\ell} = \frac{1}{R_k} \tag{1.4}$$

The thermal conductance per unit area is called the *unit thermal conductance* for conduction K_k and is given by

$$K_k = \frac{k}{\ell} \tag{1.5}$$

The thermal resistance has the units of kelvin per watt and the thermal conductance has the units of watt per kelvin.

EXAMPLE 1.1 One of the largest surfaces of a 1 m × 1 m × 0.05 m copper plate (k = 385 W/m K) is held at 100°C while the second largest surface is maintained at 0°C. Calculate the rate at which energy flows through the plate and the thermal resistance of the plate.

SOLUTION Fourier's law of heat conduction is given by

$$q_k = -kA\frac{\partial T}{\partial x}$$

or $$q_k = \frac{kA\Delta T}{\ell} = \frac{385 \times 1 \times 100}{0.05} = 770 \text{ kW}$$

The thermal resistance for conduction R_k is given by

$$R_k = \frac{\ell}{kA} = \frac{0.05}{385 \times 1} = 1.299 \times 10^{-4} \text{ K/W}$$

EXAMPLE 1.2 If 2 kW is conducted through a 3 cm thick asbestos sheet (k = 0.74 W/m K) of area 1 m^2 in cross-section, estimate the temperature difference across the surfaces of the sheet and the thermal conductance of the sheet, for conduction.

SOLUTION The integrated form of Fourier's law of heat conduction is given by

$$q_k = \frac{kA\Delta T}{\ell} \quad \text{or} \quad \Delta T = \frac{q_k \ell}{kA}$$

Substituting the values of q_k, ℓ, k and A, we get

$$\Delta T = \frac{2 \times 10^3 \times 0.03}{0.74 \times 1} = 81.1°\text{C}$$

The thermal conductance for conduction K_k is given by

$$K_k = \frac{kA}{\ell} = \frac{0.74 \times 1}{0.03} = 24.67 \text{ W/K}$$

EXAMPLE 1.3 It is necessary to insulate a kitchen oven with cork board (k = 0.043 W/m K) so that the heat losses from the oven surface does not exceed 400 W/m^2 when the inner surface of the oven is at 225°C and the outer surface of the oven is at 40°C. Determine the thickness of insulation required.

SOLUTION A schematic of a kitchen oven is shown in Fig. 1.3.

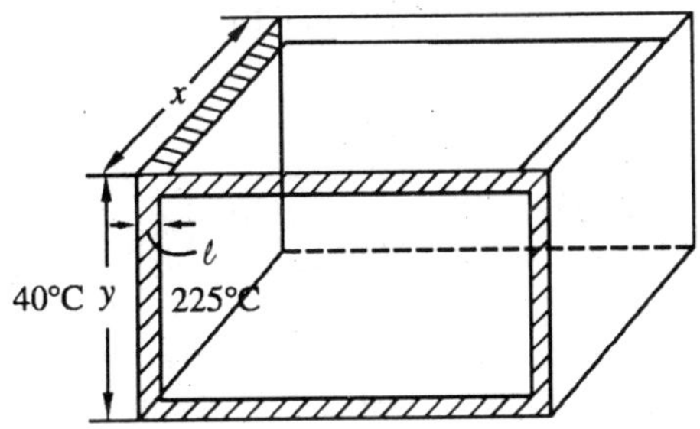

Fig. 1.3 *Schematic of a kitchen oven.*

Consider one side wall, say the left-hand side wall, of dimensions x m × y m as shown in Fig. 1.3. Fourier's law of heat conduction gives

$$\frac{q_k}{A} = \frac{k\Delta T}{\ell} \quad \text{or} \quad \frac{400}{1} = \frac{0.043 \times (225 - 40)}{\ell}$$

or $\ell = 2$ cm, where $A = xy$.

1.3.2 Radiation Heat Transfer

In conduction, energy transfers through a medium. In contrast to this mechanism heat transfer also occurs by electromagnetic radiation through a region where perfect vacuum exists. The *Stefan–Boltzmann law of thermal radiation* for a *black body* or *perfect radiator* states that the rate of radiant energy per unit area is proportional to the fourth power of the temperature of the radiator. That is,

$$q_r = \sigma A T^4 \tag{1.6}$$

where

q_r = rate of radiant energy from the surface

A = surface area of the radiator

σ = Stefan–Boltzmann constant = 5.67×10^{-8} W/m^2 K^4.

The above law was proposed by the Austrian scientist J. Stefan in 1879 based on experimental observations and in 1884 another Austrian scientist L. Boltzmann derived it theoretically. Equation (1.6) shows that any black body at a temperature $T > 0$ emits radiant energy which is independent of the conditions in the surroundings. If we consider two surfaces at temperatures T_1 and T_2, the net radiant exchange between the surfaces will be proportional to the difference in T_1^4 and T_2^4. If a black body at temperature T_1 radiates to another black body at temperature T_2 which completely surrounds it, the second black body completely absorbs the incident radiant energy and emits radiant energy which is proportional to T_2^4. Then the net rate of radiant heat transfer is given by

$$q_r = \sigma A(T_1^4 - T_2^4) \tag{1.7}$$

Real bodies do not act like an ideal radiator or a black body, but emit radiation at a lower rate than a black body. To take into account the *gray* nature of the real bodies, we introduce a factor ε, called *emissivity* which is equal to the ratio of emission from a gray surface to the emission from a perfect black surface when both the surfaces are at the same temperature. Then the net rate of radiation heat transfer from a gray body at temperature T_1 which is enclosed by a black surface at temperature T_2 is given by

$$q_r = \sigma A_1 \varepsilon_1 (T_1^4 - T_2^4) \tag{1.8}$$

In addition to this, it is possible that all the radiant energy from a surface A_1 may not reach the surface A_2 and some radiant energy may be lost to the surroundings. To take this into account it is necessary to introduce another factor related to the geometry of the surfaces. Then the net rate of radiation heat transfer between two gray bodies is given by

$$q_r = \sigma A_1 F_{1-2}(T_1^4 - T_2^4) \tag{1.9}$$

where F_{1-2} is the *view factor* or *shape factor* of surface A_1 with respect to surface A_2.

EXAMPLE 1.4 Two perfectly black surfaces are separated from each other and are enclosed such that all the radiant energy leaving the surface at 1000°C reaches the other surface maintained at 200°C. Calculate the net rate of radiation heat transfer per unit area of the first surface.

SOLUTION We know that the net rate of radiation heat transfer between two black surfaces is given by

$$q_r = \sigma A \left(T_1^4 - T_2^4\right)$$

or $\frac{q_r}{A} = \sigma\left(T_1^4 - T_2^4\right) = 5.67 \times 10^{-8}\left(1273^4 - 473^4\right) = 146 \text{ kW/m}^2$

1.3.3 Convection Heat Transfer

The rate of heat transfer by convection from a surface (with surface area A) at temperature T_s to a fluid at temperature T_∞ is given by

$$q_c = hA\left(T_s - T_\infty\right) \tag{1.10}$$

where h is the convection heat transfer coefficient. The Eq. (1.10) was proposed by Isaac Newton in 1701 and is known as the *Newton's law of cooling*. This relation was proposed as a definition for h, and is not a law for convection heat transfer. For simple systems it is possible to predict the value of h. However, for complex situations it must be determined experimentally. When a fluid is flowing over a heated plate, the velocity of the fluid adjacent to the plate will be zero as a result of viscous action. Then heat transfers from the hot plate to the adjacent layer of fluid by conduction only. Therefore, quite often the convection heat transfer coefficient is called the *film conductance*. Since convection heat transfer is a combined action of conduction and material mixing, the convection heat transfer depends on the viscosity of the fluid in addition to the thermal properties of the fluid. The approximate ranges of convection heat transfer coefficients are presented in Table 1.1.

Table 1.1 *Approximate values of convection heat transfer coefficients.*

Condition	h (W/m^2 K)
Free convection (air)	5
Free convection (water)	850
Forced convection	
Air at low speed	12
Air at high speed	70
Water	3000
Boiling water	3000–30,000
Condensing steam	4000–10,000

1.4 Combined Heat Transfer Processes

In engineering applications we usually come across situations where heat is transferred from a hot body to a cooler body by a combination of conduction, convection and radiation mechanisms. These modes of heat transfer may occur either in series or in parallel. If the temperature of the body is not sufficiently large, the radiation heat transfer may be neglected. Suppose a hot fluid is placed inside a large rectangular container which is exposed to atmospheric air. The heat transfers by convection from the hot fluid to the container wall. Then heat transfers by conduction from the inside surface to the outside surface of the container followed by convection heat transfer from the outside surface of the container to the atmospheric air. A schematic of heat transfer from the hot fluid in the container to the atmospheric air is shown in Fig. 1.4.

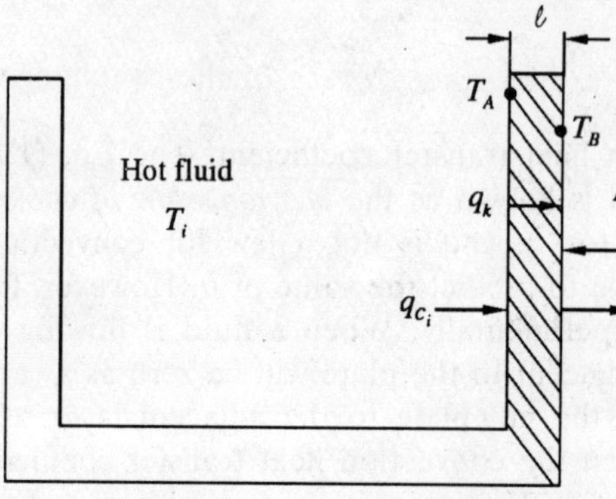

Fig. 1.4 *Schematic of heat transfer through a container wall.*

In this case, the heat transfer modes which are acting in series are convection, conduction and convection. In steady-state conditions the rate of convection heat transfer q_{c_i}, from the hot fluid to the inside surface of the container is equal to the rate of conduction heat transfer q_k, from the inside surface to the outer surface of the container wall, which is also equal to the convection heat transfer q_{c_o}, from the outer surface of the container wall to the air. The heat transfer rates can be expressed as

$$q_{c_i} = h_i A(T_i - T_A) = K_{c_i}(T_i - T_A) = \frac{T_i - T_A}{R_{c_i}} \tag{1.11}$$

$$q_k = \frac{kA(T_A - T_B)}{\ell} = K_k(T_A - T_B) = \frac{T_A - T_B}{R_k} \tag{1.12}$$

$$q_{c_o} = h_o A(T_B - T_o) = K_{c_o}(T_B - T_o) = \frac{T_B - T_o}{R_{c_o}} \tag{1.13}$$

where

h_i = convection heat transfer coefficient from the hot fluid to the inner surface of the container wall

h_o = convection heat transfer coefficient from the outer surface of the container to atmospheric air

A = area of the container wall

ℓ = thickness of the container wall

T_i = temperature of the hot fluid

T_o = temperature of the surrounding air

T_A = temperature of inner surface of the container wall

T_B = temperature of outer surface of the container wall

K_{c_i}, K_{c_o} = thermal conductance for convection at the inner and outer surfaces, respectively

R_{c_i}, R_{c_o} = thermal resistance for convection at the inner and outer surfaces, respectively

$$K_c = hA = \text{thermal conductance for convection} \tag{1.14}$$

$$R_c = \frac{1}{hA} = \text{thermal resistance for convection.} \tag{1.15}$$

In practice, the temperatures of the hot fluid and the air are known but the surface temperatures T_A and T_B are not available. They can be eliminated as given below.

$$(T_i - T_A) + (T_A - T_B) + (T_B - T_o) = qR_{c_i} + qR_k + qR_{c_o}$$

or $\quad T_i - T_o = q(R_{c_i} + R_k + R_{c_o})$

$$\text{or} \quad q = \frac{T_i - T_o}{R_{c_i} + R_k + R_{c_o}} = \frac{\Delta T}{\Sigma R} \tag{1.16}$$

where $\quad q_{c_i} = q_k = q_{c_o} = q.$

By observing the order of magnitude of the individual thermal resistances in Eq. (1.16), sometimes it is possible to ignore some resistances which leads to some simplification in the analysis of heat transfer problems.

There are certain types of problems, specifically in the analysis of heat exchangers, in which it is convenient to calculate the *overall heat transfer coefficient* U. In terms of the overall heat transfer coefficient, the rate of heat transfer q is given by

$$q = UA\Delta T \tag{1.17}$$

Comparing Eqs. (1.16) and (1.17), we get

$$UA = \frac{1}{R_{c_i} + R_k + R_{c_o}} \tag{1.18}$$

As a further example, consider the problem of heat transfer from a furnace wall. Fuel oil or fuel gas is burned inside a furnace producing high temperature gases. The heat losses from the furnace wall can be estimated by identifying the modes of heat transfer from the wall to the surrounding air. Heat flows from the furnace wall to the surrounding air by convection as well as by radiation. In this case, the mechanisms of convection and radiation act in parallel. Then the total heat transfer rate q is given by the sum of the heat transfer rates by convection (q_c), and by radiation (q_r).

1.5 Electrical Analogy to Heat Flow

If two systems obey similar relations with similar boundary conditions, then the systems are said to be analogous. That is, the equation describing the behaviour of one system can be transformed into the equation of the other system by changing the meaning of the variables. For example, the flow of direct current through an electrical resistance is analogous to the flow of heat through a thermal resistance. The rate of flow of electric charge is given by

$$\frac{dQ_e}{dt} = i = \frac{\Delta E}{R_e} \tag{1.19}$$

where

Q_e = electric charge

i = electric current

ΔE = electric potential difference

R_e = electrical resistance.

The rate of heat transfer through a thermal resistance is given by

$$q = \frac{\Delta T}{R} \tag{1.20}$$

Comparing Eqs. (1.19) and (1.20), we find that ΔT is an analogue of ΔE and q is an analogue of i. Therefore, for a given thermal circuit, one can find an equivalent electrical circuit as illustrated below.

Consider the problem of estimating the heat losses through a container wall shown in Fig. 1.4. For clarity, the temperature distribution in the physical system is shown in Fig. 1.5 (a). The electrical analogy can be used for a quantitative analysis of complex heat transfer problems encountered in practice.

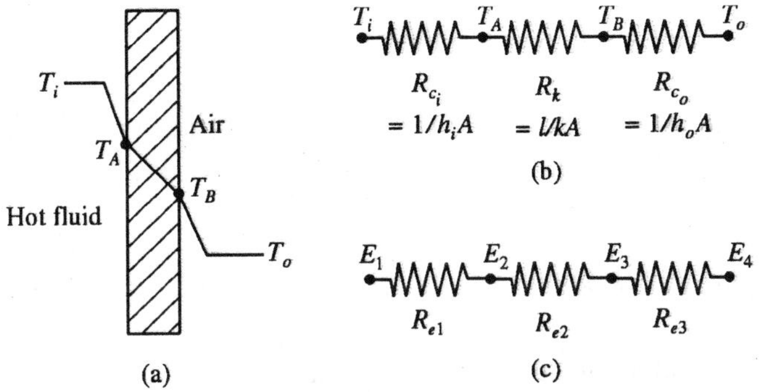

Fig. 1.5 (a) Temperature distribution in the physical system. (b) Thermal circuit for the physical system. (c) Analogous electrical circuit for the thermal circuit shown in (b).

EXAMPLE 1.5 In an oil-fired furnace, the combustion products are at 1000°C and the atmospheric air is at 25°C. The convection heat transfer coefficient between the hot combustion products and the furnace wall is 10 W/m^2 K and the convection heat transfer coefficient between the furnace wall and the atmospheric air is 5 W/m^2 K. The furnace wall is to be made with fireclay brick (k = 1.04 W/m K). Determine the maximum permissible thickness of the fireclay brick wall, if the wall temperature is not to exceed 800°C.

SOLUTION A sketch of the furnace wall along with the given data is shown in Fig. 1.6.

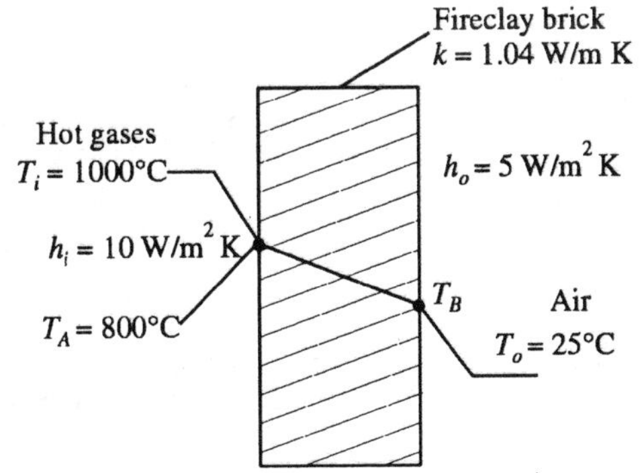

Fig. 1.6 Sketch for Example 1.5.

Let

q = rate of heat transfer from combustion products to atmospheric air, in steady-state

A = cross-sectional area of furnace wall.

In steady-state conditions (q/A) from combustion products to furnace wall is given by

$$\frac{q}{A} = h_i\,(T_i - T_A) = 10\,(1000 - 800) = 2000 \text{ W/m}^2$$

Therefore, in steady-state conditions, (q/A) from the furnace wall to atmospheric air must be equal to 2000 W/m^2, which is also equal to (q/A) for conduction through the fireclay wall. That is,

$$\frac{q}{A} = h_o(T_B - T_o) = 2000 \quad \text{or} \quad 5(T_B - 25) = 2000 \quad \text{or} \quad T_B = 425°\text{C}$$

We know that for conduction through the fireclay wall

$$\frac{q}{A} = \frac{k(T_A - T_B)}{\ell} \quad \text{or} \quad 2000 = \frac{1.04(800 - 425)}{\ell}$$

or ℓ = 0.195 m = 19.5 cm

If the fireclay wall thickness is more than 19.5 cm, the inside wall of the furnace will experience a temperature greater than 800°C.

EXAMPLE 1.6 Rework Example 1.5 by drawing an equivalent electrical circuit. Use 1 volt potential difference to represent 100°C temperature difference and a current of 1 ampere to represent a heat flux of 1000 W/m^2.

SOLUTION The physical system, its thermal circuit and equivalent electrical circuit are shown in Fig. 1.7.

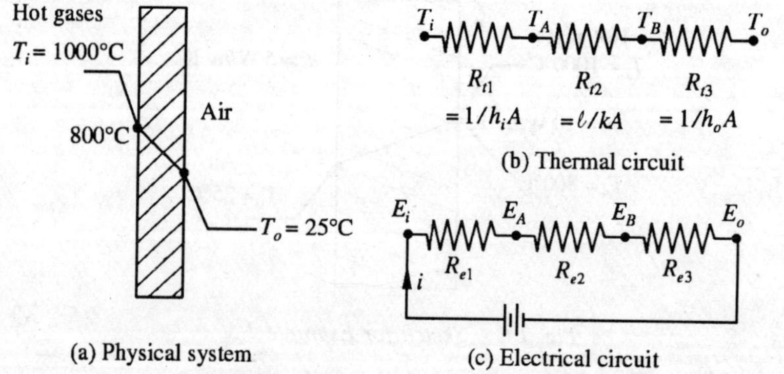

Fig. 1.7 Sketch for Example 1.6. (a) Physical system. (b) Thermal circuit. (c) Electrical circuit.

$$T_i - T_o = 1000 - 25 = 975°\text{C}$$

$$E_i - E_o = 975/100 = 9.75 \text{ V}$$

Rate of heat loss (per unit area) from combustion products to furnace wall by convection

$$= q/A = h_i(T_i - T_A) = 10 \times (1000 - 800) = 2000 \text{ W/m}^2$$

$$\text{Current through circuit} = \frac{2000}{1000} = 2 \text{ amp}$$

The total electrical resistance in the circuit

$$R_e = R_{e1} + R_{e2} + R_{e3}$$

$$i = \frac{V}{R_e} \quad \text{or} \quad R_e = \frac{V}{i} = \frac{9.75}{2} = 4.875 \text{ ohms}$$

For the thermal circuit, the thermal resistances are

$$R_{t1} = \frac{1}{h_i A} = 0.1; \quad R_{t3} = \frac{1}{h_o A} = 0.2; \quad R_{t2} = \frac{\ell}{kA}$$

$$T_i - T_A = 200°\text{C}$$

Therefore, $E_i - E_A = 2$ V

$$R_{e1} = \frac{E_i - E_A}{i} = \frac{2}{2} = 1 \text{ ohm}$$

For the thermal circuit and electrical circuit to be analogous, it is necessary that

$$\frac{R_{e1}}{R_{t1}} = \frac{R_{e3}}{R_{t3}} = \frac{R_{e2}}{R_{t2}} \tag{A}$$

Hence

$$R_{e3} = \frac{R_{e1}}{R_{t1}} R_{t3} = \frac{1 \times 0.2}{0.1} = 2 \text{ ohms}$$

Therefore $R_{e2} = R_e - R_{e1} - R_{e3} = 4.875 - 1 - 2 = 1.875$ ohms

From Eq. (A), we get

$$R_{t2} = \frac{R_{e2} \times R_{t1}}{R_{e1}} = \frac{1.875 \times 0.1}{1} = 0.1875 = \ell/kA$$

or $\ell = 0.1875 \times 1.04 = 0.195 \text{ m} = 19.5$ cm.

1.6 Summary

In thermodynamics, heat is defined as the energy transfer between a system and its surroundings, if it occurs by virtue of the temperature difference across the boundary which separates the system from its surroundings. Heat is energy in transit and it is not a property of the system. All heat transfer processes obey the laws of thermodynamics. The laws of thermodynamics can be applied to predict whether a particular change is feasible or not, and the amount of energy required for a particular change. However, the laws of thermodynamics cannot predict the rate at which the processes take place. The science of heat transfer supplements the necessary information to predict the rates of heat transfer.

There are three modes of heat transfer. They are—conduction, radiation and convection. Conduction is a mode of heat transfer from a region of high temperature to a region of low temperature within a medium (solid, liquid or gas) or between different mediums which are in direct physical contact. In conduction, energy transfer occurs through collisions between molecules in fluids or through diffusion of faster moving electrons in metals. Fourier's law of heat conduction is given by

$$q_k = -kA\frac{dT}{dx} = \frac{kA\Delta T}{\ell} = \frac{\Delta T}{(\ell/kA)}$$

where

q_k = rate of heat transfer by conduction

k = thermal conductivity of the medium

A = area of cross-section through which heat flows, measured perpendicularly to the direction of heat flow

$\frac{dT}{dx}$ = temperature gradient in the medium.

The thermal resistance for conduction, $R_k = \frac{\ell}{kA}$

where ℓ = path length for conduction.

The conductance for conduction, $K_k = \frac{1}{R_k} = \frac{kA}{\ell}$.

The mode of heat transfer from a body at high temperature to a body at low temperature, when the bodies are not in direct physical contact and are separated from each other in space is called radiation. Radiation heat transfer occurs even if perfect vacuum exists in between the bodies. The energy transferred by radiation is called radiant heat. Radiation is an electromagnetic phenomenon. The intensity of the emission depends on the temperature and nature of the surface. The Stefan–Boltzmann law of thermal radiation for a perfect radiator or black body is given by

$$q_r = \sigma A T^4$$

where

q_r = rate of radiant energy from the surface

A = surface area of the radiator

σ = Stefan–Boltzmann constant = 5.67×10^{-8} W/m^2 K^4.

If a black body at temperature T_1 is enclosed in another black surface at temperature T_2, the net rate of radiant heat transfer is given by

$$q_r = \sigma A(T_1^4 - T_2^4)$$

Real or gray bodies emit radiation at a lower rate than a black body. The emissivity ε of a gray body is defined as the ratio of emission from a gray surface to the emission from a perfect black surface at the same temperature. The net rate of radiation heat transfer from a gray surface at temperature T_1, which is enclosed by a black surface at temperature T_2, is given by

$$q_r = \sigma A_1 \varepsilon_1 (T_1^4 - T_2^4)$$

It is possible that all the radiant energy from a surface A_1 may not be intercepted by another surface A_2 and some radiant energy may be lost to the surroundings. In such cases, the net rate of radiation heat transfer is given by

$$q_r = \sigma A_1 F_{1-2} (T_1^4 - T_2^4)$$

where F_{1-2} is the view factor or shape factor of surface A_1 with respect to surface A_2.

The mode of heat transfer by the combined effect of conduction and mixing is called convection. Depending on the nature of forces which causes the mixing, convection heat transfer is classified as natural or free convection and forced convection. The rate of convection heat transfer q_c from a surface of area A at temperature T_s to a fluid at temperature T_∞ is given by Newton's law of cooling as

$$q_c = hA(T_s - T_\infty)$$

where h is the convection heat transfer coefficient which is also known as film conductance. The thermal resistance for convection (R_c) is given by

$$R_c = \frac{1}{hA}$$

The flow of direct current through an electrical resistance is analogous to the flow of heat through a thermal resistance. The two equations

$$\frac{dQ_e}{dt} = i = \frac{\Delta E}{R_e} \quad \text{and} \quad q = \frac{\Delta T}{R_t}$$

are analogous. ΔT is the analogue of ΔE and thermal resistance R_t is the analogue of electrical resistance R_e. It is possible to draw an equivalent electrical circuit from a thermal circuit. The electrical analogy can be used for a quantitative analysis of heat transfer problems.

Review Questions

1. Give a thermodynamic definition of heat.
2. Distinguish between steady-state and equilibrium.
3. Discuss the mechanisms of conduction and convection.
4. Define thermal conductivity of a material.
5. Derive the units of the Stefan–Boltzmann constant.
6. Give a few examples for natural convection and forced convection.
7. Draw the temperature profile for steady-state conduction through a material with constant thermal conductivity.
8. Draw the temperature profile for steady-state conduction through a medium, the thermal conductivity of which increases with increasing temperature.
9. What is a black body?

Problems

1.1 The inside surface of a 25 cm thick brick wall (k = 0.69 W/m K) is at 27°C while the outside surface is at 2°C. Determine the rate of heat loss per unit area through the brick wall.

1.2 A house roof is made of a 15 cm thick concrete-stone mix (k = 1.37 W/m K). On a particular day in summer, the outside surface of the concrete roof is at

55°C while the inside surface is at 30°C. Determine the rate of heat transfer per unit area through the roof.

1.3 The rate of heat transfer through a brick wall (k = 0.69 W/m K) 35 cm thick is found to be 50 W/m^2. Estimate the temperature drop in the wall in steady-state conditions.

1.4 A furnace wall is to be insulated with asbestos cement (k = 0.74 W/m K) such that the temperature drop across the insulation is 700°C when the heat flux is 2 kW/m^2. Determine the thickness of insulation.

1.5 An ice box of 50 cm × 25 cm × 25 cm is insulated with 1 cm thick polystyrene foam (k = 0.028 W/m K). If the outer surface of the ice box is at 40°C, calculate the mass of ice that melts in 4 hours, if the latent heat of fusion of water is 333.43 kJ/kg at 0°C.

1.6 A black body radiates to the surrounding black enclosure which is maintained at 300 K. Calculate the net radiation heat transfer per unit area of the black body, if it is at 400 K, 600 K, 800 K and 1000 K.

1.7 Two large perfect black surfaces are separated by a distance of 10 cm and are held at 1200°C and 200°C, respectively. Estimate the net rate of radiation energy transfer per unit area between the surfaces.

1.8 Two large perfect black surfaces separated by a distance of 20 cm are held at 800°C and 100°C, respectively. Calculate the rate of radiation heat exchange between the surfaces. Calculate the percent reduction in the rate of radiation heat transfer if another perfectly black plate is inserted in between the surfaces which are maintained at 800°C and 100°C. At steady-state conditions what will be the temperature of the centre plate?

1.9 A 50 cm × 20 cm hot plate is maintained at a temperature of 300°C while air at 25°C is blown past the plate. Under these conditions the convection heat transfer coefficient is 20 W/m^2 K. Determine the rate of heat transferred to the air from the hot plate.

1.10 The inner surface of a kettle is at 110°C while water boils in the kettle at 1 bar pressure. If the convection heat transfer is found to be 90 kW/m^2, determine the convection heat transfer coefficient. The saturation temperature of water at 1 bar is 99.6°C.

1.11 The surface temperature of the Sun is approximately 5500°C. Assuming the Sun to be a perfect radiator, estimate the radiation heat flux from the Sun.

1.12 A flowing stream of hot fluid at temperature T_1 is separated from another flowing stream of cold fluid at temperature T_2 by a metal plate of thickness ℓ, having a thermal conductivity k. Derive an expression to determine the rate of heat transfer from the hot fluid to the cold fluid, if h_1 and h_2 are the convection heat transfer coefficients on the hot fluid surface and cold fluid surface of the plate, respectively.

1.13 One side of a 10 cm thick steel plate (k = 54 W/m K) is maintained at 100°C while the second side is held at 90°C by blowing air at 25°C. Determine the convection heat transfer coefficient between the plate and air. Also calculate the heat flux through the plate.

1.14 A cubical box of inside dimensions 1 m × 1 m × 1 m is completely evacuated. The box is made of a steel plate (k = 54 W/m K) of thickness 2 cm. When a current of 10 amperes at 230 volts is continuously passed through an electrical resistance placed inside the box, the outer surface of the box reaches a steady-state temperature of 50°C. Determine the steady-state temperature of the inside surface of the box and the convection heat transfer coefficient between the outer surface of the box and the surrounding air, if the surrounding air is at 25°C.

1.15 A kitchen oven is to be designed such that the outside surface does not exceed 40°C while it is operated at the highest baking temperature of 280°C, when the kitchen air temperature is 25°C. Assume that the inside and outside convection heat transfer coefficients are 30 W/m^2 K. Determine the thickness of glass wool (k = 0.038 W/m K) insulation required for the oven.

Chapter Two

One-Dimensional Steady-State Heat Conduction

Learning Objectives

This chapter will enable the readers to

- analyze steady-state one-dimensional heat conduction in a plane wall with constant as well as with temperature dependent thermal conductivity.
- analyze steady-state one-dimensional heat transfer through a composite wall in which the thermal resistances are connected in series as well as in parallel.
- draw thermal circuits and equivalent electrical circuits for heat transfer through composite walls.
- determine the rates of heat transfer through a hollow cylinder and a multilayered cylinder.
- estimate the critical thickness of insulation of a pipe.
- analyze heat transfer through spherical shells and estimate the critical radius of insulation of a spherical vessel.
- determine the temperature distribution in a flat plate and a long cylinder with uniform heat generation.

In this chapter, one-dimensional steady-state heat conduction in a few selected simple geometrical configurations is discussed. Quite often an engineer deals with infinitely long walls through which heat flows, long hollow cylinders through which either a cold fluid or a hot fluid flows thereby transferring energy to the surroundings, and spherical shells in which a fluid at a temperature different from that of the surroundings is held. In such cases it is reasonable to assume that heat flows in only one direction and a knowledge of steady-state one-dimensional heat conduction is of considerable practical importance.

2.1 Steady-State Heat Conduction in a Plane Wall

The simplest case of one-dimensional heat flow is the conduction heat transfer through a plane wall. If the two surfaces of a plane wall of thickness ℓ are maintained at temperatures T_1 and T_2, respectively, then the rate of conduction heat transfer through a homogeneous material is given by

$$q_k = \frac{kA(T_1 - T_2)}{\ell} = \frac{\Delta T}{(\ell/kA)} = \frac{\Delta T}{R_k} \tag{2.1}$$

In general, the thermal conductivity k varies with temperature. The temperature dependence of thermal conductivity for several materials is given by

$$k(T) = k_0(1 + \beta_k T)$$

where

k_0 = thermal conductivity at zero degree kelvin

β_k = temperature coefficient of thermal conductivity

$k(T)$ = thermal conductivity at temperature T.

If the temperature range to which a material is exposed is small, it is reasonable to neglect the variation of k with temperature. The rate of conduction heat transfer is given by Fourier's law of heat conduction

$$q_k = -kA\frac{dT}{dx} \tag{2.2}$$

If energy transfer by conduction takes place in a plane whose thermal conductivity strongly depends on temperature, then Eq. (2.2) can be rewritten as

$$\frac{q_k}{A}\int_0^\ell dx = -\int_{T_1}^{T_2} k(T)dT = -\int_{T_1}^{T_2} k_0(1 + \beta_k T)dT$$

$$= -k_0\left[T + \frac{\beta_k T^2}{2}\right]_{T_1}^{T_2} = k_0\left[(T_1 - T_2) + \frac{\beta_k}{2}(T_1^2 - T_2^2)\right]$$

or

$$\frac{q_k \ell}{A} = k_0(T_1 - T_2)\left[1 + \beta_k\frac{T_1 + T_2}{2}\right]$$

or

$$q_k = (T_1 - T_2)\frac{k_o A}{\ell}\left[1 + \beta_k\frac{T_1 + T_2}{2}\right] = \frac{k_m A \Delta T}{\ell} \tag{2.3}$$

where the thermal conductivity at the mean temperature $(T_1 + T_2)/2$ is given by

$$k_m = k_0 \left[1 + \beta_k \frac{T_1 + T_2}{2}\right] \tag{2.4}$$

EXAMPLE 2.1 The thermal conductivity of pure aluminium is 214.6 W/m K and 228.5 W/m K at 200°C and 300°C, respectively. One surface of a large slab of aluminium of thickness 40 cm is exposed to 300°C while the other surface is maintained at 200°C. Assuming that the thermal conductivity of aluminium varies linearly in this temperature range, determine the rate of conduction heat transfer per unit area through the slab. Also determine the values of k_0 and β_k for aluminium.

SOLUTION If the thermal conductivity is a linear function of temperature,

$$k_m = \frac{214.6 + 228.5}{2} = 221.55 \text{ W/m K} \quad \text{at} \quad T = 250°\text{C}$$

$$q_k = \frac{\Delta T}{(\ell/k_m A)} = \frac{100}{0.4/(221.55 \times 1)} = 55.387 \text{ kW/m}^2$$

We know that

$$k(T) = k_0(1 + \beta_k T)$$

$$k(573) = 228.5 = k_0(1 + 573\beta_k) \tag{A}$$

$$k(473) = 214.6 = k_0(1 + 473\beta_k) \tag{B}$$

Subtracting Eq. (B) from Eq. (A), we get

$$13.9 = 100 k_0 \beta_k \quad \text{or} \quad k_0\beta_k = 0.139$$

$$k(573) = 228.5 = k_0 + 573 k_0 \beta_k = k_0 + 573 \times 0.139$$

or

$$k_0 = 148.853 \text{ W/m K} \quad \text{and} \quad \beta_k = 9.338 \times 10^{-4} \text{ K}^{-1}$$

2.2 Steady-State Heat Conduction in a Composite Wall

A composite wall consisting of three layers of different substances having different thicknesses and thermal conductivities is shown in Fig. 2.1. Such composite walls are commonly encountered while dealing with furnaces where the inner layer of thickness

ℓ_1 is made of firebrick (k_1). The intermediate layer of thickness ℓ_2 is made of insulating brick (k_2) which is followed by a masonry brick (k_3) of thickness ℓ_3. Suppose the two surfaces of the composite wall are maintained at temperatures T_i and T_o, respectively, and we are interested in determining the heat flow by conduction through the wall.

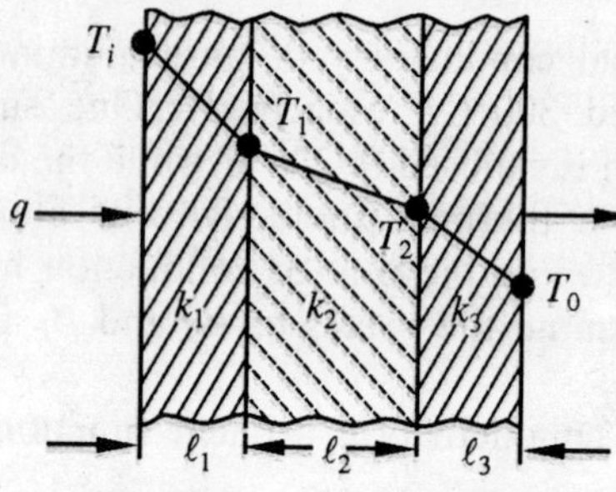

Fig. 2.1 Temperature profile in a composite wall.

In steady-state, there is a continuous heat flow from one layer to the other layer. Hence

$$q = \frac{k_1 A}{\ell_1}(T_i - T_1) = \frac{k_2 A}{\ell_2}(T_1 - T_2) = \frac{k_3 A}{\ell_3}(T_2 - T_o) \tag{2.5}$$

or

$$q = \frac{T_i - T_1}{R_1} = \frac{T_1 - T_2}{R_2} = \frac{T_2 - T_o}{R_3} \tag{2.6}$$

where

$$R_i = \ell_i / k_i A \tag{2.7}$$

= thermal resistance for conduction for layer i.

From Eq. (2.6), one can write

$$T_i - T_1 = qR_1 \tag{2.8}$$

$$T_1 - T_2 = qR_2 \tag{2.9}$$

$$T_2 - T_o = qR_3 \tag{2.10}$$

Adding Eqs. (2.8) to (2.10), we get

$$T_i - T_o = q(R_1 + R_2 + R_3) = q(\Sigma R_i)$$

or

$$q = \frac{T_i - T_o}{R_1 + R_2 + R_3} = \frac{\Delta T}{\Sigma R_i} \tag{2.11}$$

where ΔT is the overall temperature difference across the composite wall. Equation (2.11) shows that the rate of heat flow is the ratio of the overall temperature difference (driving potential for heat transfer) to the total thermal resistance. This is analogous to the flow of direct current through several electrical resistances connected in series. The thermal circuit and the equivalent electrical circuit for the physical situation shown in Fig. 2.1 is presented in Fig. 2.2.

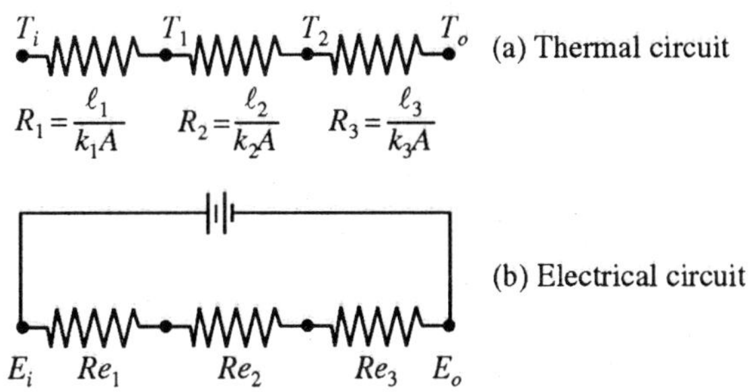

Fig. 2.2 (a) Thermal circuit. (b) Electrical circuit for the physical situation of Fig. 2.1.

EXAMPLE 2.2 A furnace wall consists of three layers. The inner layer of 10 cm thickness is made of firebrick (k = 1.04 W/m K). The intermediate layer of 25 cm thickness is made of masonry brick (k = 0.69 W/m K) followed by a 5 cm thick concrete wall (k = 1.37 W/m K). When the furnace is in continuous operation, the inner surface of the furnace is at 800°C while the outer concrete surface is at 50°C. Calculate (a) the rate of heat loss per unit area of the wall (b) the temperature at the interface of the firebrick and masonry brick and (c) the temperature at the interface of the masonry brick and concrete.

SOLUTION Let A = 1 m^2

(a) Thermal resistance of firebrick layer

$$R_1 = \frac{\ell_1}{k_1 A} = \frac{0.1}{1.04} = 0.0962 \text{ K/W}$$

Thermal resistance of masonry brick layer

$$R_2 = \frac{\ell_2}{k_2 A} = \frac{0.25}{0.69} = 0.3623 \text{ K/W}$$

Thermal resistance of concrete layer

Thermal resistance of concrete layer

$$R_3 = \frac{\ell_3}{k_3 A} = \frac{0.05}{1.37} = 0.0365 \text{ K/W}$$

$$\Sigma R_i = 0.0962 + 0.3623 + 0.0365 = 0.495 \text{ K/W}$$

$$q = \frac{\Delta T}{\Sigma R_i} = \frac{750}{0.495} = 1515 \text{ W/m}^2$$

(b) Temperature difference across firebrick layer is given by

$$qR_1 = 1515 \times 0.0962 = 145.7°\text{C}$$

Temperature difference across masonry brick layer is given by

$$qR_2 = 1515 \times 0.3623 = 548.9°\text{C}$$

Therefore, temperature at the interface of the firebrick layer and the masonry brick layer = 800 − 145.7 = 654.3°C

(c) Temperature at the interface of the masonry brick and concrete = 654.3 − 548.9 = 105.4°C

In practice, the surface temperatures are generally not known but the temperatures of the hot gases and cold air to which the surfaces are exposed can be easily estimated or measured. In such cases, it is necessary to include the convection heat transfer from the gases to the surfaces. In a furnace, the hot gases are produced by burning fuel oil or gas and it is possible to estimate the temperature of the combustion products. In general, one does not measure the temperature of the inner surface of a furnace wall. We shall assume that the combustion products are at a known constant temperature T_i in the furnace and the surroundings are at constant temperature T_o. Let h_i and h_o denote the convection heat transfer coefficients or film coefficients at the inner and outer surfaces of the furnace wall. In steady-state, there will be a constant heat flow from the hot combustion products to the surrounding air. A schematic of a furnace wall is shown in Fig. 2.3.

Since the rate of heat flow through all the sections are the same in steady-state, we can write

$$q = h_i A(T_i - T_1) = \frac{k_1 A}{\ell_1}(T_1 - T_2) = \frac{k_2 A}{\ell_2}(T_2 - T_3)$$

$$= \frac{k_3 A}{\ell_3}(T_3 - T_4) = h_o A(T_4 - T_o) \tag{2.12}$$

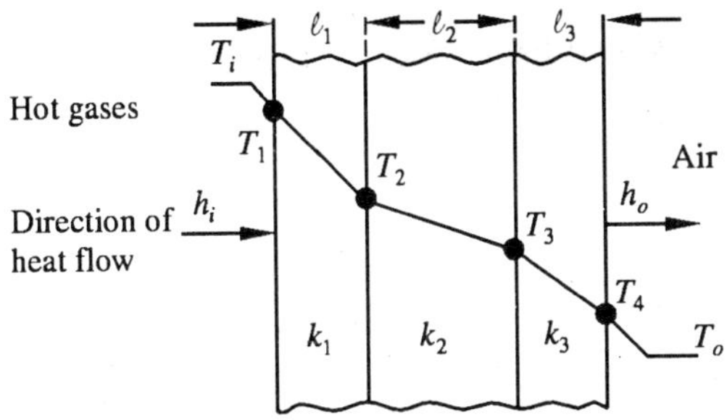

Fig. 2.3 *Temperature profile for heat flow through a furnace wall.*

or

$$q = \frac{T_i - T_1}{R_1} = \frac{T_1 - T_2}{R_2} = \frac{T_2 - T_3}{R_3} = \frac{T_3 - T_4}{R_4} = \frac{T_4 - T_o}{R_5} \tag{2.13}$$

where

R_1 = $1/h_i A$ = thermal resistance for conduction heat transfer at the inner surface

R_2 = $\ell_1/k_1 A$ = thermal resistance for conduction heat transfer in the first layer of the material with thermal conductivity k_1.

Similarly,

R_3, R_4 = thermal resistances for conduction heat transfer in the second and third layers, respectively

R_5 = $1/h_o A$ = thermal resistance for convection heat transfer at the exterior surface.

The temperature differences in Eq. (2.13) can be expressed as

$$\left.\begin{aligned} T_i - T_1 &= q\,R_1 \\ T_1 - T_2 &= q\,R_2 \\ T_2 - T_3 &= q\,R_3 \\ T_3 - T_4 &= q\,R_4 \\ T_4 - T_o &= q\,R_5 \end{aligned}\right\} \tag{2.14}$$

Adding the left-hand and right-hand sides of Eqs. (2.14), we get

$$T_i - T_o = q(R_1 + R_2 + R_3 + R_4 + R_5) = q(\Sigma R)$$

or

$$q = \frac{T_i - T_o}{R_1 + R_2 + R_3 + R_4 + R_5} = \frac{\Delta T}{\Sigma R} \tag{2.15}$$

or

$$q = \frac{T_i - T_o}{\dfrac{1}{h_i A} + \dfrac{\ell_1}{k_1 A} + \dfrac{\ell_2}{k_2 A} + \dfrac{\ell_3}{k_3 A} + \dfrac{1}{h_o A}} \tag{2.16}$$

For the physical situation shown in Fig. 2.3, the thermal circuit and the electrical circuit are shown in Fig. 2.4.

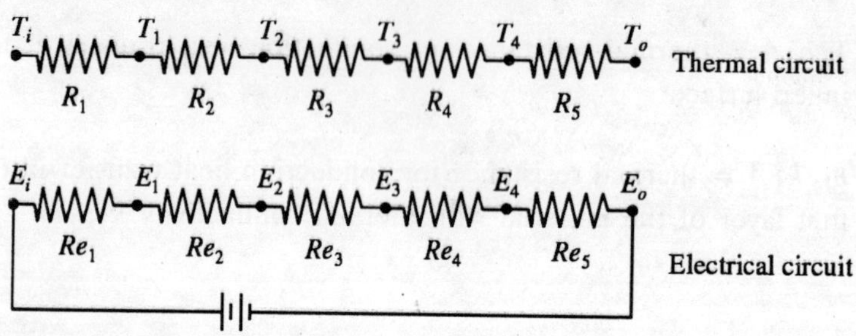

Fig. 2.4 *(a) Thermal circuit and (b) electrical circuit for the physical situation of Fig. 2.3.*

EXAMPLE 2.3 Consider the furnace wall described in Example 2.2. If the combustion products inside the furnace are at 1000°C while the atmospheric air is at 25°C, the convection heat transfer coefficients at the inner and outer surfaces of the wall are 50 W/m^2 K and 15 W/m^2 K, respectively. Determine

(a) the rate of heat transfer per unit area from the hot gases to the air.

(b) the temperature of the interior and exterior surfaces of the furnace wall.

SOLUTION (a) It is given that

$$T_i = 1000°\text{C}; \quad T_o = 25°\text{C}$$

$$h_i = 50 \text{ W/m}^2 \text{ K}; \quad h_o = 15 \text{ W/m}^2 \text{ K}$$

$$k_1 = 1.04 \text{ W/m K}; \quad k_2 = 0.69 \text{ W/m K}; \quad k_3 = 1.37 \text{ W/m K}$$

$$\ell_1 = 10 \text{ cm}; \quad \ell_2 = 25 \text{ cm}; \quad \text{and} \quad \ell_3 = 5 \text{ cm} \quad \text{[see Fig. 2.3]}$$

Let $A = 1 \text{ m}^2$, then

$$R_1 = \frac{1}{h_i A} = \frac{1}{50} = 0.02 \text{ K/W}$$

$$R_2 = \frac{\ell_1}{k_1 A} = \frac{0.1}{1.04} = 0.0962 \text{ K/W}$$

$$R_3 = \frac{\ell_2}{k_2 A} = \frac{0.25}{0.69} = 0.3623 \text{ K/W}$$

$$R_4 = \frac{\ell_3}{k_3 A} = \frac{0.05}{1.37} = 0.0365 \text{ K/W and}$$

$$R_5 = \frac{1}{h_o A} = \frac{1}{15} = 0.0667 \text{ K/W}$$

$$\Sigma R = 0.02 + 0.0962 + 0.3623 + 0.0365 + 0.0667 = 0.5817 \text{ K/W}$$

$$q = \frac{\Delta T}{\Sigma R} = \frac{975}{0.5817} = 1.6761 \text{ kW/m}^2$$

(b) We know that

$$q/A = h_i(T_i - T_1)$$

or $\quad 1676.1 = 50(1000 - T_1) \quad$ or $\quad T_1 = 966.5°\text{C}$

From Eqs. (2.14) and (2.15), we obtain

$$\frac{T_4 - T_o}{\Delta T} = \frac{R_5}{\Sigma R}$$

or $\quad T_4 - T_o = \dfrac{\Delta T \times R_5}{\Sigma R} = \dfrac{975 \times 0.0667}{0.5817}$

or $\quad T_4 - 25 = 111.8 \quad$ or $\quad T_4 = 136.8°\text{C}.$

EXAMPLE 2.4 Draw (a) the thermal circuit and (b) electrical circuit for Example 2.3. Use a potential difference of 1 volt for 10°C temperature difference and 1 ampere to represent a heat flow rate of 1 kW/m^2.

SOLUTION In Example 2.3, we found that $q = 1.6761 \text{ kW/m}^2$; $R_1 = 0.02$ K/W.

$R_2 = 0.0962$ K/W; $R_3 = 0.3623$ K/W; $R_4 = 0.0365$ K/W

and $R_5 = 0.0667$ K/W $\sum R = 0.5817$ K/W

(a) We know that $T_1 = 966.5°C$ and $T_4 = 136.8°C$ [see Example 2.3].

From Eqs. (2.14) and (2.15) we find that

$$\frac{T_1 - T_2}{\Delta T} = \frac{R_2}{\Sigma R}$$

or

$$T_1 - T_2 = \frac{\Delta T \times R_2}{\Sigma R} = \frac{975 \times 0.0962}{0.5817} = 161.2°C$$

or $T_2 = 966.5 - 161.2 = 805.3°C$

Similarly $T_2 - T_3 = \dfrac{\Delta T \times R_3}{\Sigma R} = \dfrac{975 \times 0.3623}{0.5817} = 607.3°C$

or $T_3 = 805.3 - 607.3 = 198°C$

The thermal circuit is shown in Fig. 2.5(a).

(b) Potential difference $(\Delta E) = \dfrac{975}{10} = 97.5$ volts

Current through the circuit $= i = 1.6761$ amperes

Therefore, total electrical resistance is given by

$$R_e = \frac{\Delta E}{i} = \frac{97.5}{1.6761} = 58.17 \text{ ohms}$$

Since the thermal circuit and electrical circuit are analogous, the ratio of each electrical resistance to the total electrical resistance must be equal to the corresponding ratio in the thermal circuit.

$$\frac{R_{e1}}{R_e} = \frac{R_1}{\Sigma R}$$

or $R_{e1} = \dfrac{R_1 \times R_e}{\Sigma R} = \dfrac{0.02 \times 58.17}{0.5817} = 2$ ohms

$$R_{e2} = \frac{R_2 \times R_e}{\Sigma R} = 9.62 \text{ ohms}$$

Similarly $R_{e3} = 36.23$ ohms; $R_{e4} = 3.65$ ohms and $R_{e5} = 6.67$ ohms

The electrical circuit is shown in Fig. 2.5(b).

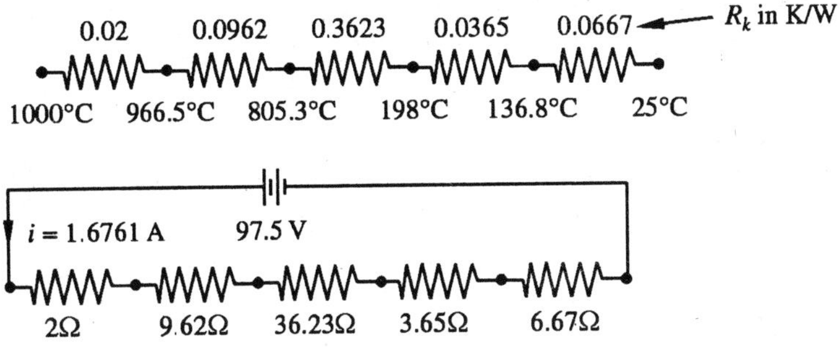

Fig. 2.5 *(a) Thermal circuit, (b) electrical circuit for Example 2.4.*

2.3 Heat Conduction in a Composite Wall with Parallel and Series Heat Flow Paths

In several practical applications one encounters a combination of parallel and series connected heat flow paths. An example of such a plane wall is shown in Fig. 2.6 in which the wall is made of three sections—the first section of thickness ℓ_1 is made of a material with thermal conductivity k_1, the intermediate section of thickness ℓ_2 is made of two different materials with thermal conductivities k_1 and k_2. The material with thermal conductivity k_1 has a breadth of b_1 while the other material has a breadth b_2. The third section is made of a material with thermal conductivity k_3.

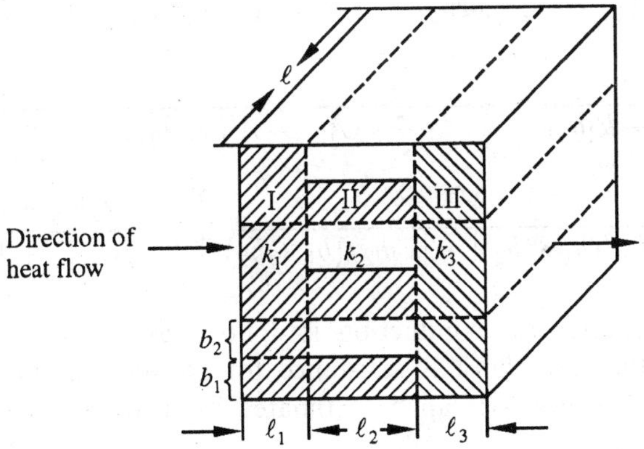

Fig. 2.6 *A composite wall with parallel and series connected heat flow paths. The intermediate layer II consists of two thermal paths in parallel—one with thermal conductivity k_1 having a breadth b_1 and the other material of breadth b_2 has thermal conductivity k_2. The composite wall has three thermal paths I, II and III connected in series.*

The thermal resistance of the intermediate layer is given by

$$\frac{1}{R_2} = \frac{1}{R_{21}} + \frac{1}{R_{22}} = \frac{R_{21} + R_{22}}{R_{21}R_{22}}$$

or

$$R_2 = \frac{R_{21}R_{22}}{R_{21} + R_{22}} \tag{2.17}$$

where

$$R_{21} = \frac{\ell_2}{k_1 A_1}$$

$$R_{22} = \frac{\ell_2}{k_2 A_2}$$

$$A_1 = b_1 \ell \quad \text{and} \quad A_2 = b_2 \ell$$

$$A = A_1 + A_2 = \ell(b_1 + b_2). \tag{2.18}$$

Substituting Eqs. (2.18) in Eq. (2.17), we get

$$R_2 = \frac{(\ell_2/k_1A_1)(\ell_2/k_2A_2)}{(\ell_2/k_1A_1) + (\ell_2/k_2A_2)} = \frac{\ell_2^2}{\ell\ell_2(k_1b_1 + k_2b_2)}$$

$$= \frac{\ell_2^2/(k_1k_2b_1b_2)}{(\ell_2/k_1\ell b_1) + (\ell_2/k_2\ell b_2)} = \frac{\ell_2^2/(k_1k_2\ell^2 b_1b_2)}{\ell_2\ell(k_1b_1 + k_2b_2)/(k_1k_2\ell^2 b_1b_2)}$$

$$= \frac{\ell_2}{\ell(k_1b_1 + k_2b_2)} = \frac{\ell_2}{k_1A_1 + k_2A_2} = \frac{\ell_2}{A[(k_1A_1/A) + (k_2A_2/A)]}$$

$$= \frac{\ell_2}{A}\left[\frac{1}{[k_1b_1/(b_1 + b_2)] + [k_2b_2/(b_1 + b_2)]}\right] = \frac{\ell_2(b_1 + b_2)}{A(k_1b_1 + k_2b_2)} \tag{2.19}$$

The thermal resistance for conduction is defined as $R_k = \frac{\ell}{kA}$. A close look at Eq. (2.19) reveals that the thermal resistance for parallel combined paths (k_1 with width b_1 and k_2 with width b_2) can be estimated by using a mean value of k which is given by

$$k_{\text{mean}} = \frac{k_1b_1 + k_2b_2}{b_1 + b_2} \tag{2.20}$$

The thermal resistance of the layers I and III are given by

$$R_1 = \frac{\ell_1}{k_1 A} \qquad \text{and} \qquad R_3 = \frac{\ell_3}{k_3 A} \tag{2.21}$$

Then the total resistance for the three layers in series is given by

$$R = R_1 + R_2 + R_3 = \frac{\ell_1}{k_1 A} + \frac{\ell_2 (b_1 + b_2)}{A(k_1 b_1 + k_2 b_2)} + \frac{\ell_3}{k_3 A} \tag{2.22}$$

The rate of heat transfer through the wall can be estimated by using the relation

$$q = \frac{\Delta T}{R}$$

EXAMPLE 2.5 A large furnace has a composite wall, the details of which are given in Fig. 2.7. The interior surface of the furnace wall is exposed to hot gases at 1000°C while the exterior surface is exposed to the atmospheric air at 25°C. If the convection heat transfer coefficients at the interior and exterior surfaces are 60 W/m^2 K and 20 W/m^2 K, respectively, calculate the rate of heat transfer, per unit area of the furnace wall, from the hot gases to the atmospheric air.

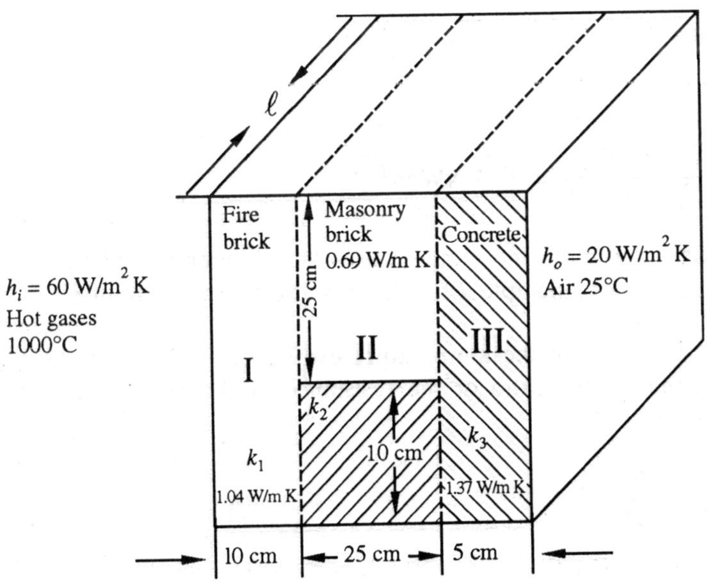

Fig. 2.7 Segment of a multilayered furnace wall.

SOLUTION Take A = 1 m^2. Thermal resistance of layer I is given by

$$R_1 = \frac{\ell_1}{k_1 A} = \frac{0.1}{1.04} = 0.0962 \text{ K/W}$$

Thermal resistance of layer II is given by

$$R_2 = \frac{\ell_2(b_1 + b_2)}{A(k_1 b_1 + k_2 b_2)} = \frac{0.25(0.1 + 0.25)}{(1.04 \times 0.1 + 0.69 \times 0.25)} = 0.3165 \text{ K/W}$$

Thermal resistance of layer III is given by

$$R_3 = \frac{\ell_3}{k_3 A} = \frac{0.05}{1.37} = 0.0365 \text{ K/W}$$

Thermal resistance for convection at the interior surface is

$$R_4 = \frac{1}{h_i A} = \frac{1}{60} = 0.0167 \text{ K/W}$$

Thermal resistance for convection at the exterior surface is

$$R_5 = \frac{1}{h_o A} = \frac{1}{20} = 0.05 \text{ K/W}$$

Total resistance $R = \Sigma R_i = 0.0962 + 0.3165 + 0.0365 + 0.0167 + 0.05 = 0.5159$ K/W

Overall temperature difference $\Delta T = 1000 - 25 = 975$°C

Therefore,

$$q = \frac{\Delta T}{R} = \frac{975}{0.5159} = 1.89 \text{ kW/m}^2$$

EXAMPLE 2.6 Estimate the temperatures of the interior and the exterior surfaces of the furnace wall of Example 2.5 and draw the thermal circuit.

SOLUTION The ratio of the temperature drop across a layer to the overall temperature is the same as the ratio of the resistance of the layer to the total resistance. Therefore,

$$\frac{T_i - T_1}{T_i - T_o} = \frac{R_4}{R}$$

or $$\frac{1000 - T_1}{1000 - 25} = \frac{0.0167}{0.5159}$$

or $T_1 = 968.4$°C

$$\frac{T_4 - T_o}{T_i - T_o} = \frac{T_4 - 25}{1000 - 25} = \frac{R_5}{R} = \frac{0.05}{0.5159}$$

or $T_4 = 119.5°\text{C}$

Interior surface temperature = $T_1 = 968.4°\text{C}$

Exterior surface temperature = $T_4 = 119.5°\text{C}$

$$R_{21} = \frac{\ell_2}{k_1 A_1} = \frac{\ell_2 A}{k_1 A A_1} = \frac{\ell_2 \ell(b_1 + b_2)}{k_1 A \ell b_1} = \frac{\ell_2(b_1 + b_2)}{k_1 A b_1} = \frac{0.25(0.1 + 0.25)}{1.04 \times 0.1}$$

or $R_{21} = 0.8413$ K/W

Similarly $R_{22} = \dfrac{\ell_2 (b_1 + b_2)}{k_2 A b_2} = \dfrac{0.25(0.1 + 0.25)}{0.69 \times 0.25} = 0.5072$ K/W

The thermal circuit is shown in Fig. 2.8.

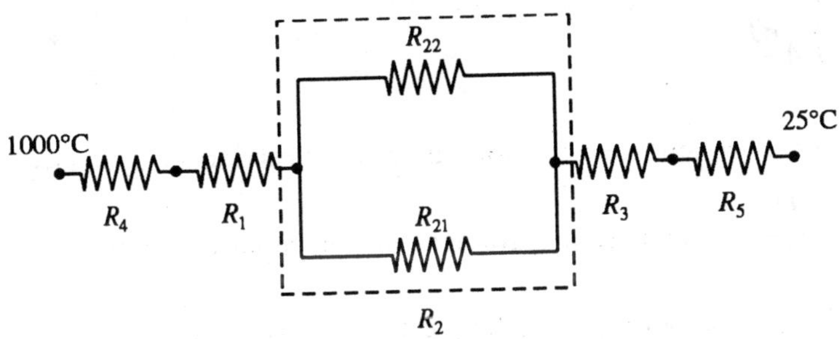

Fig. 2.8 *Thermal circuit for Example 2.6.*

2.4 Steady-State Heat Conduction in a Hollow Cylinder

Heat conduction through pipes and through pipe insulation is an example of heat conduction in a hollow cylinder. If the cylinder is sufficiently long and homogeneous and the inner and outer surfaces of the pipe are maintained at uniform and constant temperatures, then heat flows radially from the inner surface to the outer surface of the pipe. Figure 2.9 shows a sketch of radial heat flow in a hollow cylinder.

Let

T_i, T_o = temperatures of the inner and outer surfaces of the cylinder, respectively

r_i, r_o = inner and outer radii of the cylinder, respectively

ℓ = length of the pipe.

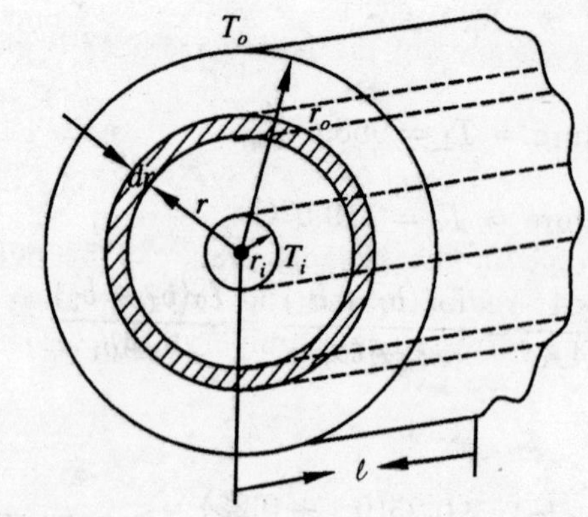

Fig. 2.9 Sketch showing radial heat flow through a hollow cylinder.

Fourier's law of heat conduction gives

$$q_k = -kA\frac{dT}{dr} \tag{2.23}$$

where $A = 2\pi r\ell$ = surface area normal to the direction of heat flow

$\frac{dT}{dr}$ = temperature gradient in the radial direction.

The rate of heat conduction [Eq. (2.23)] can be rewritten as

$$q_k = -k2\pi r\ell\frac{dT}{dr} \tag{2.24}$$

or $$\int_{T_i}^{T_o} dT = -\frac{q_k}{k2\pi\ell}\int_{r_i}^{r_o}\frac{dr}{r}$$

or $$T_i - T_o = \frac{q_k}{k2\pi\ell}\ln\left(\frac{r_o}{r_i}\right)$$

or $$q_k = \frac{2\pi k\ell(T_i - T_o)}{\ln(r_o/r_i)} = \frac{(T_i - T_o)}{[\ln(r_o/r_i)]/(2\pi k\ell)} \tag{2.25}$$

Equation (2.25) shows that the rate of conduction heat transfer in a cylindrical shell is inversely proportional to the natural logarithm of the ratio of the outer and inner radii. Equation (2.25) can be rewritten as

$$q_k = \frac{(T_i - T_o)}{[\ln(r_o/r_i)]/2\pi k\ell} = \frac{\Delta T}{R_k} \tag{2.26}$$

where R_k, the thermal resistance of the hollow cylinder is given by

$$R_k = \frac{\ln(r_o/r_i)}{2\pi k\ell} \tag{2.27}$$

The temperature profile in the cylinder wall can be determined by integrating Eq. (2.24) from the inner radius r_i and the corresponding temperature T_i to any arbitrary radius r and the corresponding temperature T. That is,

$$q_k = -2\pi r\ell k\frac{dT}{dr}$$

or

$$\int_{T_i}^{T} dT = -\frac{q_k}{2\pi k\ell}\int_{r_i}^{r}\frac{dr}{r}$$

or

$$T(r) = T_i - \frac{q_k}{2\pi k\ell}\ln(r/r_i) \tag{2.28}$$

Substituting for $q_k/(2\pi k\ell)$ from Eq. (2.25) in Eq. (2.28), we get

$$T(r) = T_i - \frac{(T_i - T_o)}{\ln(r_o/r_i)}\ln\left(\frac{r}{r_i}\right) \tag{2.29}$$

The temperature profile in the wall of a hollow cylinder is shown in Fig. 2.10.

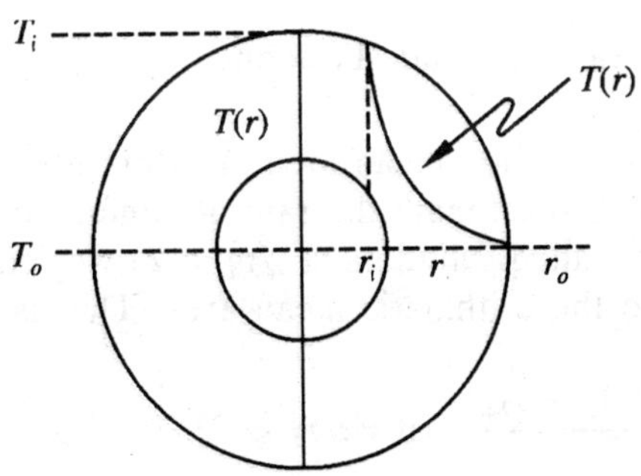

Fig. 2.10 *Temperature profile in the wall of a hollow cylinder.*

For a plane, the rate of conduction heat transfer is given by the simple expression

$$q_k = \frac{kA\Delta T}{\ell} \tag{1.2}$$

The simplicity of this expression prompts one to express the rate of conduction heat transfer through curved surfaces also in a manner similar to Eq. (1.2) by introducing a few terms. In a hollow cylinder, the length of the heat flow path is

$$\ell = (r_o - r_i) \tag{2.30}$$

Comparing Eqs. (1.2) and (2.26), we find

$$\frac{k\bar{A}\Delta T}{\ell} = \frac{k\bar{A}\Delta T}{r_o - r_i} = \frac{T_i - T_o}{[\ln(r_o/r_i)]/2\pi k\ell} = \frac{2\pi k\ell\Delta T}{\ln(r_o/r_i)}$$

or

$$\frac{k\bar{A}\Delta T}{(r_o - r_i)} = \frac{2\pi k\ell\Delta T}{\ln(r_o/r_i)} \tag{2.31}$$

or

$$\bar{A} = \frac{2\pi(r_o - r_i)\ell}{\ln(r_o/r_i)} = \frac{2\pi r_o\ell - 2\pi r_i\ell}{\ln\left(\dfrac{2\pi r_o\ell}{2\pi r_i\ell}\right)} = \frac{A_o - A_i}{\ln(A_o/A_i)} \tag{2.32}$$

where

A_i = area of the inside surface of a hollow cylinder

A_o = area of the outside surface of a hollow cylinder

$\bar{A}$ = log mean area of A_i and A_o defined by Eq. (2.32).

Thus we find that by using the log mean area $\bar{A}$ in place of A and $(r_o - r_i)$ in place of ℓ, Eq. (1.2) can be used to estimate the rate of conduction heat transfer through a hollow cylinder. For thin walled cylinders $(r_o/r_i < 2)$ we find that the log mean area is approximately equal to the arithmetic mean area. That is,

$$\bar{A} = \frac{A_o - A_i}{\ln(A_o/A_i)} \simeq \frac{A_o + A_i}{2} \quad (\text{if } r_o/r_i < 2) \tag{2.33}$$

EXAMPLE 2.7 The inner surface of a nickel-steel cylinder (k = 19 W/m K) of ID = 1 cm and OD = 7 cm is maintained at a constant temperature of 110°C, while the outer surface is held at 50°C. Calculate the heat lost per metre length of the pipe.

SOLUTION Given that $r_i = 0.5 \times 10^{-2}$ m; $r_o = 3.5 \times 10^{-2}$ m; $T_i = 110$°C; $T_o = 50$°C.

We know that the rate of conduction heat transfer in a hollow cylinder is given by

$$q_k = \frac{k\bar{A}\Delta T}{\ell} = \frac{k\bar{A}\Delta T}{(r_o - r_i)}$$

$$A_o = 2\pi r_o \ell = 2\pi(3.5 \times 10^{-2}) \times 1 = 0.2199 \text{ m}^2$$

$$A_i = 2\pi r_i \ell = 2\pi(0.5 \times 10^{-2}) \times 1 = 0.0314 \text{ m}^2$$

$$\bar{A} = \frac{A_o - A_i}{\ln(A_o/A_i)} = \frac{0.2199 - 0.0314}{\ln(0.2199/0.0314)} = 0.0969 \text{ m}^2$$

Therefore,

$$q_k = \frac{k\bar{A}\Delta T}{r_o - r_i} = \frac{19 \times 0.0969 \times (110 - 50)}{(3.5 - 0.5)10^{-2}} = 3.682 \text{ kW}$$

EXAMPLE 2.8 Calculate the temperature at r = 1, 2 and 3 cm for the cylinder of Example 2.7.

SOLUTION We know that

$$T(r) = T_i - \frac{(T_i - T_o)}{\ln(r_o/r_i)}\ln(r/r_i) = 110 - \frac{60}{\ln(3.5/0.5)}\ln(r/r_i)$$

or

$$T(r) = 110 - 30.8339\ln(r/r_i) \tag{A}$$

Substituting the value of r in Eq. (A), we get

$$T(r = 1 \text{ cm}) = 110 - 30.8339\ln\left(\frac{1}{0.5}\right) = 88.6°\text{C}$$

$$T(r = 2 \text{ cm}) = 110 - 30.8339\ln\left(\frac{2}{0.5}\right) = 67.3°\text{C}$$

$$T(r = 3 \text{ cm}) = 110 - 30.8339\ln\left(\frac{3}{0.5}\right) = 54.8°\text{C}$$

EXAMPLE 2.9 In a steam radiator, steam at 110°C is passed through a tungsten-steel (k = 54 W/m K) tube of ID = 2 cm and OD = 3 cm. The surrounding air is at 5°C. The convection heat transfer coefficients at the inner and outer surfaces of the tube are 5000 W/m^2 K and 10 W/m^2 K, respectively. Calculate the rate of heat transfer per metre length of the tube and the temperatures of the inner and outer surfaces of the tube in steady-state.

SOLUTION In steady-state, the rate of heat transfer per unit length of the tube remains constant and the same amount of heat passes from the steam to the inner surface, from the inner surface to the outer surface of the tube, and from the outer surface to the air.

$$q = h_{c_i} A_i (T_i - T_{r_i}) = \frac{T_i - T_{r_i}}{R_{c_i}} \quad \text{or} \quad T_i - T_{r_i} = \frac{q}{h_{c_i} A_i} = q R_{c_i}$$

$$q = \frac{T_{r_i} - T_{r_o}}{[\ln(r_o/r_i)/2\pi k \ell]} = \frac{T_{r_i} - T_{r_o}}{R_k} \quad \text{or} \quad T_{r_i} - T_{r_o} = q R_k$$

$$q = h_{c_o} A_o (T_{r_o} - T_o) = \frac{T_{r_o} - T_o}{R_{c_o}} \quad \text{or} \quad T_{r_o} - T_o = q R_{c_o}$$

Adding the above equations, we get

$$T_i - T_o = q(R_{c_i} + R_k + R_{c_o})$$

or

$$q = \frac{T_i - T_o}{R_{c_i} + R_k + R_{c_o}} \tag{A}$$

$$R_{c_i} = \frac{1}{h_{c_i} A_i} = \frac{1}{h_{c_i} \times 2\pi r_i \ell} = \frac{1}{5000 \times 2\pi \times 1 \times 10^{-2} \times 1} = 3.183 \times 10^{-3} \text{ K/W}$$

$$R_k = \frac{\ln(r_o/r_i)}{2\pi k \ell} = \frac{\ln(1.5/1)}{2\pi \times 54 \times 1} = 1.195 \times 10^{-3} \text{ K/W}$$

$$R_{c_o} = \frac{1}{h_{c_o} A_o} = \frac{1}{h_{c_o} 2\pi r_o \ell} = \frac{1}{10 \times 2\pi \times 1.5 \times 10^{-2} \times 1} = 1.061 \text{ K/W}$$

Total resistance

$$R = \Sigma R_i = 3.183 \times 10^{-3} + 1.195 \times 10^{-3} + 1.061 = 1.065 \text{ K/W}$$

Therefore,

$$q = \frac{T_i - T_o}{R} = \frac{110 - 5}{1.065} = 98.59 \text{ W}$$

$$T_i - T_{r_i} = qR_{c_i}$$

or $\quad 110 - T_{r_i} = 98.59 \times 3.183 \times 10^{-3}$

or $\quad T_{r_i} = 109.7°\text{C}$

Temperature of the inner surface = 109.7°C

$$T_{r_o} - T_o = qR_{c_o} \quad \text{or} \quad T_{r_o} - 5 = 98.59 \times 1.061$$

or $\quad T_{r_o} = 109.6°\text{C}$

Temperature of the outer surface = 109.6°C

2.5 Heat Conduction in a Composite Cylinder

Radial heat flow through the walls of a composite cylinder (or concentric cylinders) having different layers made of materials with different thermal conductivities is commonly encountered in several industrial applications. A typical example is the flow of heat from steam flowing through an insulated pipe. A composite cylinder having three layers made of materials with different thermal conductivities is shown in Fig. 2.11.

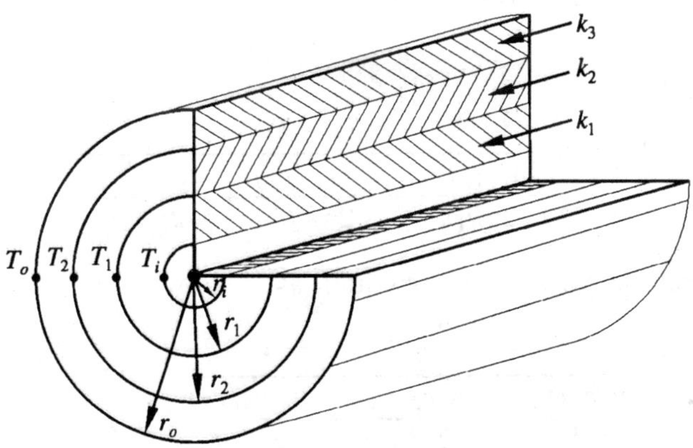

Fig. 2.11 *Sketch of a composite cylinder.*

In steady-state, the rate of heat flow through each layer is the same. For the inner layer

$$q = \frac{2\pi k_1 \ell}{\ln(r_1/r_i)}(T_i - T_1) = \frac{T_i - T_1}{R_1} \tag{2.34}$$

or

$$T_i - T_1 = qR_1 \tag{2.35}$$

For the intermediate layer

$$q = \frac{2\pi k_2 \ell}{\ln(r_2/r_1)}(T_1 - T_2) = \frac{T_1 - T_2}{R_2} \tag{2.36}$$

or

$$T_1 - T_2 = qR_2 \tag{2.37}$$

For the outer layer

$$q = \frac{2\pi k_3 \ell}{\ln(r_o/r_2)}(T_2 - T_o) = \frac{T_2 - T_o}{R_3} \tag{2.38}$$

or

$$T_2 - T_o = qR_3 \tag{2.39}$$

Adding the left-hand and right-hand sides of Eqs. (2.35), (2.37) and (2.39), we get

$$T_i - T_o = q(R_1 + R_2 + R_3) = qR \tag{2.40}$$

or

$$q = \frac{T_i - T_o}{R_1 + R_2 + R_3} = \frac{T_i - T_o}{\dfrac{\ln(r_1/r_i)}{2\pi k_1 \ell} + \dfrac{\ln(r_2/r_1)}{2\pi k_2 \ell} + \dfrac{\ln(r_o/r_2)}{2\pi k_3 \ell}} \tag{2.41}$$

Equation (2.41) can be conveniently rewritten as

$$q = UA\Delta T = \frac{\Delta T}{\dfrac{\ln(r_1/r_i)}{2\pi k_1 \ell} + \dfrac{\ln(r_2/r_1)}{2\pi k_2 \ell} + \dfrac{\ln(r_o/r_2)}{2\pi k_3 \ell}} \tag{2.42}$$

where U is the overall heat transfer coefficient. The overall heat transfer coefficient can be defined based on any area A. Since it is more practical and convenient to measure the outside diameter of the pipe, the overall heat transfer coefficient is usually defined based on the outer surface area A_o.

$$U_o A_o \Delta T = U_o(2\pi r_o \ell)\Delta T = \frac{\Delta T}{\dfrac{\ln(r_1/r_i)}{2\pi k_1 \ell} + \dfrac{\ln(r_2/r_1)}{2\pi k_2 \ell} + \dfrac{\ln(r_o/r_2)}{2\pi k_3 \ell}}$$

or

$$U_o = \frac{1}{\dfrac{r_o \ln(r_1/r_i)}{k_1} + \dfrac{r_o \ln(r_2/r_1)}{k_2} + \dfrac{r_o \ln(r_o/r_2)}{k_3}} \tag{2.43}$$

Consider a practical situation in which a hot fluid at temperature T_i is flowing through a composite cylinder consisting of two layers. Let the outer surface of the cylinder be exposed to a cold fluid at temperature T_o. The convection heat transfer coefficient at the inner and outer surfaces of the cylinder are h_i and h_o, respectively. Then, following the reasoning adopted above, we can obtain

$$q = U_o A_o \Delta T = \frac{\Delta T}{\dfrac{1}{2\pi r_i \ell h_i} + \dfrac{\ln(r_1/r_i)}{2\pi k_1 \ell} + \dfrac{\ln(r_o/r_1)}{2\pi k_2 \ell} + \dfrac{1}{2\pi r_o \ell h_o}} \tag{2.44}$$

and

$$U_o = \frac{1}{\dfrac{r_o}{r_i h_i} + \dfrac{r_o \ln(r_1/r_i)}{k_1} + \dfrac{r_o \ln(r_o/r_1)}{k_2} + \dfrac{1}{h_o}} \tag{2.45}$$

EXAMPLE 2.10 Steam at 120°C is flowing through a wrought-iron (k= 59 W/m K) tube of ID = 5 cm and OD = 7 cm which is covered with 1 cm thick asbestos (k = 0.1105 W/m K) insulation. If the convection heat transfer coefficients at the inner and outer surfaces of the tube are 200 W/m^2 K and 10 W/m^2 K, respectively, and the atmospheric air is at 25°C, estimate the rate of heat losses from steam per metre length of the tube. Assume that the steam in the tube is held at a constant temperature.

SOLUTION The rate of heat transfer from steam to air can be calculated using Eqs. (2.44) and (2.45).

$$U_o = \frac{1}{\dfrac{r_o}{r_i h_i} + \dfrac{r_o \ln(r_1/r_i)}{k_1} + \dfrac{r_o \ln(r_o/r_1)}{k_2} + \dfrac{1}{h_o}}$$

$$= \frac{1}{\dfrac{4.5}{2.5 \times 200} + \dfrac{4.5 \times 10^{-2} \ln(3.5/2.5)}{59} + \dfrac{4.5 \times 10^{-2} \ln(4.5/3.5)}{0.1105} + \dfrac{1}{10}}$$

$$= \frac{1}{9 \times 10^{-3} + 2.566 \times 10^{-4} + 0.1023 + 0.1} = \frac{1}{0.2116} = 4.7259 \text{ W/m}^2 \text{ K}$$

$$q = U_o A_o \Delta T = 4.7259 \times 2\pi \times 4.5 \times 10^{-2} \times 1 \times (120 - 25) = 127 \text{ W}$$

2.6 Critical Thickness of Insulation

For a plane, addition of an insulating layer or increasing the thickness of the insulation layer increases the thermal resistance for conduction heat transfer and hence results in lower rates of heat transfer. However, for small pipes the addition of an insulating layer may not necessarily reduce the rate of heat transfer. We know that the thermal resistance for conduction in the case of a hollow cylinder of inner radius r_i and outer radius r_o is given by

$$R_k = \frac{\ln(r_o/r_i)}{2\pi k\ell}$$

That is, increasing the value of r_o increases the thermal resistance for conduction. We also know that the thermal resistance for convection is given by

$$R_c = \frac{1}{h_o 2\pi r_o \ell}$$

That is, an increase in the value of r_o reduces the thermal resistance for convection heat transfer. Since the total thermal resistance for heat transfer is the sum of R_k and R_c, it is possible to increase the heat losses from the pipe by adding a small layer of insulation. If the thickness of insulation is further increased, the heat losses from the pipe gradually increase to a higher value than the heat loss from a bare pipe. The thickness of insulation $(r_o - r_i)$, which yields the maximum heat transfer rate from a pipe is called the *critical thickness of insulation.* If the thickness of insulation is less than the critical thickness of insulation, the heat transfer rate from an insulated pipe will be more than that of a bare pipe. On the other hand, if the thickness of insulation is greater than the critical thickness of insulation, the heat transfer rate from the insulated pipe reduces and becomes less than that from bare pipe. This information is widely used for insulating the current-carrying wires and cables to increase the heat transfer rate which in turn increases the current-carrying capacity of the wire.

Consider a pipe with outer radius r_o held at constant temperature T_i as shown in Fig. 2.12. The pipe is insulated with a material of thermal conductivity k, to a thickness of $(r_o - r_i)$. The outer surface of the insulation is exposed to a fluid at constant temperature T_o. The convection heat transfer coefficient between the outer surface and fluid is h_o.

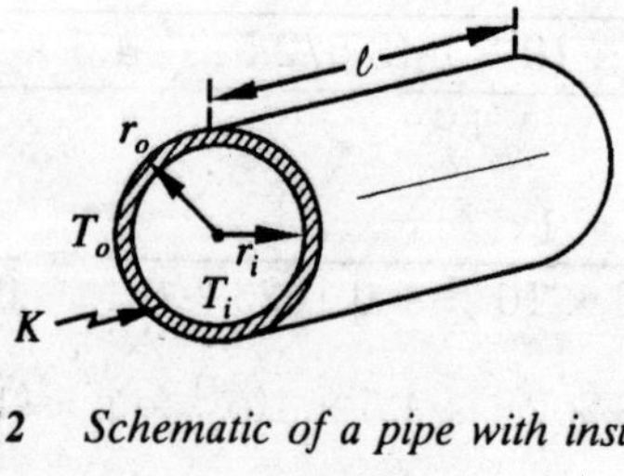

Fig. 2.12 *Schematic of a pipe with insulation.*

Thermal resistance for conduction is given by

$$R_k = \frac{\ln(r_o/r_1)}{2\pi k\ell}$$

Thermal resistance for convection is given by $R_c = \dfrac{1}{h_o 2\pi r_o \ell}$

Total resistance for heat transfer $R = R_k + R_c$

or

$$R = \frac{\ln(r_o/r_1)}{2\pi k\ell} + \frac{1}{h_o 2\pi r_o \ell} \tag{2.46}$$

and

$$q = \frac{T_i - T_o}{R} = \frac{\Delta T}{R} \tag{2.47}$$

For a fixed value of ΔT the rate of heat transfer q will be maximum when the total resistance R given by Eq. (2.46) is minimum. Therefore, the critical thickness of insulation can be determined by minimizing R with respect to r_o, the outer radius of insulation. That is,

$$\frac{dR}{dr_o} = \frac{d}{dr_o}\left[\frac{\ln(r_o/r_1)}{2\pi k\ell} + \frac{1}{h_o 2\pi r_o \ell}\right] = 0$$

or $\quad \dfrac{1}{kr_o} - \dfrac{1}{h_o r_o^2} = 0$

or

$$r_c = r_o = \frac{k}{h_o} \tag{2.48}$$

where r_c is the critical radius of insulation.

$$\frac{d^2R}{dr_o^2} = \frac{d}{dr_o}\left[\frac{1}{kr_o} - \frac{1}{h_o r_o^2}\right] = -\frac{1}{kr_o^2} + \frac{2}{h_o r_o^3} = \frac{1}{r_o^2}\left(\frac{2}{h_o r_o} - \frac{1}{k}\right)$$

$$\left(\frac{d^2R}{dr_o^2}\right)_{r_o=r_c} = \frac{1}{r_c^2}\left(\frac{2}{k} - \frac{1}{k}\right) = \frac{1}{r_c^2} > 0$$

Since dR/dr_o=0 and d^2R/dr_o^2>0, R is minimum and hence q is maximum at $r_o = r_c$.

The variation of the thermal resistance with the outer radius of insulation is shown in Fig. 2.13.

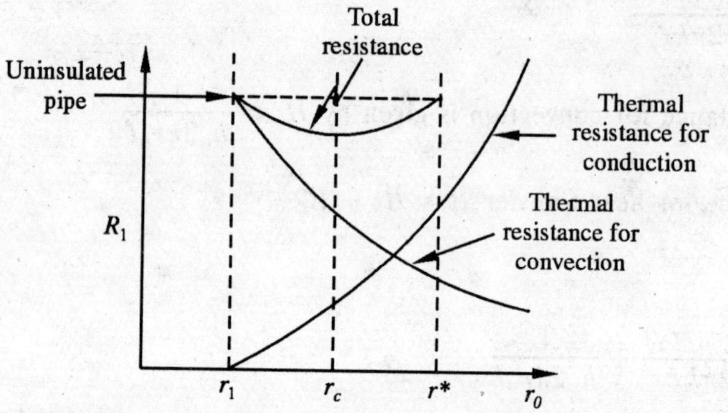

Fig. 2.13 Sketch illustrating the variation of resistance with the radius of insulation. At r^*, the total resistance is the same as that of an uninsulated pipe.

If the outer radius of the pipe, r_1, is less than r_c, then the rate of heat transfer increases by the addition of insulation. On the other hand, if $r_1 > r_c$ the rate of heat transfer decreases by the addition of insulation. In electrical conductors, the current-carrying capacity is controlled by the rate of dissipation of *ohmic heat*. The addition of insulation to current-carrying conductors provides electrical insulation and improves the current-carrying capacity.

EXAMPLE 2.11 An electrical wire of 10 m length and 1 mm diameter dissipates 200 W in air at 25°C. The convection heat transfer coefficient between the wire surface and air is 15 W/m^2 K. (a) Determine the temperature of the wire. (b) If the wire is insulated with mica (k = 0.5815 W/m K) such that its outer diameter increases to 3 mm, determine the temperature of the wire.

SOLUTION (a) Let

T_i = temperature of the wire

D = outer diameter of the wire

ℓ = length of the wire.

Then, $q = hA\Delta T = h(\pi D\ell)\Delta T$

or $200 = 15\pi \times 1 \times 10^{-3} \times 10(T_i - 25)$

or $T_i = 449.4°\text{C}$

(b) Thermal resistance for conduction due to insulation is given by

$$R_1 = \frac{\ln(r_o/r_i)}{2\pi k\ell}$$

or $$R_1 = \frac{\ln(3/1)}{2\pi \times 0.5815 \times 10} = 0.03 \text{ K/W}$$

Thermal resistance for convection is given by

$$R_2 = \frac{1}{hA} = \frac{1}{h\pi D_o \ell}$$

or $$R_2 = \frac{1}{15\pi \times 3 \times 10^{-3} \times 10} = 0.7074 \text{ K/W}$$

Total resistance is given by

$$R = R_1 + R_2 = 0.03 + 0.7074 = 0.7374 \text{ K/W}$$

$$q = \frac{\Delta T}{R}$$

or $$\Delta T = qR = 200 \times 0.7374 = 147.5°\text{C}$$

$$T_i - 25 = 147.5 \quad \text{or} \quad T_i = 172.5°\text{C}$$

EXAMPLE 2.12 Calculate the critical radius of insulation for the electrical wire of Example 2.11. Also determine the temperature of the wire if it is insulated to the critical thickness of insulation.

SOLUTION We know that the critical radius of insulation r_c is given by

$$r_c = \frac{k}{h} = \frac{0.5815}{15} = 0.0388 \text{ m} = 38.8 \text{ mm}$$

At the critical radius of insulation, the thermal resistance for convection R_2 is given by

$$R_2 = \frac{1}{hA} = \frac{1}{h(2\pi r_c)\ell} = \frac{1}{15 \times 2\pi \times 0.0388 \times 10} = 0.0273 \text{ K/W}$$

Thermal resistance for conduction due to insulation is given by

$$R_1 = \frac{ln(r_o/r_i)}{2\pi k\ell} = \frac{\ln(38.8/1)}{2\pi \times 0.5815 \times 10} = 0.1001 \text{ K/W}$$

Total resistance is given by

$$R = R_1 + R_2 = 0.0273 + 0.1001 = 0.1274 \text{ K/W}$$

$$q = \frac{\Delta T}{R} \quad \text{or} \quad \Delta T = qR$$

or $\quad T_i - 25 = 200 \times 0.1274 = 25.5°\text{C}$

or $\quad T_i = 50.5°\text{C}$

2.7 Heat Conduction in a Spherical Shell

Of all the geometrical configurations, a sphere has the largest volume per unit surface area. Since heat losses are dependent on surface area, a hollow spherical vessel is usually used to store fluids at low temperature (liquid nitrogen) to minimize heat losses and hence to reduce the rate of evaporation. If the interior and exterior surfaces of a spherical vessel are maintained at constant temperatures, the conduction through the spherical shell is a one-dimensional steady-state process. A schematic of a ollow sphere illustrating the nomenclature is shown in Fig. 2.14. The rate of conduction heat transfer is given by Fourier's law of heat conduction

$$q_k = -kA\frac{dT}{dr} = -k(4\pi r^2)\frac{dT}{dr} \tag{2.49}$$

or $\quad \dfrac{q_k}{4\pi k}\displaystyle\int_{r_i}^{r_o} \frac{dr}{r^2} = -\int_{T_i}^{T_o} dT$

or $\quad \dfrac{q_k}{4\pi k}\left(\dfrac{1}{r_i} - \dfrac{1}{r_o}\right) = T_i - T_o$

or $$q_k = \frac{4\pi r_o r_i k(T_i - T_o)}{r_o - r_i} = \frac{\Delta T}{R_k} \tag{2.50}$$

The thermal resistance for conduction for a hollow sphere is given by

$$R_k = \frac{r_o - r_i}{4\pi r_o r_i k} \tag{2.51}$$

The outer surface area $A_o = 4\pi r_o^2$.

The inner surface area $A_i = 4\pi r_i^2$.

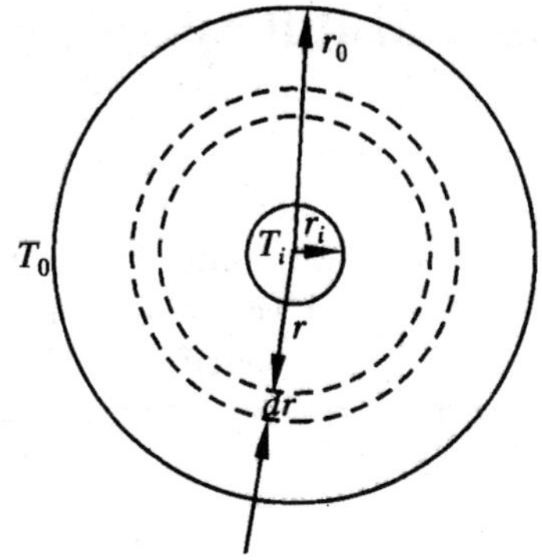

Fig. 2.14 *Schematic of a hollow sphere of inner radius r_i and outer radius r_o.*

Hence $\bar{A} = \sqrt{A_o A_i} = 4\pi r_o r_i$ (2.52)

where $\bar{A}$ is the geometric mean area. Substituting Eq. (2.52) in Eq. (2.50), we get

$$q_k = \frac{k\bar{A}\Delta T}{r_o - r_i} = \frac{k\bar{A}\Delta T}{\ell} \tag{2.53}$$

where $\ell = r_o - r_i$ = thickness through which the heat is conducted. Equation (2.53) is similar to the heat conduction equation through a plane wall [see Eq. (1.2)]. Substituting Eq. (2.52) in Eq. (2.51), we get

$$R_k = \frac{r_o - r_i}{\bar{A}k} = \frac{\ell}{\bar{A}k} \tag{2.54}$$

EXAMPLE 2.13 A spherical vessel made of carbon-steel (k = 54 W/m K) with ID = 50 cm is filled with liquid nitrogen at its normal boiling point 77 K. If the outer surface of the vessel is maintained at 97 K by blowing air, estimate the rate of conduction heat transfer through the vessel, assuming that the inner surface of the vessel is at a constant temperature of 77 K.

SOLUTION We know that the rate of conduction heat transfer through a spherical shell is given by

$$q_k = \frac{4\pi r_o r_i k(T_i - T_o)}{r_o - r_i} \tag{A}$$

Substituting the given values in Eq. (A), we get

$$q_k = \frac{4\pi(0.55/2)(0.50/2) \times 54 \times (77 - 97)}{(0.55 - 0.50)/2} = 37.32 \text{ kW}$$

2.8 Heat Conduction in a Multilayered Spherical Shell

The spherical vessels used in industries have to withstand high pressure. For this purpose multilayered spherical vessels are fabricated. To reduce heat losses, the spherical vessels are insulated. Hence, we study the heat conduction in multilayered spherical vessels. Consider the spherical vessel shown in Fig. 2.15. The inner and outer radii of the vessel are r_i and r_o, respectively. The spherical shell of thickness $(r_1 - r_i)$ is made of material with thermal conductivity k_1, while the spherical shell of thickness $(r_o - r_1)$ is made of a different material with thermal conductivity k_2. Let

T_i = temperature of the fluid in the vessel

T_o = temperature of the surrounding air

h_i = convection heat transfer coefficient at the inner surface

h_o = convection heat transfer coefficient at the outer surface.

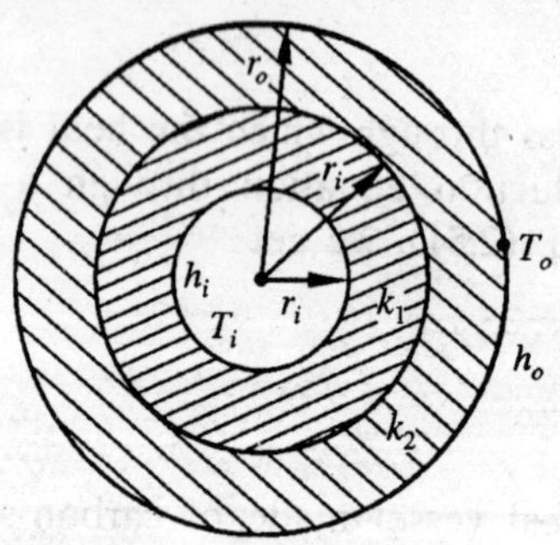

Fig. 2.15 Sketch of a two-layered hollow spherical vessel.

In steady-state, the rate of heat transfer is given by

$$q = \frac{\Delta T}{R} \tag{2.55}$$

where the total thermal resistance is given by

$$R = R_{c1} + R_{k1} + R_{k2} + R_{c2} \tag{2.56}$$

Thermal resistance for convection at the inner surface is given by

$$R_{c1} = \frac{1}{h_i A_i} = \frac{1}{h_i 4\pi r_i^2} \tag{2.57}$$

Thermal resistance due to the first layer is given by

$$R_{k1} = \frac{r_1 - r_i}{4\pi r_1 r_i k_1} \tag{2.58}$$

Thermal resistance due to the second layer is given by

$$R_{k2} = \frac{r_o - r_1}{4\pi r_o r_1 k_2} \tag{2.59}$$

Thermal resistance for convection at the outer surface is given by

$$R_{c2} = \frac{1}{h_o A_o} = \frac{1}{h_o 4\pi r_o^2} \tag{2.60}$$

Substituting Eqs. (2.56)–(2.60) in Eq. (2.55), we get

$$q = \frac{T_i - T_o}{\dfrac{1}{h_i 4\pi r_i^2} + \dfrac{r_1 - r_i}{4\pi r_1 r_i k_1} + \dfrac{r_o - r_1}{4\pi r_o r_1 k_2} + \dfrac{1}{h_o 4\pi r_o^2}} \tag{2.61}$$

EXAMPLE 2.14 Liquid ammonia at −20°C is stored in a spherical vessel of inner and outer radii 1 m and 1.05 m, respectively. The inner shell of 3 cm thickness is made of nickel-steel (k = 19 W/m K) and the outer shell of thickness 2 cm is made of carbon-steel (k = 54 W/m K). The vessel is exposed to atmospheric air at 30°C. The inside and outside film coefficients are 100 W/m^2 K and 10 W/m^2 K, respectively. Calculate the rate of heat losses from liquid ammonia.

SOLUTION We know that the rate of heat transfer is given by

$$q = \frac{\Delta T}{R} = \frac{30 - (-20)}{R} = \frac{50}{R}$$

$$R_{c_i} = \frac{1}{h_i 4\pi r_i^2} = \frac{1}{100 \times 4\pi (1)^2} = 7.958 \times 10^{-4} \text{ K/W}$$

$$R_{k1} = \frac{r_1 - r_i}{4\pi r_1 r_i k_1} = \frac{0.03}{4\pi \times 1.03 \times 1 \times 19} = 1.22 \times 10^{-4} \text{ K/W}$$

$$R_{k2} = \frac{r_o - r_i}{4\pi r_o r_1 k_2} = \frac{0.02}{4\pi \times 1.05 \times 1.03 \times 54} = 2.725 \times 10^{-5} \text{ K/W}$$

$$R_{c_o} = \frac{1}{h_o 4\pi r_o^2} = \frac{1}{10 \times 4\pi \times (1.05)^2} = 7.2179 \times 10^{-3} \text{ K/W}$$

$$R = R_{c_i} + R_{k1} + R_{k2} + R_{c_o} = 8.163 \times 10^{-3} \text{ K/W}$$

Therefore,

$$q = \frac{50}{8.163 \times 10^{-3}} = 6.125 \text{ kW}$$

2.9 Critical Radius of Insulation for a Hollow Sphere

To determine the critical radius of insulation, let us consider a spherical vessel of inner radius r_i and outer radius r_1. Let the thermal conductivity of the vessel material be k_1. If the vessel is insulated with a material of thermal conductivity k to a radius r_o, then the rate of heat transfer from the fluid inside the vessel to the surrounding fluid [see Fig. 2.15] is given by Eq. (2.61). The rate of heat transfer will be maximum if the denominator (total resistance, R) of Eq. (2.61) is minimum. Then the critical radius of insulation can be obtained from

$$\frac{dR}{dr_o} = \frac{d}{dr_o}\left[\frac{1}{h_i 4\pi r_i^2} + \frac{r_1 - r_i}{4\pi r_1 r_i k_1} + \frac{r_o - r_1}{4\pi r_o r_1 k} + \frac{1}{h_o 4\pi r_o^2}\right] = 0$$

or $$\frac{1}{k r_o^2} - \frac{2}{h_o r_o^3} = 0 \quad \text{or} \quad r_c = \frac{2k}{h_o} \tag{2.62}$$

$$\left[\frac{d^2R}{dr_o^2}\right]_{r_o = r_c} = \left[-\frac{2}{k r_o^3} + \frac{6}{h_o r_o^4}\right]_{r_o = r_c} = \frac{1}{k r_c^3} > 0$$

Hence at the critical radius of insulation r_c, the heat losses will be maximum for a spherical vessel.

EXAMPLE 2.15 A hollow spherical vessel of ID = 19 cm and OD = 20 cm contains a hot fluid. The fluid is to be cooled by exposing the vessel to a surrounding cold fluid when the outside film coefficient is 10 W/m^2 K. If the vessel is to be lagged by mica sheet (k = 0.5815 W/m K), determine the thickness of insulation so that the rate of heat transfer from the hot fluid is maximum.

SOLUTION We know that the critical radius of insulation for a hollow sphere is given by

$$r_c = \frac{2k}{h_o} = \frac{2 \times 0.5815}{10} = 0.1163 \text{ m} = 11.63 \text{ cm.}$$

Thickness of insulation = $r_c - r_o = 11.63 - 10 = 1.63$ cm.

2.10 Steady-State Heat Conduction in Systems with Heat Sources

Quite often an engineer has to analyze heat transfer in systems like chemical reactors, nuclear reactors and electrical heaters in which heat is generated. The dissipation of heat from such systems is quite essential to prevent break down of the equipment. In this section steady-state, one-dimensional heat conduction in a flat plate and in a cylinder with homogeneous heat generation is discussed.

2.10.1 Flat Plate with Uniform Heat Generation

Consider a large, homogeneous flat plate in which heat is generated uniformly. A differential element of the flat plate is shown in Fig. 2.16.

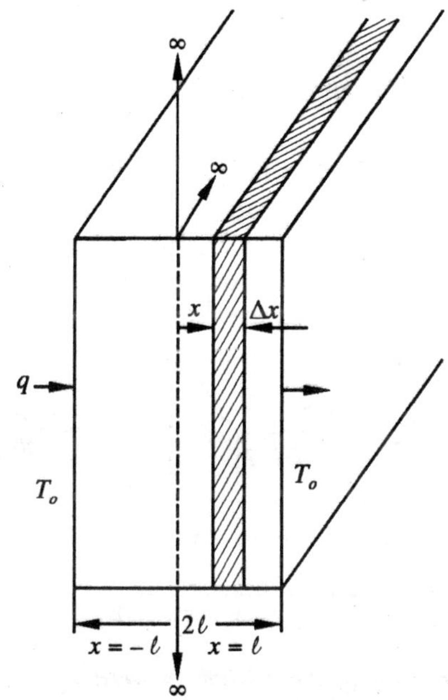

Fig. 2.16 Sketch of a large flat plate with homogeneous heat generation.

Let

A = cross-sectional area of the flat plate, normal to the direction of heat flow

2ℓ = thickness of the plate

$\dot{q}$ = rate of heat generation per unit volume of the plate

T_o = temperature at the faces $x = 0$ and $x = 2\ell$.

In steady-state, the energy balance across the differential element gives

$$\begin{bmatrix}\text{Rate of energy transfer}\\ \text{by conduction}\\ \text{into the element}\end{bmatrix} + \begin{bmatrix}\text{Rate of energy}\\ \text{generation}\\ \text{in the element}\end{bmatrix} = \begin{bmatrix}\text{Rate of energy transfer}\\ \text{by conduction}\\ \text{out of the element}\end{bmatrix}$$

or

$$-kA\left(\frac{dT}{dx}\right)_x + \dot{q}(A\Delta x) = -kA\left(\frac{dT}{dx}\right)_{x+\Delta x} \tag{2.63}$$

According to the mean value theorem, the temperature gradient at $(x+\Delta x)$ is given by

$$\left(\frac{dT}{dx}\right)_{x+\Delta x} = \left(\frac{dT}{dx}\right)_x + \frac{d}{dx}\left(\frac{dT}{dx}\right)\Delta x \tag{2.64}$$

The relation between the temperature gradients at x and $(x+\Delta x)$ given by Eq. (2.64) is shown in Fig. 2.17.

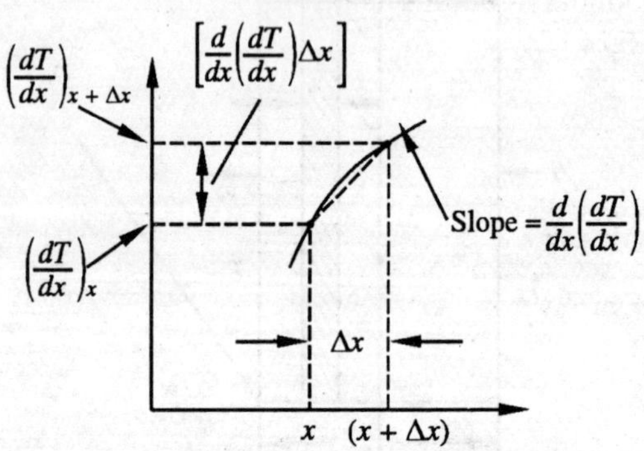

Fig. 2.17 Graphical representation of the relation between the temperature gradients at x and $(x+\Delta x)$.

Substituting Eq. (2.64) in Eq. (2.63), we get

$$-kA\left(\frac{dT}{dx}\right)_x + \dot{q}(A\Delta x) = -kA\left[\left(\frac{dT}{dx}\right)_x + \frac{d}{dx}\left(\frac{dT}{dx}\right)\Delta x\right]$$

or

$$-kA\left(\frac{dT}{dx}\right)_x + \dot{q}(A\Delta x) = -kA\left(\frac{dT}{dx}\right)_x - kA\left(\frac{d^2T}{dx^2}\right)\Delta x$$

or

$$\frac{d^2T}{dx^2} = -\frac{\dot{q}}{k} \tag{2.65}$$

The solution of Eq. (2.65) is obtained by two successive integrations:

$$\frac{dT}{dx} = -\frac{\dot{q}}{k}x + c_1 \tag{2.66}$$

$$T = -\frac{\dot{q}x^2}{2k} + c_1x + c_2 \tag{2.67}$$

where c_1 and c_2 are constants, the values of which can be obtained from the following boundary conditions:

At the left face: $x = -\ell, \quad T = T_o$ (2.68)

At the right face: $x = \ell; \quad T = T_o$ (2.69)

Substituting the boundary conditions given by Eqs. (2.68) and (2.69) in Eq. (2.67), we get

$$T_o = \frac{-\dot{q}\ell^2}{2k} - c_1\ell + c_2 \tag{2.70}$$

$$T_o = \frac{-\dot{q}\ell^2}{2k} + c_1\ell + c_2 \tag{2.71}$$

From Eqs. (2.70) and (2.71), we get

$$c_1 = 0 \quad \text{and} \quad c_2 = T_o + \frac{\dot{q}\ell^2}{2k} \tag{2.72}$$

Substituting the values of c_1 and c_2 from Eqs. (2.72) in Eq. (2.67), we get the temperature profile in the plate as

$$T = -\frac{\dot{q}x^2}{2k} + T_o + \frac{\dot{q}\ell^2}{2k}$$

or

$$T - T_o = \frac{\dot{q}}{2k}(\ell^2 - x^2) \tag{2.73}$$

The temperature profile given by Eq. (2.73) is shown in Fig. 2.18.

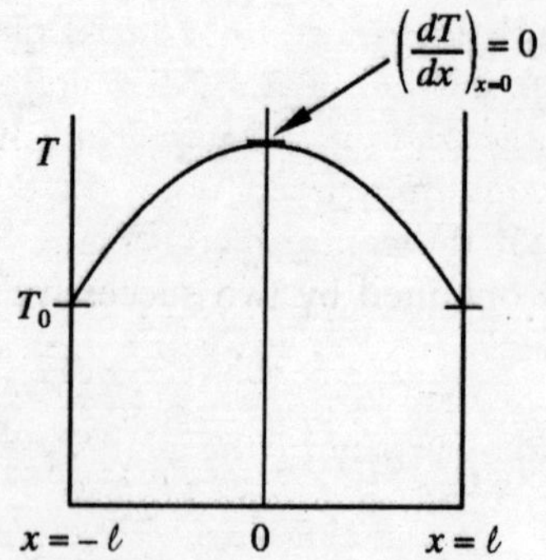

Fig. 2.18 Temperature profile in a homogeneous flat plate with uniform heat generation for steady-state, one-dimensional heat conduction.

It can be observed from Eqs. (2.66) and (2.73) and Fig. 2.18 that at the centre of the plate ($x = 0$) the temperature gradient is zero and the temperature is maximum. They are given by

$$\left(\frac{dT}{dx}\right)_{x=0} = 0 \tag{2.74}$$

and

$$(T - T_o)_{\max} = \frac{\dot{q}\ell^2}{2k} \tag{2.75}$$

EXAMPLE 2.16 A 20 cm thick nickel (k = 90 W/m K) plate is used as an element for a heater. Heat is uniformly generated in the plate at a rate of 100 kW/m^3 and both the surfaces of the plate are maintained at 120°C. Determine the temperature at the centre of the plate.

SOLUTION We know that the temperature will be maximum at the centre of the plate and is given by

$$(T - T_o)_{\max} = \frac{\dot{q}\ell^2}{2k} \tag{A}$$

Substituting the known values of $T_o, \dot{q}, \ell$ and k, we obtain

$$T - 120 = \frac{100 \times 10^3 \times (0.1)^2}{2 \times 90} = 5.6$$

or

$$T = 125.6°\text{C}$$

EXAMPLE 2.17 A flat nickel-steel (k = 19 W/m K) plate of 6 cm thickness is used as a heater by passing electric current, and heat is uniformly generated in the plate at the rate of 100 kW/m^3. If the plate is immersed in a flowing fluid at 25°C and the film coefficient under these conditions is 50 W/m^2 K, determine the temperature at the surface of the plate, and at the centre of the plate.

SOLUTION Let

T_o = temperature at the surface of the plate

T_∞ = temperature of the fluid

h_o = film coefficient at the surface.

In steady-state, the heat generated in one half of the plate passes through one surface and eventually from the surface to the fluid. This condition gives

$$\dot{q}A\ell = -kA\left(\frac{dT}{dx}\right)_{x=\ell} = -kA\left(\frac{dT}{dx}\right)_{x=-\ell} = hA(T_o - T_\infty)$$

or

$$T_o - T_\infty = \frac{\dot{q}\ell}{h} = \frac{100 \times 10^3 \times 0.03}{50} = 60$$

or $T_o = T_\infty + 60 = 25 + 60 = 85°\text{C}$

The maximum temperature attained at the centre of the plate is given by

$$(T - T_o)_{\text{max}} = \frac{\dot{q}\ell^2}{2k}$$

or

$$T - 85 = \frac{100 \times 10^3 \times (0.03)^2}{2 \times 19} = 2.4$$

or $T = 87.4°\text{C}$

2.10.2 Long Solid Cylinder with Uniform Heat Generation

An electric heating element in which heat is generated by passing electric current is an example of a solid cylinder with uniform heat generation. Consider an annular element of a solid cylindrical section with uniform heat generation shown in Fig. 2.19.

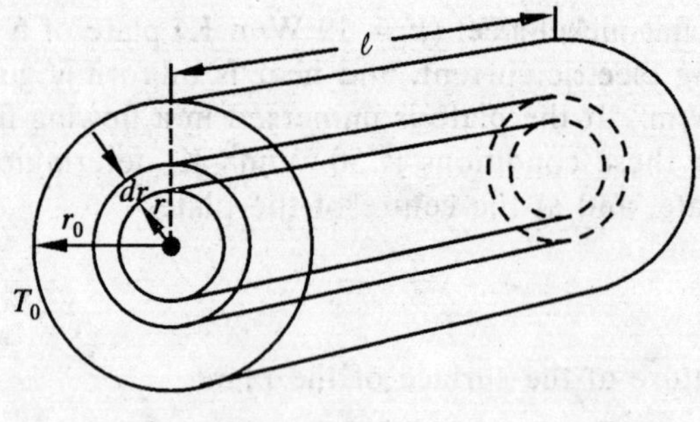

Fig. 2.19 Sketch of a long solid cylinder with uniform heat generation.

Let

$\dot{q}$ = rate of heat generation per unit volume

T_s = temperature at the outer surface of the cylinder

ℓ = length of the cylinder.

If the cylinder is long, heat transfer takes place radially. Then the energy balance for the cylindrical shell of inner radius r and outer radius $(r + dr)$ gives

$$\begin{bmatrix} \text{Rate of energy} \\ \text{conduction} \\ \text{into the shell} \end{bmatrix} + \begin{bmatrix} \text{Rate of energy} \\ \text{generation} \\ \text{in the shell} \end{bmatrix} = \begin{bmatrix} \text{Rate of energy} \\ \text{conduction} \\ \text{out of the shell} \end{bmatrix} \tag{2.76}$$

The rate of heat flow $(q)_{r+dr}$ is related to the rate of heat flow $(q)_r$ as

$$(q)_{r+dr} = (q)_r + \left[\frac{d}{dr}(q)_r\right] dr \tag{2.77}$$

Substituting Eq. (2.77) in Eq. (2.76), we get

$$-k2\pi r\ell\left(\frac{dT}{dr}\right)_r + \dot{q}2\pi r\ell dr = -k2\pi r\ell\left(\frac{dT}{dr}\right)_r - k2\pi\ell\frac{d}{dr}\left(r\frac{dT}{dr}\right) dr$$

or

$$\dot{q}r = -k\frac{d}{dr}\left(r\frac{dT}{dr}\right) \tag{2.78}$$

Integrating Eq. (2.78), we get

$$\frac{\dot{q}r^2}{2} = -kr\frac{dT}{dr} + c_1 \tag{2.79}$$

Substituting the boundary condition

$$\frac{dT}{dr} = 0 \qquad at \qquad r = 0 \tag{2.80}$$

we get $c_1 = 0$. Integrating once again, we obtain

$$T = -\frac{\dot{q}r^2}{4k} + c_2 \tag{2.81}$$

Applying the second boundary condition that

$$T = T_o \text{ at } r = r_o \tag{2.82}$$

we get

$$c_2 = T_o + \frac{\dot{q}r_o^2}{4k} \tag{2.83}$$

Therefore,

$$T = T_o + \frac{\dot{q}r_o^2}{4k} - \frac{\dot{q}r^2}{4k} = T_o + \frac{\dot{q}r_o^2}{4k}\left[1 - (r/r_o)^2\right] \tag{2.84}$$

The maximum temperature occurs at the centre. That is, when $r = 0$ we get

$$T_{\max} = T_o + \frac{\dot{q}r_o^2}{4k} \tag{2.85}$$

EXAMPLE 2.18 A long cylindrical nickel-steel (k = 19 W/m K) rod of 4 mm diameter is used as a heating element. By passing current through the rod heat is uniformly generated at the rate of 2 GW/m^3. If the outer surface of the rod is maintained at 110°C, calculate the maximum temperature at the centre of the rod.

SOLUTION The maximum temperature reached at the centre of a cylinder is given by

$$T_{\max} = T_o + \frac{\dot{q}r_o^2}{4k} = 110 + \frac{2 \times 10^9 \times (2 \times 10^{-3})^2}{4 \times 19} = 110 + 105.3$$

or

$$T_{\max} = 215.3°\text{C}$$

EXAMPLE 2.19 It is proposed to use a 1 m long nickel (k = 90 W/m K) rod as a heating element to dissipate 25 kW in a fluid at 30°C. In the envisaged operating conditions the convection heat transfer coefficient between the rod and the fluid is 750 W/m^2 K. Determine the diameter of the rod if the maximum allowed temperature in the heating element is 900°C.

SOLUTION Rate of heat generation

$$\dot{q} = \frac{25 \times 10^3}{\pi r_o^2 \ell} \text{ W/m}^3$$

The maximum temperature at the centre of the rod is given by

$$T_{\max} = T_o + \frac{\dot{q} r_0^2}{4k}$$

or $$900 = T_o + \left(\frac{25 \times 10^3}{\pi r_0^2 \times 1}\right) \frac{r_0^2}{4 \times 90} = T_o + 22.1$$

or $T_o = 877.9$°C

In steady-state,

the rate of heat generation = the rate of heat dissipation by convection

or $$25 \times 10^3 = h(\pi d \ell)(T_o - 30) = 750 \pi d \times 1(877.9 - 30)$$

or $d = 12.5$ mm

2.11 Summary

If two surfaces of a plane wall of thickness ℓ are maintained at temperatures T_1 and T_2, respectively, then the rate of conduction heat transfer through a homogeneous material is given by

$$q_k = \frac{kA}{\ell}(T_1 - T_2) = \frac{\Delta T}{(\ell/kA)} = \frac{\Delta T}{R_k}$$

If the thermal conductivity of the material varies linearly with temperature, the rate of conduction heat transfer through a plane wall is given by

$$q_k = \frac{\Delta T}{(\ell/k_m A)}$$

where k_m is the thermal conductivity of the material at the mean temperature $(T_1 + T_2)/2$.

The rate of conduction heat transfer through a composite wall is given by

$$q = \frac{\Delta T}{\sum R_i}$$

where ΔT is the overall temperature difference across the wall and $\sum R_i$ is the sum of the thermal resistances for conduction offered by different layers. The thermal resistances for conduction R_k and convection R_c are given by

$$R_k = \frac{\ell}{kA} \quad \text{and} \quad R_c = \frac{1}{hA}$$

where h is the convection heat transfer coefficient or film coefficient between the surface and the surrounding fluid. The total thermal resistance R is given by $R = \sum R_i$ for thermal paths connected in series and

$$\frac{1}{R} = \sum \frac{1}{R_i}$$

for thermal paths connected in parallel.

The rate of heat transfer through a hollow cylinder of inner radius r_i at temperature T_i and outer radius r_o at temperature T_o is given by

$$q = \frac{T_i - T_o}{\dfrac{\ln(r_o/r_i)}{2\pi k\ell}} = \frac{\Delta T}{R_k}$$

where ℓ is the length of the cylinder. The thermal resistance for conduction for a hollow cylinder is given by

$$R_k = \frac{\ln(r_o/r_i)}{2\pi k\ell}$$

The temperature distribution in the cylindrical shell is given by

$$T(r) = T_i - \frac{(T_i - T_o)}{\ln(r_o/r_i)} \ln(r/r_i)$$

The maximum temperature occurs at the centre. That is when $r = 0$, we get

$$T_{\max} = T_o + \frac{\dot{q}r_0^2}{4k}$$

The expression for the rate of conduction heat transfer through a plane wall can be used for a hollow cylinder also if the path length of heat transfer ℓ is changed to $(r_o - r_i)$ and the log mean area $\bar{A}$ is used in place of A. The log mean area $\bar{A}$ is given by

$$\bar{A} = \frac{A_o - A_i}{\ln(A_o/A_i)}$$

In a composite cylinder having three different layers, the rate of conduction heat transfer is given by

$$q = \frac{\Delta T}{\dfrac{\ln(r_1/r_i)}{2\pi k_1 \ell} + \dfrac{\ln(r_2/r_1)}{2\pi k_2 \ell} + \dfrac{\ln(r_o/r_2)}{2\pi k_3 \ell}}$$

The rate of heat transfer from a fluid flowing through a two-layered cylinder to the surrounding fluid (with convection heat transfer included) can be estimated by the relation

$$q = U_o A_o \Delta T = \frac{\Delta T}{\dfrac{1}{2\pi r_i \ell h_i} + \dfrac{\ln(r_1/r_i)}{2\pi k_1 \ell} + \dfrac{\ln(r_o/r_1)}{2\pi k_2 \ell} + \dfrac{1}{2\pi r_o \ell h_o}}$$

and the overall heat transfer coefficient U_o based on the outer area is given by

$$U_o = \frac{1}{\dfrac{r_o}{r_i h_i} + \dfrac{r_o \ln(r_1/r_i)}{k_1} + \dfrac{r_o \ln(r_o/r_1)}{k_2} + \dfrac{1}{h_o}}$$

For small diameter pipes, the addition of insulation may enhance the rate of heat transfer. The rate of heat transfer will be maximum if the pipe is insulated to the critical radius (r_c) of insulation. The critical radius of insulation for a cylindrical pipe is given by

$$r_c = \frac{k}{h_o}$$

The rate of conduction heat transfer through a spherical shell of inner radius r_i and outer radius r_o is given by

$$q = \frac{4\pi r_o r_i k(T_i - T_o)}{r_o - r_i} = \frac{\Delta T}{R_k} = \frac{k\bar{A}\Delta T}{\ell}$$

where $R_k = \dfrac{r_o - r_i}{4\pi r_o r_i k}$ = thermal resistance for conduction for a hollow sphere

$\ell = r_o - r_i$ = path length for heat transfer

$\bar{A} = \sqrt{A_o A_i} = 4\pi r_o r_i$ = geometric mean area.

The rate of heat transfer from a fluid inside a hollow spherical vessel made of two layers to a fluid outside the spherical vessel is given by

$$q = \frac{T_i - T_o}{\dfrac{1}{h_i 4\pi r_i^2} + \dfrac{r_1 - r_i}{4\pi r_1 r_i k_1} + \dfrac{r_o - r_1}{4\pi r_o r_1 k_2} + \dfrac{1}{h_o 4\pi r_0^2}}$$

The critical radius (r_c) of insulation for a hollow sphere is given by

$$r_c = \frac{2k}{h_o}$$

The temperature distribution in a large, homogeneous flat plate with uniform heat generation is given by

$$T - T_o = \frac{\dot{q}}{2k}(\ell^2 - x^2)$$

where

T = temperature at plane x

T_o = surface temperature

2ℓ = thickness of the plate

$\dot{q}$ = rate of heat generation per unit volume of the plate.

The mid-plane of the plate will attain the maximum temperature.

The temperature distribution in a long solid cylinder with uniform heat generation is given by

$$T - T_o = \frac{\dot{q} r_0^2}{4k}\left[1 - (r/r_o)^2\right]$$

and the maximum temperature attained at the centre is given by

$$T_{\max} = T_o + \frac{\dot{q} r_0^2}{4k}$$

Review Questions

1. What is meant by one-dimensional, steady-state heat conduction?
2. Sketch the temperature distribution for conduction heat transfer through a thick flat plate of homogeneous material of constant thermal conductivity, the two opposite surfaces of which are maintained at temperature T_1 and T_2, respectively ($T_1 > T_2$).
3. If the thermal conductivity of a material increases linearly with temperature, sketch the temperature distribution for steady-state conduction heat transfer through a flat plate, the two opposite surfaces of which are maintained at different constant temperatures.
4. The rate of conduction heat transfer through a flat plate of homogeneous material with constant thermal conductivity is given by

$$q = \frac{kA\Delta T}{\ell}$$

   If the thermal conductivity varies linearly with temperature as $k = k_o(1 + \beta_k T)$, where β_k is constant, and it is desired to use the same expression for estimating the rate of conduction heat transfer, by replacing k with k_m, obtain a relation between k_m and T.
5. What is meant by thermal resistance?
6. A plane wall is made in two different layers of different materials of the same thickness. The thermal conductivities of the two materials differ by a factor of three. The two surfaces of the wall are maintained at two different constant temperatures. Does the rate of conduction heat transfer through the wall depend on which layer is exposed to the higher temperature? Explain why or why not.
7. One surface of a thick plane wall of homogeneous material is exposed to a high temperature gas at T_i, while the other surface is exposed to a fluid at low temperature T_o. If the convection heat transfer coefficients are h_i and h_o on the high temperature and low temperature sides, respectively, what will be the total thermal resistance for heat flow?
8. What will be the rate of conduction heat transfer if the inside surface of a hollow cylinder is maintained at T_i while the outer surface is maintained at T_o? The inner and outer radii of the cylinder are r_i and r_o, respectively.
9. Express the thermal resistance for conduction for a hollow cylinder in terms of inner and outer radii.
10. Sketch the temperature profile in the wall of a hollow cylinder if the inner and outer surfaces are maintained at two different constant temperatures.
11. If it is desired to use the relation $q = (kA\Delta T)/\ell$ (which expresses the rate of conduction heat transfer through a plane wall) for conduction heat transfer through a hollow cylinder in which the cross-sectioned area changes with radius, how should we define A and ℓ?

12. What is meant by overall heat transfer coefficient?
13. Express the overall heat transfer coefficient, based on the outer area for the transfer of heat from a hot fluid flowing in a pipe to a cold fluid to which the pipe is exposed, in terms of film coefficients and radii of the pipe.
14. Addition of an insulating layer to a plane wall reduces the rate of heat transfer. Will this be always true in the case of a pipe or a cylinder?
15. What is meant by critical radius of insulation?
16. What is the critical radius of insulation for a cylinder?
17. In the case of a pipe, is it possible that the rate of heat transfer for an insulated pipe will be identical with that of the bare pipe? Explain with the help of a diagram showing heat transfer rate as a function of insulation thickness.
18. What is the thermal resistance for conduction for a hollow sphere?
19. How does one redefine A and ℓ for the relation $q = (kA\Delta T)/\ell$ to be valid for conduction heat transfer in a hollow sphere?
20. What is the critical radius of insulation for a hollow sphere?
21. Sketch the temperature distribution in a flat plate, with uniform heat generation, the surfaces of which are maintained at the same constant temperature.
22. If both the surfaces of a long flat plate with uniform heat generation are maintained at the same constant temperature, which point of the plate will attain maximum temperature?
23. Specify the location of maximum temperature in a long solid cylinder, with uniform heat generation, the outer surface of which is maintained at constant temperature.
24. Specify the expression to describe the temperature distribution in a long solid cylinder, with uniform heat generation, the outer surface of which is maintained at constant temperature.

Problems

2.1 A concrete wall (k = 1.37 W/m K) of 10 cm thickness is to be plastered with gypsum (k = 0.48 W/m K) so that the heat losses from the wall do not exceed 500 W/m^2 when the inner and outer surfaces of the wall are at 110°C and 40°C, respectively. Determine the thickness of plastering to be added to the concrete wall.

2.2 When the two surfaces of a plane wall of 4 cm thickness of unknown material are maintained at 80°C and 40°C, respectively, the centre plane was found to be at 50°C and the heat flow through the wall was 8 kW/m^2. Determine the thermal conductivity of the material as a function of temperature.

2.3 An oven wall is formed by 0.5 cm thick chrome-nickel (k = 19 W/m K) supported by 1 cm thick asbestos (k = 0.1105 W/m K) sheet. When the oven is operated in steady-state, the inner surface of the chrome-nickel sheet is at 300°C while the outer asbestos surface is at 40°C. Calculate the rate of heat transfer per unit area of the oven wall.

2.4 A domestic oven has a composite wall formed by 0.5 cm thick chrome-nickel (k = 19 W/m K) sheet supported by 1 cm thick asbestos (k = 0.1105 W/m K) sheet. In steady-state operation the hot gases inside the oven are at 350°C while the atmospheric air is at 30°C. The convection heat transfer coefficients at the inside surface and outside surface of the oven are 100 W/m^2 K and 15 W/m^2 K, respectively. Determine the rate of heat losses per unit area, through the oven wall.

2.5 Draw the thermal circuit for Problem 2.4.

2.6 A furnace has a composite wall, the details of which are presented in Fig. 2.20. Determine the rate of heat transfer per metre length of the composite wall.

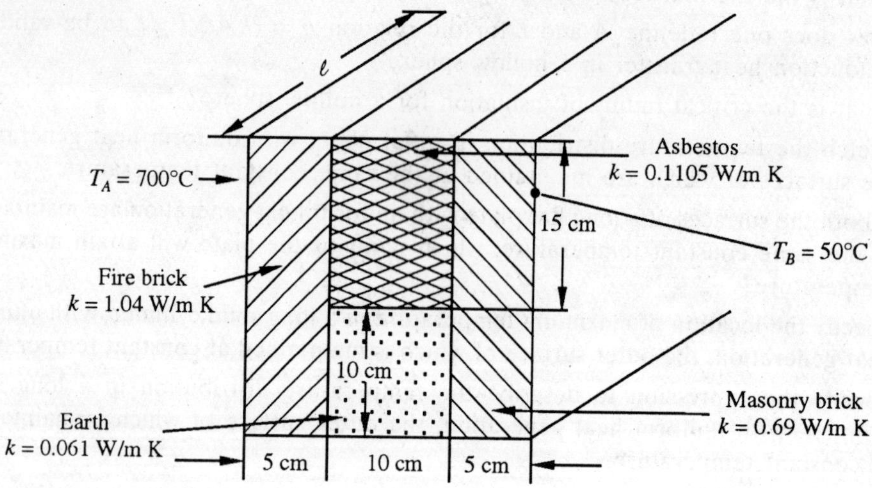

Fig. 2.20 *Sketch for Problem 2.6.*

2.7 The composite wall of Problem 2.6 is used in the construction of a furnace. The hot gases in the furnace are at 850°C, while the air on the outer side of the furnace is at 30°C. If the inside and outside convective heat transfer coefficients are 70 W/m^2 K and 15 W/m^2 K, determine the rate of heat transfer per metre length of the wall, from the hot gases to air.

2.8 The thermal conductivity of a material is determined by using the experimental setup shown in Fig. 2.21. A current of 15 amperes at 220 V is passed through the electrical resistance and the temperatures of the hot and cold surfaces are measured by the thermocouples. At steady-state, the temperatures are recorded as 150°C and 50°C. Determine the thermal conductivity of the material.

2.9 The experimental setup shown in Fig. 2.21 is used in measuring the thermal conductivity of a different material. The thermocouples recorded the temperatures as 150°C and 50°C while the surrounding air is at 30°C. The convection heat transfer coefficient between the outer surface and the surrounding air under these conditions is 20 W/m^2 K. The current passing through the electrical resistance could not be measured as the ammeter failed. Determine the thermal conductivity of the material.

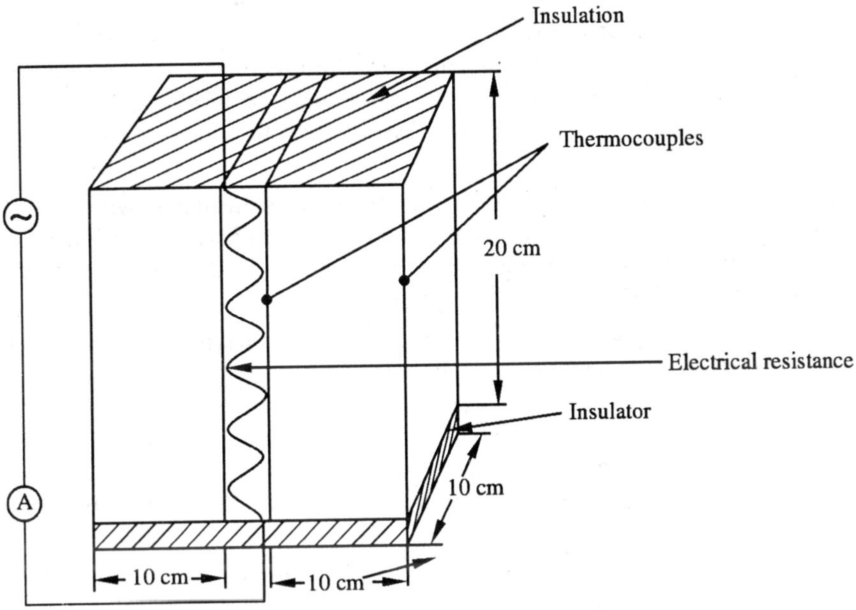

Fig. 2.21 Sketch for Problem 2.8.

2.10 It is required to reduce the heat loss from a furnace wall by doubling the thickness of the insulating brick work. Initially the temperatures of the inner and outer surfaces of the insulating brick are 480°C and 180°C, respectively. The atmospheric air is at 30°C. Calculate the percentage decrease in heat loss because of doubling of the thickness of insulating brick.

2.11 A steel pipe of 5 cm OD is covered with 0.5 cm asbestos (k = 0.1105 W/m K) followed by 0.5 cm thick glass wool (k = 0.038 W/m K) insulation. It was found that the insulation temperature reached a steady-state value of 40°C, while the pipe surface is at 150°C and the atmospheric air to which the insulated pipe is exposed is at 25°C. Determine the temperature at the interface of the asbestos and glass wool and the film coefficient between the glass wool and atmospheric air.

2.12 The inner surface of a thick invar (k = 10.7 W/m K) cylinder of ID = 2 cm and OD = 12 cm is held at 150°C, while the outer surface is at 50°C. Calculate the temperatures at r = 2 cm, 3 cm, 4 cm and 5 cm.

2.13 In a steam radiator saturated steam at 120°C flows through a carbon-steel (k = 54 W/m K) tube of 1.5 cm ID and 2.5 cm OD. The radiator is placed in a room in which the air is at 5°C. The convection heat transfer coefficient between the condensing steam and tube is 4000 W/m^2 K. Calculate the temperature of the inner and outer surfaces of the tube and the rate of heat transfer from steam to air per metre length of the tube.

2.14 Steam at 125°C is to be transported through a carbon-steel pipe of ID = 5 cm and OD = 6 cm. To reduce heat losses it is to be insulated with asbestos (k= 0.1105 W/m K) so that the rate of heat loss from steam to the atmospheric air is not more than 50 W per metre length of the pipeline. The atmospheric air is at 25°C and the film coefficient between air and the insulation is 10 W/m^2 K. Determine the thickness of insulation. Assume that the outer surface of the steel pipe is at 125°C.

2.15 A current-carrying wire of 3 mm diameter is to be insulated with rubber (k= 0.163 W/m K) to increase the current-carrying capacity of the wire. If the convection heat transfer coefficient from the rubber surface to air is 10 W/m^2 K, determine the critical thickness of insulation.

2.16 A carbon-steel (k = 54 W/m K) pipe of OD = 6 cm carrying steam at 300°C is to be insulated to reduce heat losses. The outer surface of the pipe may be assumed to be at 300°C. It is desired to use 1 cm thick asbestos (k = 0.1105 W/m K) and 1 cm thick glass wool (k = 0.038 W/m K) insulation. The atmospheric air is at 30°C and the film coefficient between the insulating layer and air is 10 W/m^2 K. Determine which insulating material should be used as the first layer of insulation.

2.17 An electrical heater uses 5 m length and 2 mm diameter wire to dissipate 500 W in air at 30°C. The convection heat transfer coefficient between the wire surface and air is 20 W/m^2 K. (a) Determine the temperature of the wire. (b) Determine the temperature of the wire if 3 mm thick rubber (k = 0.163 W/m^2 K) insulation is used, assuming that the film coefficient between the rubber surface and air is 20 W/m^2 K. (c) Calculate the temperature of the wire if the thickness of rubber insulation is equal to the critical thickness of insulation.

2.18 Freon-12 at −40°C is flowing through a copper tube of 1 cm OD. The atmospheric air is at 30°C and the film coefficient between the tube surface and air is 10 W/m^2 K. An engineer wanted to reduce the heat losses and added 6 mm thick asbestos (k = 0.1105 W/m K) insulation to the tube. Determine the percent reduction in the heat loss due to the added insulation. Assume that the outer surface of the tube is at −40°C.

2.19 Rework Problem 2.18 if the thickness of insulation is 9 mm.

2.20 A thick walled spherical vessel of ID = 30 cm and OD = 40 cm made of bronze (k = 26 W/m K) is filled with saturated steam at 120°C. If the outer surface of the vessel is maintained at 90°C, determine the rate of heat transfer from the vessel, assuming that the inner surface is at 120°C.

2.21 If the inner and outer surfaces of a spherical vessel of inside and outside radii r_i and r_o are maintained at T_i and T_o, respectively, determine the temperature profile in the wall of the vessel.

2.22 A spherical vessel of ID = 40 cm and OD = 45 cm is covered with 5 cm thick plaster (k = 0.48 W/m K). The thermal conductivity of the vessel material is 73 W/m K. If the inner surface of the vessel is kept at 150°C while the outer surface is exposed to air at 30°C, determine the rate of heat transfer from the vessel. The outside film coefficient is 5 W/m^2 K.

2.23 A small spherical vessel of OD = 6 cm is covered with asbestos (k = 0.1105 W/m K) and left in the atmospheric air at 30°C. The film coefficient between air and asbestos is 5 W/m^2 K. If it is desired to maximize the heat transfer rate from the contents of the vessel to the air, determine the thickness of asbestos cover needed.

2.24 An electrical furnace uses a 10 cm thick chrome-nickel plate (k = 19 W/m K) as a heating element. Heat is uniformly generated in the plate at a rate of 10 MW/m^3 and both the surfaces of the plate are maintained at 250°C. Determine the temperature at the centre of the plate.

2.25 If the plate of Problem 2.24 is immersed in a fluid at 50°C and the film coefficient between the plate and fluid is 2 kW/m^2 K, determine the maximum temperature attained in the plate.

2.26 The heating element used in an electrical oven can be treated as a long cylinder with uniform heat generation. If a tungsten-steel (k = 54 W/m K) rod of 1 cm diameter and 1 m length is used as a heating element with a uniform heat generation of 10 MW/m^3 and the air surrounding the heating element is at 120°C, calculate the surface temperature of the rod and the temperature at the centre of the rod assuming that the film coefficient between air and the heating element is 100 W/m^2 K.

2.27 It is desired to use 1 metre long tungsten-steel rod (k = 54 W/m K) as a heating element to dissipate 100 kW in a fluid at 10°C. If the maximum allowed temperature in the rod is 750°C, determine the diameter of the heating element, assuming that the film coefficient between the heating element and the fluid is 2000 W/m^2 K.

2.28 Suppose the outer surface of a sphere of radius r_o, with uniform heat generation, is maintained at temperature T_o. Derive a relation to determine the temperature distribution in the sphere.

2.29 A 3 mm diameter nickel-chromium (k = 19 W/m K) wire of 1 metre length has been used as a heating element by applying 30 V. The outer surface of the wire is maintained at 100°C. Calculate the temperature at the centre of the wire. The electrical resistivity of the wire is $6 \times 10^{-5} \Omega$ cm.

2.30 An aluminium-alloy (k = 200 W/m K) electrical cable of 3 cm diameter carrying an electric current of 200 A is exposed to air at 30°C. The resistivity of the cable is 3×10^{-6} ohm cm. Calculate the temperature at the centre of the cable, if the outside surface is at 60°C.

Chapter Three

Heat Transfer from Extended Surfaces

Learning Objectives

This chapter will enable the readers to

- understand the meaning of extended surfaces or fins.
- familiarize himself or herself with the types of fins in common use.
- analyze the heat transfer through fins of uniform cross-section.
- determine the temperature distribution and heat transferred through an infinitely long fin.
- predict the temperature distribution and heat transfer through a fin of finite length with insulated end.
- calculate the temperature distribution and rate of heat transfer through a fin of finite length with heat losses through the end.
- define and calculate fin efficiency.

3.1 Extended Surfaces or Fins

The heat conducted through a body must be subsequently transferred to a colder fluid by convection. For example, the heat conducted through a furnace wall is dissipated into the surrounding air by convection. Whenever it is desired to enhance the rate of heat transfer, it is a common practice to extend the surface by adding fins. In a heat exchanger, heat is transferred by convection from the hot fluid to the tube, transferred by conduction through the tube wall followed by convection from the tube surface to the colder fluid. Hence an analysis of the combined conduction-convection heat transfer is quite important from a practical viewpoint. When a solid of relatively small cross-sectional area protrudes from a large body into a fluid at different temperature, the protruding surface is called an *extended surface* or *fin*. The addition of fins increases the surface area for heat transfer. The engine cylinder and radiators of automobiles, the evaporators and condensers of refrigerators are common examples where extended surfaces or fins are extensively used. The fins may have different configurations. Some of the widely used fins are shown in Fig. 3.1.

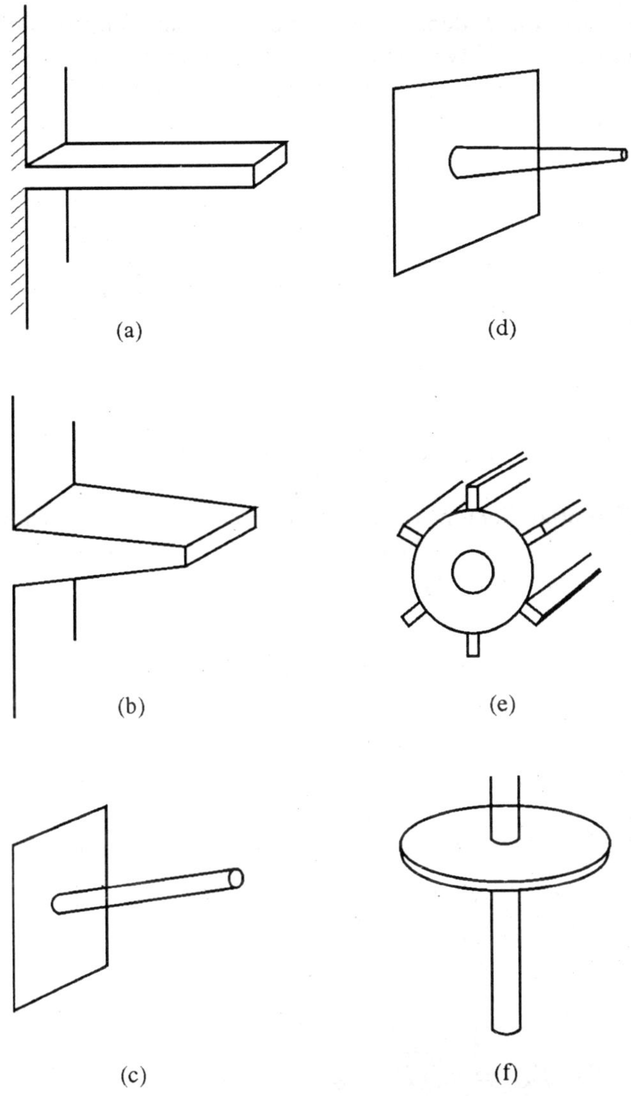

Fig. 3.1 *(a) Longitudinal fin of rectangular profile. (b) Longitudinal fin of trapezoidal profile. (c) Pin fin or cylindrical spine. (d) Truncated conical fin. (e) Cylindrical tube with fins of rectangular profile. (f) Cylindrical tube with radial fin of rectangular profile.*

3.2 Fins of Uniform Cross-Section

Consider a pin fin having the shape of a cylindrical rod, the base of which is attached to a wall at surface temperature T_s as shown in Fig. 3.2. The fin is exposed to a fluid at temperature T_∞. Assume that the fin is homogeneous and the thermal conductivity of the fin material is k. Let h denote the convection heat transfer coefficient between the fin surface and the fluid. We shall also assume that at any given location of the fin, the temperature is uniform. In other words, there is no temperature gradient in the radial direction and the heat flow is one-dimensional.

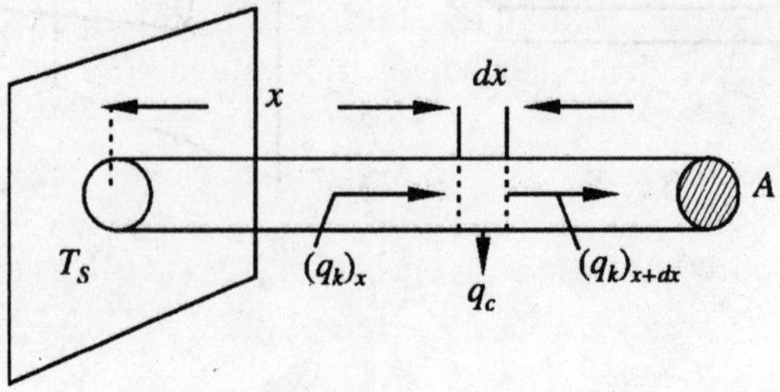

Fig. 3.2 Schematic of a pin fin.

Consider a differential element of the fin of thickness dx, at a distance x from the base. At steady-state, the rate of heat flow into the element is equal to the rate of heat flow out of the element. That is,

$$\begin{bmatrix}\text{Rate of heat flow}\\ \text{by conduction into}\\ \text{the element at } x\end{bmatrix} = \begin{bmatrix}\text{Rate of heat flow}\\ \text{by conduction out of}\\ \text{the element at } x+dx\end{bmatrix} + \begin{bmatrix}\text{Rate of heat flow}\\ \text{by convection from}\\ \text{the surface between}\\ x \text{ and } x+dx\end{bmatrix}$$

or $\quad [q_k]_x = [q_k]_{x+dx} + q_c$

or $\quad -kA\dfrac{dT}{dx} = -kA\dfrac{dT}{dx} + \dfrac{d}{dx}\left(-kA\dfrac{dT}{dx}\right) dx + hPdx\,(T - T_\infty)$

or

$$\frac{d^2T}{dx^2} = \frac{hP}{kA}(T - T_\infty) = m^2(T - T_\infty) \tag{3.1}$$

where

$$m = \sqrt{\frac{hP}{kA}} \tag{3.2}$$

Pdx = surface area between x and $x + dx$, which is in contact with the fluid at T_∞

P = perimeter of the fin.

The general solution of Eq. (3.1) is given by

$$T - T_\infty = C_1 e^{mx} + C_2 e^{-mx} \tag{3.3}$$

The values of the constants C_1 and C_2 are determined from the boundary conditions. The first boundary condition is that the temperature of the base of the fin is equal to the temperature of the surface to which it is attached. That is,

$$T = T_s \quad \text{at} \quad x = 0 \tag{3.4}$$

Substituting the boundary condition given by Eq. (3.4) in Eq. (3.3), we get

$$T_s - T_\infty = C_1 + C_2 \tag{3.5}$$

The second boundary condition depends on the nature of the problem. Let us consider the following three different cases and obtain the temperature distribution in the fin.

3.3 Infinitely Long Fin

If the fin is infinitely long, the temperature at the end of the fin ($x = \infty$) is equal to the temperature of the surrounding fluid. That is, the second boundary condition is given by

$$T = T_\infty \quad \text{at} \quad x = \infty \tag{3.6}$$

Substituting the boundary condition given by Eq. (3.6) in Eq. (3.3), we get

$$T_\infty - T_\infty = 0 = C_1 e^{m\infty}$$

or $\quad C_1 = 0$

Hence

$$C_2 = T_s - T_\infty$$

Therefore, the temperature distribution in an infinitely long fin is given by

$$T - T_\infty = (T_s - T_\infty)e^{-mx} \tag{3.7}$$

The temperature distribution in an infinitely long fin is shown in Fig. 3.3.

The rate of heat transfer from the fin to the fluid can be estimated by any one of the following two methods.

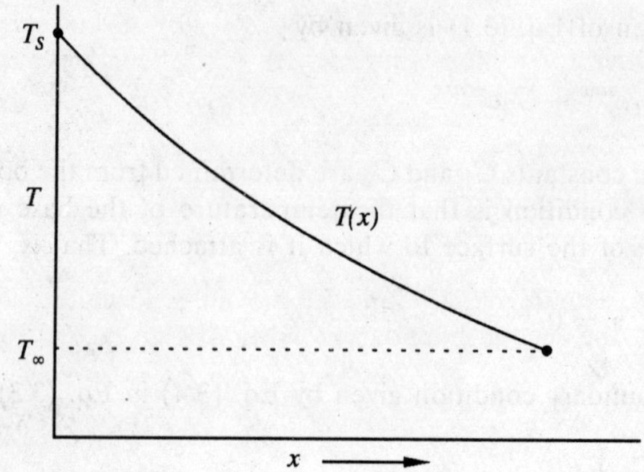

Fig. 3.3 Temperature distribution in an infinitely long fin.

METHOD-1

The heat which flows from the body to the base of the fin ($x = 0$) must be transferred to the fluid by convection from the total surface of the fin. If $\bar{h}$ denotes the average value of the film coefficient, then

$$q = \left[-kA\frac{dT}{dx}\right]_{x=0} = \int_0^\infty \bar{h}(Pdx)(T - T_\infty) \tag{3.8}$$

Differentiating Eq. (3.7) and substituting in Eq. (3.8), we get

$$q = -kA\left[-m(T_s - T_\infty)e^{-m(0)}\right]_{x=0} = kAm(T_s - T_\infty) = kA\sqrt{\frac{hP}{kA}}(T_s - T_\infty)$$

or

$$q = \sqrt{hPkA}(T_s - T_\infty) \tag{3.9}$$

METHOD-2

The rate of heat flow from the fin can also be determined by evaluating the convection heat transfer.

$$q = \int_0^\infty \bar{h}P(T - T_\infty)dx = \int_0^\infty \bar{h}P(T_s - T_\infty)e^{-mx}dx$$

$$= \left[-\frac{\bar{h}P}{m}(T_s - T_\infty)e^{-mx}\right]_0^\infty = \frac{\bar{h}P(T_s - T_\infty)}{\sqrt{(\bar{h}P/kA)}}$$

$$= \sqrt{\bar{h}PkA}(T_s - T_\infty) \tag{3.10}$$

The temperature distribution and the rate of heat flow given by Eqs. (3.7) and (3.9) can be used in practice for a fin of finite length if the length is very large compared to its cross-sectional area.

3.4 Finite Length Fin with Insulated End

If the fin is of finite length, but the end of the fin is insulated so that there is no heat loss through it, the second boundary condition is given by

$$\frac{dT}{dx} = 0 \quad \text{at} \quad x = \ell \tag{3.11}$$

Differentiating the general solution given by Eq. (3.3), and substituting the boundary condition (3.11), we get

$$\left[\frac{dT}{dx}\right]_{x=\ell} = mC_1e^{m\ell} - mC_2e^{-m\ell} = 0$$

or

$$C_2 = C_1e^{2m\ell} \tag{3.12}$$

We know that

$$T_s - T_\infty = C_1 + C_2 \tag{3.5}$$

Therefore,

$$T_s - T_\infty = C_1 + C_1e^{2m\ell} = C_1(1 + e^{2m\ell})$$

or

$$C_1 = \frac{T_s - T_\infty}{1 + e^{2m\ell}} \tag{3.13}$$

and

$$C_2 = \frac{(T_s - T_\infty)e^{2m\ell}}{1 + e^{2m\ell}} = \frac{T_s - T_\infty}{1 + e^{-2m\ell}} \tag{3.14}$$

Substituting the values of C_1 and C_2 given by Eqs. (3.13) and (3.14) in the general solution [Eq. (3.3)], we obtain the complete solution or the temperature distribution as

$$\frac{T - T_\infty}{T_s - T_\infty} = \frac{e^{mx}}{1 + e^{2m\ell}} + \frac{e^{-mx}}{1 + e^{-2m\ell}}$$

$$\frac{T - T_\infty}{T_s - T_\infty} = \frac{e^{mx}e^{-m\ell}}{e^{-m\ell} + e^{m\ell}} + \frac{e^{-mx}e^{m\ell}}{e^{m\ell} + e^{-m\ell}}$$

$$= \frac{e^{-m(\ell-x)}}{e^{m\ell} + e^{-m\ell}} + \frac{e^{m(\ell-x)}}{e^{m\ell} + e^{-m\ell}} = \frac{e^{m(\ell-x)} + e^{-m(\ell-x)}}{e^{m\ell} + e^{-m\ell}} \tag{3.15}$$

or

$$\frac{T - T_\infty}{T_s - T_\infty} = \frac{\cosh[m(\ell - x)]}{\cosh(m\ell)} \tag{3.16}$$

where the hyperbolic functions are defined as

$$\sinh(x) = \frac{e^x - e^{-x}}{2} \tag{3.17}$$

$$\cosh(x) = \frac{e^x + e^{-x}}{2} \tag{3.18}$$

$$\tanh(x) = \frac{\sinh(x)}{\cosh(x)} = \frac{e^x - e^{-x}}{e^x + e^{-x}} \tag{3.19}$$

The rate of heat flow from the fin can be estimated from the relation

$$q = -\left[kA\frac{dT}{dx}\right]_{x=0} \tag{3.20}$$

Differentiating Eq. (3.15) we get

$$\left[\frac{dT}{dx}\right]_{x=0} = (T_s - T_\infty)\left[\frac{me^{m(0)}}{1 + e^{2m\ell}} - \frac{me^{-m(0)}}{1 + e^{-2m\ell}}\right]$$

$$= m(T_s - T_\infty)\left[\frac{1}{1 + e^{2m\ell}} - \frac{1}{1 + e^{-2m\ell}}\right]$$

$$= m(T_s - T_\infty)\left[\frac{e^{-m\ell}}{e^{m\ell} + e^{-m\ell}} - \frac{e^{m\ell}}{e^{m\ell} + e^{-m\ell}}\right]$$

$$= -m(T_s - T_\infty)\left[\frac{e^{m\ell} - e^{-m\ell}}{e^{m\ell} + e^{-m\ell}}\right]$$

or

$$\left[\frac{dT}{dx}\right]_{x=0} = -m(T_s - T_\infty)\tanh(m\ell) \tag{3.21}$$

Substituting Eq. (3.21) in Eq. (3.20), we get

$$q = mkA(T_s - T_\infty)\tanh(m\ell) = \sqrt{hPkA}(T_s - T_\infty)\tanh(m\ell) \tag{3.22}$$

3.5 Finite Length Fin with Convection Heat Transfer at the End

If the fin is of finite length and the end is insulated (it loses heat from its end by convection), then the rate of heat flow by conduction to the face at $x = \ell$ must be equal to the rate of heat flow by convection from the end to the fluid. In other words, the second boundary condition is given by

$$\left[-kA\frac{dT}{dx}\right]_{x=\ell} = hA(T_{x=\ell} - T_\infty) \tag{3.23}$$

The temperature distribution in a fin of finite length, which loses heat by convection from the end, is shown in Fig. 3.4.

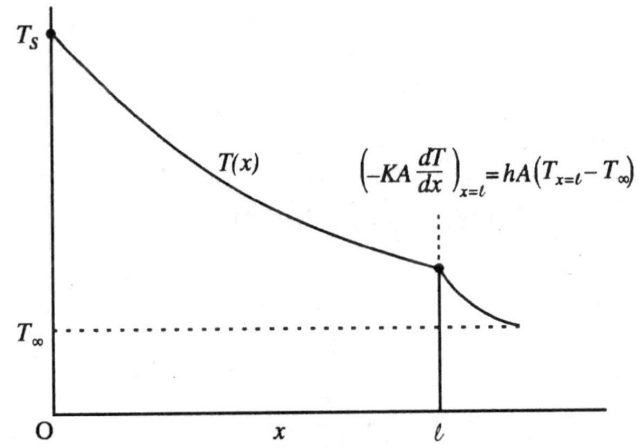

Fig. 3.4 *Temperature distribution in a finite length fin with convection heat transfer at the end.*

The general solution of Eq. (3.1) is given by

$$T - T_\infty = C_1 e^{mx} + C_2 e^{-mx} \tag{3.3}$$

Application of the first boundary condition gives

$$T_s - T_\infty = C_1 + C_2 \tag{3.5}$$

At $x = \ell$, Eq. (3.3) gives

$$T_{x=\ell} - T_\infty = C_1 e^{m\ell} + C_2 e^{-m\ell} \tag{3.24}$$

$$\left[\frac{dT}{dx}\right]_{x=\ell} = mC_1e^{m\ell} - mC_2e^{-m\ell} \tag{3.25}$$

Substituting Eqs. (3.24) and (3.25) in the boundary condition (3.23), we get

$$-km(C_1e^{m\ell} - C_2e^{-m\ell}) = h(C_1e^{m\ell} + C_2e^{-m\ell}) \tag{3.26}$$

Solving Eqs. (3.5) and (3.26), we get

$$C_1 = \frac{(T_s - T_\infty)\left(1 - \dfrac{h}{km}\right)e^{-m\ell}}{(e^{m\ell} + e^{-m\ell}) + \dfrac{h}{km}(e^{m\ell} - e^{-m\ell})}$$

$$C_2 = \frac{(T_s - T_\infty)\left(1 - \dfrac{h}{km}\right)e^{-m\ell}}{(e^{m\ell} + e^{-m\ell}) + \dfrac{h}{km}(e^{m\ell} - e^{-m\ell})} \tag{3.27}$$

Substituting the values of C_1 and C_2 from Eqs. (3.26) and (3.27) in Eq. (3.3), we obtain the temperature distribution as

$$\frac{T - T_\infty}{T_s - T_\infty} = \frac{\cosh[m(\ell - x)] + \dfrac{h}{km}\sinh[m(\ell - x)]}{\cosh(m\ell) + \dfrac{h}{km}\sinh(m\ell)} \tag{3.28}$$

The rate of heat transfer from the fin can be found from Eqs. (3.20) and (3.28) and is given by

$$q = \sqrt{hPkA}(T_s - T_\infty)\frac{\sinh(m\ell) + \dfrac{h}{km}\cosh(m\ell)}{\cosh(m\ell) + \dfrac{h}{km}\sinh(m\ell)} \tag{3.29}$$

3.6 Fin Effectiveness

For a plane surface area, the thermal resistance for convection is equal to $(1/hA)$. Addition of fins to a plane increases the surface area for convection and hence reduces the thermal resistance for convection. However, the thermal resistance for conduction, over that portion of the original surface at which the fins are added, increases. Hence, addition of fins will not always increase the rate of heat transfer. The effectiveness of a fin in transferring energy is expressed in terms of *fin efficiency* η_f, which is defined as

$$\eta_f = \frac{\text{Actual heat transferred}}{\text{heat which would be transferred if the entire fin area were at base temperature}}$$

or

$$\eta_f = \frac{\text{Actual heat transferred}}{hP\ell(T_s - T_\infty)} \tag{3.30}$$

For an infinitely long fin (discussed in section 3.3), the fin efficiency is given by

$$\eta_f = \frac{1}{m\ell} \tag{3.31}$$

For a fin of finite length with insulated end, the fin efficiency is given by

$$\eta_f = \frac{\tanh(m\ell)}{m\ell} \tag{3.32}$$

and for a fin of finite length with heat transfer by convection from the end, the fin efficiency is given by

$$\eta_f = \frac{1}{m\ell}\left[\frac{\sinh(m\ell) + \dfrac{h}{km}\cosh(m\ell)}{\cosh(m\ell) + \dfrac{h}{km}\sinh(m\ell)}\right] \tag{3.33}$$

For a fin of uniform cross-sectional area, the limiting condition can be obtained by differentiating the rate of heat flow with respect to fin length for given values of h, k, P and A and setting the result equal to zero. Now, let us consider a finite length fin with heat transfer by convection from the end. The rate of heat transfer for such a fin is given by

$$q = \sqrt{hPkA}(T_s - T_\infty)\frac{\sinh(m\ell) + \dfrac{h}{km}\cosh(m\ell)}{\cosh(m\ell) + \dfrac{h}{km}\sinh(m\ell)} \tag{3.29}$$

$$q = \sqrt{hPkA}(T_s - T_\infty)\frac{\tanh(m\ell) + \dfrac{h}{km}}{1 + \dfrac{h}{km}\tanh(m\ell)}$$

Then $$\frac{dq}{d\ell} = \frac{\left[1+\frac{h}{km}\tanh(m\ell)\right]\frac{m}{\cosh^2(m\ell)} - \left[\tanh(m\ell)+\frac{h}{km}\right]\frac{h}{k\cosh^2(m\ell)}}{\left[1+\frac{h}{km}\tanh(m\ell)\right]^2} = 0$$

or $$m\left[1+\frac{h}{km}\tanh(m\ell)\right] - \frac{h}{k}\left[\tanh(m\ell)+\frac{h}{km}\right] = 0$$

or

$$m^2 = \frac{h^2}{k^2} \quad \text{or} \quad \frac{hP}{kA} = \frac{h^2}{k^2} \quad \text{or} \quad \frac{hA}{kP} = 1 \tag{3.34}$$

The rate of heat transfer can be expected to increase by the addition of fins only if $(hA/kP) \leq 1$. However, the addition of fins is not justified unless $h < 0.25\ kP/A$.

Consider a rectangular fin, shown in Fig. 3.5, of length ℓ, width b and thickness δ. Then, perimeter $P = 2(b+\delta) \simeq 2b$ (since width $b \gg \delta$)

Cross-sectional area $A = b\delta$

$$m = \sqrt{\frac{hP}{kA}} = \sqrt{\frac{h2b}{kb\delta}} = \sqrt{\frac{2h}{k\delta}}$$

The limiting condition is $m^2 = \dfrac{h^2}{k^2}$

or

$$\frac{2h}{k\delta} = \frac{h^2}{k^2} \quad \text{or} \quad \frac{1}{h} = \frac{\delta}{2k} \tag{3.35}$$

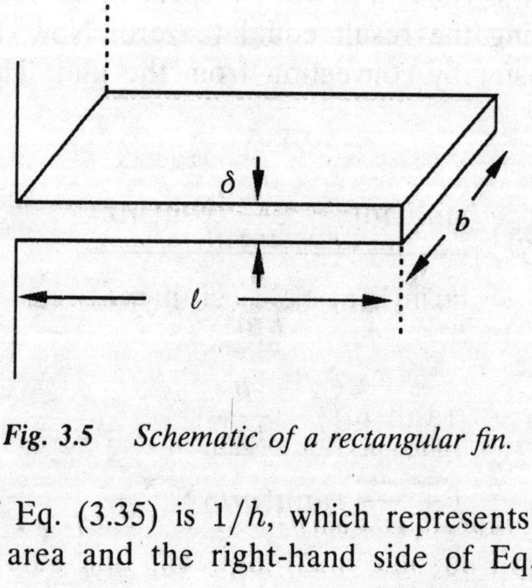

Fig. 3.5 Schematic of a rectangular fin.

The left-hand side of Eq. (3.35) is $1/h$, which represents the thermal resistance for convection per unit area and the right-hand side of Eq. (3.35) is $\delta/2k$, which represents the thermal resistance for conduction per unit area of a wall, the thickness of which is one half of the thickness of the fin. The rate of heat transfer from a fin increases if $(hA/kP) \leq 1$. This implies that $(h\delta/2k) \leq 1$ for a rectangular fin.

From the above considerations it is apparent that fins would be effective if

1. k is large. Hence, fins are usually made of aluminium which has high thermal conductivity.
2. h is small. Therefore, fins are usually used to increase the rate of heat transfer to a gas medium and they are less effective when the medium is a liquid.
3. δ is small. Therefore, a large number of thin and closely spaced fins are more effective than fewer and thicker fins.

EXAMPLE 3.1 A plane wall is fitted with an aluminium (k = 204 W/m K) pin fin of 1 cm diameter and 30 cm length. The fin base temperature is 300°C and the pin fin is in contact with air at 30°C. The convection heat transfer coefficient between the fin surface and air is 10 W/m^2 K. Assuming that the fin is infinitely long, calculate the temperature at 5 cm, 15 cm and 25 cm from the base and the rate of heat transfer from the fin.

SOLUTION Perimeter of the fin

$$P = \pi d = \pi \times 1 \times 10^{-2} = 3.14 \times 10^{-2} \text{ m}$$

Cross-sectional area

$$A = \frac{\pi}{4}d^2 = \frac{\pi}{4} \times (1 \times 10^{-2})^2 = 0.7854 \times 10^{-4} \text{ m}^2$$

$$m = \sqrt{\frac{hP}{kA}} = \sqrt{\frac{10 \times 3.14 \times 10^{-2}}{204 \times 0.785 \times 10^{-4}}} = 4.428$$

The temperature distribution in an infinitely long fin is given by

$$T - T_\infty = (T_s - T_\infty)e^{-mx} = (300 - 30)e^{-4.428\,x}$$

When x = 0.05 m

$$T - T_\infty = 270e^{-4.428 \times 0.05} = 216.4°\text{C}$$

or T (5 cm) = 246.4°C

When $x = 0.15$ m, $T - T_\infty = 139°\text{C}$ or T (15 cm) = 169°C

When $x = 0.25$ m, $T - T_\infty = 270\,e^{-4.428 \times 0.25} = 89.2°\text{C}$

or T (25 cm) = 119.2°C

Rate of heat transfer from the fin is given by

$$q = \sqrt{hPkA}(T_s - T_\infty)$$

$$= \sqrt{10 \times 3.14 \times 10^{-2} \times 204 \times 0.7854 \times 10^{-4}}(300 - 30) = 19.15 \text{ W}$$

EXAMPLE 3.2 An aluminium (k = 204 W/m K) rod 2 cm in diameter and 20 cm long protrudes from a wall which is maintained at 300°C. The end of the rod is insulated and the surface of the rod is exposed to air at 30°C. The convection heat transfer coefficient between the rod's surface and air is 10 W/m^2 K. Calculate the heat lost by the rod. Also calculate the temperature of the rod at a distance of 10 cm from the wall.

SOLUTION The aluminium rod acts as a fin of finite length with insulated end.

Perimeter of fin, $P = \pi d = \pi \times 2 \times 10^{-2} = 6.283 \times 10^{-2}$ m

Surface area of fin, A is given by

$$A = \frac{\pi}{4}d^2 = \frac{\pi}{4} \times 4 \times 10^{-4} = 3.14 \times 10^{-4} \text{ m}^2$$

$$m = \sqrt{\frac{hP}{kA}} = \sqrt{\frac{10 \times 6.283 \times 10^{-2}}{204 \times 3.14 \times 10^{-4}}} = 3.132$$

Rate of heat transfer,

$$q = \sqrt{hPkA}(T_s - T_\infty)\tanh(m\ell)$$

$$= \sqrt{10 \times 6.283 \times 10^{-2} \times 204 \times 3.14 \times 10^{-4}}(300 - 30)\tanh(3.132 \times 0.2)$$

$$= 30 \text{ W}$$

The temperature distribution is given by

$$\frac{T - T_\infty}{T_s - T_\infty} = \frac{\cosh\{m(\ell - x)\}}{\cosh(m\ell)}$$

Therefore, T at 10 cm from the base is given by

$$\frac{T - 30}{300 - 30} = \frac{\cosh(3.132 \times 0.1)}{\cosh(3.132 \times 0.2)} = 0.8726$$

or $T = 265.6°\text{C}$

EXAMPLE 3.3 A rectangular fin of length 30 cm, width 30 cm and thickness 2 mm is attached to a surface at 300°C. The fin is made of aluminium (k = 204 W/m K) and is exposed to air at 30°C. The fin end is uninsulated and can lose heat through its end also. The convection heat transfer coefficient between the fin surface and air is 15 W/m^2 K. Calculate the temperature of the fin at (a) 10 cm (b) 20 cm and (c) 30 cm from the base and (d) the rate of heat transfer from the fin.

SOLUTION (a) Perimeter of fin, $P = 2(b + \delta) = 2(0.3 + 0.002) = 0.604$ m

Cross-sectional area of fin, $A = b\delta = 0.3 \times 0.002 = 0.0006$ m^2

$$m = \sqrt{\frac{hP}{kA}} = \sqrt{\frac{15 \times 0.604}{204 \times 0.0006}} = 8.603$$

The temperature distribution in the fin is given by

$$\frac{T - T_\infty}{T_s - T_\infty} = \frac{\cosh\{m(\ell - x)\} + \dfrac{h}{km}\sinh\{m(\ell - x)\}}{\cosh(m\ell) + \dfrac{h}{km}\sinh(m\ell)}$$

$$= \frac{\cosh\{8.603(0.3 - x)\} + \dfrac{15}{204 \times 8.603}\sinh\{8.603(0.3 - x)\}}{\cosh(8.603 \times 0.3) + \dfrac{15}{204 \times 8.603}\sinh(8.603 \times 0.3)}$$

or
$$\frac{T - T_\infty}{300 - 30} = \frac{\cosh\{8.603(0.3 - x)\} + 8.547 \times 10^{-3}\sinh\{8.603(0.3 - x)\}}{6.6985}$$

when x = 0.1 m

$$\frac{T - 30}{270} = \frac{\cosh(8.603 \times 0.2) + 8.547 \times 10^{-3}\sinh(8.603 \times 0.2)}{6.6985}$$

or $T(x = 0.1 \text{ m}) = 147.2°\text{C}$

(b) When x = 0.2 m

$$\frac{T - 30}{270} = \frac{\cosh(8.603 \times 0.1) + 8.547 \times 10^{-3}\sinh(8.603 \times 0.1)}{6.6985}$$

or $T(x = 0.2 \text{ m}) = 86.5°\text{C}$

(c) When $x = 0.3$ m

$$\frac{T-30}{270} = \frac{\cosh(0) + 8.547 \times 10^{-3} \sinh(0)}{6.6985}$$

or $T(x = 0.3 \text{ m}) = 70.3°\text{C}$

(d) The rate of heat transfer from the fin is given by

$$q = \sqrt{hPkA}(T_s - T_\infty)\frac{\sinh(m\ell) + \dfrac{h}{km}\cosh(m\ell)}{\cosh(m\ell) + \dfrac{h}{km}\sinh(m\ell)}$$

$$\sqrt{hPkA}(T_s - T_\infty) = \sqrt{15 \times 0.604 \times 204 \times 0.0006}(300 - 30) = 284.33$$

$$\frac{\sinh(m\ell) + \dfrac{h}{km}\cosh(m\ell)}{\cosh(m\ell) + \dfrac{h}{km}\sinh(m\ell)}$$

$$= \frac{\sinh(8.603 \times 0.3) + \dfrac{15}{204 \times 8.603}\cosh(8.603 \times 0.3)}{\cosh(8.603 \times 0.3) + \dfrac{15}{204 \times 8.603}\sinh(8.603 \times 0.3)} = 0.9888$$

Therefore, $q = 284.33 \times 0.9888 = 281.1$ W.

EXAMPLE 3.4 Calculate the fin efficiency for the fins described in Examples 3.1 to 3.3.

SOLUTION (a) Example 3.1 refers to an infinitely long fin. For an infinitely long fin, the fin efficiency is given by

$$\eta_f = \frac{1}{m\ell} = \frac{1}{4.428 \times 0.3} = 0.7528$$

$m = 4.428$ and $\ell = 0.3$ m [see Example 3.1]

(b) Example 3.2 deals with a finite length fin with insulated end. The efficiency of a fin of finite length with insulated end is given by

$$\eta_f = \frac{\tanh(m\ell)}{m\ell} = \frac{\tanh(3.132 \times 0.2)}{3.132 \times 0.2} = 0.8869$$

($m = 3.132$ and $\ell = 0.2$ m, see Example 3.2)

(c) Example 3.3 deals with a fin of finite length with convection heat transfer from the end. The efficiency of such a fin is given by

$$\eta_f = \frac{1}{m\ell}\left[\frac{\sinh(m\ell) + \dfrac{h}{km}\cosh(m\ell)}{\cosh(m\ell) + \dfrac{h}{km}\sinh(m\ell)}\right]$$

$$= \frac{1}{8.603 \times 0.3}\left[\frac{\sinh(8.603 \times 0.3) + \dfrac{15}{204 \times 8.603}\cosh(8.603 \times 0.3)}{\cosh(8.603 \times 0.3) + \dfrac{15}{204 \times 8.603}\sinh(8.603 \times 0.3)}\right]$$

$$= 0.3875 \times 0.9888 = 0.3832$$

EXAMPLE 3.5 The temperature of steam flowing through a 50 mm carbon-steel pipe is measured with the help of a mercury-in-glass thermometer immersed in an oil-filled steel well (k = 54 W/m K). A schematic of the thermometer well is shown in Fig. 3.6. The heat transfer coefficient between steam and the thermometer well is 1500 W/m^2 K. If the surface temperature of the pipe is 50°C and the thermometer reading is 250°C, determine the true temperature of steam.

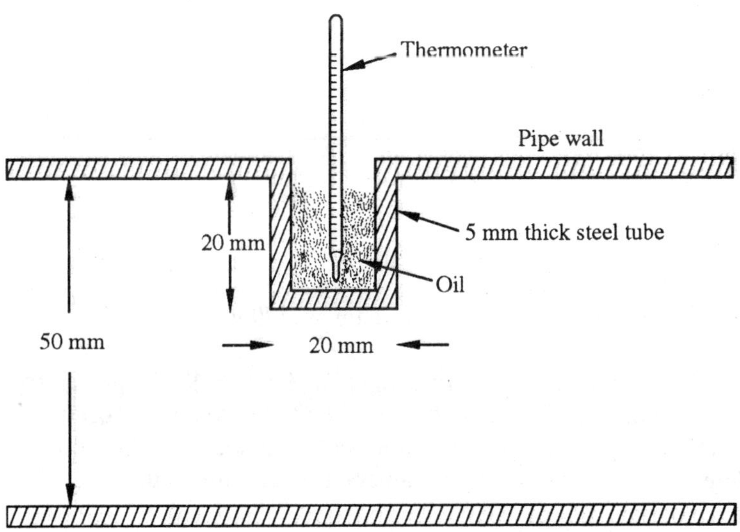

Fig. 3.6 Schematic of a thermometer well.

SOLUTION The thermometer well is a hollow rod protruding into steam at temperature T_∞. As the heat flows from the steam to the cooler pipe wall through the thermometer well, the thermometer reads the temperature at the tip of the thermometer well but does not indicate the true temperature of the steam. The thermometer

well can be treated as a fin of finite length with heat transfer from the end.
Perimeter of fin,

$$P = \pi d = \pi \times 10 \times 10^{-3} = 3.1416 \times 10^{-2} \text{ m}$$

Cross-sectional area of fin,

$$A = \frac{\pi}{4}(d_o^2 - d_i^2) = \frac{\pi}{4}\left[(20 \times 10^{-3})^2 - (10 \times 10^{-3})^2\right] = 0.2356 \times 10^{-3} \text{ m}^2$$

$$m = \sqrt{\frac{hP}{kA}} = \sqrt{\frac{1500 \times 3.1416 \times 10^{-2}}{54 \times 0.2356 \times 10^{-3}}} = 60.86$$

$$m\ell = 60.86 \times 20 \times 10^{-3} = 1.217$$

The temperature distribution in a fin of finite length with heat transfer from the end is given by

$$\frac{T - T_\infty}{T_s - T_\infty} = \frac{\cosh[m(\ell - x)] + \dfrac{h}{km}\sinh[m(\ell - x)]}{\cosh(m\ell) + \dfrac{h}{km}\sinh(m\ell)}$$

When $x = \ell$, we get

$$\frac{T - T_\infty}{T_s - T_\infty} = \frac{250 - T_\infty}{50 - T_\infty} = \frac{1}{\cosh(1.217) + \dfrac{1500}{54 \times 60.86}\sinh(1.217)} = 0.3937$$

or $\quad T_\infty = 379.9°\text{C}$

Therefore, the true temperature of the steam = 379.9°C.

EXAMPLE 3.6 One end of a very long nickel (k = 90 W/m K) rod of 25 mm diameter is maintained at 500°C and the rest of the rod is exposed to air at 30°C. If the convection heat transfer coefficient between air and the rod's surface is 10 W/m^2 K, determine the rate of heat transfer through the rod.

SOLUTION The rod acts as an infinitely long fin whose base is at 500°C. The rate of heat transfer from an infinitely long fin is given by

$$q = \sqrt{hPkA}(T_s - T_\infty)$$

$$= \sqrt{10 \times \pi \times 25 \times 10^{-3} \times 90 \times \frac{\pi}{4}(25 \times 10^{-3})^2}(500 - 30) = 87.5 \text{ W}$$

3.7 Summary

The heat conducted through a body is subsequently transferred to a fluid by convection. Hence, an understanding of the combined conduction-convection heat transfer is important from a practical viewpoint. When a solid of relatively small cross-sectional area protrudes from the surface of a large body into a fluid at different temperature, the protruding body is called an extended surface or fin. The fins may have different configurations. Longitudinal fins of rectangular cross-section, cylindrical tubes with fins of rectangular profile or with radial fins, cylindrical spine fins or pin fins, and truncated conical fins are in wide use.

If the base of a fin of uniform cross-section A and perimeter P is held at temperature T_s and the fin is exposed to a fluid at temperature T_∞, the temperature distribution in the fin is given by

$$T - T_\infty = C_1 e^{mx} + C_2 e^{-mx}$$

where

$$m = \sqrt{\frac{hP}{kA}}$$

h = convection heat transfer coefficient between the fin surface and fluid

k = thermal conductivity of fin material

T = temperature of the fin at a distance x from the base.

The constants C_1 and C_2 depend on the boundary conditions. The first boundary condition is

$$T = T_s \quad \text{at} \quad x = 0$$

The second boundary condition and the temperature distribution depend on the type of the fin.

For an infinitely long fin, the second boundary condition is $T = T_\infty$ at $x = \infty$. The temperature distribution in the fin is given by

$$\frac{T - T_\infty}{T_s - T_\infty} = e^{-mx}$$

and the rate of heat transfer from the fin is given by

$$q = \sqrt{hPkA}(T_s - T_\infty)$$

For a fin of finite length with an insulated end, the second boundary condition is given by

$$\frac{dT}{dx} = 0 \quad \text{at} \quad x = \ell$$

where ℓ is the length of the fin. The temperature distribution in the fin is given by

$$\frac{T - T_\infty}{T_s - T_\infty} = \frac{\cosh[m(\ell - x)]}{\cosh(m\ell)}$$

and the rate of heat transfer from the fin is given by

$$q = \sqrt{hPkA}(T_s - T_\infty)\tanh(m\ell)$$

For a fin of finite length ℓ with convection heat transfer from the end, the second boundary condition is giver by

$$\left[-kA\frac{dT}{dx}\right]_{x=\ell} = hA(T_{x=\ell} - T_\infty)$$

The temperature distribution in the fin is given by

$$\frac{T - T_\infty}{T_s - T_\infty} = \frac{\cosh[m(\ell - x)] + \dfrac{h}{km}\sinh[m(\ell - x)]}{\cosh(m\ell) + \dfrac{h}{km}\sinh(m\ell)}$$

and the rate of heat transfer from the fin is given by

$$q = \sqrt{hPkA}(T_s - T_\infty)\frac{\sinh(m\ell) + \dfrac{h}{km}\cosh(m\ell)}{\cosh(m\ell) + \dfrac{h}{km}\sinh(m\ell)}$$

The fin efficiency η_f is defined as the ratio of the *actual heat transferred by the fin to the heat which would be transferred if the entire fin area were at the base temperature.*

For an infinitely long fin

$$\eta_f = \frac{1}{m\ell}$$

For a finite length fin with insulated end

$$\eta_f = \frac{\tanh(m\ell)}{m\ell}$$

For a fin of finite length with convection heat transfer from the end

$$\eta_f = \frac{1}{m\ell}\left[\frac{\sinh(m\ell) + \dfrac{h}{km}\cosh(m\ell)}{\cosh(m\ell) + \dfrac{h}{km}\sinh(m\ell)}\right]$$

The fins would be effective if
(a) they are made of material having high thermal conductivity.
(b) the convection heat transfer coefficient is low and
(c) the thickness of the fin is very small.

Review Questions

1. What is an extended surface?
2. What is the importance of neglecting radial heat flow in fins?
3. What is the mode of heat transfer through a fin?
4. Name a few appliances (domestic) which use fins.
5. Name a few commonly used fins.
6. Define fin efficiency.
7. Why is aluminium used as a fin material?
8. In which medium – gas or liquid – will the use of fins be more effective?
9. Is it advisable to use fins on the surface exposed to condensing steam?
10. Is it better to use a large number of closely spaced thin fins or a small number of thick fins?

Problems

3.1 Consider an infinitely long fin of rectangular cross-section and derive the relation to determine the temperature distribution in the fin.

3.2 One end of a very long copper (k = 386 W/m K) rod of 5 cm diameter is kept in a furnace and is maintained at 600°C, while the surface of the rod is exposed to air at 25°C. If the heat transfer coefficient between the surface and air is 10 W/m^2 K, calculate the rate of heat transfer through the rod. Also estimate the temperature of the rod at 20 cm, 40 cm, and 60 cm as measured from the hot end.

3.3 A rod fin of 2 cm diameter and 50 cm length is to be attached to a surface at 400°C, while the fin surface is exposed to air at 25°C. The convection heat transfer coefficient between the fin surface and air is 15 W/m^2 K. Compare the temperature distributions for three fin materials (a) aluminium (k = 204 W/m K) (b) nickel (k = 90 W/m K) and (c) nickel-steel (k = 19 W/m K).

3.4 A very long aluminium pin fin of 2 cm diameter is attached to a surface at 300°C, while the fin surface is exposed to an environment at 25°C. Calculate the rate of heat transfer from the fin if the medium is (a) flowing liquid with a heat transfer coefficient of 200 W/m^2 K. (b) flowing air with h = 20 W/m^2 K. (c) stagnant air with h = 5 W/m^2 K.

3.5 An aluminium rectangular fin of length 50 cm, width 20 cm and thickness 2 mm is attached to a wall at 350°C and the end of the fin is insulated. The fin surface is exposed to an environment at 25°C with h = 10 W/m^2 K. Calculate the temperature of the fin
(a) at a distance of 25 cm from the base and
(b) at the insulated end.
(c) Also calculate the rate of heat transfer from the fin.

3.6 One end of a long, 2 cm diameter metal rod of unknown thermal conductivity is maintained at 400°C by keeping it in a furnace, while the rod's surface is exposed to air at 25°C. The convection heat transfer coefficient from the surface is 10 W/m^2 K. Once the steady-state is reached, the temperature of the rod at a distance of 60 cm from the hot end is 100°C. Determine the thermal conductivity of the metal.

3.7 One end of a long copper rod of 3 cm diameter is maintained at 400°C, while the rest of the rod is exposed to an environment at 25°C. In steady-state, the temperature of the rod at a distance of 50 cm from the hot end is found to be 120°C. Determine the average value of the convection heat transfer coefficient from the surface to the environment.

3.8 One end of a long copper (k = 386 W/m K) rod of 3 cm diameter is maintained at a constant temperature by placing it in a furnace, while the rest of the rod is exposed to an environment at 25°C. In steady-state, the temperatures at two different locations 25 cm apart along the length of the rod are found to be 200°C and 100°C, respectively. Calculate the heat transfer coefficient from the rod's surface to the environment.

3.9 One end of a wrought-iron (k = 59 W/m K) rod of 30 cm length and 3 cm diameter is attached to a wall at 300°C, while the second end is insulated. The surface of the rod is exposed to an environment at 25°C and the convection heat transfer coefficient from the surface to the environment is 15 W/m^2 K. Determine (a) the temperature of the insulated end and (b) the rate of heat transfer from the rod.

3.10 The handle of a ladle used for pouring molten metal is 50 cm long and is made of 5 cm × 5 cm solid carbon-steel (k = 54 W/m K). The end of the handle is covered with a thick layer of insulating material. To reduce the temperature at the point of grip, it is proposed to form a hollow handle of the same size by using 1 cm thick carbon-steel plates. If the surrounding air is at 30°C and the convection heat transfer coefficient is 10 W/m^2 K, determine the percent reduction in the temperature at the grip point.

3.11 The temperature distribution in a fin of finite length with heat transfer from the end is given by Eq. (3.28). Show that the rate of heat transfer from the fin is given by Eq. (3.29).

3.12 One of the ends of two long rods of the same diameter, one made of copper (k = 386 W/m K) and the other of nickel-steel (k = 19 W/m K), are attached to a surface which is maintained at a high temperature. The surfaces of both the rods are exposed to the same environment. At steady-state, the temperature of the copper rod at a distance of 50 cm from the hot end is found to be 100°C. At what distance from the hot end would the same temperature be attained in the nickel-steel rod?

3.13 One end of an aluminium rod of 2 cm diameter and 50 cm length is attached to a wall at 200°C, and the rest of the rod is exposed to an environment at 25°C. The other end of the rod is losing heat to the environment. If the heat transfer coefficient to the environment is 5 W/m^2 K, determine (a) the temperature at the end of the rod and (b) the rate of heat transfer from the rod.

3.14 Calculate the efficiency of the fin described in Problem 3.5.

3.15 An aluminium rod of 2 cm diameter and 30 cm length is used as a fin. The heat transfer coefficient from the fin to the environment is 10 W/m^2 K. Calculate the fin efficiency assuming that the fin is
(a) of finite length but the end is insulated, and
(b) of finite length and loses heat through its end also.

3.16 The temperature of steam flowing through a pipe is measured as 150°C with a thermometer placed in a thermometer well. A sketch of the thermometer well is shown in Fig. 3.6. The pipe surface is at 50°C and the convection heat transfer coefficient between the thermometer well and steam is 600 W/m^2 K. The thermal conductivity of the thermometer well is 54 W/m K. Estimate the true temperature of steam, assuming that the thermometer well is an infinitely long fin.

3.17 Rework Problem 3.16 assuming that the thermometer well is a finite length fin with an insulated end.

3.18 Rework Problem 3.16 assuming that the thermometer well is a finite length fin with heat losses from the end also.

3.19 Two walls A and B are maintained at temperatures T_A and T_B, respectively. One end of a metal rod of length ℓ is embedded in wall A, while the other end is fixed to wall B. The rod loses heat by convection to the environment at T_∞. Derive an expression to determine (a) the temperature distribution in the rod and (b) the total heat lost by the rod.

3.20 Derive an expression to determine the heat transferred from the wall A described in Problem 3.19.

3.21 Two walls 1 m apart are kept at 200°C and 100°C, respectively. If the walls are connected by a 2 cm diameter brass (k = 111 W/m K) rod and the rod loses heat by convection (h = 10 W/m^2 K) to the ambient air at 25°C, determine the total heat transferred from the brass rod to air.

CHAPTER FOUR

TWO- AND THREE-DIMENSIONAL STEADY-STATE HEAT CONDUCTION

Learning Objectives

This chapter will enable the readers to

- know the various methods of analysis which can be applied to treat heat conduction.
- derive the general heat conduction equation in cartesian coordinates and in cylindrical coordinates.
- deduce Fourier, Poisson and Laplace equations.
- obtain analytical solutions to two-dimensional steady-state heat conduction problems in some simple cases.
- obtain approximate solutions to two-dimensional heat conduction problems by graphical methods (flux plot).
- predict conduction shape factors for several systems.
- develop electrical analogues of steady-state two-dimensional heat conduction problems.
- understand and apply the relaxation method of analysis for steady-state two-dimensional heat conduction problems.

4.1 Methods of Analysis

In chapters 2 and 3, steady-state heat conduction was discussed in systems in which the temperature gradient and heat flow were treated as functions of only one variable. In many practical situations the temperature gradient depends on more than one variable and hence it is necessary to develop the more general heat conduction equation. In this chapter, a few methods for analyzing heat conduction in two- and three-dimensional systems will be discussed. Since the analysis of three-dimensional systems is more complex and involved, we shall focus our attention on the analysis of two-dimensional systems to illustrate the methods of analysis.

Heat conduction in two- and three-dimensional systems can be treated by (a) analytical (b) graphical (c) analogical and (d) numerical methods. A complete treatment of analytical solutions requires a good knowledge of Fourier series, Bessel functions,

Legendre polynomials and complex variables. As this book is primarily meant for an introductory course on heat transfer, we shall focus our attention on the analytical solution of a relatively simple system to illustrate the method. The graphical, analogical and numerical methods do not require a knowledge of advanced mathematics, and they provide reasonably accurate solutions which are quite useful in an engineering analysis. Moreover, an analytical solution may not be feasible for many of the problems encountered by an engineer. In such situations, the approximate solutions can be obtained by adopting the graphical or numerical methods. In this chapter, we shall discuss the graphical, analogical and numerical methods of analysis.

4.2 General Heat Conduction Equation in Cartesian Coordinates

In this section we shall derive the general equation which governs the temperature distribution in a body. For this purpose consider a small rectangular parallelopiped element of dimensions dx, dy and dz, of the body as shown in Fig. 4.1.

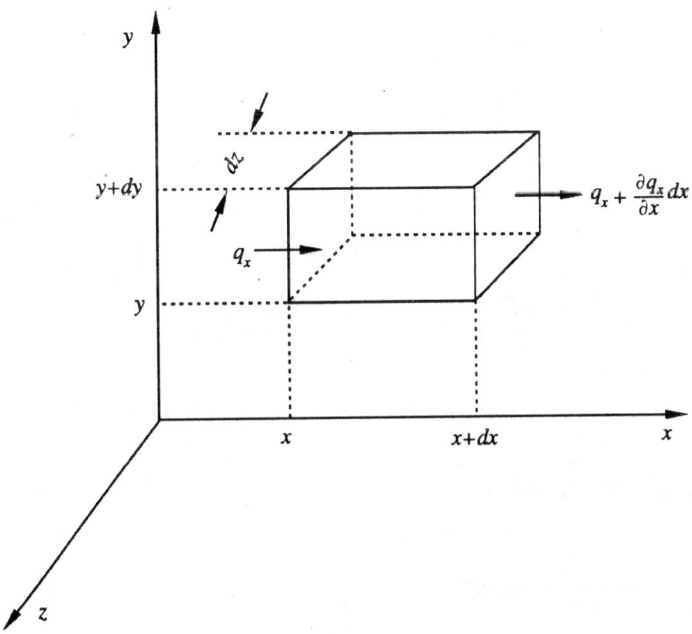

Fig. 4.1 Small element of dimensions dx, dy and dz of a body.

During a time interval dt the energy balance equation for the element is given by

$$\begin{bmatrix}\text{Heat flow into}\\ \text{the element}\\ \text{during time } dt\end{bmatrix} - \begin{bmatrix}\text{Heat flow out}\\ \text{of the element}\\ \text{during time } dt\end{bmatrix} + \begin{bmatrix}\text{Heat generated}\\ \text{in the element}\\ \text{during time } dt\end{bmatrix} = \begin{bmatrix}\text{Change in the}\\ \text{energy of the}\\ \text{element during}\\ \text{time } dt\end{bmatrix}$$

or

$$(q_x+q_y+q_z)dt+\dot{q}(dxdydz)dt-(q_{x+dx}+q_{y+dy}+q_{z+dz})\,dt=\rho c(dxdydz)dT \quad (4.1)$$

where

q_x = rate of heat conduction into the element across the left face in the x-direction

$$= -k(dydz)\frac{\partial T}{\partial x}$$

q_{x+dx} = rate of heat conduction out of the element across the right face in the x-direction

$$= q_x+\frac{\partial q_x}{\partial x}dx = -k(dydz)\frac{\partial T}{\partial x}+\frac{\partial}{\partial x}\left[-k(dydz)\frac{\partial T}{\partial x}\right]dx$$

$$= \left[-k\frac{\partial T}{\partial x}+\frac{\partial}{\partial x}\left(-k\frac{\partial T}{\partial x}\right)dx\right]dydz$$

c = specific heat of the body

ρ = density of the body.

Then

$$q_x - q_{x+dx} = \frac{\partial}{\partial x}\left[k\frac{\partial T}{\partial x}\right]dxdydz = k\frac{\partial^2 T}{\partial x^2}dxdydz \quad (4.2)$$

Similarly for the y and z directions, we get

$$q_y - q_{y+dy} = k\frac{\partial^2 T}{\partial y^2}dxdydz \quad (4.3)$$

$$q_z - q_{z+dz} = k\frac{\partial^2 T}{\partial z^2}dxdydz \quad (4.4)$$

Substituting Eqs. (4.2)–(4.4) in Eq. (4.1) and dividing throughout by $dxdydzdt$, we get

$$k\left[\frac{\partial^2 T}{\partial x^2}+\frac{\partial^2 T}{\partial y^2}+\frac{\partial^2 T}{\partial z^2}\right]+\dot{q}=\rho c\frac{\partial T}{\partial t}$$

or

$$\frac{\partial^2 T}{\partial x^2}+\frac{\partial^2 T}{\partial y^2}+\frac{\partial^2 T}{\partial z^2}+\frac{\dot{q}}{k}=\left(\frac{\rho c}{k}\right)\frac{\partial T}{\partial t}=\frac{1}{\alpha}\frac{\partial T}{\partial t} \quad (4.5)$$

where

$$\alpha = \frac{k}{\rho c} = \text{thermal diffusivity (has the units m}^2\text{/s)} \tag{4.6}$$

Equation (4.5) is the required general heat conduction equation which governs the temperature distribution and conduction heat flow in an isotropic solid.

If the body has no heat sources ($\dot{q} = 0$), Eq. (4.5) reduces to the *Fourier equation* given by

$$\frac{\partial^2 T}{\partial x^2} + \frac{\partial^2 T}{\partial y^2} + \frac{\partial^2 T}{\partial z^2} = \frac{1}{\alpha}\frac{\partial T}{\partial t} \tag{4.7}$$

If the system is in steady-state but has heat sources, then Eq. (4.5) reduces to the *Poisson equation* given by

$$\frac{\partial^2 T}{\partial x^2} + \frac{\partial^2 T}{\partial y^2} + \frac{\partial^2 T}{\partial z^2} + \frac{\dot{q}}{k} = 0 \tag{4.8}$$

If the system is in steady-state and has no heat sources, Eq. (4.5) reduces to the *Laplace equation* given by

$$\frac{\partial^2 T}{\partial x^2} + \frac{\partial^2 T}{\partial y^2} + \frac{\partial^2 T}{\partial z^2} = 0 \tag{4.9}$$

4.3 General Heat Conduction Equation in Cylindrical Coordinates

There are several situations in heat conduction which can be handled more conveniently in the cylindrical coordinate system. Consider a small cylindrical volume element, having the coordinates r, θ and z, of the body as shown in Fig. 4.2.

The volume of the element is $rd\theta drdz$. The energy balance on the volume element can be written as

$$\begin{bmatrix}\text{Change in the}\\ \text{energy of the}\\ \text{element during}\\ \text{time } dt\end{bmatrix} = \begin{bmatrix}\text{Heat flow}\\ \text{into the ele-}\\ \text{ment during}\\ \text{time } dt\end{bmatrix} - \begin{bmatrix}\text{Heat flow}\\ \text{out of the}\\ \text{element dur-}\\ \text{ing time } dt\end{bmatrix} + \begin{bmatrix}\text{Heat gene-}\\ \text{rated in the}\\ \text{element dur-}\\ \text{ing time } dt\end{bmatrix}$$

or

$$\rho c(rd\theta drdz)dT = (q_z + q_r + q_\theta)dt - [q_{z+dz} + q_{r+dr} + q_{\theta+d\theta}] + \dot{q}(rd\theta drdz)dt \tag{4.10}$$

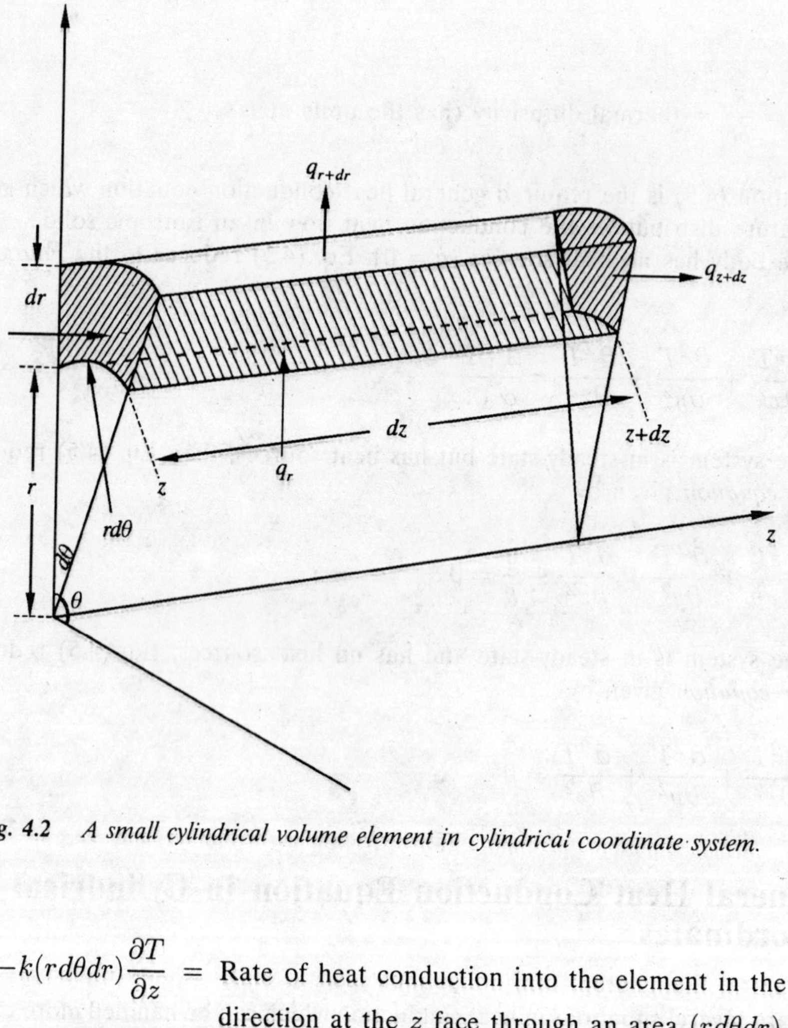

Fig. 4.2 A small cylindrical volume element in cylindrical coordinate system.

where

$q_z = -k(rd\theta dr)\dfrac{\partial T}{\partial z}$ = Rate of heat conduction into the element in the axial direction at the z face through an area $(rd\theta dr)$

$$q_{z+dz} = q_z + \frac{\partial q_z}{\partial z}$$

= Rate of heat conduction out of the element in the axial direction at the $(z+dz)$ face

$$= -k(rd\theta dr)\frac{\partial T}{\partial z} + \frac{\partial}{\partial z}\left[-k(rd\theta dr)\frac{\partial T}{\partial z}\right] dz$$

or

$$q_z - q_{z+dz} = \frac{\partial}{\partial z}\left[k(rd\theta dr)\frac{\partial T}{\partial z}\right] dz = k(rd\theta drdz)\frac{\partial^2 T}{\partial z^2} \tag{4.11}$$

$$q_r = -k(rd\theta dz)\frac{\partial T}{\partial r}$$

= Heat flow into the element in the radial direction at the r face through an area $(rd\theta dz)$

$$q_{r+dr} = q_r + \frac{\partial q_r}{\partial r}dr$$

= Heat flow out of the element in the radial direction at the $(r+dr)$ face through the area $(r+dr)d\theta dz$

$$= -k(rd\theta dz)\frac{\partial T}{\partial r} + \frac{\partial}{\partial r}\left[-k(rd\theta dz)\frac{\partial T}{\partial r}\right] dr$$

or

$$q_r - q_{r+dr} = k(rd\theta dzdr)\frac{\partial^2 T}{\partial r^2} + k(d\theta dzdr)\frac{\partial T}{\partial r} \tag{4.12}$$

$$q_\theta = -k(drdz)\left(\frac{\partial T}{rd\theta}\right)$$

= Heat flow into the element in the circumferential direction at the θ face through an area $(drdz)$

$$q_{\theta+d\theta} = q_\theta + \frac{\partial q_\theta}{rd\theta}(rd\theta)$$

= Heat flow out of the element in the circumferential direction at $(\theta + d\theta)$ face

$$= -k(drdz)\frac{\partial T}{rd\theta} + \frac{\partial}{rd\theta}\left[-k(drdz)\frac{\partial T}{rd\theta}\right] rd\theta$$

or

$$q_\theta - q_{\theta+d\theta} = k(drdzrd\theta)\frac{\partial^2 T}{r^2 d\theta^2} \tag{4.13}$$

Heat generated in the element during time $dt = \dot{q}(rd\theta drdz)dt$ (4.14)

Change in the energy of the element during time $dt = \rho c(rd\theta drdz)\frac{\partial T}{\partial t}$ (4.15)

Substituting Eqs. (4.11)–(4.15) in Eq. (4.10) and dividing throughout with $k(rd\theta dr dz)dt$, we get

$$\frac{\partial^2 T}{\partial z^2}+\frac{\partial^2 T}{\partial r^2}+\frac{1}{r}\frac{\partial T}{\partial r}+\frac{1}{r^2}\frac{\partial^2 T}{\partial \theta^2}+\frac{\dot{q}}{k}=\frac{\rho c}{k}\frac{\partial T}{\partial t}=\frac{1}{\alpha}\frac{\partial T}{\partial t} \tag{4.16}$$

Equation (4.16) is the required general heat conduction equation in cylindrical coordinates.

If the cylinder is long compared to the diameter, the heat flows in the radial direction only (one-dimensional), and in the absence of heat sources ($\dot{q}=0$) and the system in steady-state, Eq. (4.16) reduces to

$$\frac{d^2 T}{dr^2}+\frac{1}{r}\frac{dT}{dr}=0 \tag{4.17}$$

In the spherical coordinate system (r, θ, ϕ) the general heat conduction equation is given by

$$\frac{\partial^2 T}{\partial r^2}+\frac{2}{r}\frac{\partial T}{\partial r}+\frac{1}{r^2\sin^2\theta}\frac{\partial^2 T}{\partial \phi^2}+\frac{1}{r^2\sin\theta}\frac{\partial}{\partial \theta}\left(\sin\theta\frac{\partial T}{\partial \theta}\right)+\frac{\dot{q}}{k}=\frac{1}{\alpha}\frac{\partial T}{\partial t} \tag{4.18}$$

If there are no heat sources in the body and the system is in steady-state, and the heat flow is in the radial direction (one-dimensional), Eq. (4.18) reduces to

$$\frac{d^2 T}{dr^2}+\frac{2}{r}\frac{dT}{dr}=0 \tag{4.19}$$

4.4 Analytical Solution of Two-Dimensional Heat Conduction in Rectangular Plates

A relatively simple problem to illustrate the analytical solution of the heat conduction equation is the steady-state heat conduction in a thin rectangular plate with no internal heat generation. Let us consider a thin rectangular plate of size $\ell \times b$ which is insulated at the top and bottom faces as shown in Fig. 4.3. For a thin plate which is insulated at the top and bottom faces, there is no temperature gradient in the z direction. Then the temperature is a function of x and y only. That is,

$$T = T(x, y) \quad \text{and} \quad \frac{\partial T}{\partial z}=0$$

Let us assume that the three faces $x = 0, x = \ell$ and $y = 0$ are held at a constant temperature T_1 and the upper face $y = b$ has a temperature distribution such as a

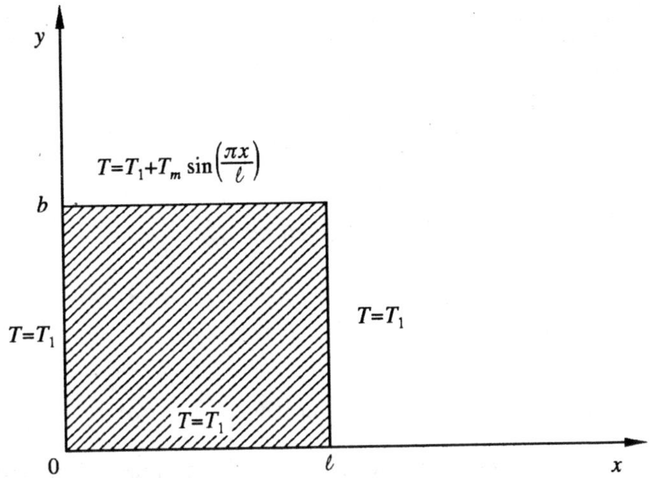

Fig. 4.3 Rectangular plate with boundary conditions.

sine wave distribution impressed on it. Then the general heat conduction equation (4.5) reduces to

$$\frac{\partial^2 T}{\partial x^2} + \frac{\partial^2 T}{\partial y^2} = 0 \tag{4.20}$$

Equation (4.20) is a linear homogeneous partial differential equation and one can use the method of separation of variables to solve it. The solution of this equation is assumed to be of the form

$$T(x, y) = X(x)Y(y) \tag{4.21}$$

Substituting Eq. (4.21) in Eq. (4.20), we get

$$-\frac{1}{X}\frac{d^2X}{dx^2} = \frac{1}{Y}\frac{d^2Y}{dy^2} \tag{4.22}$$

The left-hand side of Eq. (4.22) is a function of x only, while the RHS is a function of y only. If we change x while holding y constant, the LHS changes but the RHS does not change. But Eq. (4.22) demands that the LHS must be equal to the RHS which is a constant. Hence the LHS is independent of x. Similarly the RHS is independent of y. Therefore, both the LHS and RHS must be equal to a constant, say λ^2. Then Eq. (4.22) reduces to two ordinary differential equations

$$\frac{d^2X}{dx^2} + \lambda^2 X = 0 \tag{4.23}$$

$$\frac{d^2Y}{dy^2} - \lambda^2 Y = 0 \tag{4.24}$$

The general solution to Eq. (4.23) is given by

$$X(x) = c_1 \cos \lambda x + c_2 \sin \lambda x \tag{4.25}$$

and the general solution to Eq. (4.24) is given by

$$Y(y) = c_3 e^{-\lambda y} + c_4 e^{\lambda y} \tag{4.26}$$

Hence

$$T = X(x)Y(y) = (c_1 \cos \lambda x + c_2 \sin \lambda x)(c_3 e^{-\lambda y} + c_4 e^{\lambda y}) \tag{4.27}$$

The values of the constants c_1–c_4 depend on the boundary conditions. To simplify the algebra, let us define the new variable θ as

$$\theta = T - T_1 \tag{4.28}$$

In terms of the new variable, the boundary conditions are

$$T = T_1 \quad \text{or} \quad \theta = 0 \quad \text{at} \quad y = 0 \tag{4.29}$$

$$T = T_1 \quad \text{or} \quad \theta = 0 \quad \text{at} \quad x = 0 \tag{4.30}$$

$$T = T_1 \quad \text{or} \quad \theta = 0 \quad \text{at} \quad x = \ell \tag{4.31}$$

$$T = T_1 + T_m \sin(\pi x/\ell) \quad \text{or} \quad \theta = T_m \sin(\pi x/\ell) \quad \text{at} \quad y = b \tag{4.32}$$

Applying the first boundary condition (4.29) in Eq. (4.27), we get

$$0 = (c_1 \cos \lambda x + c_2 \sin \lambda x)(c_3 + c_4) \tag{4.33}$$

Application of the second boundary condition (4.30) to Eq. (4.27) gives

$$0 = c_1(c_3 e^{-\lambda y} + c_4 e^{\lambda y}) \tag{4.34}$$

Application of the third and fourth boundary conditions give

$$0 = (c_1 \cos \lambda\ell + c_2 \sin \lambda\ell)(c_3 e^{-\lambda y} + c_4 e^{\lambda y}) \tag{4.35}$$

$$T_m \sin(\pi x/\ell) = (c_1 \cos \lambda x + c_2 \sin \lambda x)(c_3 e^{-\lambda b} + c_4 e^{\lambda b}) \tag{4.36}$$

Equation (4.33) is satisfied only if $c_3 = -c_4$ and Eq. (4.34) is satisfied only if $c_1 = 0$. Substituting the values of c_1 and c_4 in Eq. (4.35), we get

$$0 = c_2 \sin(\lambda\ell) c_3 (e^{-\lambda y} - e^{\lambda y}) = 2c_2 c_3 \sin(\lambda\ell) \sinh(\lambda y) \tag{4.37}$$

Equation (4.37) is satisfied only if $\sin(\lambda\ell) = 0$ or $\lambda = n\pi/\ell$, where n = 1,2,3, etc. The value $n = 0$ is not acceptable since θ will be zero all the time when $n = 0$. There exists different solutions for different values of n with different constants c_n. Then the general solution is given as a sum of these solutions. Thus, the general solution is given by

$$\theta = T - T_1 = \sum_{n=1}^{\infty} c_n \sin\left(\frac{n\pi x}{\ell}\right) \sinh\left(\frac{n\pi y}{\ell}\right) \tag{4.38}$$

where the constants c_2 and c_3 have been combined.

Now, applying the last boundary condition (4.36), we get

$$T_m \sin\left(\frac{\pi x}{\ell}\right) = \sum_{n=1}^{\infty} c_n \sin\left(\frac{n\pi x}{\ell}\right) \sinh\left(\frac{n\pi b}{\ell}\right) \tag{4.39}$$

Equation (4.39) shows that for $n = 1$

$$c_1 = \frac{T_m}{\sinh(\pi b/\ell)} \tag{4.40}$$

and $c_n = 0$ for $n > 1$. Therefore, the final solution is given by

$$T - T_1 = T_m \frac{\sin(\pi x/\ell)\sinh(\pi y/\ell)}{\sinh(\pi b/\ell)} \tag{4.41}$$

The temperature distribution given by Eq. (4.41) is shown in Fig. 4.4.

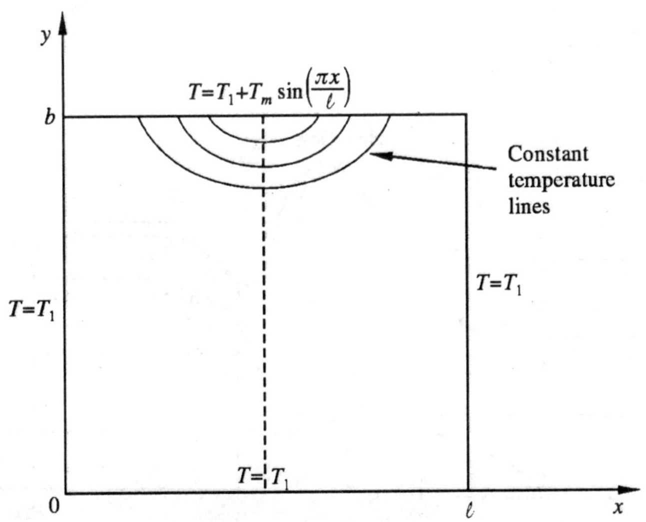

Fig. 4.4 Constant temperature lines in a rectangular plate, three faces of which are at temperature T_1 while the top face is impressed with temperature distribution given by $T = T_1 + T_m \sin(\pi x/\ell)$.

The analytical solutions are quite useful if they can be obtained. However, in practice one has to deal with different geometries and different types of boundary conditions which are not as simple as the one illustrated above.

4.5 Graphical Analysis

An approximate solution of the Laplace equation for a two-dimensional body can be obtained through a graphical analysis. The graphical analysis is simple if the boundaries of the body are at constant temperature, but it can also be used to analyze the heat transfer through a body which loses heat by convection or radiation through its surface.

In the graphical analysis one constructs a network of isotherms and constant heat flow lines. The constant heat flow lines are analogous to stream lines in fluid flow. The heat flow lines are tangent to the direction of heat flow and hence no heat flows across the heat flow lines. The isotherms and heat flow lines appear as shown in Fig. 4.5.

Consider a plane wall, one surface of which is maintained at temperature T_1 while the second surface is held at T_2. The heat flow lines are perpendicular to the wall surfaces and the isotherms are parallel to the faces of the wall. Now let us consider the corner section of a furnace wall. Then the curvilinear section (marked as A in Fig. 4.5) formed by two adjacent isotherms and two adjacent heat flow lines is shown in Fig. 4.6.

The rate of heat flow through one flow tube (per unit depth) shown in Fig. 4.6 is given by

$$q = -k\Delta y(1)\frac{\Delta T}{\Delta x} \tag{4.42}$$

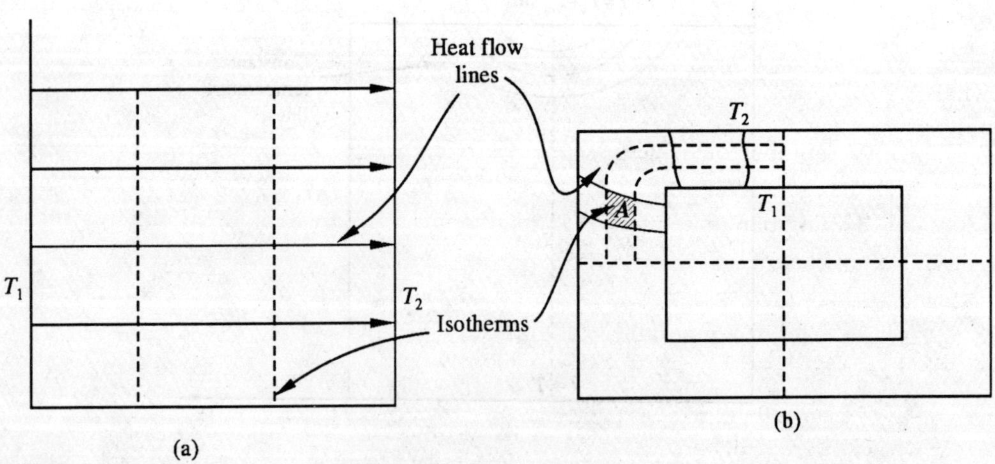

Fig. 4.5 Isotherms and heat flow lines in (a) a plane wall. (b) a corner section of a furnace wall.

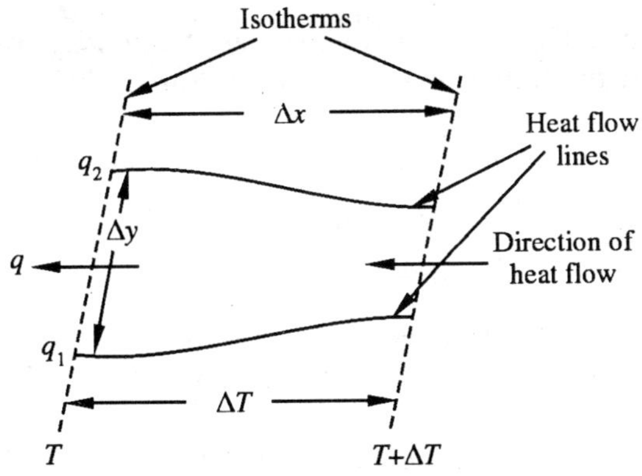

Fig. 4.6 *Curvilinear section formed by isotherms and heat flow lines.*

Suppose the heat flow lines are equally spaced so that the same amount of heat flows in each section. Then Eq. (4.42) gives

$$\frac{-q}{k\Delta T} = \frac{\Delta y}{\Delta x} \tag{4.43}$$

That is, if the temperature difference (ΔT) between any two adjacent isotherms is the same, the ratio $\Delta y/\Delta x$ is constant. In other words, in a two-dimensional heat flow region, the network of isotherms and heat flow lines consists of rectangles. It is convenient, from a practical point of view, to choose $\Delta x = \Delta y$. Then Eq. (4.43) gives $q = -k\Delta T$, irrespective of the size of the squares. Except for a plane wall, the elements of a heat flow network will not be geometrically perfect squares and they are called curvilinear squares. The total rate of heat transfer is given by

$$q = \frac{M}{N} k\Delta T = Sk\Delta T \tag{4.44}$$

where

M = number of heat flow lanes (or channels)

N = number of temperature increments (or the number of squares in a channel)

ΔT = overall temperature difference

S = conduction shape factor = M/N.

To calculate the rate of heat transfer by this graphical method we have to construct curvilinear squares and count the number of heat flow tubes and the number of temperature increments. This method is known as *flux plotting*. In making a flux plot we should ensure that the following conditions are satisfied:

1. The heat flow lines are perpendicular to the isothermal surfaces.
2. The isotherms are perpendicular to insulated surfaces.
3. The isotherms and heat flow lines are perpendicular to each other.
4. The heat flow lines leading to a corner of an isothermal boundary bisect the angle between the surfaces of the boundary at the corner.

EXAMPLE 4.1 A furnace wall is made of fireclay brick (k = 1.04 W/m K). In steady-state the inside and outside surfaces of the wall are at 550°C and 50°C, respectively. A schematic of the furnace wall is shown in Fig. 4.7. Determine the conduction shape factor and the rate of heat conduction per metre length of the furnace.

SOLUTION By observing the symmetry of the furnace wall, one can choose only one-fourth of the furnace wall for flux plotting. The overall temperature difference between the inside and outside surfaces is ΔT = 500°C. Let us choose 5 temperature increments of 100°C each. That is, N = 5. Figure 4.8 shows one fourth of the furnace wall. The surface ABC is isothermal at 550°C and the surface DEF is isothermal at 50°C. The surfaces AD and CF coincide with the lines of symmetry and hence there is no heat flow across these surfaces. Therefore, they can be treated as insulated faces. These insulated faces can be used as boundaries of heat flow tubes.

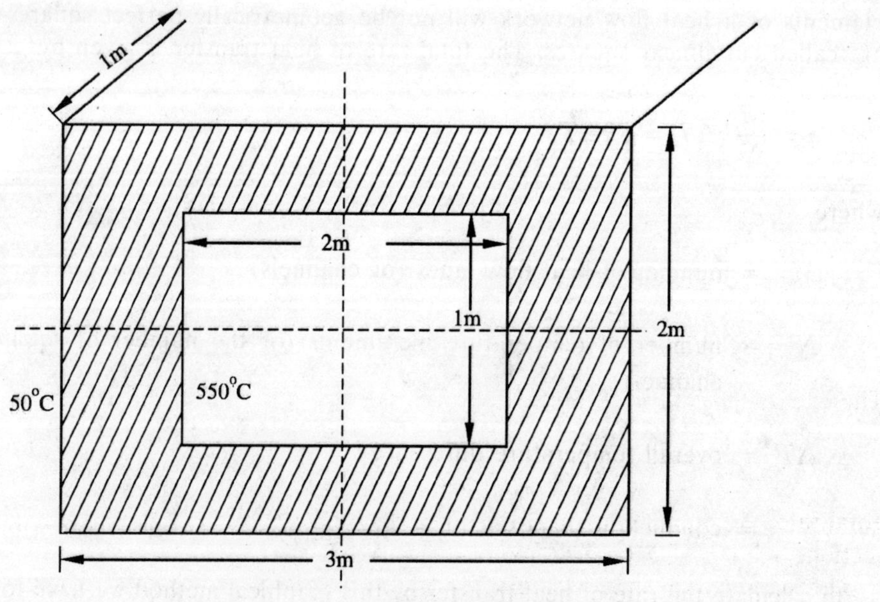

Fig. 4.7 Schematic of a furnace.

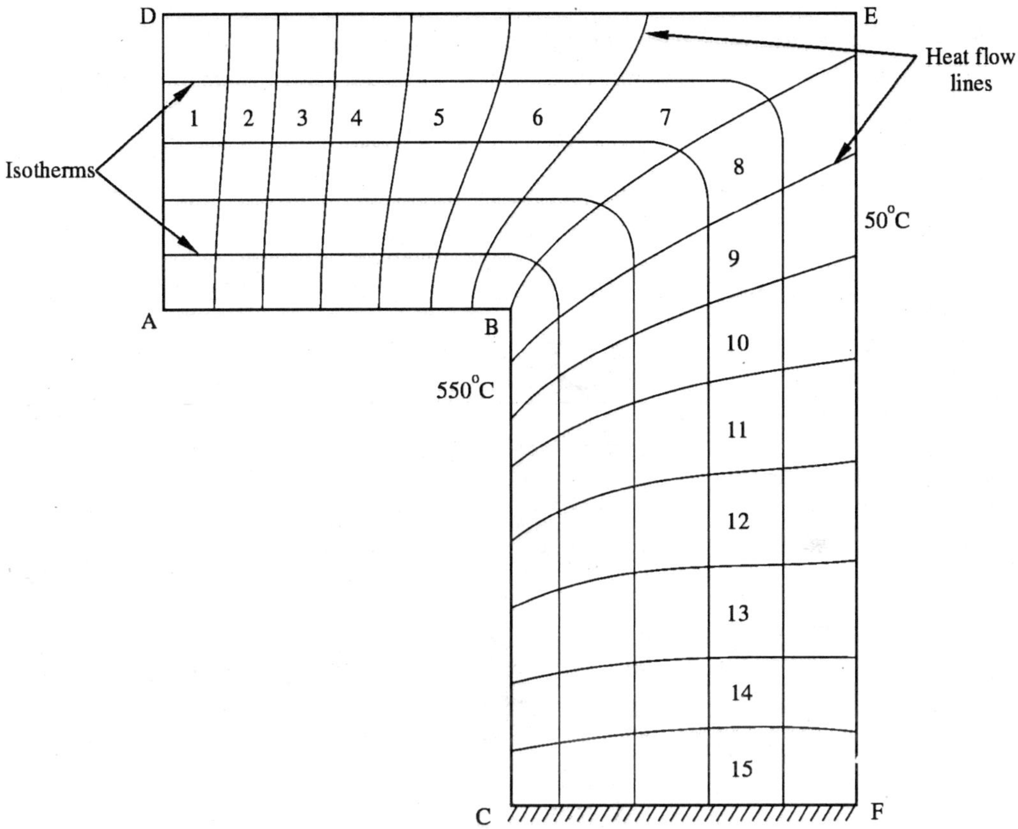

Fig. 4.8 *Sketch showing the construction of flux plot.*

Next, draw the isotherms such that they are perpendicular to the insulated faces. Bisect the corners at A and E, the lines being the beginnings of heat flow tubes. After a few trials we can draw curvilinear squares as shown in Fig. 4.8.

The number of squares per flow tube or the number of heat flow lanes $M = 15$.

Therefore, the conduction shape factor,

$$S = M/N = 15/5 = 3.$$

The rate of heat flow,

$$q = Sk(T_1 - T_2) = 3 \times 1.04(550 - 50) = 1560 \text{ W}.$$

Total rate of heat conduction from 1 metre length of furnace (i.e. for four symmetrical sections) = $4 \times 1560 = 6240$ W.

The conduction shape factors for a few selected geometrical configurations of practical interest are presented in the Table 4.1.

Table 4.1 Conduction shape factors for a few selected systems.

Physical system	Schematic	Shape factor	Restriction
Isothermal cylinder of radius r buried in semi-infinite medium having isothermal surface	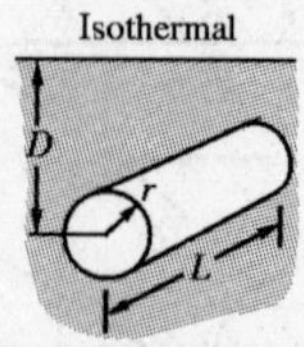	$\dfrac{2\pi L}{\cosh^{-1}(D/r)}$ $\dfrac{2\pi L}{\ln(D/r)}$	$L \gg r$ $L \gg r$ $D > 3r$
Isothermal sphere of radius r buried in infinite medium		$4\pi r$	
Isothermal sphere of radius r buried in semi-infinite medium having isothermal surface $\Delta T = T_{\text{surf}} - T_{\text{far field}}$	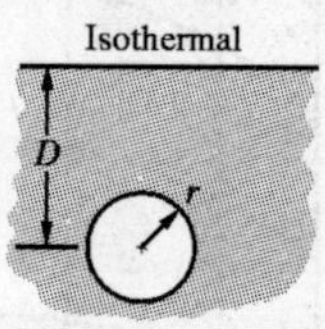	$\dfrac{4\pi r}{1-(r/2D)}$	
Conduction between two isothermal cylinders of length L buried in infinite medium	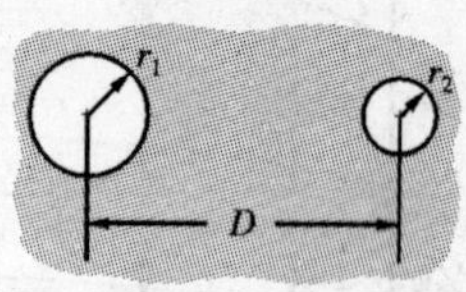	$\dfrac{2\pi L}{\cosh^{-1}\left(\dfrac{D^2-r_1^2-r_2^2}{2r_1r_2}\right)}$	$L \gg r$ $L \gg D$
Buried cube in infinite medium, L on a side		$8.24\ L$	
Isothermal cylinder of radius r placed in semi-infinite medium as shown	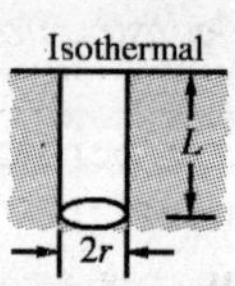	$\dfrac{2\pi L}{\ln(2L/r)}$	$L \gg 2r$
Isothermal rectangular parallelopiped buried in semi-infinite medium having isothermal surface	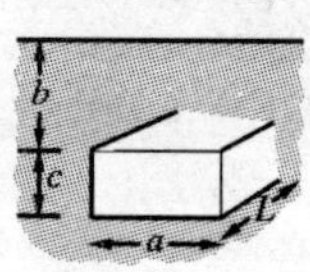	$1.685L\left[\log\left(1+\dfrac{b}{a}\right)\right]^{-0.59} \times \left(\dfrac{b}{c}\right)^{-0.078}$	see Ref.7
Plane wall	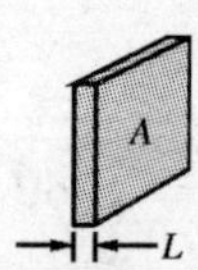	$\dfrac{A}{L}$	One-dimensional heat flow
Hollow cylinder, length L		$\dfrac{2\pi L}{\ln(r_o/r_i)}$	$L \gg r$

Continues

Table 4.1 *Continued*

Physical system	Schematic	Shape factor	Restriction
Hollow sphere	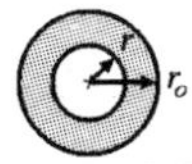	$\dfrac{4\pi r_o r_i}{r_o - r_i}$	
Thin horizontal disk buried in semi-infinite medium with isothermal surface	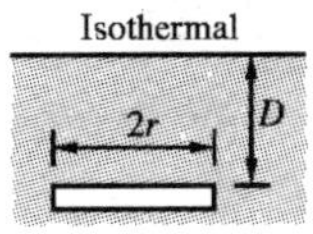	$4r$ $8r$ $\dfrac{4\pi r}{\pi/2 - \tan^{-1}(r/2D)}$	$D = 0$ $D \gg 2r$ $D/2r > 1$ $\tan^{-1}(r/2D)$ in radians
Hemisphere buried in semi-infinite medium $\Delta T = T_{\text{sphere}} - T_{\text{far field}}$	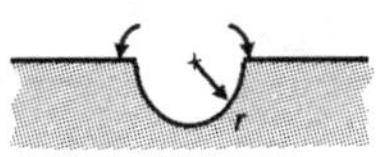	$2\pi r$	
Isothermal sphere buried in semi-infinite medium with insulated surface	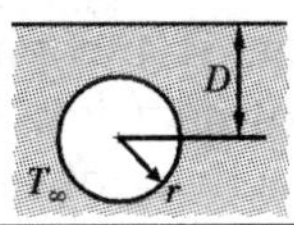	$\dfrac{4\pi r}{1 + r/2D}$	
Two isothermal spheres buried in infinite medium	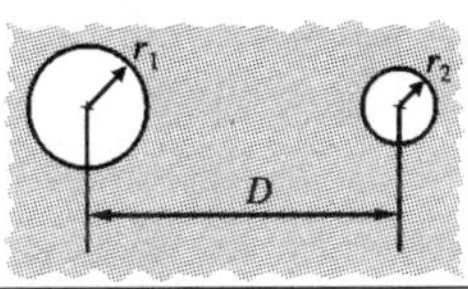	$\dfrac{4\pi r_2}{\dfrac{r_2}{r_1}\left[1 - \dfrac{(r_1/D)^4}{1-(r_2/D)^2}\right] - \dfrac{2r_2}{D}}$	$D > 5r_{\max}$
Thin rectangular plate of length L, buried in semi-infinite medium having isothermal surface	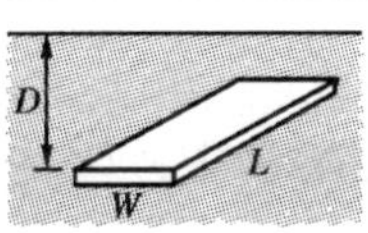	$\dfrac{\pi W}{\ln(4W/L)}$ $\dfrac{2\pi W}{\ln(4W/L)}$ $\dfrac{2\pi W}{\ln(2\pi D/L)}$	$D = 0$ $W > L$ $D \gg W$ $W > L$ $W \gg L$ $D > 2W$
Parallel disks buried in infinite medium	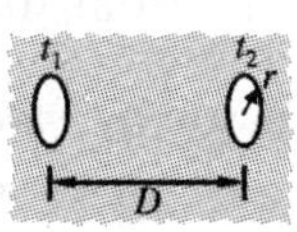	$\dfrac{4\pi r}{\left[\dfrac{\pi}{2} - \tan^{-1}(r/D)\right]}$	$D > 5r$ $\tan^{-1}(r/D)$ in radians
Eccentric cylinders of length L	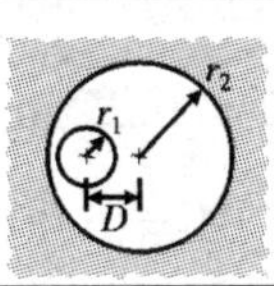	$\dfrac{2\pi L}{\cosh^{-1}\left(\dfrac{r_1^2 + r_2^2 - D^2}{2r_1 r_2}\right)}$	$L \gg r_2$
Cylinder centered in a square of length L	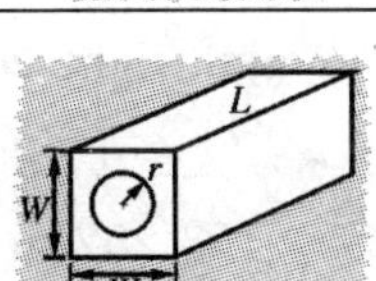	$\dfrac{2\pi L}{\ln(0.54W/r)}$	$L \gg W$

Continues

Table 4.1 Continued

Physical system	Schematic	Shape factor	Restriction
Horizontal cylinder of length L centered in infinite plate	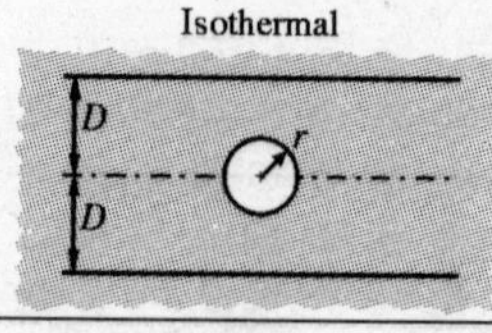	$\dfrac{2\pi L}{\ln(4D/r)}$	
This horizontal disk buried in semi-infinite medium with adiabatic surface $\Delta T = T_{\text{disk}} - T_{\text{far field}}$	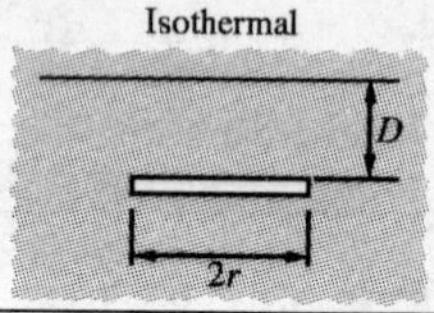	$\dfrac{4\pi r}{\pi/2 + \tan^{-1}(r/2D)}$	$D/2r > 1$ $\tan^{-1}(r/2D)$ in radians

EXAMPLE 4.2 A horizontal pipe of 10 cm radius and 5 m length is buried in a masonry brick work at a depth of 30 cm. If the pipe wall is at 60°C while the earth's surface is at 5°C, calculate the rate of heat loss by the pipe, by estimating the conduction shape factor. The thermal conductivity of masonry brick is 0.69 W/m K.

SOLUTION The shape factor S is given by

$$S = \frac{2\pi\ell}{\cosh^{-1}(z/r)} = \frac{2\pi \times 5}{\cosh^{-1}(30/10)} = 17.82$$

Rate of heat transfer by conduction,

$$q = kS\Delta T = 0.69 \times 17.82 \times 55 = 676.3 \text{ W}$$

EXAMPLE 4.3 An oven of size 50 cm × 50 cm × 50 cm is covered by 1 cm thick asbestos sheet (k = 0.1105 W/m K). While the oven is in use, the inside surface temperature is 300°C and the outer surface is at 40°C. Estimate the conduction shape factor and the rate of conduction heat transfer from the oven.

SOLUTION We calculate the total shape factor by calculating the shape factors for plane walls, edges and corners.

Shape factor for plane wall $= A/\ell = (0.5 \times 0.5)/0.01 = 25$ m

Shape factor for edge $= 0.54\ell = 0.54 \times 0.5 = 0.27$ m

Shape factor for corner $= 0.15\ell = 0.15 \times 0.01 = 0.0015$ m

The cubical box has six walls, twelve edges and eight corners. Therefore, the total shape factor, S is given by

$$S = 6 \times 25 + 12 \times 0.27 + 8 \times 0.0015 = 153.252 \text{ m}$$

Rate of heat conduction, $q = kS\Delta T = 0.1105 \times 153.252(300 - 40) = 4403$ W

4.6 Analogical Method of Analysis

When two different physical phenomena can be described mathematically by the same equation, the two phenomena are said to be mathematically analogous. The variables of one system are called analogues of the corresponding variables of the other system. In a two-dimensional case, the Laplace equation for heat conduction is given by

$$\frac{\partial^2 T}{\partial x^2} + \frac{\partial^2 T}{\partial y^2} = 0 \tag{4.45}$$

The potential distribution in an electric field is given by

$$\frac{\partial^2 E}{\partial x^2} + \frac{\partial^2 E}{\partial y^2} = 0 \tag{4.46}$$

where E is the electric potential. Equations (4.45) and (4.46) are similar. Then, the temperature distribution in a two-dimensional body having no internal heat generation and in steady-state is analogous to the voltage distribution in an electric field. The temperature $T(x, y)$ is the analogue of voltage $E(x, y)$. Then constant voltage lines in an electric field are analogous to isotherms in a heat flow, and electric current corresponds to heat flow.

There are many situations in heat transfer for which analytical solutions cannot be obtained. To obtain the solutions through experimentation would be too expensive and time consuming. In such cases it is possible to obtain experimental solutions in an analogous system and the results can be translated in terms of the heat transfer problem.

Analogue field plotter

An experimental arrangement for measuring the electric potential distribution in a two-dimensional system is shown in Fig. 4.9. Such experimental arrangements are called *analogue field plotters*. A thin sheet of electrically conducting paper of relatively high resistivity is cut to an exact geometric model of the two-dimensional heat conduction system. Boundary conditions corresponding to constant temperature in the heat flow model are obtained by connecting copper strips or by painting highly conducting silver paint on the surface of the paper in the electrical analogue model. Insulated surfaces in the heat flow model correspond to plain edges of the conducting paper.

An electrical potential, proportional to the overall temperature difference $(T_1 - T_2)$ is applied to the copper strips. Then lines of constant voltage (for a few selected values) are traced on the paper by using a movable stylus and voltmeter. The constant potential lines correspond to the isotherms. Since the heat flow lines are perpendicular to the isotherms, the heat flow lines are drawn in free hand so that one obtains curvilinear squares. The heat flow lines can also be traced by changing the copper strips (isothermal faces) to plain edges (insulated faces) and vice versa and then repeat the measurements to trace a new set of constant voltage lines.

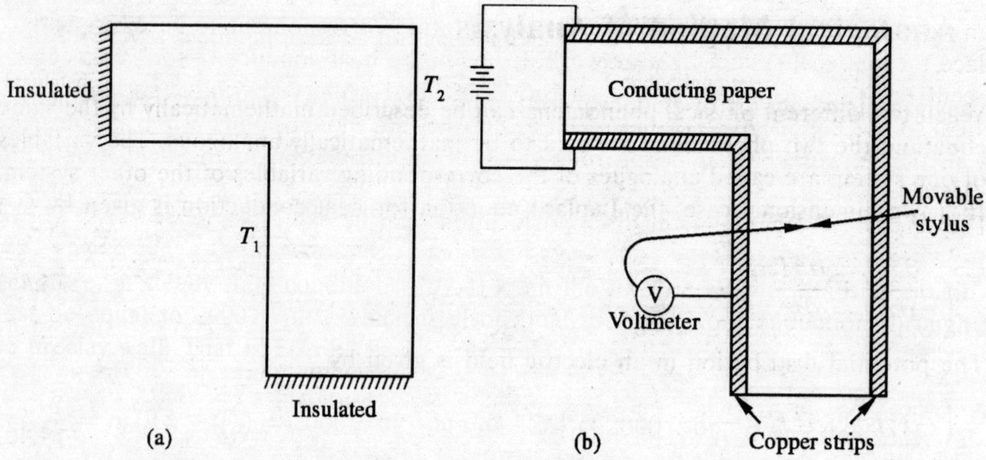

Fig. 4.9 (a) Heat flow model. (b) Analog field plotter.

4.7 Numerical Method of Analysis

In the numerical method of solving a heat transfer problem described by a differential equation, we approximate the differential equation and boundary conditions by a set of algebraic equations. We subdivide the two-dimensional system into a number of small but finite squares as shown in Fig. 4.10 and assign a reference number to each.

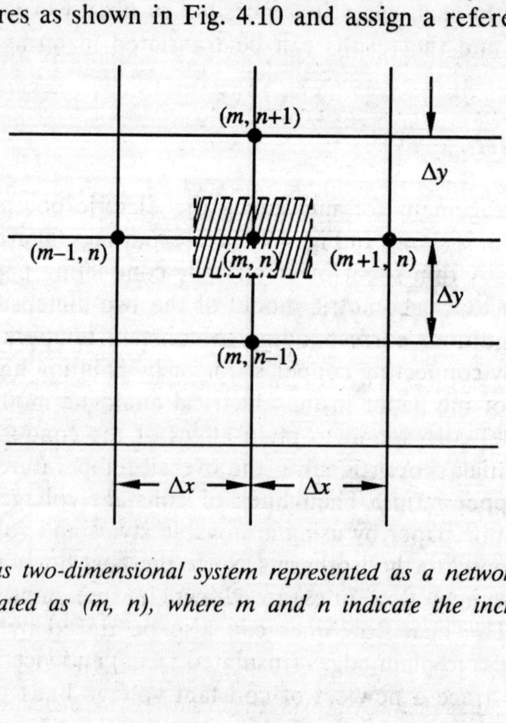

Fig. 4.10 A continuous two-dimensional system represented as a network of nodal points. The nodal points are designated as (m, n), where m and n indicate the increments in the x and y directions, respectively.

Then we assume that each square is at the temperature corresponding to its centre and replace the physical system by a network of imaginary heat conducting rods between the centres or nodal points. Now each rod is assigned a thermal conductance which is equal to the conductance of the material between the nodal points. Then the heat flow in the network of conducting rods gives an approximation to the heat flow in the physical system.

The temperature gradients at different locations can be expressed as

$$\left(\frac{\partial T}{\partial x}\right)_{m+\frac{1}{2},n} \simeq \frac{T_{m+1,n} - T_{m,n}}{\Delta x} \tag{4.47}$$

$$\left(\frac{\partial T}{\partial x}\right)_{m-\frac{1}{2},n} \simeq \frac{T_{m,n} - T_{m-1,n}}{\Delta x} \tag{4.48}$$

$$\left(\frac{\partial^2 T}{\partial x^2}\right)_{m,n} \simeq \frac{\left(\frac{\partial T}{\partial x}\right)_{m+\frac{1}{2},n} - \left(\frac{\partial T}{\partial x}\right)_{m-\frac{1}{2},n}}{\Delta x} = \frac{T_{m+1,n} + T_{m-1,n} - 2T_{m,n}}{(\Delta x^2)} \tag{4.49}$$

$$\left(\frac{\partial T}{\partial y}\right)_{m,n+\frac{1}{2}} \simeq \frac{T_{m,n+1} - T_{m,n}}{\Delta y} \tag{4.50}$$

$$\left(\frac{\partial T}{\partial y}\right)_{m,n-\frac{1}{2}} \simeq \frac{T_{m,n} - T_{m,n-1}}{\Delta y} \tag{4.51}$$

$$\left(\frac{\partial^2 T}{\partial y^2}\right)_{m,n} \simeq \frac{\left(\frac{\partial T}{\partial y}\right)_{m,n+\frac{1}{2}} - \left(\frac{\partial T}{\partial y}\right)_{m,n-\frac{1}{2}}}{\Delta y} = \frac{T_{m,n+1} + T_{m,n-1} - 2T_{m,n}}{(\Delta y^2)} \tag{4.52}$$

Then the Laplace equation

$$\frac{\partial^2 T}{\partial x^2} + \frac{\partial^2 T}{\partial y^2} = 0$$

can be written in the finite difference approximation as

$$\frac{T_{m+1,n} + T_{m-1,n} - 2\,T_{m,n}}{(\Delta x)^2} + \frac{T_{m,n+1} + T_{m,n-1} - 2\,T_{m,n}}{(\Delta y)^2} = 0 \tag{4.53}$$

If $\Delta x = \Delta y$, then Eq. (4.53) reduces to

$$T_{m+1,n} + T_{m-1,n} + T_{m,n+1} + T_{m,n-1} - 4\,T_{m,n} = 0 \tag{4.54}$$

Since the material is homogeneous, having the same value of thermal conductivity, the heat flow is directly related to the temperature difference. Equation (4.54) thus states that the net heat flow into any node is zero at steady-state. To obtain the temperature distribution in the system one should apply Eq. (4.54) to each nodal point and obtain a system of equations (one equation for every nodal point). The resultant system of equations can be solved to obtain the temperatures of the nodal points. Once the temperatures of the nodal points are determined, the heat flow can be calculated by summing up the rates of heat flow through the fictitious rods terminating either at the inner or at the outer surfaces. That is,

$$q = -\sum k(\Delta y)\frac{\Delta T}{\Delta x} \tag{4.55}$$

The heat flows calculated at the inner and outer surfaces may differ if one chooses a coarse grid. However, if one chooses a sufficiently fine grid, the two values of heat flow will be the same. In general, it is advisable to calculate the two values of heat flow (based on the inner surface and outer surface) and take the arithmetic average of the two values.

When a two-dimensional body is subdivided to form a grid, the interior nodal points are surrounded by other nodal points, but the nodal points at a boundary may have lesser material associated with them. Therefore, the nodal equation for a boundary nodal point depends on the geometry as well as the boundary condition. The nodal equations for a few typical geometrical conditions and boundary conditions are given below.

(a) Nodal point at the surface in contact with a fluid

Consider the nodal point (m, n) shown in Fig. 4.11 which at the surface is in contact with a fluid. The energy balance on nodal point (m, n) gives

$$-k(\Delta y)\frac{T_{m,n} - T_{m-1,n}}{\Delta x} - k\left(\frac{\Delta x}{2}\right)\frac{T_{m,n} - T_{m,n+1}}{\Delta y} - k\left(\frac{\Delta x}{2}\right)\frac{T_{m,n} - T_{m,n-1}}{\Delta y}$$
$$= h(\Delta y)(T_{m,n} - T_\infty) \tag{4.56}$$

If $\Delta x = \Delta y$, Eq. (4.56) reduces to

$$T_{m,n}\left(\frac{h}{k}\Delta x + 2\right) - \frac{h}{k}\Delta x T_\infty - \frac{1}{2}\left(2T_{m-1,n} + T_{m,n+1} + T_{m,n-1}\right) = 0$$

or

$$T_0\left(2 + \frac{h}{k}\Delta x\right) - \frac{h}{k}\Delta x T_\infty - \frac{1}{2}\left(2T_1 + T_2 + T_4\right) = 0 \tag{4.57}$$

where the central node is renamed as the zeroth node which is surrounded by four other nodes as shown in Fig. 4.11. The node 3 is not present, because the zeroth node is on the surface.

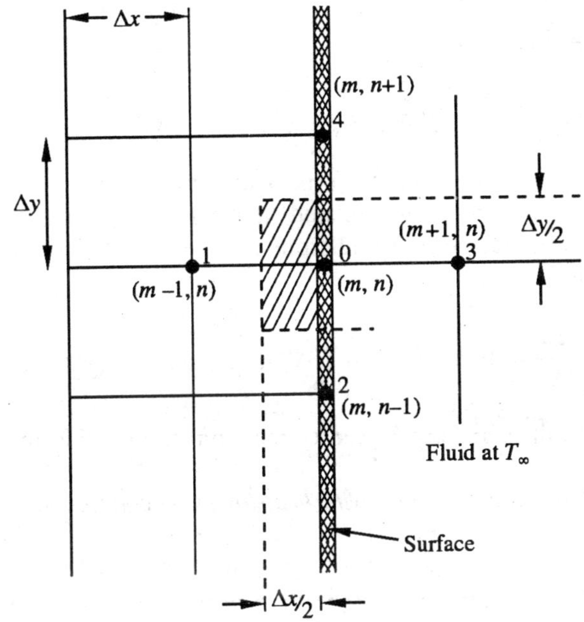

Fig. 4.11 *Nodal point on a surface with convection heat transfer.*

(b) Nodal point at the insulated surface

A nodal point at an insulated surface is shown in Fig. 4.12. In this case there is no heat flow from nodal point 2 (fictitious) to nodal point 0. However, heat flows from

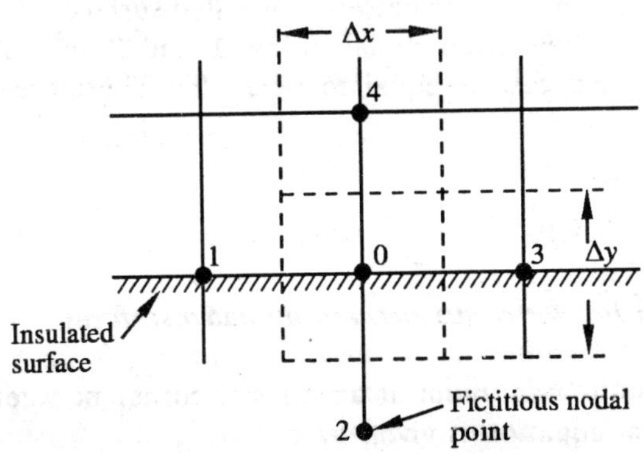

Fig. 4.12 *Nodal point at an insulated surface.*

nodal points 1, 3 and 4 to nodal point 0. The area for heat transfer between nodal points 1 and 0 and between the nodal points 3 and 0 is only $\Delta x/2$. Then the nodal equation is given by

$$\frac{1}{2}(T_1 + T_3) + T_4 - 2T_0 = 0 \tag{4.58}$$

(c) Nodal point at an isothermal boundary

Figure 4.13 (a) shows a nodal point on an isothermal boundary. The nodal points 0, 2 and 4 are at the same temperature and hence no heat flows from 2 and 4 to 0. Heat flows from the nodal point 3 and from outside to the nodal point 0. Therefore, the nodal equation is given by

$$T_3 - T_0 + q'\frac{\Delta x}{k} = 0 \tag{4.59}$$

where q' is the rate of heat flow per unit area from the outside to the nodal point 0.

(d) Nodal point at an extreme corner with heat losses by convection

A nodal point at an extreme corner and in contact with a fluid at T_∞ is shown in Fig. 4.13(b). In this case, the nodal points 1 and 2 are fictitious. Then the nodal equation is given by

$$\frac{1}{2}(T_3 + T_4) + \frac{h}{k}\Delta x T_\infty - \left(\frac{h}{k}\Delta x + 1\right) T_0 = 0 \tag{4.60}$$

(e) Nodal point at an extreme corner between insulated surfaces

A nodal point at an extreme corner between insulated surfaces is shown in Fig. 4.13(c). In this case, heat can flow from nodal points 1 and 2 to the nodal point 0 and the conductance in each case is equal to $(k\Delta x/2)$. Therefore, the nodal equation is given by

$$\frac{1}{2}(T_1 + T_2) - T_0 = 0 \tag{4.61}$$

(f) Nodal point at an interior corner between insulated surfaces

Figure 4.13 (d) shows a nodal point at an interior corner between insulated surfaces. In this case the nodal equation is given by

$$\frac{1}{2}(T_1 + T_2) + (T_3 + T_4) - 3T_0 = 0 \tag{4.62}$$

(g) Nodal point at an interior corner with heat losses by convection

A nodal point at an interior corner whose surfaces are in contact with a fluid at T_∞ is shown in Fig. 4.13(e). In this case, the nodal equation is given by

$$\frac{1}{2}(T_1 + T_2) + (T_3 + T_4) + \frac{h}{k}\Delta x T_\infty - \left(3 + \frac{h}{k}\Delta x\right) T_0 = 0 \tag{4.63}$$

(h) Interior nodal point near a curved boundary

Figure 4.13(f) shows an interior nodal point near a curved boundary. In this case the nodal equation is given by

$$\frac{2T_1}{a(a+1)} + \frac{2T_2}{b+1} + \frac{2T_3}{a+1} + \frac{2T_4}{b(b+1)} - 2\left(\frac{1}{a} + \frac{1}{b}\right) T_0 = 0 \tag{4.64}$$

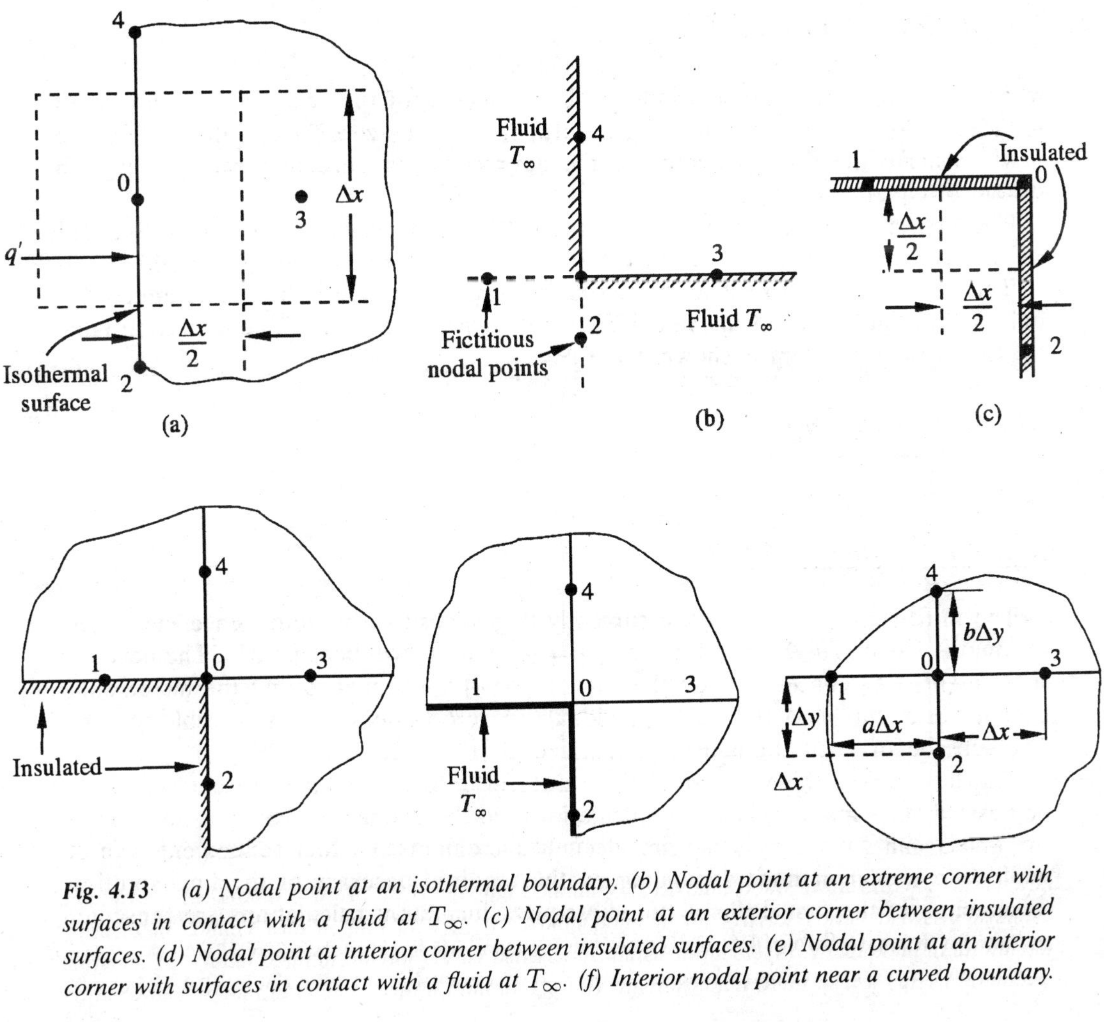

Fig. 4.13 (a) Nodal point at an isothermal boundary. (b) Nodal point at an extreme corner with surfaces in contact with a fluid at T_∞. (c) Nodal point at an exterior corner between insulated surfaces. (d) Nodal point at interior corner between insulated surfaces. (e) Nodal point at an interior corner with surfaces in contact with a fluid at T_∞. (f) Interior nodal point near a curved boundary.

4.8 Relaxation Method

The relaxation method is a numerical method for solving a set of algebraic equations and was first used by Sir Richard Southwell. The basic principles of the relaxation method are illustrated below by solving the following pair of algebraic equations:

$$-15x + 16y = -279$$

$$5x - 24y = -145$$

To obtain a solution of the above equations, we rewrite the given equations as

$$R_1 = -15x + 16y + 279$$

$$R_2 = 5x - 24y + 145$$

where R_1 and R_2 are called residuals. The objective of the relaxation method is to reduce the residuals systematically to zero or close to zero. The values of x and y which make the residuals zero are the desired values since they satisfy the given system of equations.

The first step in the solution is to choose the initial values for x and y. Suppose we choose $x = 0$ and $y = 0$. Then the residuals are $R_1 = 279$ and $R_2 = 145$.

The second step is to set up an *Operation table* which will indicate the effect of a unit positive increment in the values of x and y on the residuals. The operation table for the present problem is shown Table 4.2.

Table 4.2 *Operation table*

Increment	ΔR_1	ΔR_2
$\Delta x = 1$	-15	5
$\Delta y = 1$	16	-24

The third step is to reduce the currently largest residual to zero. Since the largest residual is $R_1 = 279, \Delta x = 19$ makes $R_1 = -6$ but R_2 will become 240. The next and subsequent steps concentrate on the currently largest residual to find the incremental changes in x and y to reduce the residuals to zero or as close as possible to zero. The subsequent calculations are summarized in tabular form in Table 4.3.

The last row of the table gives the values of x and y, which are correct to the nearest whole number. Then the calculations can be refined to obtain a solution to the first decimal place by using first decimal increments. Further refinements can be made in a similar manner depending on the required accuracy of the final solution. A summary of the first decimal and subsequent increment calculations are presented in tabular form in Table 4.4.

Table 4.3 Relaxation table - I

x	y	R_1	R_2
0	0	279	145
$\Delta x = 19$		-6	240
	$\Delta y = 10$	154	0
$\Delta x = 11$		-11	55
	$\Delta y = 3$	37	-17
$\Delta x = 3$		-8	-2
$x = 33$	$y = 13$	-8	-2

Table 4.4 Relaxation table - II

x	y	R_1	R_2	
33	13	-8	-2	
$\Delta x = -0.5$		-0.5	-4.5	
	$\Delta y = -0.2$	-3.7	0.3	
$\Delta x = -0.2$		-0.7	-0.7	
$x = 32.3$	$y = 12.8$	-0.7	-0.7	A
	$\Delta y = -0.03$	-1.18	0.02	
$\Delta x = -0.08$		0.02	-0.38	
	$\Delta y = -0.02$	-0.3	0.10	
$\Delta x = -0.02$		0	0	
$x = 32.2$	$y = 12.75$	0	0	B

The row marked A shows the solutions which are correct up to the first decimal place. The row marked B in the table above shows that the residuals are zero. Hence it gives the final values of x and y as 32.2 and 12.75, respectively.

Shortcuts in relaxation method

The relaxation method (explained above) can also be applied to complex problems. However, by modifying the method one can save considerable effort and time. These modifications are called shortcuts.

1. The first shortcut is to assign some initial values to the variables, instead of assigning zero values. The initial values can be guessed based on a close look at the physical problem.
2. The second shortcut, called *over-relaxation* is used to change the residuals from a positive value to a negative value and vice versa, by choosing larger increments. However, one cannot quantify the size of the increment to choose. The choice of the increment size may vary from person to person analyzing the problem.
3. The third shortcut, called *block operation*, is to change each of the variables by the

same amount. The changes in the residuals due to the individual unit increments in the variables are added to obtain the effect of the *unit block operation*. This unit block operation can be used at any stage of calculation, but it is quite valuable in the initial step of the calculation. To apply this unit block operation in the initial step we add the residuals and divide the sum by the effect of the unit block operation to obtain the size of the increment. If one uses the increment size thus obtained in a block operation it will reduce the sum of the residuals to zero and the initial step considerably reduces the subsequent steps to obtain the final solution.

The method of applying the unit block operation is illustrated below by considering the earlier example. The operation table for the earlier example (Table 4.2) is given in Table 4.5.

Table 4.5 *Initial step increment size calculation table.*

Increment	R_1	R_2
$\Delta x = 1$	− 15	5
$\Delta y = 1$	16	−24
$\Delta x = 1;\ \Delta y = 1$	1	−19

Sum of residuals = 1 − 19 = −18 = effect of unit block operation.

Sum of residuals in the initial stage (with $x = 0$ and $y = 0$) = 279 + 145 = 424 (see Table 4.3)

$$\text{Step size in the initial step} = \frac{424}{18} \simeq 23$$

Now we can use a step size of 23 for both the variables x and y to apply the unit block operation. Table 4.6 illustrates the application of unit block operation.

Table 4.6 *Application of unit block operation*

x	y	R_1	R_2
0	0	279	145
$\Delta x = 23$	$\Delta y = 23$	302	−292

The subsequent calculations can be carried out as in Table 4.3.

EXAMPLE 4.4 Determine the temperature distribution and the rate of heat flow per metre length of a long furnace, the cross-section of which is shown in Fig. 4.14. Assume that the inside surface temperature is 750°C, the outside surface temperature is 50°C and the thermal conductivity of the furnace wall (refractory clay) is 1.035 W/m K.

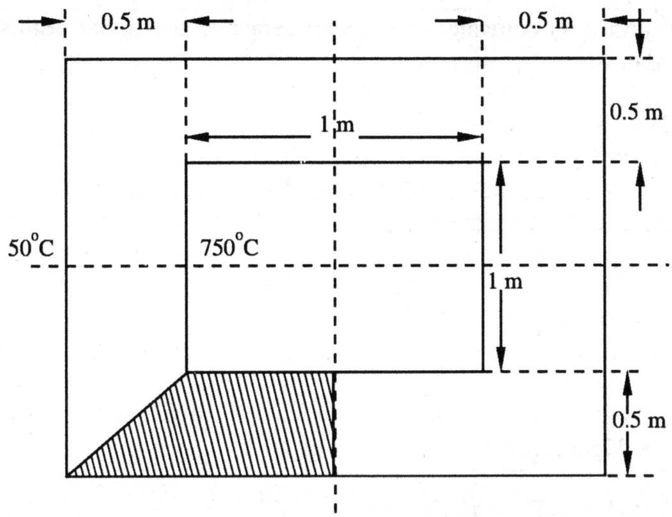

Fig. 4.14 *Schematic of a furnace cross-section.*

SOLUTION We can observe the symmetry of the furnace wall. We need to consider only a one-eighth section of the furnace wall which is shown in Fig. 4.15. To reduce the amount of calculations, let us choose a grid size of 0.25 m × 0.25 m and number the nodal points as 1, 2,...,12. The nodal points 5, 6 and 7 lie on the inside surface and are at 750°C and the nodal points 8, 9,...,12 lie on the outside surface and are at 50°C. We need to estimate the temperatures of the nodal points 1, 2, 3 and 4 only. The nodal points 2′ and 9′ represent the mirror images (about the line of symmetry) of the nodal points 2 and 9. Nodal point 3′ is an image of nodal point 3.

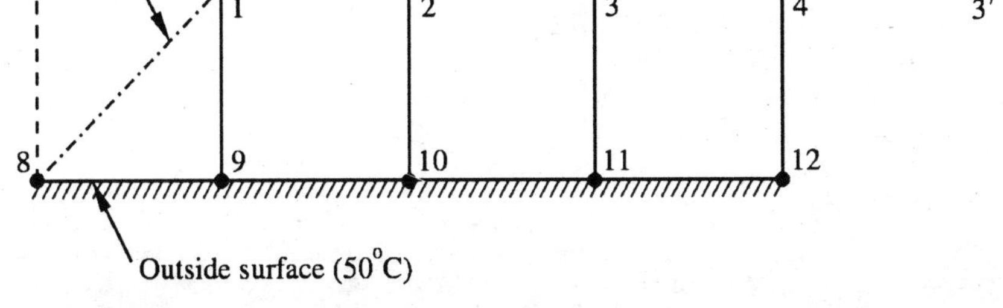

Fig. 4.15 *Network of nodal points for relaxation calculations.*

The nodal point 1 is located on the line of symmetry and hence the influence of nodal points 2′ and 9′ is equivalent to the influence of the nodal points 2 and 9. The residual for nodal point 1 is given by

$$R_1 = 2(T_2 + T_9) - 4T_1 \tag{A}$$

Nodal points 2 and 3 are interior points and Eq. (4.54) applies to them. Therefore, the residuals R_2 and R_3 are given by

$$R_2 = T_1 + T_{10} + T_3 + T_7 - 4T_2 \tag{B}$$

$$R_3 = T_2 + T_{11} + T_4 + T_6 - 4T_3 \tag{C}$$

The nodal point 4 is an interior point for which the nodal point 3′ is a mirror image of nodal point 3. Hence the residual R_4 is given by

$$R_4 = 2T_3 + T_{12} + T_5 - 4T_4 \tag{D}$$

Let us make the following initial guess for the temperatures $T_1, ..., T_4$:

$$T_1 = T_2 = T_3 = T_4 = 400°\text{C}$$

Substituting the known values of the temperatures, i.e.,

$$T_9 = T_{10} = T_{11} = T_{12} = 50°\text{C}$$

$$T_5 = T_6 = T_7 = 750°\text{C}$$

in Eqs. (A) to (D), we get

$$R_1 = 2(T_2 + 50) - 4T_1 = 50 + 2T_2 - 4T_1 \tag{E}$$

$$R_2 = T_1 + T_{10} + T_3 + T_7 - 4T_2 = T_1 + 50 + T_3 + 750 - 4T_2$$

or

$$R_2 = 800 + T_1 + T_3 - 4T_2 \tag{F}$$

$$R_3 = T_2 + T_{11} + T_4 + T_6 - 4T_3 = T_2 + 50 + T_4 + 750 - 4T_3$$

or

$$R_3 = 800 + T_2 + T_4 - 4T_3 \tag{G}$$

$$R_4 = 2T_3 + T_{12} + T_5 - 4T_4 = 2T_3 + 50 + 750 - 4T_4$$

or

$$R_4 = 800 + 2T_3 - 4T_4 \tag{H}$$

Operation table and initial step size calculation

T_1	R_1	T_2	R_2	T_3	R_3	T_4	R_4
400	−750	400	0	400	0	400	0
$\Delta T_1 = 1$	−4		1				
$\Delta T_2 = 1$	2		−4		1		
$\Delta T_3 = 1$			1		−4		2
$\Delta T_4 = 1$					1		−4
$\Delta T_1, \Delta T_2, \Delta T_3, \Delta T_4$ = 1	−2		−2		−2		−2

Sum of residuals = $-2 - 2 - 2 - 2 = -8$ = effect of unit block operation.

Sum of residuals with guess values of $T_1, ..., T_4 = -750$.

Step size = $\dfrac{750}{8} \simeq 94$.

Therefore, let us choose a block step size = 94°C.

In the relaxation table (Table 4.7), row 1 shows the initial assumed values of T_1, T_2, T_3, T_4 and the values of R_1, R_2, R_3 and R_4 obtained from Eqs. (E) to (H) by using the assumed values of T_1 to T_4. Row 2 shows the increment in temperature through a block change. Then the residuals are calculated and shown in Row 3. The largest value of residual R_1 (which is encircled) can be reduced by changing T_1 by -145 which is ΔT_1, and is shown in Row 4. Row 5 shows the revised values of residuals when T_1 is changed. At this stage the largest residual is $R_3 = 188$. Row 6 shows the increment in T_3. The change in the value of T_3 changes the residuals R_2, R_3 and R_4 and the revised values are shown in Row 7. Then the largest residual is $R_4 = 288$. The calculations are continued as shown in the relaxation table. Row 10 shows the current temperatures obtained by adding the increments to the assumed values and Row 11 shows the values of the current residuals, which are checked through calculations by using Eqs. (E) to (H). Then the unit block operation calculations are repeated to obtain $\Delta T = 22$. Rows 12 to 19 show the calculations for increments in T_2, T_3 and T_4. Rows 20 and 21 show the current values of T_1 to T_4 and the residuals R_1 to R_4, respectively. The residuals in Row 21 are checked again through calculations. Rows 22 to 25 show similar calculation results for the next block operation. Finally, Row 26 shows the final values of T_1 to T_4 and R_1 to R_4 . Since the residuals are close to zero the calculations are discontinued and we obtain the temperatures (which are correct to units place) at the nodal points 1 to 4 as $T_1 = 185$°C; $T_2 = 342$°C; $T_3 = 384$°C; and $T_4 = 392$°C.

Table 4.7 Relaxation table

S.No		T_1	R_1	T_2	R_2	T_3	R_3	T_4	R_4
1	Initial	400	−750	400	0	400	0	400	0
2	Block (−94)	−94		−94		−94		−94	
3			−562		188		188		188
4		−145							
5			18		43				
6						+50			
7					93		−12		288
8								+72	
9							60		0
10		161		306		356		378	
11			18		93		60		0
12	Block (+22)	+22		+22		+22		+22	
13			−26		49		16		−44
14				+12					
15			−2		1		28		
16								−11	
17							17		0
18						+4			
19					5		1		8
20		183		340		382		389	
21			−2		5		1		8
22	Block(+2)	+2		+2		+2		+2	
23			2		1		−3		4
24								+1	
25							−2		0
26		185	2	342	1	384	−2	392	0

The rate of heat flow from the furnace is given by the sum of heat flow rates through the nodal points terminating at the inside surface.

$$q = 8 \sum \frac{kA\Delta T}{\ell} = \frac{8k(1 \times 0.25)}{0.25}\left[(T_7 - T_2) + (T_6 - T_3) + \frac{1}{2}(T_5 - T_4)\right]$$

$$= 8 \times 1.035 \left[(750 - 342) + (750 - 384) + \frac{1}{2}(750 - 392)\right]$$

$$= 8 \times 1.035 \times 953 = 7891 \text{ W}$$

where A = area normal to the direction of heat flow.

Note that there are eight identical sections in the furnace wall and one-half of the energy transferred from nodal point 5 to nodal point 4 is included in one section.

4.9 Summary

Heat conduction in two- and three-dimensional systems can be analyzed by (a) analytical (b) graphical (c) analogical and (d) numerical methods. A complete treatment of analytical solutions requires a knowledge of higher mathematics, but relatively simple systems can be analyzed without much difficulty. The graphical, analogical and numerical methods do not require a knowledge of higher mathematics and at the same time provide a reasonably accurate solution.

The general heat conduction equation in Cartesian coordinates is given by

$$\frac{\partial^2 T}{\partial x^2} + \frac{\partial^2 T}{\partial y^2} + \frac{\partial^2 T}{\partial z^2} + \frac{\dot{q}}{k} = \frac{1}{\alpha}\frac{\partial T}{\partial t}$$

where $\alpha = (k/\rho c)$ = thermal diffusivity.

If the body has no internal heat generation, it reduces to the Fourier equation

$$\frac{\partial^2 T}{\partial x^2} + \frac{\partial^2 T}{\partial y^2} + \frac{\partial^2 T}{\partial z^2} = \frac{1}{\alpha}\frac{\partial T}{\partial t}$$

If the system is in steady-state, but has internal heat generation, the general heat conduction equation reduces to the Poisson equation

$$\frac{\partial^2 T}{\partial x^2} + \frac{\partial^2 T}{\partial y^2} + \frac{\partial^2 T}{\partial z^2} + \frac{\dot{q}}{k} = 0$$

If the system is in steady-state and has no internal heat generation, the heat conduction equation reduces to the Laplace equation

$$\frac{\partial^2 T}{\partial x^2} + \frac{\partial^2 T}{\partial y^2} + \frac{\partial^2 T}{\partial z^2} = 0$$

The heat conduction equation in the cylindrical coordinates is given by

$$\frac{\partial^2 T}{\partial z^2} + \frac{\partial^2 T}{\partial r^2} + \frac{1}{r}\frac{\partial T}{\partial r} + \frac{1}{r^2}\frac{\partial^2 T}{\partial \theta^2} + \frac{\dot{q}}{k} = \frac{1}{\alpha}\frac{\partial T}{\partial t}$$

In the spherical coordinates, the heat conduction equation is given by

$$\frac{\partial^2 T}{\partial r^2} + \frac{2}{r}\frac{\partial T}{dr} + \frac{1}{r^2 \sin^2\theta}\frac{\partial^2 T}{\partial \phi^2} + \frac{1}{r^2 \sin\theta}\frac{\partial}{\partial \theta}\left(\sin\theta\frac{\partial T}{\partial \theta}\right) + \frac{\dot{q}}{k} = \frac{1}{\alpha}\frac{\partial T}{\partial t}$$

If the three sides of a thin rectangular plate are held at constant temperature T_1 and a temperature distribution of $T = T_1 + T_m \sin(\frac{\pi x}{\ell})$, where ℓ is the length of the plate, is impressed on the top face which is at a height b, then the temperature distribution in the plate can be obtained analytically as

$$T - T_1 = T_m \frac{\sin\left(\frac{\pi x}{\ell}\right) \sinh\left(\frac{\pi y}{\ell}\right)}{\sinh\left(\frac{\pi b}{\ell}\right)}$$

An approximate solution of the Laplace equation for a two-dimensional body can be obtained graphically by constructing a network of isotherms and heat flow lines. The isotherms and heat flow lines are perpendicular to each other. This method is known as flux plotting. In a flux plot, the isotherms and heat flow lines form curvilinear squares. The ratio of the number of heat flow lanes to the number of squares in a heat flow tube (or the number of temperature increments) is called the conduction shape factor S. The rate of conduction heat transfer through a body is then given by $q = kS\Delta T$, where ΔT is the overall temperature difference.

If two different physical phenomena can be described mathematically by the same equation, the two phenomena are said to be mathematically analogous. The variables of one system are called analogues of the corresponding variables of the other system. In a two-dimensional case, the Laplace equation for heat conduction is analogous to the potential distribution in an electric field. The temperature is analogous to the voltage, and the thermal resistance is analogous to the electrical resistance. For many practical situations the analytical solutions to heat transfer problems cannot be obtained. In such cases an experimental electrical analogue of the heat transfer problem can be devised to obtain the solution.

In the numerical method of analysis of heat transfer problems, the differential equation and the boundary conditions are expressed as a set of algebraic equations. These algebraic equations can be solved by the relaxation technique. Over-relaxation and unit block operation are powerful shortcuts which can be introduced in the relaxation to save considerable effort and time in the analysis.

Review Questions

1. What are the methods of analyzing heat conduction in two- and three-dimen sional systems?
2. What are the difficulties associated with the analytical method of analysis of two- and three-dimensional heat conduction problems?
3. What is the advantage associated with the graphical and numerical methods of analysis of two- and three-dimensional heat conduction problems?
4. Define thermal diffusivity.
5. State the general conduction equation in Cartesian coordinates.

6. Under what conditions does the general heat conduction equation reduce to the Fourier equation?
7. State the conditions under which the general heat conduction equation reduces to the Poisson equation.
8. Specify the conditions under which the general heat conduction equation reduces to the Laplace equation.
9. What is the assumption involved in making use of the separation of variables method for solving the Laplace equation?
10. Is it possible for the heat flow lines and isotherms to be parallel? If not, specify the reason.
11. What advantage is obtained by choosing $\Delta x = \Delta y$ in the graphical method of analyzing the heat conduction equation?
12. Define the conduction shape factor.
13. When do we say that two different physical phenomena are analogous?
14. Is the Laplace equation valid for determining the voltage distribution in an electric field?
15. Name the electrical analogues of thermal resistance, temperature and the rate of heat flow.
16. What is the basic procedure in setting up a numerical solution to a two-dimensional conduction problem?
17. Having developed the finite difference equations for a conduction problem, what methods can be used to obtain a solution?
18. What purpose does an *operation table* serve in the relaxation technique?
19. Name a few shortcuts which can be introduced in the relaxation technique to save considerable effort and time.
20. What is meant by *block operation* in the relaxation technique?

Problems

4.1 The relationships between the cylindrical and Cartesian coordinates are given by $x = r\cos\theta$, $y = r\sin\theta$ and $z = z$. Transform the general heat conduction equation given by Eq. (4.5) into the cylindrical coordinates by employing the standard coordinate transformation technique.

4.2 Starting from the general heat conduction equation in rectangular coordinates, derive the temperature profile for steady-state heat conduction through a plane wall of homogeneous material with no internal heat generation.

4.3 Consider a long hollow cylinder of inside radius r_i and outside radius r_0. In steady-state, the inside and outside surfaces of the cylinder are at temperatures T_i and T_0, respectively. Assume that the heat flows in the radial direction only.

Starting from the general heat conduction equation in cylindrical coordinates, derive a relation to determine the temperature distribution in the cylinder.

4.4 Consider a flat plate of thickness 2ℓ which has uniformly distributed heat sources. Assume that the heat flow is one-dimensional and the plate is in steady-state. Starting from the one-dimensional Poisson equation derive a relation to determine the temperature distribution in the plate.

4.5 Consider a long solid circular cylinder with uniform internal heat generation. Starting from the general heat conduction equation in cylindrical coordinates, derive a relation to determine the temperature distribution in the cylinder.

4.6 The heat conduction equation for a two-dimensional body with no internal heat generation and in steady-state is

$$\frac{\partial^2 T}{\partial x^2} + \frac{\partial^2 T}{\partial y^2} = 0$$

Consider the thin rectangular plate shown in Fig. 4.3 along with the specified boundary conditions. Apply the separation of variables method to solve the above equations and write down the general solutions for (a) the constant $\lambda^2 = 0$ and (b) the constant $\lambda^2 < 0$. [see Eqs. (4.23) and (4.24).] Judge whether the solutions satisfy the boundary conditions or not.

4.7 The sides and bottom of a long mica pad (k = 0.5815 W/m^2 K) of 2 cm width and 5 cm height are bound to a metal plate at 150°C, and the top surface of the mica pad is impressed with a temperature distribution $T = 150 + 300\sin(\pi x/0.02)$, where x is in metres and T is in °C. Determine the rate of heat flow per metre length of the mica pad.

4.8 A thin rectangular plate $0 \le x \le a; 0 \le y \le b$ has the following temperature distribution on its boundary:

$$x = 0;\ T = 300 \text{ K}$$

$$x = a;\ T = 300 + 100\sin\left(\frac{\pi y}{b}\right)$$

$$y = 0;\ T = 300 \text{ K}$$

$$y = b;\ T = 300 + 200\sin\left(\frac{\pi x}{a}\right)$$

Determine (a) the steady-state temperature distribution and (b) the centre temperature if $a = b$.

4.9 A pipe of 10 cm OD passes through the centre of a concrete slab of cross-section 60 cm × 60 cm. The outer surface of the pipe is maintained at 130°C while the outer surface of the concrete slab is at 30°C. Using the flux plot, estimate the heat loss per metre length of the pipe.

4.10 A furnace wall constructed with fireclay brick (k= 1.04 W/m^2 K) appears as shown in Fig. 4.16. Using the flux plot, determine the conduction shape factor for the wall.

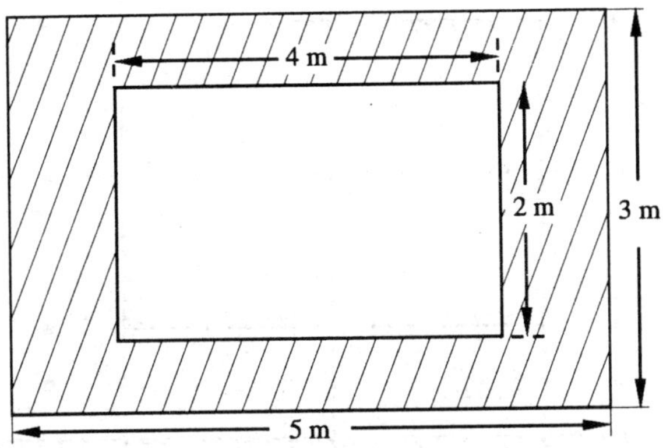

Fig. 4.16 *Schematic of a furnace wall.*

4.11 An oven in a bakery has inside dimensions of 2 m width, 1 m height and 1 m depth. The walls are made of 10 cm thick masonry brick. At steady operating conditions the inside temperature is 180°C while the outside temperature is 40°C. Estimate (a) the conduction shape factor and (b) heat loss from the oven.

4.12 Consider the one-dimensional fin shown in Fig. 4.17 and obtain the nodal equation for the nodal point 3.

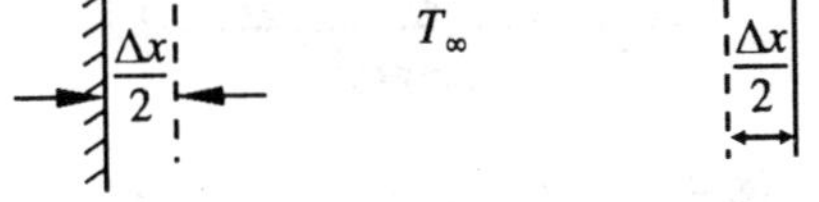

Fig. 4.17 *Circular pin fin of diameter D.*

4.13 Derive the nodal equation for the nodal point 6 shown in Fig. 4.17.

4.14 Consider the nodal point 0 near a curved boundary shown in Fig. 4.13(f) and derive the nodal equation.

4.15 Figure 4.18 shows 16 nodal points. Make use of the symmetry of the system and determine the temperatures (to an accuracy of 0.1°C) through the relaxation technique.

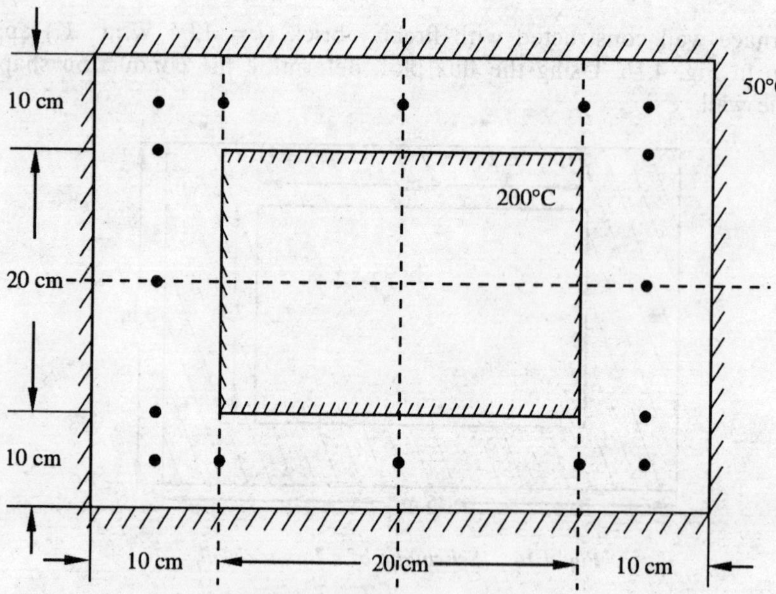

Fig. 4.18 *Sketch for Problem 4.15.*

4.16 Estimate the temperatures at points 1, 2, 3 and 4 shown in Fig. 4.19 using the relaxation technique.

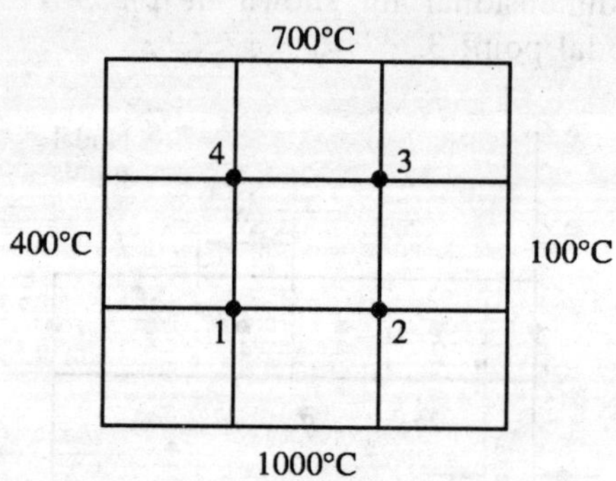

Fig. 4.19 *Sketch for Problem 4.16.*

4.17 A pin fin of 1 cm diameter and 5 cm length from the base appears as shown in Fig. 4.17. If $T_s = 300°C, T_\infty = 30°C$ and the convection heat transfer coefficient is 10 W/m^2 K, calculate the steady-state temperatures at the nodes. Also calculate the rate of heat loss if the thermal conductivity of the fin material is 25 W/m K.

Chapter Five

Unsteady-State Heat Conduction

Learning Objectives

This chapter will enable the readers to

- analyze transient heat transfer problems using lumped heat capacity analysis.
- draw an electrical analogy for transient heat flow.
- extend the lumped heat capacity analysis to composite systems.
- obtain an analytical solution to transient heat flow in an infinite plate.
- analyze transient heat flow through infinite plates, infinite cylinders and spheres by using Heisler charts.
- analyze transient heat flow through two- and three-dimensional bodies.
- solve problems on transient heat flow in a semi-infinite solid
    - when the surface temperature suddenly changes.
    - when the surface is exposed to a constant heat flux.
    - with convection heat transfer at the surface.

5.1 Introduction

When a solid body is suddenly exposed to an environment which is at a temperature different from that of the body, some time must elapse before the body can attain a steady-state condition. In the transient heating and cooling of bodies, the internal energy of the body changes with time and such a change should be taken into account while analyzing transient or unsteady-state heat transfer problems. An engineer encounters many unsteady-state heat flow problems in practice. To analyze a transient heat conduction problem, one can use the analytical method for solving the general heat conduction equation. The simplest problems in unsteady-state heat conduction are those in which the temperature is a function of time only and it is uniform through out the system. First, we shall analyze such systems.

5.2 Transient Heat Flow in Systems with Negligible Internal Resistance

In several systems we find that the internal conduction resistance is very small compared to the external thermal resistance between the surface of the body and the environment. In such situations, the body will be at a uniform temperature. Suppose a hot steel ball is suddenly immersed in a cold liquid. Then the temperature distribution in the ball depends on the thermal conductivity of the ball material and the convection heat transfer from the surface of the ball to the fluid. If the conduction resistance in the ball is very small compared to the convection resistance for heat transfer from the surface of the ball to the fluid, then the steel ball will be at a uniform temperature. The analysis of systems in which the system is assumed to be at uniform temperature is called *lumped heat capacity analysis*.

A measure of the relative importance of the internal conduction resistance is expressed by the ratio of internal conduction resistance (ℓ/kA) to the external convection resistance $(1/hA)$. This ratio is called the *Biot number* or *Biot modulus*, i.e.,

$$Bi = \frac{\text{Internal conduction resistance}}{\text{External convection resistance}} = \frac{\ell/kA}{1/hA} = \frac{h\ell}{k} \tag{5.1}$$

It is reasonable to assume that a body is at uniform temperature if the Biot number is less than 0.1 (that is, if the internal conduction resistance is less than 10 percent of the external convection resistance.)

Consider a small metal casting of volume V, density ρ and specific heat c at temperature T_0 which is suddenly immersed in a large quantity of fluid at T_∞. Since the thermal conductivity of a metal is very high, it is reasonable to assume that there is no temperature gradient within the metal. The energy balance over a small time interval dt on the metal casting gives

$$\left(\begin{array}{l}\text{Change in the internal energy}\\ \text{of the casting during time } dt\end{array}\right) = \left(\begin{array}{l}\text{net heat flow from the casting}\\ \text{to the fluid during time } dt\end{array}\right)$$

or

$$-\rho V c dT = hA(T - T_\infty)dt \tag{5.2}$$

where A is the surface area of the casting. Equation (5.2) can be rewritten as

$$\frac{dT}{T - T_\infty} = \frac{d(T - T_\infty)}{T - T_\infty} = -\frac{hA}{\rho V c}dt \tag{5.3}$$

Equation (5.3) can be solved along with the initial condition $T = T_0$ at $t = 0$ to obtain the solution

$$\ln\left(\frac{T - T_\infty}{T_0 - T_\infty}\right) = \frac{-hAt}{\rho V c}$$

or

$$\frac{T - T_\infty}{T_0 - T_\infty} = \exp\left(\frac{-hAt}{\rho Vc}\right) = \exp\left(-\frac{t}{\tau}\right) \tag{5.4}$$

The quantity $(\rho Vc/hA)$ has the dimensions of time and is called the time constant τ. That is,

$$\tau = \rho Vc/hA = R_{th}C_{th} \tag{5.5}$$

$$= \text{(thermal resistance for convection)} \times \text{(thermal capacity of the body)}$$

Equation (5.4) gives the temperature–time history of the cooling casting. This equation is similar to the equation describing the voltage–time history in an electrical system consisting of a capacitor C_e and resistance R_e in series. In an electrical network, the time constant $\tau = R_eC_e$. The analogy between a thermal system and an electrical R_e–C_e circuit is shown in Fig. 5.1.

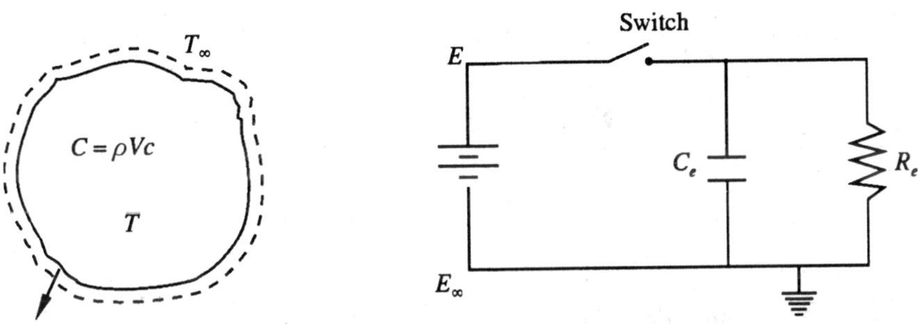

$$R_{th} = \frac{1}{hA}$$

$$q = \frac{T - T_\infty}{R_{th}} = \frac{T - T_\infty}{(1/hA)} = -C_{th}\frac{dT}{dt} = -\rho VC\frac{dT}{dt}$$

$$\frac{T - T_\infty}{T_0 - T_\infty} = \exp\left(-\frac{1}{R_{th}C_{th}}t\right)$$

$$i = \frac{E - E_\infty}{R_e} = -C_e\frac{dE}{dt}$$

$$\frac{E - E_\infty}{E_0 - E_\infty} = \exp\left(-\frac{1}{R_eC_e}t\right)$$

Fig. 5.1 *Analogy between the cooling of a casting and the discharging of a capacitor.*

In the thermal system, the energy is stored in the casting, while in the electrical system the electric charge is stored in the capacitor. The flow of energy in the thermal system is analogous to the flow of electric charge or current. When the switch in the electrical system is closed, the capacitor is initially charged to the voltage E_0. When the switch is opened, the capacitor discharges through the resistor. The temperature–time or voltage–time history of a cooling casting (or a capacitor) is shown in Fig. 5.2. It should be noted that at time $t = \tau$, the temperature difference $(T - T_\infty)$ has a value of 36.8 per cent of the initial temperature difference $(T_0 - T_\infty)$.

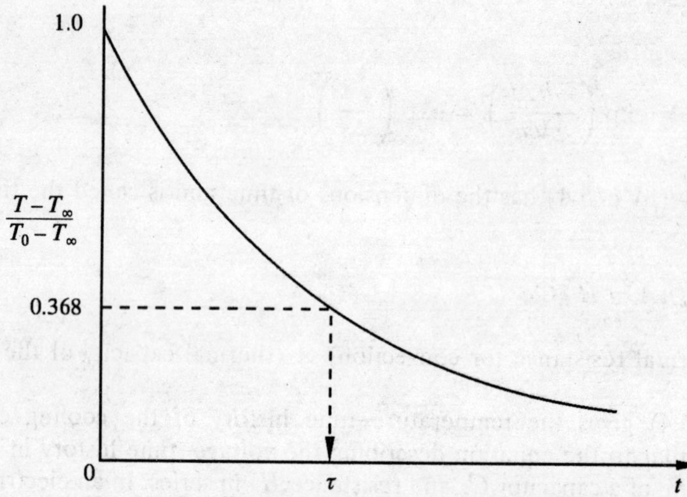

Fig. 5.2 Temperature–time history of a cooling casting.

The argument $\left(\frac{hAt}{\rho Vc}\right)$ of the exponential term in Eq. (5.4) can be rewritten as

$$\frac{hAt}{\rho Vc} = \frac{ht}{\rho c\ell}\left(\frac{\ell k}{\ell k}\right) = \left(\frac{h\ell}{k}\right)\left(\frac{k}{\rho c}\frac{t}{\ell^2}\right) = \left(\frac{h\ell}{k}\right)\left(\frac{\alpha t}{\ell^2}\right) = (Bi)(Fo) \tag{5.6}$$

where

$$\ell = \frac{V}{A} = \text{characteristic length} \tag{5.7}$$

$$\alpha = \frac{k}{\rho c} = \text{thermal diffusivity} \tag{5.8}$$

$$Fo = \frac{\alpha t}{\ell^2} = \text{Fourier number or Fourier modulus} \tag{5.9}$$

Then Eq. (5.4) can be rewritten as

$$\frac{T - T_\infty}{T_0 - T_\infty} = \exp\left(-\frac{hAt}{\rho Vc}\right) = \exp[-(Bi)(Fo)] \tag{5.10}$$

or

$$\ln\left(\frac{T - T_\infty}{T_0 - T_\infty}\right) = -(Bi)(Fo) \tag{5.11}$$

Hence a plot of $\ln\left(\frac{T - T_\infty}{T_0 - T_\infty}\right)$ versus $(Bi)(Fo)$ yields a straight line with a slope of -1 as shown in Fig. 5.3.

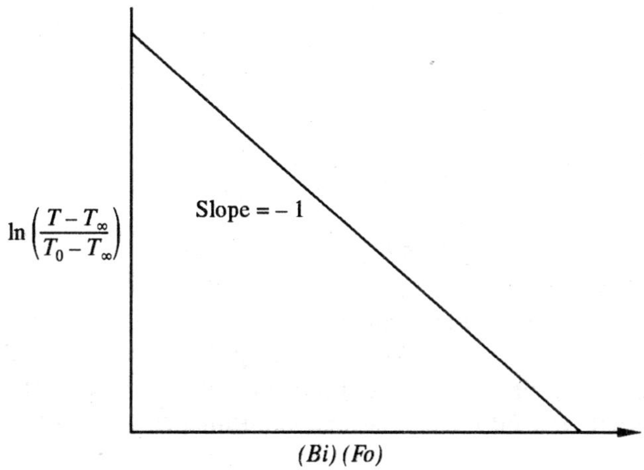

Fig. 5.3 *Plot of* $\ln\left(\frac{T - T_\infty}{T_0 - T_\infty}\right)$ *versus the product of* Bi *and* Fo *numbers.*

Sometimes one may be interested in determining the change in the energy of the system during a given time interval. This can be determined by evaluating the instantaneous rate of heat flow at time t and integrating it from $t = 0$ to the required time t, as shown below.

The instantaneous rate of heat flow at time t

$$= q = \rho V c \frac{dT}{dt} \tag{5.12}$$

We know that the temperature variation with time is givcn by

$$\frac{T - T_\infty}{T_0 - T_\infty} = \exp\left(\frac{-hAt}{\rho V c}\right) \tag{5.4}$$

or

$$\frac{dT}{dt} = (T_\infty - T_0)\left(\frac{hA}{\rho V c}\right)\exp\left(\frac{-hAt}{\rho V c}\right) \tag{5.13}$$

Substituting Eq. (5.13) in Eq. (5.12), we get

$$q = (hA)(T_\infty - T_0)\exp\left(\frac{-hAt}{\rho V c}\right) \tag{5.14}$$

Then the quantity of heat transferred Q, in the time interval $t = 0$ to time t is given by

$$Q = \int_0^t q dt = \int_0^t (hA)(T_\infty - T_0) \exp\left(\frac{-hAt}{\rho Vc}\right) dt$$

or

$$Q = \left(\frac{\rho Vc}{hA}\right)(hA)(T_\infty - T_0)\left[1 - \exp\left(\frac{-hAt}{\rho Vc}\right)\right]$$

$$= \rho VchA\,[1 - \exp(-(Bi)(Fo))] \tag{5.15}$$

EXAMPLE 5.1 A carbon-steel (k= 54 W/m K; ρ = 7833 kg/m^3; c = 0.465 kJ/kg K) ball of 10 cm diameter which is initially at a uniform temperature of 625°C is suddenly immersed in a large quantity of oil at 25°C. If the convection heat transfer coefficient between the steel ball and oil is 10 W/m^2 K, determine the time required for the ball to attain a temperature of 100°C.

SOLUTION The Biot number is given by

$$Bi = \frac{h\ell}{k} = \frac{h(V/A)}{k} = \frac{h}{k}\left(\frac{4}{3}\pi r^3\right)\frac{1}{4\pi r^2} = \frac{hr}{3k}$$

or $$Bi = \frac{10 \times 0.05}{3 \times 54} = 0.0031$$

Since $Bi \ll 0.1$, we can use the lumped heat capacity analysis.

Thermal diffusivity is given by

$$\alpha = \frac{k}{\rho c} = \frac{54}{7833 \times 0.465 \times 10^3} = 1.4826 \times 10^{-5}$$

We know that the temperature variation with time is given by

$$\ln\left(\frac{T - T_\infty}{T_0 - T_\infty}\right) = -(Bi)(Fo)$$

or $$\ln\left(\frac{100 - 25}{625 - 25}\right) = -(0.0031)(Fo)$$

or $$(Fo) = 670.8 = \frac{\alpha t}{\ell^2} = \frac{1.4826 \times 10^{-5}\, t}{(r/3)^2} = \frac{1.4826 \times 10^{-5}\, t}{(0.05/3)^2}$$

or $t = 12,568 \text{ s} = 3.49$ hours.

EXAMPLE 5.2 A constantan wire of 1 mm diameter originally at 25°C is suddenly exposed to air at 250°C. If the convection heat transfer coefficient is 3 W/m^2 K, determine the time required for the wire to reach a temperature of 225°C.

SOLUTION From Appendix (Table A1), we read the following values for constantan:

$$k = 22.7 \text{ W/m K}; \ \alpha = 0.612 \times 10^{-5} \text{ m}^2\text{/s}$$

Characteristic length is given by

$$\ell = \frac{V}{A} = \frac{\pi r^2 \ell}{2\pi r \ell} = \frac{r}{2} = \frac{0.5 \times 10^{-3}}{2} = 0.25 \times 10^{-3} \text{ m}$$

$$Bi = \frac{h\ell}{k} = \frac{3 \times 0.25 \times 10^{-3}}{22.7} = 3.304 \times 10^{-5}$$

The temperature–time history is given by

$$\ln\left(\frac{T - T_\infty}{T_0 - T_\infty}\right) = -(Bi)(Fo)$$

Therefore,

$$\ln\left(\frac{225 - 250}{25 - 250}\right) = -3.304 \times 10^{-5}(Fo)$$

or $$Fo = \frac{\alpha t}{\ell^2} = 66502$$

or $$\frac{0.612 \times 10^{-5}\, t}{(0.25 \times 10^{-3})^2} = 66502$$

or $t = 679 \text{ s} = 11.32$ minutes

5.3 Two Lumped Heat Capacity System

The lumped heat capacity method of analysis can also be applied to composite systems. If the size and the type of material of each system are known, one can determine the heat capacity of each system constituting the composite system. Then it is necessary to know the thermal resistance connecting the individual systems of the composite system. Consider a container filled with a fluid at temperature T_0 which is exposed to an environment at temperature T_∞ as shown in Fig. 5.4. Then the fluid can be treated as system 1 and the container can be treated as system 2. The thermal capacitance of system 1 is $\rho_1 V_1 c_1$ and the thermal capacitance of system 2 is $\rho_2 V_2 c_2$. Let

h_1 = convection heat transfer coefficient between the fluid and the inner surface of the container

h_2 = convection heat transfer coefficient between the outer surface of the container and the environment.

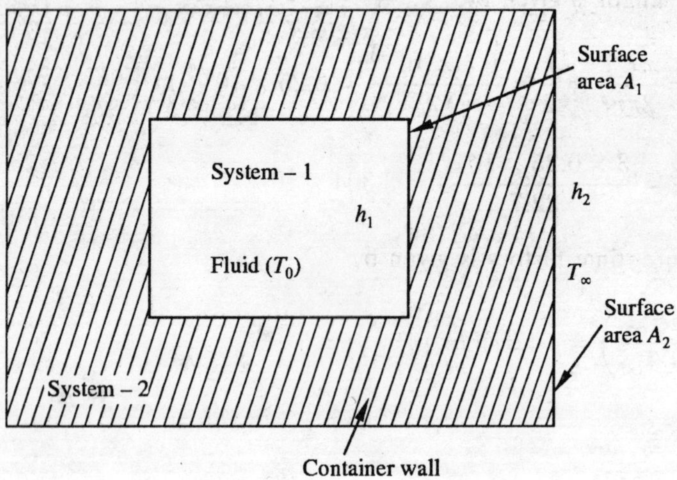

Fig. 5.4 *Two lumped heat capacity system.*

Initially, the fluid and the container are at temperature T_0. When the container is exposed to an environment at T_∞, the temperature T_1 of the fluid (system 1) and the temperature T_2 of the container (system 2) change with time due to convection heat transfer from the fluid to the environment through the container wall.

Energy balance on the fluid in the container gives

$$-\rho_1 V_1 c_1 \frac{dT_1}{dt} = h_1 A_1 (T_1 - T_2) \tag{5.16}$$

Energy balance on the container gives

$$-\rho_2 V_2 c_2 \frac{dT_2}{dt} = -h_1 A_1 (T_1 - T_2) + h_2 A_2 (T_2 - T_\infty) \tag{5.17}$$

Thus we have two simultaneous linear differential equations which can be solved to obtain the temperature–time history of the fluid and container. These equations must satisfy the following boundary conditions:

$$T_1 = T_0 \qquad \text{at} \qquad t = 0 \tag{5.18}$$

$$T_2 = T_0 \qquad \text{at} \qquad t = 0 \tag{5.19}$$

Equations (5.16) and (5.17) can be rearranged as

$$\frac{dT_1}{dt} + \frac{h_1 A_1}{\rho_1 V_1 c_1}(T_1 - T_2) = 0 \tag{5.20}$$

$$\frac{dT_2}{dt} - \frac{h_1 A_1}{\rho_2 V_2 c_2}(T_1 - T_2) + \frac{h_2 A_2}{\rho_2 V_2 c_2}(T_2 - T_\infty) = 0 \tag{5.21}$$

For convenience let

$$m_1 = \frac{h_1 A_1}{\rho_1 V_1 c_1} \tag{5.22}$$

$$m_2 = \frac{h_1 A_1}{\rho_2 V_2 c_2} \tag{5.23}$$

$$m_3 = \frac{h_2 A_2}{\rho_2 V_2 c_2} \tag{5.24}$$

Then Eq. (5.20) can be rewritten as

$$\frac{dT_1}{dt} + m_1(T_1 - T_2) = 0$$

or

$$T_2 = \frac{1}{m_1}\frac{dT_1}{dt} + T_1 \tag{5.25}$$

or

$$\frac{dT_2}{dt} = \frac{1}{m_1}\frac{d^2T_1}{dt^2} + \frac{dT_1}{dt} \tag{5.26}$$

Equation (5.21) can be rewritten as

$$\frac{dT_2}{dt} - m_2 T_1 + m_2 T_2 + m_3 T_2 - m_3 T_\infty = 0$$

or

$$\frac{dT_2}{dt} = m_2 T_1 + m_3 T_\infty - (m_2 + m_3) T_2 \tag{5.27}$$

From Eqs. (5.26) and (5.27), we get

$$\frac{1}{m_1}\frac{d^2T_1}{dt^2} + \frac{dT_1}{dt} = m_2T_1 + m_3T_\infty - (m_2 + m_3)T_2 \tag{5.28}$$

Substitute Eq. (5.25) in Eq. (5.28) to obtain

$$\frac{1}{m_1}\frac{d^2T_1}{dt^2} + \frac{dT_1}{dt} = m_2T_1 + m_3T_\infty - (m_2 + m_3)\left(\frac{1}{m_1}\frac{dT_1}{dt} + T_1\right)$$

or

$$\frac{d^2T_1}{dt^2} + (m_1 + m_2 + m_3)\frac{dT_1}{dt} + m_1m_3T_1 = m_1m_3T_\infty \tag{5.29}$$

The general solution of Eq. (5.29) is given by

$$T_1 = T_\infty + c_1\exp(k_1t) + c_2\exp(k_2t) \tag{5.30}$$

where

$$k_1 = \frac{-(m_1 + m_2 + m_3) + \sqrt{(m_1 + m_2 + m_3)^2 - 4m_1m_3}}{2} \tag{5.31}$$

$$k_2 = \frac{-(m_1 + m_2 + m_3) - \sqrt{(m_1 + m_2 + m_3)^2 - 4m_1m_3}}{2} \tag{5.32}$$

The constants c_1 and c_2 are to be determined by using the boundary conditions (5.18) and (5.19). Substituting these boundary conditions in Eq. (5.16) we get

$$\frac{dT_1}{dt} = 0 \quad \text{at} \quad t = 0 \tag{5.33}$$

Thus, the two boundary conditions which are to be used in determining the values of c_1 and c_2 are given by Eqs. (5.18) and (5.33). Substituting the boundary condition (5.18) in Eq. (5.30), we get

$$T_0 = T_\infty + c_1 + c_2 \quad \text{or} \quad c_1 + c_2 = T_0 - T_\infty \tag{5.34}$$

Substituting the boundary condition (5.33) in Eq. (5.30) we get

$$0 = k_1c_1 + k_2c_2 \tag{5.35}$$

From Eqs. (5.34) and (5.35), we get

$$c_1 = \frac{k_2(T_0 - T_\infty)}{k_2 - k_1} \tag{5.36}$$

and

$$c_2 = \frac{k_1(T_\infty - T_0)}{k_2 - k_1} \tag{5.37}$$

Substituting the values of c_1 and c_2 in Eq. (5.30), we get the final solution as

$$\frac{T_1 - T_\infty}{T_0 - T_\infty} = \frac{k_2}{k_2 - k_1}\exp(k_1 t) - \frac{k_1}{k_2 - k_1}\exp(k_2 t) \tag{5.38}$$

The solution for T_2 can be obtained by substituting the value of T_1 obtained from Eq. (5.38) in Eq. (5.16).

The electrical analogy for the two lumped heat capacity system is shown in Fig. 5.5. When the switch is closed the two capacitors C_1 and C_2 are charged to the same potential difference $(E_0 - E_\infty)$.

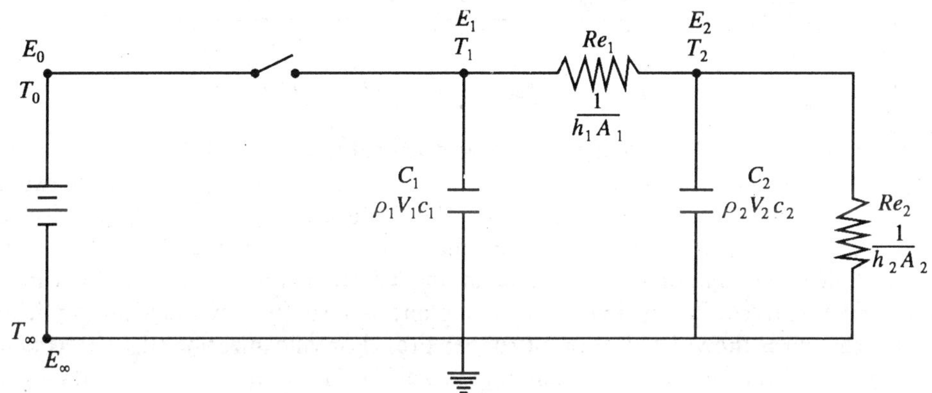

Fig. 5.5 Electrical analogue of a two lumped heat capacity system.

5.4 Transient Heat Flow in an Infinite Plate

The analytical method of solving transient heat flow problems in which the thermal resistance within the system is negligible is discussed in sections 5.2 and 5.3 by treating the thermal capacity of the entire system as a lumped parameter. This approach is quite satisfactory if the Biot modulus is less than 0.1. In this section we shall discuss the analytical method of solving the general heat conduction equation by considering a simple case.

Consider a large flat plate of thickness ℓ which is initially at a uniform temperature T_i. One side of the plate is insulated and the other side is suddenly exposed to a large mass of fluid at temperature T_∞ as shown in Fig. 5.6. Then heat flows from the fluid to the plate. We are interested in determining the time dependent temperature distribution in the plate and the rate of heat flow from the fluid to the plate.

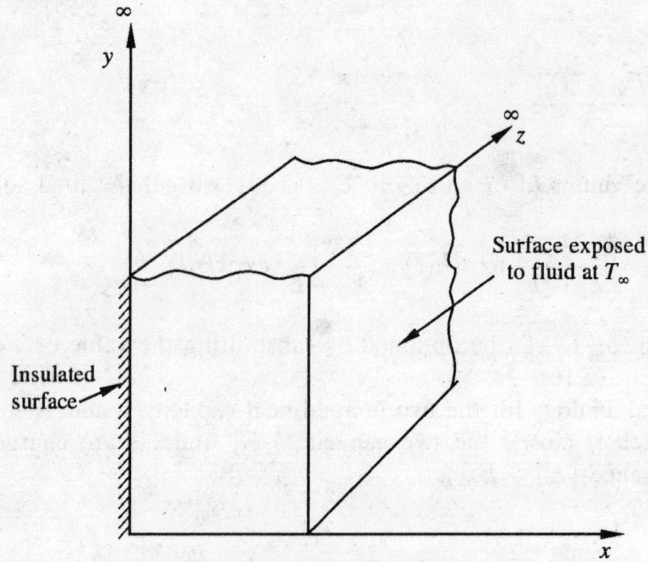

Fig. 5.6 Schematic of an infinite plate.

The solution to this problem has several practical applications. Suppose an infinite plate of thickness 2ℓ which is initially at uniform temperature T_i is suddenly exposed to a fluid at temperature T_∞. Since both the surfaces are exposed to uniform and equal temperatures, the temperature profile would be symmetric about the centre plane. Hence the temperature gradient at the centre plane must be zero, which is equivalent to the condition that the centre plane is insulated. A practical application of this problem is the vulcanization of rubber tires. For vulcanization, the tire is placed in a mould and the surfaces are maintained at an elevated temperature till the central layer reaches a specified temperature.

The plate is assumed to be very long in the y and z directions. Then the temperature will be uniform in any yz plane at a specified value of x. Then the general heat conduction equation, in the absence of internal heat generation, reduces to

$$\frac{\partial^2 T}{\partial x^2} = \frac{1}{\alpha}\frac{\partial T}{\partial t} \tag{5.39}$$

This equation may be rewritten in a more convenient form by introducing the variable θ, defined as

$$\theta = T - T_\infty \tag{5.40}$$

Then Eq. (5.39) reduces to

$$\frac{\partial^2 \theta}{\partial x^2} = \frac{1}{\alpha}\frac{\partial \theta}{\partial t} \tag{5.41}$$

The initial and boundary conditions are given by

$$\theta = \theta_i = T_i - T_\infty \quad \text{at} \quad t = 0 \tag{5.42}$$

$$\frac{\partial \theta}{\partial x} = 0 \quad \text{at} \quad x = 0 \tag{5.43}$$

$$-\frac{\partial \theta}{\partial x} = \frac{h}{k}(T_{x=\ell} - T_\infty) = \frac{h}{k}\theta_{x=\ell} \quad \text{at} \quad x = \ell \tag{5.44}$$

Since the temperature is a function of time (t) and distance (x), we assume a product solution of the form

$$\theta(x,t) = X(x)\tau(t) \tag{5.45}$$

Then

$$\frac{\partial \theta}{\partial t} = X\frac{d\tau}{dt} \quad \text{and} \quad \frac{\partial^2 \theta}{\partial x^2} = \tau\frac{\partial^2 X}{\partial x^2} \tag{5.46}$$

Substituting Eqs. (5.46) in Eq. (5.41), we get

$$\tau\frac{\partial^2 X}{\partial x^2} = \frac{1}{\alpha}X\frac{\partial \tau}{\partial t} \quad \text{or} \quad \frac{1}{\alpha\tau}\frac{\partial \tau}{\partial t} = \frac{1}{X}\frac{\partial^2 X}{\partial x^2} \tag{5.47}$$

Now we observe that the left-hand side of Eq. (5.47) is a function of time only and is independent of x. Similarly, the right-hand side of Eq. (5.47) is a function of x only and is independent of t. Since both the left-hand side and right-hand side are equal, they do not depend on either x or t and must be equal to a constant k. Then we can write

$$\frac{1}{\alpha\tau}\frac{\partial \tau}{\partial t} = \frac{1}{X}\frac{\partial^2 X}{\partial x^2} = k \tag{5.48}$$

Equation (5.48) gives two ordinary and linear differential equations with constant coefficients as

$$\frac{\partial \tau}{\partial t} = \alpha k \tau \tag{5.49}$$

$$\frac{\partial^2 X}{\partial x^2} = kX \tag{5.50}$$

The value of the constant k must be determined from the initial and boundary conditions. The form of the solution to Eqs. (5.49) and (5.50) depend on the sign of k. The sign of k can be determined by an inspection of the boundary conditions. The general solution of Eq. (5.49) is given by

$$\tau(t) = c_1 \exp(\alpha k t) \tag{5.51}$$

Equation (5.51) shows that the temperature of the plate keeps on increasing with increasing t if $k > 0$, which is not correct. Hence we exclude the possibility that $k > 0$. Therefore, k cannot take a positive value. On the other hand, if $k = 0$, Eq. (5.51) shows that the temperature of the plate is constant, which is incorrect. Hence, we exclude the possibility that $k = 0$. Therefore, we conclude that k must be a negative quantity. For convenience let us write $k = -\lambda^2$. Then the solution of Eqs. (5.49) and (5.50) are given by

$$\tau(t) = c_1 \exp(-\alpha\lambda^2 t) \tag{5.52}$$

$$X(x) = c_2 \cos(\lambda x) + c_3 \sin(\lambda x) \tag{5.53}$$

Therefore, the product solution which gives the temperature of the plate as a function of distance and time is given by

$$\begin{aligned}\theta(x,t) &= c_1 \exp(-\alpha\lambda^2 t)\,[c_2 \cos(\lambda x) + c_3 \sin(\lambda x)] \\ &= [A\cos(\lambda x) + B\sin(\lambda x)]\exp(-\alpha\lambda^2 t)\end{aligned} \tag{5.54}$$

where $A = c_1 c_2$ and $B = c_1 c_3$. The values of the constants A, B and λ are to be determined from the boundary conditions.

In order to apply the boundary condition (5.43) we take the partial derivative of θ with respect to x and equate it to zero, to obtain

$$\begin{aligned}\left(\frac{\partial\theta}{\partial x}\right)_{x=0} &= \left[A\lambda\sin(\lambda x) + B\lambda\cos(\lambda x)\exp(-\alpha\lambda^2 t)\right]_{x=0} = 0 \\ &= B\lambda\cos(0)\exp(-\alpha\lambda^2 t) = 0\end{aligned}$$

or

$$B = 0 \tag{5.55}$$

Then Eq. (5.54) reduces to

$$\theta(x,t) = A\cos(\lambda x)\exp(-\alpha\lambda^2 t) \tag{5.56}$$

Application of the second boundary condition (5.44) gives

$$-\left(\frac{\partial\theta}{\partial x}\right)_{x=\ell} = A\lambda\sin(\lambda\ell)\exp(-\alpha\lambda^2 t) = \frac{h}{k}A\cos(\lambda\ell)\exp(-\alpha\lambda^2 t)$$

or $\quad A\lambda\sin(\lambda\ell) = \dfrac{h}{k}A\cos(\lambda\ell)$

or

$$\cot(\lambda\ell) = \frac{k\lambda}{h} = \frac{k(\lambda\ell)}{h\ell} = \frac{\lambda\ell}{Bi} \tag{5.57}$$

An infinite number of values of λ satisfy the Eq. (5.57) and these values of λ are called *characteristic values or eigen values.* The numerical values of λ can be obtained by a graphical solution of Eq. (5.57). Suppose one plots $\cot(\lambda\ell)$ and $(\frac{\lambda\ell}{Bi})$ against $\lambda\ell$. The values of λ at the points of intersection of these curves are the required eigen values as shown in Fig. 5.7.

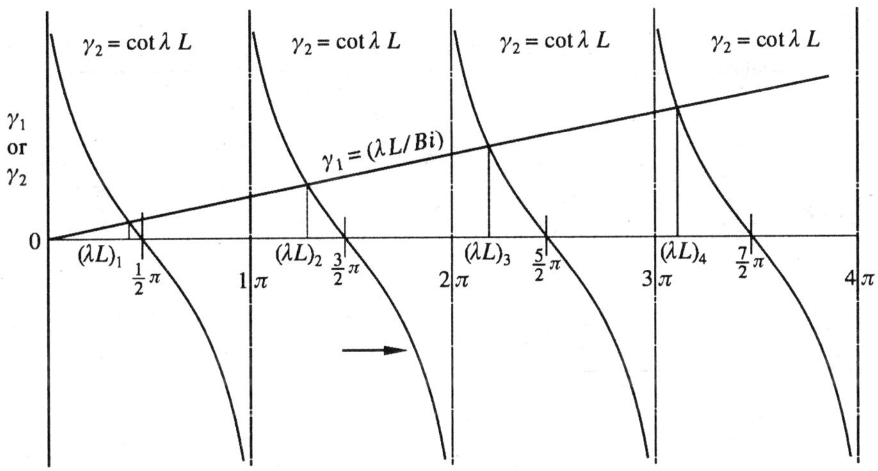

Fig. 5.7 Graphical solution of Eq. (5.57).

If $\ell = 1$, we get the characteristic values as

$$\lambda_1 = 0.86Bi; \quad \lambda_2 = 3.43Bi; \quad \lambda_3 = 6.44Bi; \text{ etc.}$$

The value of $\lambda = 0$ is not considered because it gives the trivial solution $\theta = 0$. A particular solution of Eq. (5.56) corresponds to each value of λ and each value of λ gives a different value for the constant A. To identify the values of λ and A we shall use a subscript n for λ and A. Then the complete solution of Eq. (5.56) is given by the sum of the solutions corresponding to each eigen value. That is,

$$\theta(x,t) = \sum_{n=1}^{\infty} A_n \cos(\lambda_n x) \exp(-\alpha\lambda_n^2 t) \tag{5.58}$$

The constants A_n are determined by substituting the initial condition (5.42) in the temperature distribution given by Eq. (5.58). That is,

$$\theta(x,0) = 0 = \sum_{n=1}^{\infty} A_n \cos(\lambda_n x) \tag{5.59}$$

Multiply both sides of Eq. (5.59) with $\cos(\lambda_m x)$ and integrate between 0 and ℓ, where λ_m is any characteristic value of λ. Then, we get

$$\int_0^\ell \theta_0 \cos(\lambda_m x)dx = \sum A_n \int_0^\ell \cos(\lambda_n x)\cos(\lambda_m x)dx$$

or

$$\frac{\theta_0 \sin(\lambda_m \ell)}{\lambda_m} = A_n \int_0^\ell \cos(\lambda_n x)\cos(\lambda_m x)dx \tag{5.60}$$

$$\int_0^\ell \cos(\lambda_n x)\cos(\lambda_m x)dx = \frac{\lambda_n \sin(\lambda_n \ell)\cos(\lambda_m \ell) - \lambda_m \sin(\lambda_m \ell)\cos(\lambda_n \ell)}{2\ell(\lambda_m^2 - \lambda_n^2)} \tag{5.61}$$

But we know that

$$\cot(\lambda\ell) = \frac{k\lambda}{h} \tag{5.57}$$

or

$$\frac{\cot(\lambda_m \ell)}{\lambda_m} = \frac{k}{h} = \frac{\cot(\lambda_n \ell)}{\lambda_n}$$

or

$$\lambda_n \frac{\cos(\lambda_m \ell)}{\sin(\lambda_m \ell)} = \lambda_m \frac{\cos(\lambda_n \ell)}{\sin(\lambda_n \ell)}$$

or

$$\lambda_n \sin(\lambda_n \ell)\cos(\lambda_m \ell) = \lambda_m \sin(\lambda_m \ell)\cos(\lambda_n \ell) \tag{5.62}$$

Substituting Eq. (5.62) in Eq. (5.61), we find that

$$\int_0^\ell \cos(\lambda_n x)\cos(\lambda_m x)dx = 0 \qquad \text{if } m \neq n \tag{5.63}$$

$$\begin{aligned}
\int_0^\ell \cos(\lambda_n x)\cos(\lambda_m x)dx &= \int_0^\ell \cos^2(\lambda_n x)dx \qquad \text{if } m = n \\
&= \int_0^\ell \frac{1+\cos(2\lambda_n x)}{2}dx \\
&= \frac{\ell}{2} + \frac{\sin(\lambda_n \ell)\cos(\lambda_n \ell)}{2\lambda_n}
\end{aligned} \tag{5.64}$$

Substituting Eq. (5.64) in Eq. (5.60), we get

$$\frac{\theta_0 \sin(\lambda_n \ell)}{\lambda_n} = \frac{A_n}{2\lambda_n}[\ell\lambda_n + \sin(\lambda_n \ell)\cos(\lambda_n \ell)]$$

or

$$A_n = \frac{2\theta_0 \sin(\lambda_n \ell)}{\ell\lambda_n + \sin(\lambda_n \ell)\cos(\lambda_n \ell)} \tag{5.65}$$

Substituting the value of A_n from Eq. (5.65) in Eq. (5.58), we obtain the general solution as

$$\theta(x,t) = 2\theta_0 \sum_{n=1}^{\infty} \frac{\sin(\lambda_n \ell)\cos(\lambda_n x)\exp(-\alpha\lambda_n^2 t)}{\ell\lambda_n + \sin(\lambda_n \ell)\cos(\lambda_n \ell)} \tag{5.66}$$

or

$$\frac{T - T_\infty}{T_i - T_\infty} = 2\sum_{n=1}^{\infty} \frac{\sin(\lambda_n \ell)\cos(\lambda_n x)\exp(-\alpha\lambda_n^2 t)}{\ell\lambda_n + \sin(\lambda_n \ell)\cos(\lambda_n \ell)} \tag{5.67}$$

At any instant, the rate of heat transfer per unit area, from the fluid to the surface of the plate is given by

$$\frac{q}{A} = -k\left(\frac{\partial T}{\partial x}\right)_{x=\ell} \tag{5.68}$$

The temperature gradient at the surface of the plate at any given time t is obtained by differentiating Eq. (5.67) with respect to x.

$$\left(\frac{\partial T}{\partial x}\right)_{x=\ell} = (T_i - T_\infty)2\sum_{n=1}^{\infty} \frac{-\lambda_n \sin(\lambda_n \ell)\sin(\lambda_n \ell)\exp(-\alpha\lambda_n^2 t)}{\ell\lambda_n + \sin(\lambda_n \ell)\cos(\lambda_n \ell)} \tag{5.69}$$

Therefore,

$$\frac{q}{A} = 2k(T_i - T_\infty)\sum_{n=1}^{\infty} \frac{\lambda_n \sin^2(\lambda_n \ell)\exp(-\alpha\lambda_n^2 t)}{\ell\lambda_n + \sin(\lambda_n \ell)\cos(\lambda_n \ell)} \tag{5.70}$$

Then the change in the energy of the plate or the total amount of heat transferred per unit area during the time $t = 0$ to $t = t$ is given by

$$Q = \int_0^t \frac{q}{A}dt = 2k(T_i - T_\infty)\int_0^t \sum_{n=1}^{\infty} \frac{\lambda_n \sin^2(\lambda_n \ell)\exp(-\alpha\lambda_n^2 t)}{\ell\lambda_n + \sin(\lambda_n \ell)\cos(\lambda_n \ell)}dt$$

or

$$Q = \frac{2k(T_i - T_\infty)}{\alpha} \sum_{n=1}^{\infty} \frac{\sin^2(\lambda_n \ell)\left[1 - \exp(-\alpha\lambda_n^2 t)\right]}{\lambda_n \left[\ell\lambda_n + \sin(\lambda_n \ell)\cos(\lambda_n \ell)\right]} \tag{5.71}$$

The temperature distribution in the plate and the amount of heat transferred at any given time [see Eq. (5.67) and (5.71)] are expressed in the form of infinite series. These expressions have been evaluated and the results presented in the form of charts by Heisler. The expressions for temperature distribution and the amount of heat transferred can be expressed in terms of dimensionless quantities. For this purpose, let us write

$$\lambda_n = \delta_n/\ell \tag{5.72}$$

Then Eq. (5.57) can be rewritten as

$$\cot(\lambda_n \ell) = \cot\delta_n = \frac{\lambda_n \ell}{Bi} = \frac{\delta_n}{Bi}$$

or

$$\delta_n \tan\delta_n = Bi \tag{5.73}$$

That is, δ_n is a function of the dimensionless quantity Biot modulus. Then Eq. (5.67) can be rewritten as

$$\frac{T - T_\infty}{T_i - T_\infty} = 2\sum_{n=1}^{\infty} \frac{\sin(\delta_n)\cos\left(\dfrac{\delta_n x}{\ell}\right)\exp\left(\dfrac{-\alpha t\delta_n^2}{\ell^2}\right)}{\delta_n + \sin(\delta_n)\cos(\delta_n)} \tag{5.74}$$

We know that

$$Fo = \frac{\alpha t}{\ell^2}$$

Hence, Eq. (5.74) gives

$$\frac{T - T_\infty}{T_i - T_\infty} = \text{Function of } (Bi,\ Fo,\ x/\ell) \tag{5.75}$$

Considering T_∞ as the reference temperature, the initial energy of the plate, per unit area, is given by

$$Q_i = c\rho\frac{V}{A}(T_i - T_\infty) \tag{5.76}$$

where V is the volume of the plate.

In order to make Eq. (5.71) dimensionless, we can divide both sides of Eq. (5.71) by Q_i and obtain

$$\frac{Q}{Q_i} = 2\sum_{n=1}^{\infty} \frac{\sin^2 \delta_n \left[1 - \exp\left(\frac{-\alpha t \delta_n^2}{\ell^2}\right)\right]}{\delta_n(\delta_n + \sin \delta_n \cos \delta_n)}$$

or

$$\frac{Q}{Q_i} = \text{Function of } (Bi, Fo, x/\ell) \tag{5.77}$$

The Heisler charts provide the dimensionless temperature ratio and heat flow ratio as functions of dimensionless quantities Bi, Fo and x/ℓ.

5.5 Heisler Charts for Transient Heat Flow

The analytical method of solving the general heat conduction equation in an infinite plate is discussed in section 5.4. The general solution is found to be in the form of an infinite series. The analytical solutions of transient heat flow problems with boundary conditions of practical importance have been obtained by Heisler and the results are presented in the form of charts. The most important geometries are

1. Infinite plates or plates whose thickness is small compared to the other dimensions.
2. Infinite cylinders or cylinders whose diameter is small compared to the length, and
3. Spheres.

The calculations for the construction of Heisler charts were performed by truncating the infinite series solutions after a few terms. Hence the charts can be used if the Fourier modulus is greater than 0.2. For problems of practical importance, $1/Bi$ lies between 0 and 100. Hence, the Heisler charts are available for $Fo > 0.2$ and $0 < (1/Bi) < 100$. In this section we shall discuss the application of the Heisler charts in the analysis of transient heat flow problems.

5.5.1 Infinite Flat Plate

The Heisler charts for an infinite flat plate of thickness 2ℓ which is initially at a uniform temperature T_i are presented in Figs. 5.8–5.11. The centre plane temperature at $x = 0$ is represented by T_0. At time $t = 0$, the two surfaces of the plate are exposed to a fluid at T_∞. The temperature distribution in the plate is presented in Figs. 5.8–5.10.

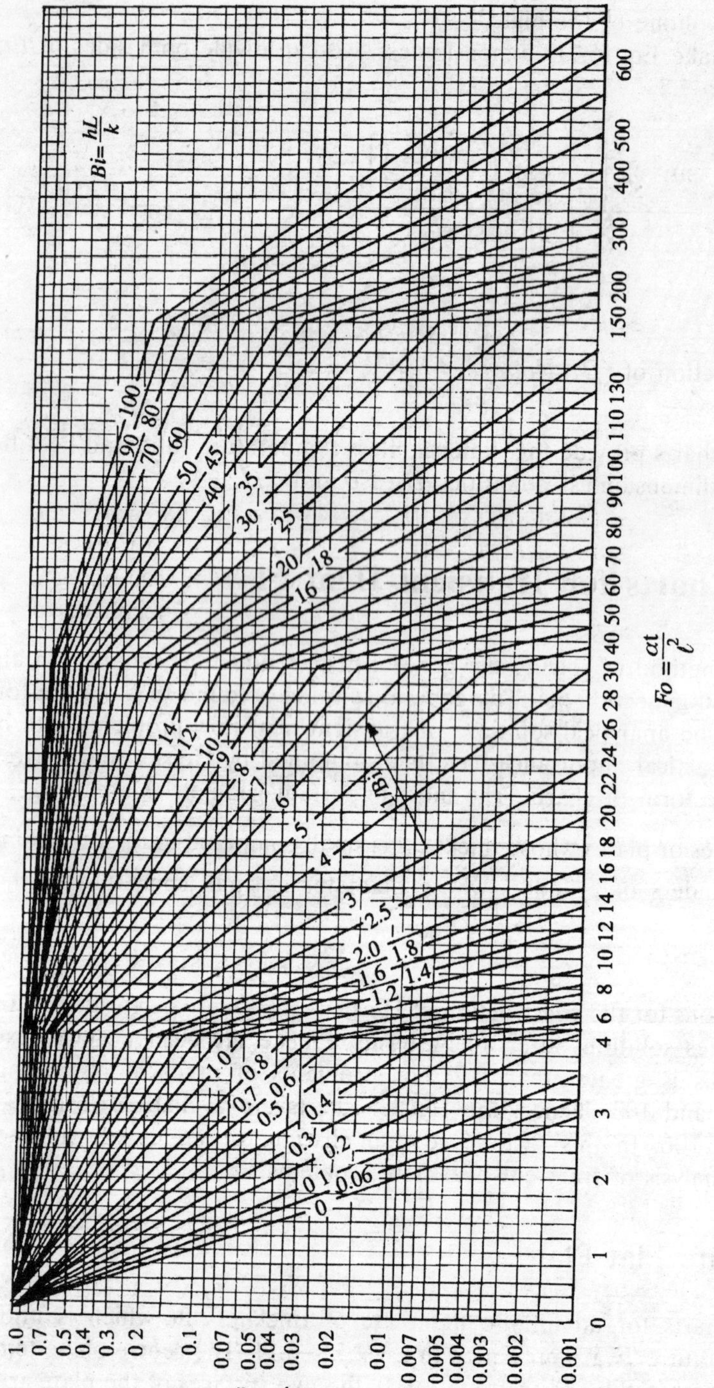

Fig. 5.8 *Mid-plane temperature for an infinite flat plate of thickness 2ℓ for $0 < Fo < 700$.*

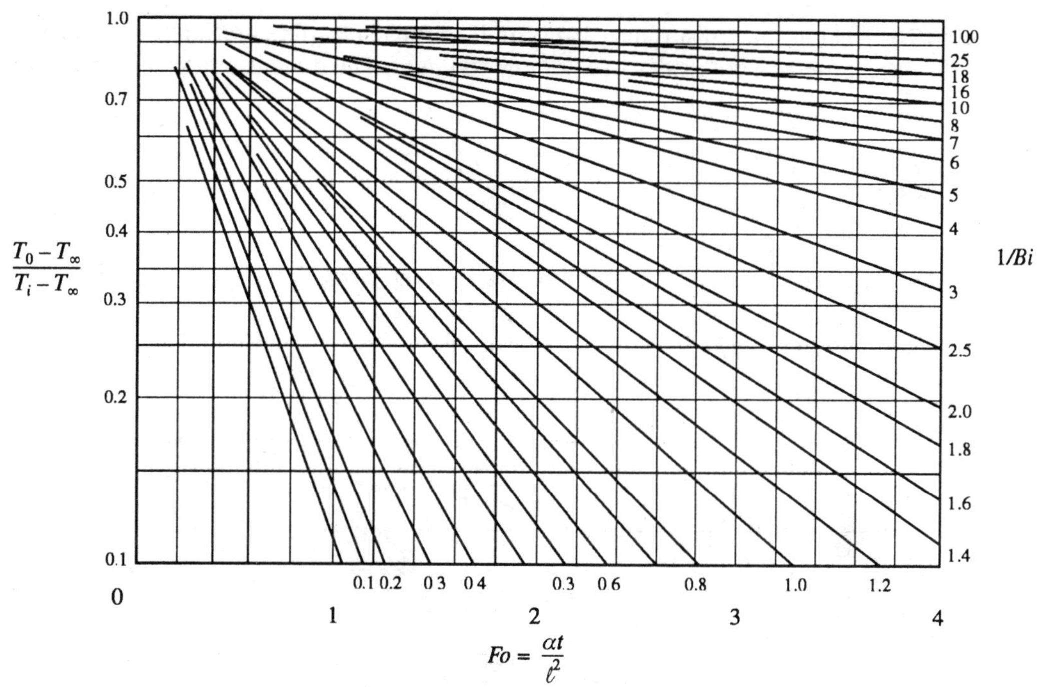

Fig. 5.9 Mid-plane temperature for an infinite flat plate of thickness 2ℓ *for* $0 < Fo < 4$.

Figure 5.8 presents the dimensionless temperature ratio $(T_0 - T_\infty)/(T_i - T_\infty)$ at the centre plane of the plate as a function of Fo and Bi in the range $0 < Fo < 700$, whereas Fig. 5.9 presents the same for clarity in an expanded scale for $0 < Fo < 4$. The temperature of the plate at any location other than the centre plane can be calculated as

$$\frac{T - T_\infty}{T_i - T_\infty} = \left(\frac{T_0 - T_\infty}{T_i - T_\infty}\right)\left(\frac{T - T_\infty}{T_0 - T_\infty}\right) \tag{5.78}$$

The dimensionless temperature ratio $(T - T_\infty)/(T_0 - T_\infty)$ is shown as a function of the dimensionless quantities Bi and (x/ℓ) in Fig. 5.10. The dimensionless heat transfer ratio Q/Q_0 for an infinite flat plate of thickness 2ℓ is presented as a function of $Fo(Bi)^2$ and Bi in Fig. 5.11.

EXAMPLE 5.3 A masonry brick wall ($\rho = 1600$ kg/m^3; $c = 0.84$ kJ/kg K; $k = 0.69$ W/m K; and $\alpha = 5.2 \times 10^{-7}$ m^2/s) of 25 cm thickness and originally at 30°C is suddenly exposed on one side to hot gases at 630°C and the second side is insulated. The convection heat transfer coefficient on the hot gas side is 20 W/m^2 K. Determine the time required for the insulated surface to attain a temperature of 270°C and the heat transferred to the wall per unit area till that instant.

SOLUTION The insulated surface corresponds to the centre-line of a wall of thickness 2ℓ, which is exposed on both sides to a hot environment. The dimensionless temperature ratio at the insulated surface is given by

$$\left(\frac{T - T_\infty}{T_i - T_\infty}\right)_{x=\ell} = \frac{T_0 - T_\infty}{T_i - T_\infty} = \frac{270 - 630}{30 - 630} = 0.6$$

$$\alpha = 5.2 \times 10^{-7} \text{ m}^2/\text{s}; \quad \ell = 0.25 \text{ m}; \quad k = 0.69 \text{ W/m K}; \quad h = 20 \text{ W/m}^2 \text{ K}$$

The Heisler charts given in Figs. 5.8–5.11 can be used for a solution of this problem.

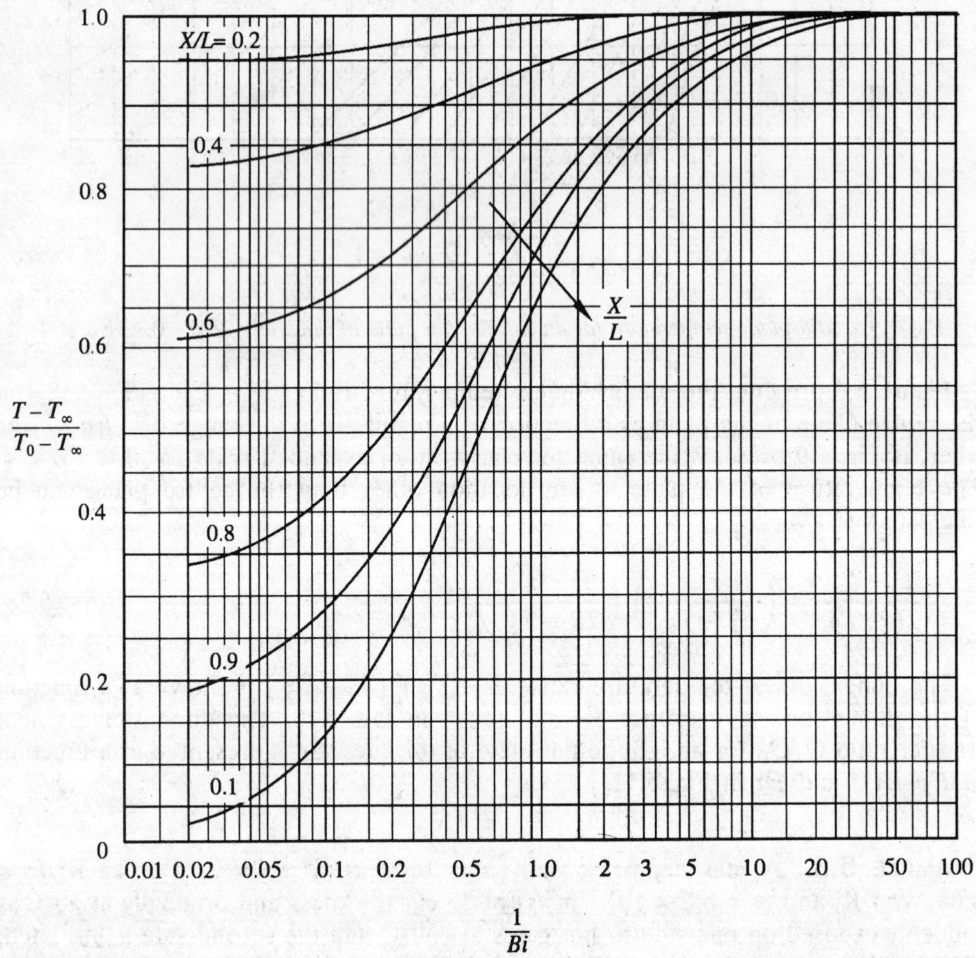

Fig. 5.10 Temperature as a function of mid-plane temperature in an infinite flat plate of thickness 2ℓ.

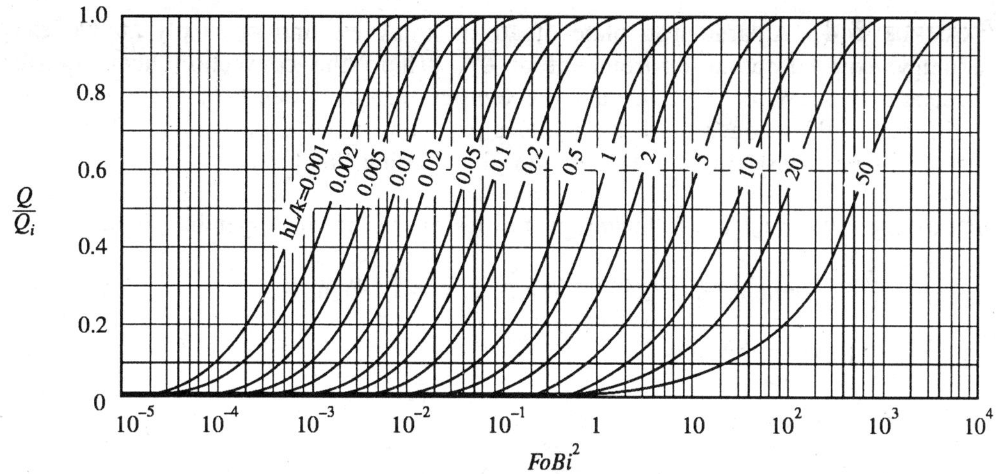

Fig. 5.11 *Dimensionless heat loss Q/Q_0 for an infinite flat plate of thickness 2ℓ with time.*

$$\frac{1}{Bi} = \frac{k}{h\ell} = \frac{0.69}{20 \times 0.25} = 0.138 \quad \text{or} \quad Bi = 7.25$$

From Fig. 5.9 we read

$$Fo = 0.35 \text{ at } 1/Bi = 0.138 \quad \text{and} \quad \frac{T_0 - T_\infty}{T_i - T_\infty} = 0.6$$

Therefore, $Fo = \dfrac{\alpha t}{\ell^2} = 0.35$

$$\text{or} \quad t = \frac{0.35\ell^2}{\alpha} = \frac{0.35(0.25)^2}{5.2 \times 10^{-7}} = 42067 \text{ s} = 11.685 \text{ hours}$$

$$Fo(Bi)^2 = 0.35(7.25)^2 = 18.4$$

Then read the value of $\dfrac{Q}{Q_i} = 0.22$ at $Bi = 7.25$ and $Fo(Bi)^2 = 18.4$ from Fig. 5.11.

$$\frac{Q}{Q_i} = 0.22 \quad \text{or} \quad Q = 0.22Q_i = 0.22c\rho\frac{V}{A}(T_i - T_\infty)$$

or

$$Q = 0.22c\rho(2\ell)(T_i - T_\infty)$$

$$= 0.22 \times 0.84 \times 10^3 \times 1600 \times 0.5 \times 600 = 88.704 \text{ MJ}$$

EXAMPLE 5.4 A large iron plate of 10 cm thickness and originally at 800°C is suddenly exposed to an environment at 0°C where the convection heat transfer coefficient is 50 W/m^2 K. Calculate the temperature at a depth of 4 cm from one of the faces 100 seconds after the plate is exposed to the environment. How much energy has been lost per unit area of the plate during this time?

SOLUTION We read the following properties of iron from Appendix (Table A1).

$$\rho = 7897 \text{ kg/m}^3; \quad c = 0.452 \text{ kJ/kg K}; \quad k = 73 \text{ W/m K}$$

$$\alpha = 2.034 \times 10^{-5} \text{ m}^2\text{/s}$$

The given conditions are

$$T_i = 800°\text{C}; \quad T_\infty = 0°\text{C}; \quad h = 50 \text{ W/m}^2 \text{ K}$$

$$t = 100 \text{ s}; \quad 2\ell = 0.1 \text{ m} \quad \text{and} \quad x = 0.04 \text{ m}$$

This problem can be conveniently solved using the Heisler charts presented in Figs. 5.8–5.11.

$$Bi = \frac{h\ell}{k} = \frac{50 \times 0.05}{73} = 0.034 \quad \text{or} \quad \frac{1}{Bi} = 29.2$$

$$Fo = \frac{\alpha t}{\ell^2} = \frac{2.034 \times 10^{-5} \times 100}{(0.05)^2} = 0.814$$

$$\frac{x}{\ell} = \frac{0.04}{0.05} = 0.8$$

We read the value of

$$\frac{T_0 - T_\infty}{T_i - T_\infty} = 0.92 \quad \text{at } 1/Bi = 29.2 \quad \text{and} \quad Fo = 0.814$$

from Fig. 5.9. Then read the value

$$\frac{T - T_\infty}{T_0 - T_\infty} = 0.9 \quad \text{at} \quad \frac{1}{Bi} = 29.2 \quad \text{and} \quad \frac{x}{\ell} = 0.8 \qquad \text{from Fig. 5.10.}$$

Then

$$T - T_\infty = \left(\frac{T - T_\infty}{T_0 - T_\infty}\right)\left(\frac{T_0 - T_\infty}{T_i - T_\infty}\right)(T_i - T_\infty) = 0.9 \times 0.92 \times 800 = 662.4°\text{C}$$

or $T = 662.4°\text{C}$

$$Fo(Bi)^2 = 0.814(0.034)^2 = 9.4 \times 10^{-4}$$

Then read $Q/Q_i = 0.02$ at $(Fo)(Bi)^2 = 9.4 \times 10^{-4}$ and $Bi = 0.034$ from Fig. 5.11.

$$Q_i = \frac{c\rho V}{A}(T_i - T_\infty) = c\rho(2\ell)(T_i - T_\infty)$$

$$= 0.452 \times 10^3 \times 7897 \times 0.1 \times 800 = 285.56 \text{ MJ}$$

Therefore, $Q = 0.02\,Qi = 5.71$ MJ

EXAMPLE 5.5 A large aluminium plate of thickness 20 cm originally at a uniform temperature of 350°C is suddenly exposed to an environment at 30°C. The convection heat transfer coefficient between the plate and the environment is 1000 W/m^2 K. Determine the time required for the centre-line temperature to drop to 126°C.

SOLUTION The properties of aluminium are given by [see Appendix (Table A1)]

$$\rho = 2707 \text{ kg/m}^3; \quad c = 0.896 \text{ kJ/kg K};$$

$$k = 204 \text{ W/m K}; \quad \alpha = 8.418 \times 10^{-5} \text{ m}^2\text{/s}$$

Given data is

$$2\ell = 0.2 \text{ m}; \quad h = 1000 \text{ W/m}^2 \text{ K}; \quad T_i = 350°\text{C}$$

$$T_\infty = 30°\text{C} \quad \text{and} \quad T_0 = 126°\text{C}$$

Therefore,

$$\frac{T_0 - T_\infty}{T_i - T_\infty} = \frac{126 - 30}{350 - 30} = 0.3$$

$$\frac{1}{Bi} = \frac{k}{h\ell} = \frac{204}{1000 \times 0.1} = 2.04$$

Then we read $Fo = 3$ at $\dfrac{1}{Bi} = 2.04$ and $\dfrac{T_0 - T_\infty}{T_i - T_\infty} = 0.3$ from Fig. 5.9.

Hence, $Fo = \dfrac{\alpha t}{\ell^2} = 3$

or

$$t = \frac{3\ell^2}{\alpha} = \frac{3(0.1)^2}{8.418 \times 10^{-5}} = 356.4 \text{ s} = 5.94 \text{ minutes}$$

5.5.2 Infinitely Long Cylinder

The solutions to the transient heat flow problems are also available in the form of Heisler charts for an infinitely long cylinder, that is, for a cylinder the diameter of which is small compared to the length. These charts are presented in Figs. 5.12–5.15. The initial and boundary conditions for which the charts are applicable are: (i) The cylinder is initially at a uniform temperature T_i. (ii) At time $t = 0$, the cylinder of radius r_0 is suddenly exposed to an environment at T_∞ with convection heat transfer coefficient h. The centre-line temperature is represented as T_0. The characteristic dimension is r_0 (in place of ℓ) for evaluation of Bi and Fo.

EXAMPLE 5.6 A long aluminium cylinder of radius 10 cm which is initially at a uniform temperature of 350°C is suddenly exposed to an environment at 30°C. The convection heat transfer coefficient between the cylinder's surface and the environment is 1000 W/m² K. Determine the time required for the axis of the cylinder to attain a temperature of 126°C.

SOLUTION For aluminium, $k = 204$ W/m K; $\alpha = 8.418 \times 10^{-5}$ m²/s

It is given that

$$T_0 = 126°\text{C}; \quad T_\infty = 30°\text{C}; \quad T_i = 350°\text{C}$$

$$r_0 = 0.1 \text{ m} \quad \text{and} \quad h = 1000 \text{ W/m}^2\text{ K}$$

Therefore,

$$\frac{T_0 - T_\infty}{T_i - T_\infty} = \frac{126 - 30}{350 - 30} = 0.3$$

$$\frac{k}{hr_0} = \frac{204}{1000 \times 0.1} = 2.04$$

From Fig. 5.12, we read

$$\frac{\alpha t}{r_0^2} = 1.55 \quad \text{at} \quad \frac{T_0 - T_\infty}{t_i - T_\infty} = 0.3 \quad \text{and} \quad \frac{k}{hr_0} = 2.04.$$

Hence,

$$t = \frac{1.55\, r_0^2}{\alpha} = \frac{1.55 \times (0.1)^2}{8.418 \times 10^{-5}} = 184 \text{ s}$$

The axis of the cylinder reaches a temperature of 126°C after 184 s.

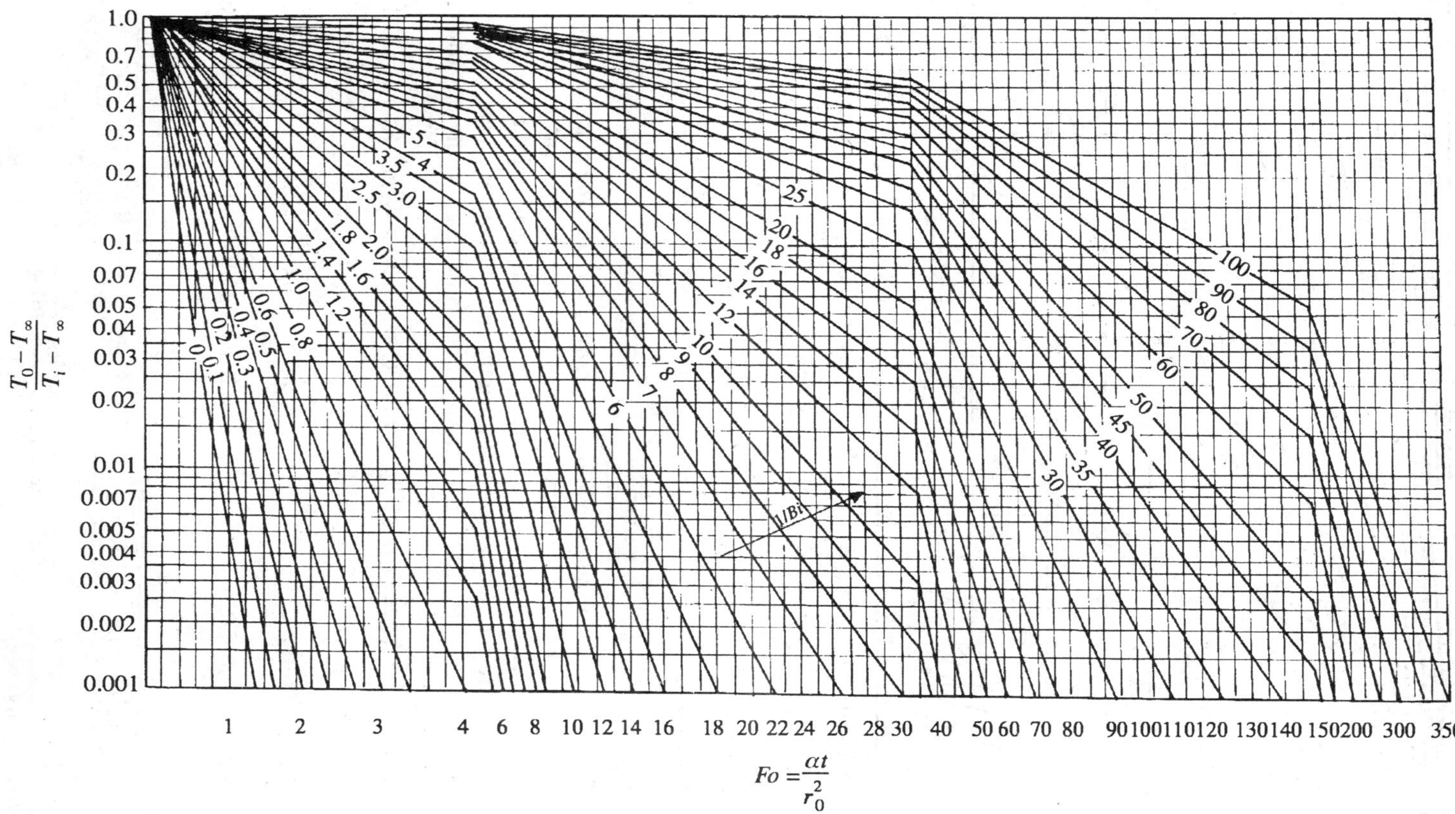

***Fig. 5.12** The dimensionless temperature ratio* $\frac{T_0 - T_\infty}{T_i - T_\infty}$ *at the axis of an infinitely long cylinder of radius* r_0.

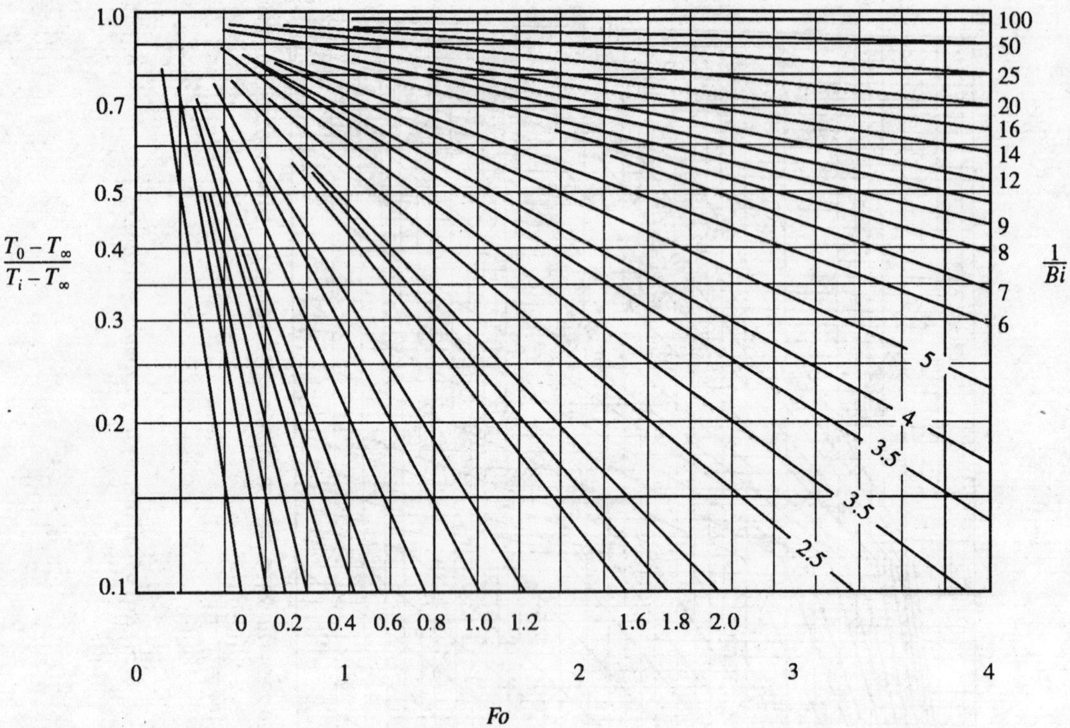

Fig. 5.13 Enlarged chart for the dimensionless temperature ratio $\left(\frac{T_0 - T_\infty}{T_i - T_\infty}\right)$ *at the axis of an infinitely long cylinder of radius* r_0 *in the range* $0 < Fo < 4$.

EXAMPLE 5.7 A long aluminium cylinder of radius 10 cm which is initially at a uniform temperature of 350°C is suddenly exposed to an environment at 30°C. The convection heat transfer coefficient between the cylinder's surface and the environment is 600 W/m^2 K. Calculate the temperature at a radius of 6 cm and the heat transferred per metre length of the cylinder 120 s after the cylinder is exposed to the environment.

SOLUTION Given data: $T_i = 350°\text{C}$; $T_\infty = 30°\text{C}$; $h = 600$ W/m^2 K; $r_0 = 0.1$ m; $r = 0.06$ m; $t = 120$ s

We know that for aluminium

$$k = 204 \text{ W/m K}; \ \alpha = 8.418 \times 10^{-5} \text{ m}^2\text{/s}; \ \rho = 2707 \text{ kg/m}^3 \text{ and } c = 0.896 \text{ kJ/kg K}$$

Then $\dfrac{\alpha t}{r_0^2} = \dfrac{8.418 \times 10^{-5} \times 120}{(0.1)^2} = 1.01$

$$\frac{1}{Bi} = \frac{k}{hr_0} = \frac{204}{600 \times 0.1} = 3.4; \qquad \frac{r}{r_0} = \frac{0.06}{0.1} = 0.6$$

First, we determine the temperature at the axis of the cylinder by using Fig. 5.13. We read

$$\frac{T_0 - T_\infty}{T_i - T_\infty} = 0.65 \text{ at } \frac{\alpha t}{r_0^2} = 1.01 \text{ and } \frac{k}{hr_0} = 3.4 \text{ from Fig. 5.13.}$$

Then we read from Fig. 5.14

$$\frac{T - T_\infty}{T_0 - T_\infty} = 0.93 \quad \text{at} \quad \frac{k}{hr_0} = 3.04 \quad \text{and} \quad \frac{r}{r_0} = 0.6$$

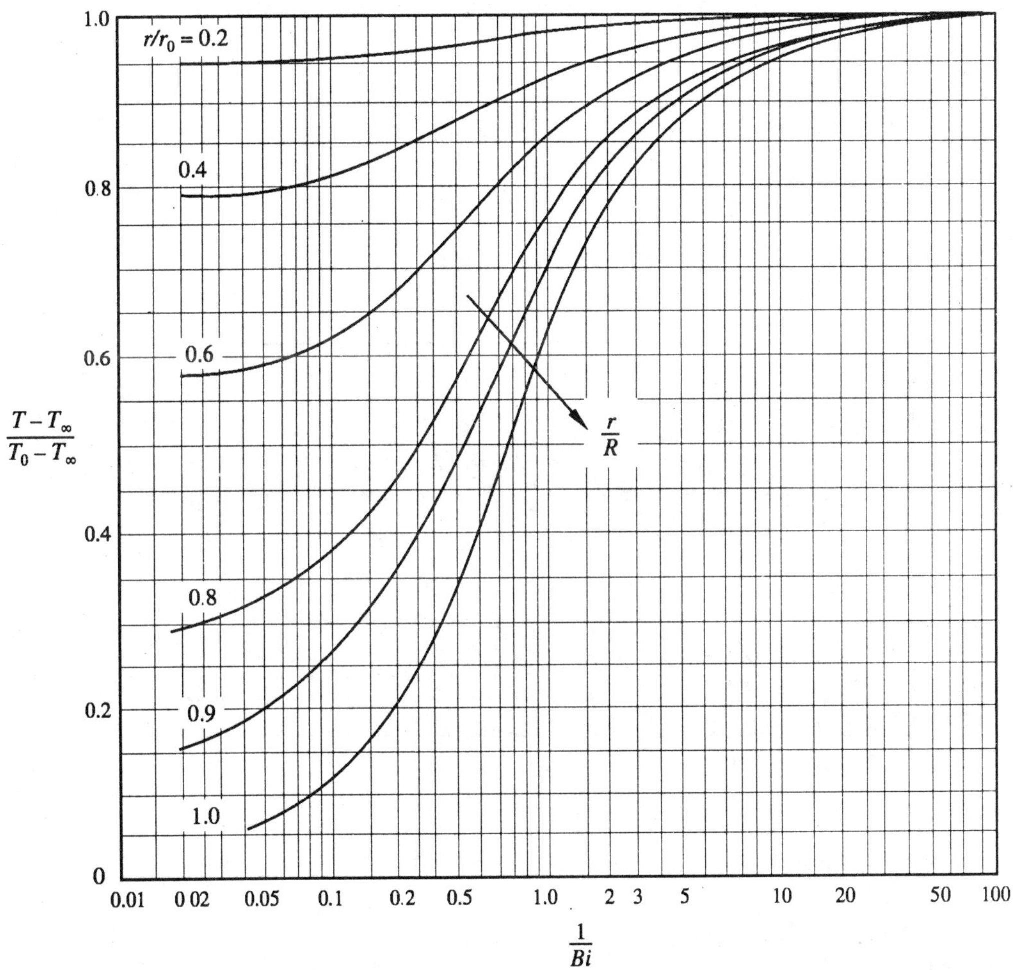

Fig. 5.14 *The temperature ratio* $\left(\frac{T-T_\infty}{T_0-T_\infty}\right)$ *for an infinitely long cylinder of radius* r_0 *as a function of* r/r_0.

Therefore,

$$\frac{T - T_\infty}{T_i - T_\infty} = \left(\frac{T_0 - T_\infty}{T_i - T_\infty}\right)\left(\frac{T - T_\infty}{T_0 - T_\infty}\right) = 0.65 \times 0.93 = 0.6045$$

or $$\frac{T - 30}{350 - 30} = 0.6045 \quad \text{or} \quad T = 223.4°\text{C}$$

That is, the temperature at $r = 6$ cm is 223.4°C.

$$(Fo)(Bi)^2 = \frac{h^2 \alpha t}{k^2} = \frac{(600)^2 \times 8.418 \times 10^{-5} \times 120}{(204)^2} = 0.087$$

$$Bi = \frac{hr_0}{k} = \frac{1}{3.4} = 0.294$$

Then we read from Fig. 5.15

$$\frac{Q}{Q_i} = 0.38 \quad \text{at} \quad (Fo)(Bi)^2 = 0.087 \quad \text{and} \quad Bi = 0.294$$

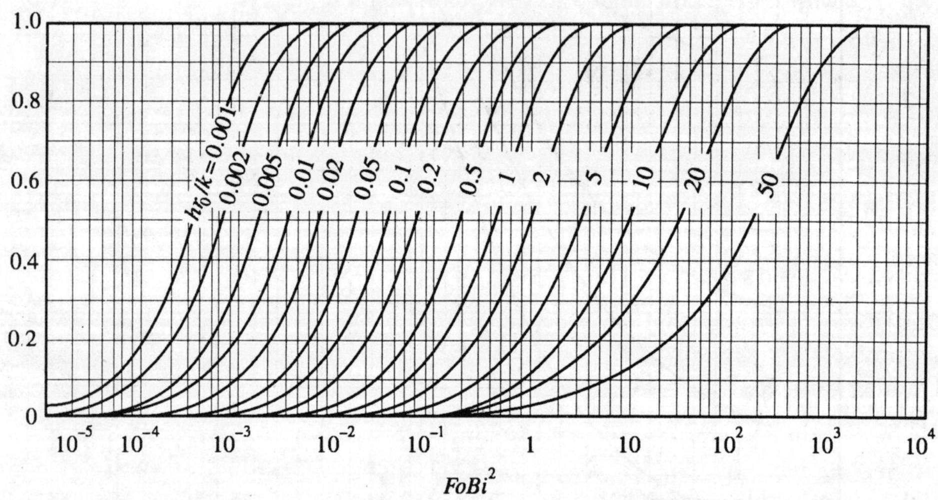

Fig. 5.15 Dimensional heat transfer ratio Q/Q_i for an infinitely long cylinder of radius r_0.

For unit length of cylinder,

$$\frac{Q_i}{\ell} = \frac{\rho c V}{\ell} = \rho c \pi r_0^2 (T_i - T_\infty)$$

$$= 2707 \times 0.896 \times 10^3 \times \pi(0.1)^2(350 - 30) = 24.384 \text{ MJ}$$

Hence, heat transferred per unit length of the cylinder $= 24.384 \times 10^6 \times 0.38 =$ 9.266 MJ/m.

5.5.3 Sphere

The basic differential equation (for transient heat flow) to be solved for obtaining temperature distribution in a sphere is given by

$$\frac{\partial T}{\partial t} = \alpha \left(\frac{\partial^2 T}{\partial r^2} + \frac{2}{r^2} \frac{\partial T}{\partial r} \right) \tag{5.79}$$

The solution of the above equation can be obtained by assuming a product solution similar to that described in section 5.4 for an infinite flat plate. The solution to the above problem is available in the form of Heisler charts which are presented in Figs. 5.16–5.19. The initial and boundary conditions for which the charts are applicable are:

1. The sphere is initially at a uniform temperature T_i.
2. At time $t = 0$ the sphere of radius r_0 is suddenly exposed to a fluid/ environment at T_∞ with convection heat transfer coefficient h.
3. The centre temperature at $r = 0$ is denoted by T_0 and the characteristic dimension is r_0 for the evaluation of Bi and Fo.

EXAMPLE 5.8 A steel sphere of radius 10 cm which is initially at a uniform temperature of 630°C is suddenly immersed in an oil bath at 30°C with convection heat transfer coefficient 500 W/m^2 K. How long will it take for the centre of the sphere to reach a temperature of 90°C ?

SOLUTION We know that for steel, $k = 54$ W/m K and $\alpha = 1.474 \times 10^{-5}$ m^2/s. This problem can be solved by using the Heisler charts presented in Figs. 5.16 and 5.17. It is given that $T_0 = 90$°C; $T_\infty = 30$°C; $T_i = 630$°C.

Therefore,

$$\frac{T_0 - T_\infty}{T_i - T_\infty} = \frac{90 - 30}{630 - 30} = 0.1$$

$$Bi = \frac{hr_0}{k} = \frac{500 \times 0.1}{54} = 0.9259 \quad \text{or} \quad \frac{1}{Bi} = 1.08$$

We read from Fig. 5.17

$$Fo = \frac{\alpha t}{r_0^2} = 1.125 \quad \text{at} \quad \frac{T_0 - T_\infty}{T_i - T_\infty} = 0.1 \quad \text{and} \quad \frac{1}{Bi} = 1.08$$

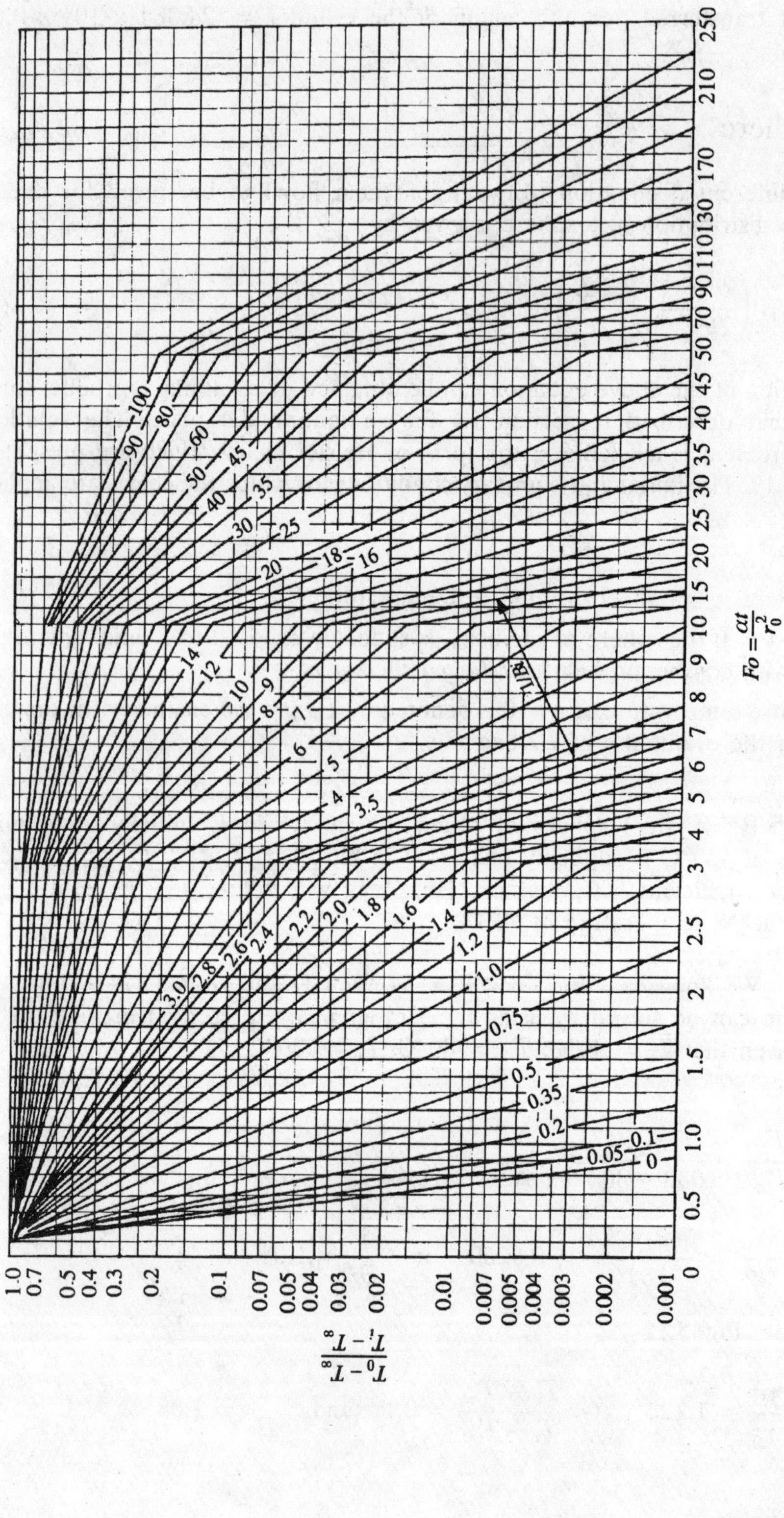

Fig. 5.16 *The dimensionless temperature ratio* $\frac{T_0 - T_\infty}{T_i - T_\infty}$ *at the centre of a sphere of radius* r_0 *with time.*

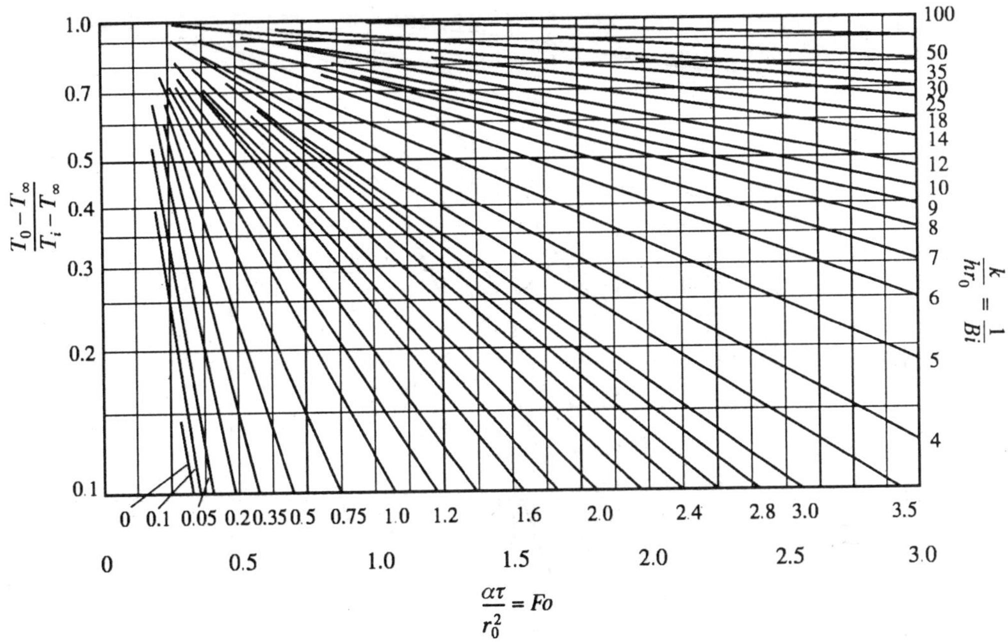

Fig. 5.17 Enlarged chart for the dimensionless temperature ratio $\frac{T_0 - T_\infty}{T_i - T_\infty}$ *at the centre of a sphere of radius* r_0 *in the range* $0 < Fo < 3$.

Therefore,

$$\frac{\alpha t}{r_0^2} = 1.125 = \frac{1.474 \times 10^{-5} t}{(0.1)^2}$$

or

$$t = 7632 \text{ s} = 2.12 \text{ hours.}$$

EXAMPLE 5.9 A steel sphere of radius 6 cm which is initially at a uniform temperature of 325°C is suddenly exposed to an environment at 25°C, with convection heat transfer coefficient 500 W/m^2 K. Calculate the temperature at a radius 3.6 cm and the heat transferred 100 seconds after the sphere is exposed to the environment.

SOLUTION This problem can be conveniently solved by using the Heisler charts presented in Figs. 5.16–5.19. First let us determine the centre temperature. We know that for steel,

$$\rho = 7833 \text{ kg/m}^3; \qquad c = 0.465 \text{ kJ/kg K}$$

$$k = 54 \text{ W/m K} \quad \text{and} \quad \alpha = 1.474 \times 10^{-5} \text{ m}^2\text{/s}$$

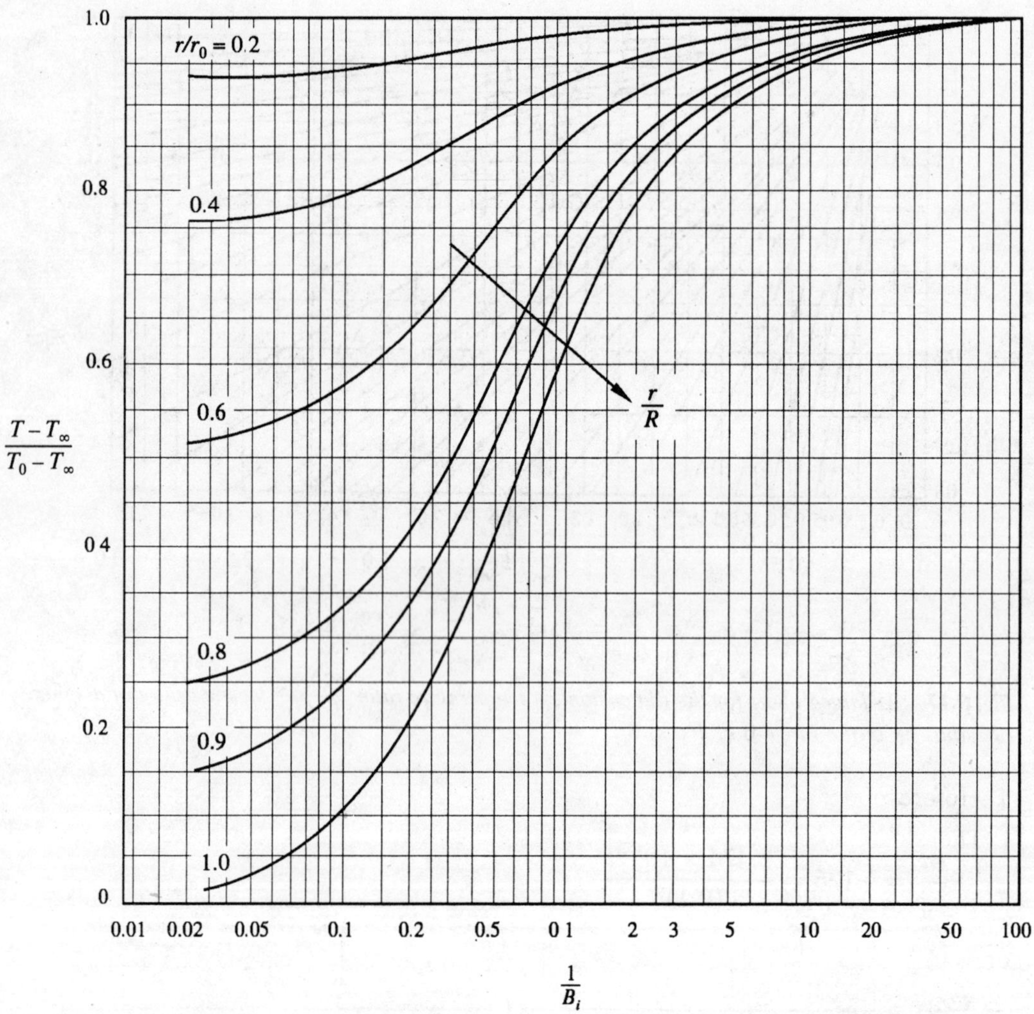

Fig. 5.18 *The temperature ratio* $\dfrac{T - T_\infty}{T_0 - T_\infty}$ *for a sphere of radius* r_0 *as a function of* $\dfrac{r}{r_0}$.

It is given that

$$r_0 = 6 \text{ cm}; \quad r = 3.6 \text{ cm}; \quad T_i = 325°\text{C}$$

$$T_\infty = 25°\text{C}; \quad h = 500 \text{ W/m}^2 \text{ K} \quad \text{and} \quad t = 100 \text{ s}$$

Then

$$Bi = \frac{hr_0}{k} = \frac{500 \times 0.06}{54} = 0.5555 \quad \text{and} \quad \frac{1}{Bi} = 1.8$$

$$Fo = \frac{\alpha t}{r_0^2} = \frac{1.474 \times 10^{-5} \times 100}{(0.06)^2} = 0.409 \quad \text{and} \quad \frac{r}{r_0} = \frac{3.6}{6} = 0.6$$

We read from Fig. 5.17

$$\frac{T_0 - T_\infty}{T_i - T_\infty} = 0.7 \quad \text{at} \quad Fo = 0.409 \quad \text{and} \quad \frac{1}{Bi} = 1.8$$

Then we read from Fig. 5.18

$$\frac{T - T_\infty}{T_0 - T_\infty} = 0.9 \quad \text{at} \quad \frac{1}{Bi} = 1.8 \quad \text{and} \quad \frac{r}{r_0} = 0.6$$

Therefore,

$$\frac{T - T_\infty}{T_i - T_\infty} = \left(\frac{T_0 - T_\infty}{T_i - T_\infty}\right)\left(\frac{T - T_\infty}{T_0 - T_\infty}\right) = 0.7 \times 0.9 = 0.63$$

or $$\frac{T - 25}{325 - 25} = 0.63 \quad \text{or} \quad T = 214°\text{C}$$

Hence, the temperature (at r = 3.6 cm and t = 100 s) = 214°C

$$(Fo)(Bi)^2 = 0.409(0.5555)^2 = 0.126$$

Then we read from Fig. 5.19 $Q/Q_i = 0.45$ at $(Fo)(Bi)^2 = 0.126$ and $Bi = 0.5555$ from Fig. 5.19.

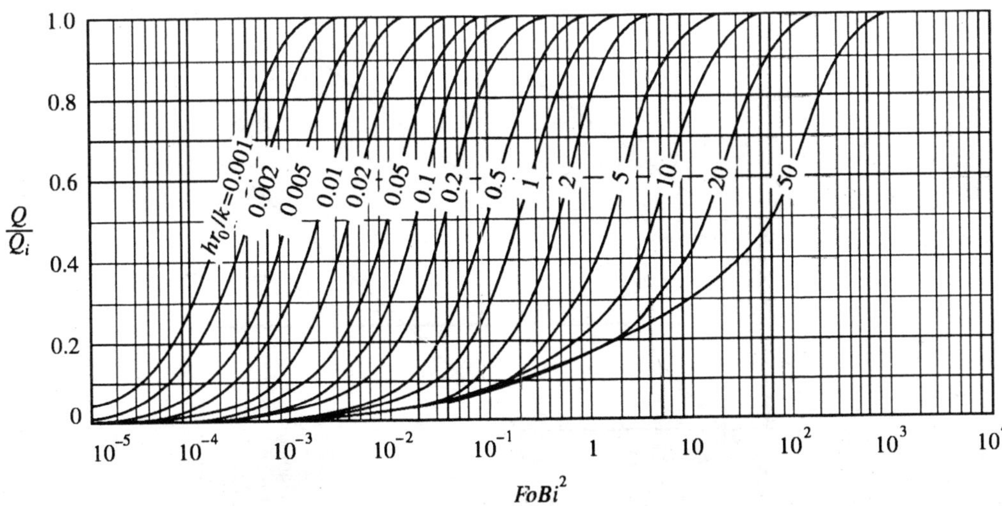

Fig. 5.19 *The dimensionless heat transfer ratio* Q/Q_i *for a sphere of radius* r_0.

$$Q_i = \rho c V(T_i - T_\infty) = \rho c \frac{4}{3}\pi r_0^3 (T_i - T_\infty)$$

$$= 7833 \times 0.465 \times 10^3 \times \frac{4}{3}\pi(0.06)^3(325 - 25) = 0.9877 \text{ MJ}$$

Therefore,

$$Q = 0.45 Q_i = 0.45 \times 0.9877 \times 10^6 = 0.445 \text{ MJ}$$

Hence heat transferred from the sphere in 100 s after the sphere is exposed to the environment = 358700 J.

5.6 Transient Heat Flow in Two- and Three-Dimensional Bodies

So far, we have discussed the use of Heisler charts in the analysis of transient heat flow in an infinite flat plate of thickness 2ℓ, in a long cylinder and a sphere. Many practical problems encountered by an engineer deal with transient heat flow through bodies in which the height and depth of a flat plate are not very large compared to its thickness, or a cylinder whose length is not very large compared to its diameter. In such cases the heat flow is not one-dimensional and it is necessary to consider additional space coordinates in expressing the heat flow equation. In this section we shall show the method of analysis for transient heat flow in two- and three-dimensional bodies.

Consider the rectangular bar shown in Fig. 5.20, which is initially at a uniform temperature T_i. At time $t = 0$, the bar is exposed to an environment at T_∞, with the convection heat transfer coefficient h.

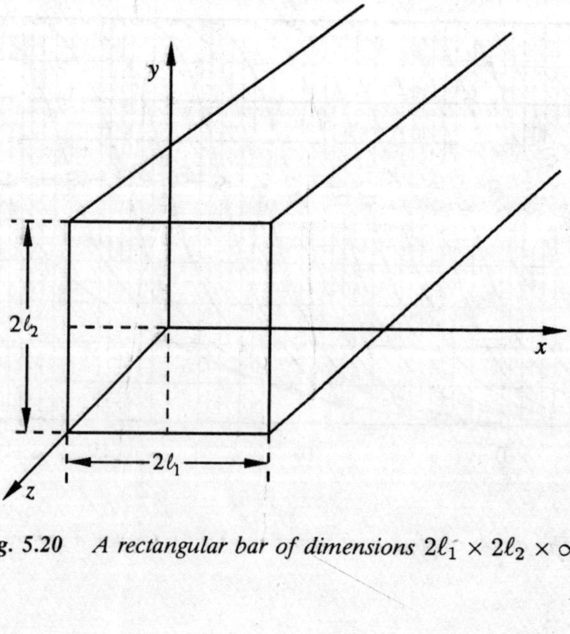

Fig. 5.20 A rectangular bar of dimensions $2\ell_1 \times 2\ell_2 \times \infty$.

The rectangular bar extends from $-\ell_1 < x < \ell_1$ and $-\ell_2 < y < \ell_2$ and extends to infinity in the z-direction. The differential equation governing the heat flow in such a bar is given by

$$\frac{\partial T}{\partial t} = \alpha\left(\frac{\partial^2 T}{\partial x^2} + \frac{\partial^2 T}{\partial y^2}\right) \tag{5.80}$$

Equation (5.80) can be solved by using the method of separation of variables. For this purpose, we can assume a product solution of the form

$$T(x, y, t) = X(x)Y(y)\tau(t) \tag{5.81}$$

The infinite rectangular bar shown in Fig. 5.20 can be formed from two infinite plates of thickness $2\ell_1$ and $2\ell_2$, respectively. For the two infinite plates, the differential equations governing the temperature distribution are given by:

For plate 1 of thickness $2\ell_1$,

$$\frac{\partial T_1}{\partial t} = \alpha\frac{\partial^2 T_1}{\partial x^2} \tag{5.82}$$

For plate 2 of thickness $2\ell_2$,

$$\frac{\partial T_2}{\partial t} = \alpha\frac{\partial^2 T_2}{\partial y^2} \tag{5.83}$$

The solutions of Eqs. (5.82) and (5.83) are of the form

$$T_1 = T_1(x, t) \tag{5.84}$$

and

$$T_2 = T_2(y, t)$$

The product solution for the rectangular bar given by Eq. (5.81) can be obtained as a simple product of the solutions given by Eqs. (5.84). That is,

$$T(x, y, t) = T_1(x, t)T_2(y, t) \tag{5.85}$$

From Eq. (5.85) we get

$$\frac{\partial^2 T}{\partial x^2} = T_2\frac{\partial^2 T_1}{\partial x^2} \tag{5.86}$$

$$\frac{\partial^2 T}{\partial y^2} = T_1\frac{\partial^2 T_2}{\partial y^2} \tag{5.87}$$

$$\frac{\partial T}{\partial t} = T_1 \frac{\partial T_2}{\partial t} + T_2 \frac{\partial T_1}{\partial t} \tag{5.88}$$

From Eqs. (5.82), (5.83) and (5.88) we get

$$\frac{\partial T}{\partial t} = T_1 \alpha \frac{\partial^2 T_2}{\partial y^2} + T_2 \alpha \frac{\partial^2 T_1}{\partial x^2} \tag{5.89}$$

Substituting Eqs. (5.86),(5.87) and (5.89) in Eq. (5.80), we get

$$T_2 \alpha \frac{\partial^2 T_1}{\partial x^2} + T_1 \alpha \frac{\partial^2 T_2}{\partial y^2} = \alpha \left(T_2 \frac{\partial^2 T_1}{\partial x^2} + T_1 \frac{\partial^2 T_2}{\partial y^2} \right) \tag{5.90}$$

Equation (5.90) shows that the assumed product solution given by Eq. (5.85) satisfies the differential equation [Eq. (5.80)] for a rectangular bar. This implies that the temperature distribution in a rectangular bar is given by the product of the temperature distributions in two infinite plates as

$$\left(\frac{T - T_\infty}{T_i - T_\infty} \right)_{\text{bar}} = \left(\frac{T - T_\infty}{T_i - T_\infty} \right)_{\text{plate 1}} \left(\frac{T - T_\infty}{T_i - T_\infty} \right)_{\text{plate 2}} \tag{5.91}$$

Similarly, the solution for a three-dimensional block shown in Fig. 5.21 can be expressed as a product of three infinite plates — plate 1 of thickness $2\ell_1$, plate 2 of thickness $2\ell_2$ and plate 3 of thickness $2\ell_3$ — as

$$\left(\frac{T - T_\infty}{T_i - T_\infty} \right)_{\text{block}} = \left(\frac{T - T_\infty}{T_i - T_\infty} \right)_{\text{plate 1}} \left(\frac{T - T_\infty}{T_i - T_\infty} \right)_{\text{plate 2}} \left(\frac{T - T_\infty}{T_i - T_\infty} \right)_{\text{plate 3}} \tag{5.92}$$

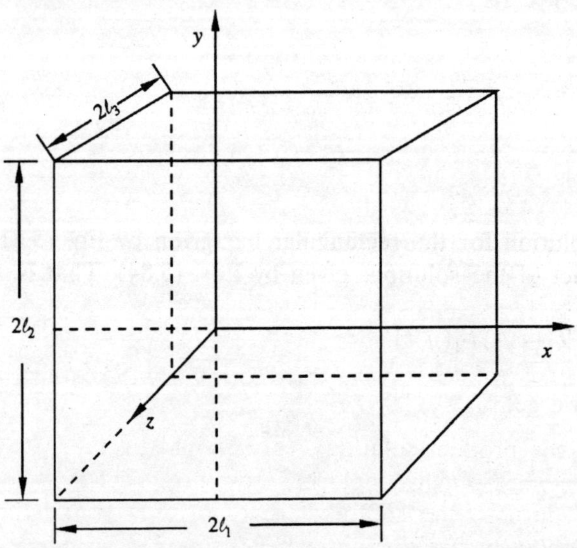

Fig. 5.21 A three-dimensional block of size $2\ell_1 \times 2\ell_2 \times 2\ell_3$.

Similarly, a solution for a cylinder of finite length shown in Fig. 5.22 can be expressed as a product of solutions of an infinitely long cylinder and an infinite plate having a thickness equal to the length of the cylinder.

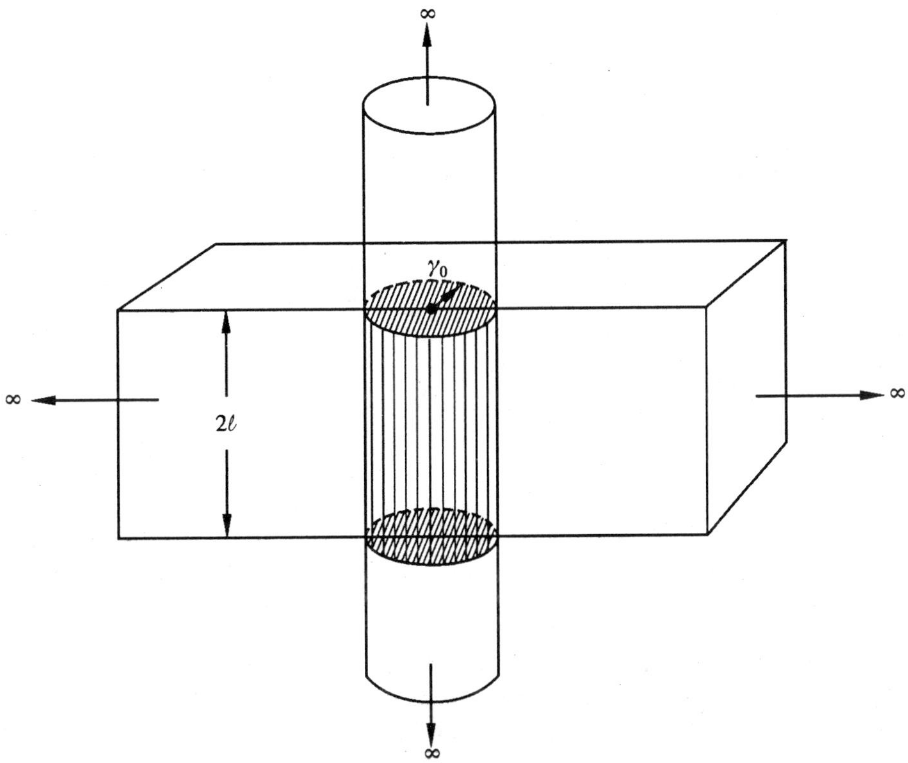

Fig. 5.22 *A cylinder of radius r_0 and length 2ℓ is treated as an intersection of an infinitely long cylinder of radius r_0 and an infinite flat plate of thickness 2ℓ.*

Hence

$$\left(\frac{T - T_\infty}{T_i - T_\infty}\right)_{2\ell \text{ long cylinder}}$$

$$= \left(\frac{T - T_\infty}{T_i - T_\infty}\right)_{\text{infinite cylinder}} \left(\frac{T - T_\infty}{T_i - T_\infty}\right)_{2\ell \text{ thick infinite plate}} \tag{5.93}$$

While using the product solutions, care should be taken to satisfy the boundary conditions of the one-dimensional heat flow problems to which the Heisler charts apply. For this purpose the coordinate axes of a two- and three-dimensional body should be chosen such that they lie along the axes of symmetry and intersect at the centre of the body.

The heat transfer from a two- and three-dimensional body can be obtained by superimposing the heat transfer solutions from a one-dimensional body. The heat transfer from a two-dimensional body which is obtained as an intersection of two one-dimensional bodies 1 and 2 is given by

$$\left(\frac{Q}{Q_i}\right)_{\text{2-D body}} = \left(\frac{Q}{Q_i}\right)_1 + \left(\frac{Q}{Q_i}\right)_2 \left[1 - \left(\frac{Q}{Q_i}\right)_1\right] \tag{5.94}$$

where the subscripts 1 and 2 refer to the two intersecting one-dimensional bodies 1 and 2 respectively. For a three-dimensional body obtained as an intersection of three one-dimensional bodies, the heat transfer is given by

$$\left(\frac{Q}{Q_i}\right)_{\text{3-D body}} = \left(\frac{Q}{Q_i}\right)_1 + \left(\frac{Q}{Q_i}\right)_2 \left[1 - \left(\frac{Q}{Q_i}\right)_1\right] + \left(\frac{Q}{Q_i}\right)_3 \left[1 - \left(\frac{Q}{Q_i}\right)_1\right]\left[1 - \left(\frac{Q}{Q_i}\right)_2\right] \tag{5.95}$$

EXAMPLE 5.10 A large rectangular iron bar of cross-section 10 cm × 6 cm which is initially at a uniform temperature of 630°C is suddenly exposed to an environment at 30°C, with a convection heat transfer coefficient 500 W/m^2 K. Calculate the temperature at the center of the bar 100 s after it is exposed to the environment.

SOLUTION We know that for iron, k = 73 W/m K and $\alpha = 2.034 \times 10^{-5}$ m^2/s. It is given that T_i =630°C; T_∞ = 30°C; t = 100 s and h = 500 W/m^2 K. The rectangular iron bar can be considered as a two-dimensional body obtained as an intersection of (1) an infinite flat plate of thickness ($2\ell_1$) 10 cm and (2) an infinite flat plate of thickness ($2\ell_2$) 6 cm. That is, for plate 1, ℓ_1 = 5 cm and for plate 2, ℓ_2 = 3 cm.

For plate 1

$$(Fo)_1 = \frac{\alpha t}{\ell_1^2} = \frac{2.034 \times 10^{-5} \times 100}{(0.05)^2} = 0.8136$$

$$(Bi)_1 = \frac{h\ell_1}{k} = \frac{500 \times 0.05}{73} = 0.3425$$

or $$\left(\frac{1}{Bi}\right)_1 = 2.92$$

Then we read from Fig. 5.9

$$\left(\frac{T_0 - T_\infty}{T_i - T_\infty}\right)_1 = 0.85 \quad \text{at} \quad Fo = 0.8136 \quad \text{and} \quad \left(\frac{1}{Bi}\right) = 2.92$$

For plate 2

$$(Fo)_2 = \frac{\alpha t}{\ell_2^2} = \frac{2.034 \times 10^{-5} \times 100}{(0.03)^2} = 2.26$$

$$(Bi)_2 = \frac{h\ell_2}{k} = \frac{500 \times 0.03}{73} = 0.2055$$

or $$\left(\frac{1}{Bi}\right)_2 = 4.87$$

Then we read from Fig. 5.9

$$\left(\frac{T_0 - T_\infty}{T_i - T_\infty}\right)_2 = 0.68 \quad \text{at} \quad Fo = 2.26 \quad \text{and} \quad \left(\frac{1}{Bi}\right) = 4.87$$

Therefore,

$$\left(\frac{T_0 - T_\infty}{T_i - T_\infty}\right)_{\text{bar}} = \left(\frac{T_0 - T_\infty}{T_i - T_\infty}\right)_{\text{plate 1}} \left(\frac{T_0 - T_\infty}{T_i - T_\infty}\right)_{\text{plate 2}} = 0.85 \times 0.68 = 0.578$$

or

$$\frac{T_0 - 30}{630 - 30} = 0.578 \quad \text{or} \quad T_0 = 376.8°\text{C}$$

EXAMPLE 5.11 An iron beam of rectangular cross-section of size 30 cm × 20 cm is used in the construction of a building. Initially, the beam is at a uniform temperature of 30°C. Owing to an accidental fire, the beam is suddenly exposed to hot gases at 730°C, with the convection heat transfer coefficient 100 W/m^2 K. Calculate the time required for the centre plane of the beam to reach a temperature of 310°C.

SOLUTION The rectangular iron beam can be considered as an intersection of infinite plate 1 with thickness $2\ell_1$ = 30 cm and a second infinite plate 2 with thickness $2\ell_2$ = 20 cm. We know that for iron, k = 73 W/m K and α = 2.034 ×10^{-5} m^2/s. It is given that

$$T_i = 30°\text{C}; \quad T_\infty = 730°\text{C}; \quad T_0 = 310°\text{C} \quad \text{and} \quad h = 100 \text{ W/m}^2 \text{ K}$$

$$\left(\frac{T_0 - T_\infty}{T_i - T_\infty}\right)_{\text{beam}} = \left(\frac{T_0 - T_\infty}{T_i - T_\infty}\right)_{\text{plate 1}} \left(\frac{T_0 - T_\infty}{T_i - T_\infty}\right)_{\text{plate 2}}$$

$$\left(\frac{T_0 - T_\infty}{T_i - T_\infty}\right)_{\text{beam}} = \frac{310 - 730}{30 - 730} = 0.6$$

The solution cannot be obtained directly, but requires trial and error calculations. We assume some value for t to determine the centre-line temperature for plates 1 and 2 and hence to determine the centre-line temperature of the beam. The value of t at which

$$\left(\frac{T_0 - T_\infty}{T_i - T_\infty}\right)_{\text{beam}} = 0.6$$

is satisfied is the required time.

Assume $t = 900$ s.

For plate 1

$$(Bi)_1 = \frac{h\ell_1}{k} = \frac{100 \times 0.15}{73} = 0.2055 \quad \text{or} \quad \left(\frac{1}{Bi}\right)_1 = 4.87$$

$$(Fo)_1 = \frac{\alpha t}{\ell_1^2} = \frac{2.034 \times 10^{-5} \times 900}{(0.15)^2} = 0.814$$

Then we read from Fig. 5.9

$$\left(\frac{T_0 - T_\infty}{T_i - T_\infty}\right)_{\text{plate 1}} = 0.79 \quad \text{at } Fo = 0.814 \text{ and } 1/Bi = 4.87.$$

For plate 2

$$(Bi)_2 = \frac{h\ell_2}{k} = \frac{100 \times 0.1}{73} = 0.137$$

or

$$\left(\frac{1}{Bi}\right)_2 = 7.3$$

$$(Fo)_2 = \frac{\alpha t}{\ell_2^2} = \frac{2.034 \times 10^{-5} \times 900}{(0.1)^2} = 1.83$$

Then we read from Fig. 5.9

$$\left(\frac{T_0 - T_\infty}{T_i - T_\infty}\right)_{\text{plate 2}} = 0.66$$

at $Fo = 1.83$ and $1/Bi = 7.3$.

Therefore,

$$\left(\frac{T_0 - T_\infty}{T_i - T_\infty}\right)_{\text{beam}} = 0.79 \times 0.66 = 0.5214$$

Since the calculated value of

$$\left(\frac{T_0 - T_\infty}{T_i - T_\infty}\right) = 0.5214$$

is less than 0.6, the assumed value of t is large. Hence, we assume a second value of t which is less than 900 s. Now, let us assume $t = 800$ s. Then

$$(Fo)_1 = \frac{\alpha t}{\ell_1^2} = \frac{2.034 \times 10^{-5} \times 800}{(0.15)^2} = 0.723$$

$$(Fo)_2 = \frac{\alpha t}{\ell_2^2} = \frac{2.034 \times 10^{-5} \times 800}{(0.1)^2} = 1.628$$

We read the values of $(T_0 - T_\infty)/(T_i - T_\infty)$ for plates 1 and 2 from Fig. 5.9 as

$$\left(\frac{T_0 - T_\infty}{T_i - T_\infty}\right)_{\text{plate 1}} = 0.84$$

at $Fo = 0.723$ and $1/Bi = 4.87$.

$$\left(\frac{T_0 - T_\infty}{T_i - T_\infty}\right)_{\text{plate 2}} = 0.7$$

at $Fo = 1.628$ and $1/Bi = 7.3$.

Therefore,

$$\left(\frac{T_0 - T_\infty}{T_i - T_\infty}\right)_{\text{beam}} = 0.84 \times 0.7 = 0.588$$

The calculated value of $(T_0 - T_\infty)/(T_i - T_\infty)$ is still less than 0.6 and hence the assumed value of t is still large. Now, let us assume $t = 720$ s. Then

$$(Fo)_1 = \frac{\alpha t}{\ell_1^2} = \frac{2.034 \times 10^{-5} \times 720}{(0.15)^2} = 0.65$$

$$(Fo)_2 = \frac{\alpha t}{\ell_2^2} = \frac{2.034 \times 10^{-5} \times 720}{(0.1)^2} = 1.465$$

We read the values of $(T_0 - T_\infty)/(T_i - T_\infty)$ for plates 1 and 2 from Fig. 5.9 as

$$\left(\frac{T_0 - T_\infty}{T_i - T_\infty}\right)_{\text{plate 1}} = 0.85$$

at $Fo = 0.65$ and $1/Bi = 4.87$.

$$\left(\frac{T_0 - T_\infty}{T_i - T_\infty}\right)_{\text{plate 2}} = 0.7$$

at $Fo = 1.465$ and $1/Bi = 7.3$.

Therefore,

$$\left(\frac{T_0 - T_\infty}{T_i - T_\infty}\right)_{\text{beam}} = 0.85 \times 0.7 = 0.595$$

The calculated value of $(T_0 - T_\infty)/(T_i - T_\infty)$ is 0.595 which is close to the required value of 0.6. Let us assume that the value is in agreement. Hence, the time required for the center of the beam to reach 310°C is 720 s.

EXAMPLE 5.12 Calculate the amount of energy transferred per unit length in 100 seconds from the rectangular iron bar of Example 5.10.

SOLUTION We know that the amount of energy transferred from a two-dimensional body is given by

$$\left(\frac{Q}{Q_i}\right)_{\text{2-D body}} = \left(\frac{Q}{Q_i}\right)_1 + \left(\frac{Q}{Q_i}\right)_2 \left[1 - \left(\frac{Q}{Q_i}\right)_1\right]$$

We know that for plate 1, $(Fo)_1 = 0.8136$ and $(Bi)_1 = 0.3425$ or $(1/Bi)_1 = 2.92$ and for plate 2, $(Fo)_2 = 2.26$ and $(Bi)_2 = 0.2055$ or $(1/Bi)_2 = 4.87$ (see Example 5.10). Therefore,

$$\left[(Fo)(Bi)^2\right]_1 = 0.8136 \times (0.3425)^2 = 0.0954$$

$$\left[(Fo)(Bi)^2\right]_2 = 2.26 \times (0.2055)^2 = 0.0954$$

Then we read from Fig. 5.11 $(Q/Q_i)_1 = 0.28$ at $(Fo)(Bi)^2 = 0.0954$ and $Bi = 0.3425$ and $(Q/Q_i)_2 = 0.39$ at $(Fo)(Bi)^2 = 0.0954$ and $Bi = 0.2055$. Hence,

$$\left(\frac{Q}{Q_i}\right)_{\text{bar}} = 0.28 + 0.39(1 - 0.28) = 0.5608$$

$$\left(\frac{Q_i}{\ell}\right)_{\text{bar}} = \frac{\rho c V}{\ell}(T_i - T_\infty) = \frac{\rho c \times 2\ell_1 \times 2\ell_2 \times \ell}{\ell}(T_i - T_\infty) = 4\rho c \ell_1 \ell_2 (T_i - T_\infty)$$

$$= 4 \times 7897 \times 0.452 \times 10^3 \times 0.05 \times 0.03(630 - 30) = 12.85 \text{ MJ}$$

where ρ = 7897 kg/m^3 and c = 0.452 kJ/kg K for iron.

Therefore,

$$\left(\frac{Q}{\ell}\right)_{\text{bar}} = 0.5608 \times 12.85 \times 10^6 = 7.206 \text{ MJ}$$

EXAMPLE 5.13 A short nickel-steel cylinder 10 cm in diameter and 30 cm in length, originally at a uniform temperature of 530°C, is suddenly exposed to an environment at 30°C, with convection heat transfer coefficient 500 W/m^2 K. Calculate the temperature at the center of the cylinder 600 s after it is exposed to the environment.

SOLUTION The short cylinder can be considered as an intersection of an infinite flat plate of thickness 2ℓ = 30 cm and an infinitely long cylinder of radius 5 cm. To obtain the temperature distribution, the Heisler charts for infinite flat plate and infinite cylinder can be used.

$$\left(\frac{T_0 - T_\infty}{T_i - T_\infty}\right)_{\text{finite cylinder}} = \left(\frac{T_0 - T_\infty}{T_i - T_\infty}\right)_{\text{plate}} \left(\frac{T_0 - T_\infty}{T_i - T_\infty}\right)_{\text{cylinder}}$$

For nickel-steel, k = 19 W/m K and α = 0.526 × 10^{-5} m^2/s.

For the flat plate ℓ = 15 cm

$$(Bi)_{\text{plate}} = \frac{h\ell}{k} = \frac{500 \times 0.15}{19} = 3.95 \quad \text{or} \quad \left(\frac{1}{Bi}\right)_{\text{plate}} = 0.25$$

$$(Fo)_{\text{plate}} = \frac{\alpha t}{\ell^2} = \frac{0.526 \times 10^{-5} \times 600}{(0.15)^2} = 0.14$$

From the Heisler chart for a flat plate at $Fo = 0.14$ and $1/Bi = 0.25$, we obtain

$$\left(\frac{T_0 - T_\infty}{T_i - T_\infty}\right) = 0.95$$

For the cylinder

$$(Bi)_{\text{cylinder}} = \frac{hr_0}{k} = \frac{500 \times 0.05}{19} = 1.32 \quad \text{or} \quad \left(\frac{1}{Bi}\right)_{\text{cylinder}} = 0.76$$

$$(Fo)_{\text{cylinder}} = \frac{\alpha t}{r_0^2} = \frac{0.526 \times 10^{-5} \times 600}{(0.05)^2} = 1.26$$

From the Heisler chart for an infinite cylinder, we get

$$\left(\frac{T_0 - T_\infty}{T_i - T_\infty}\right)_{\text{cylinder}} = 0.12 \quad \text{at} \quad Fo = 1.26 \quad \text{and} \quad \frac{1}{Bi} = 0.76$$

Therefore,

$$\left(\frac{T_0 - T_\infty}{T_i - T_\infty}\right)_{\text{finite cylinder}} = 0.95 \times 0.12 = 0.114$$

or $$\frac{T_0 - 30}{530 - 30} = 0.114 \quad \text{or} \quad T_0 = 87°\text{C}$$

Hence, the centre of the cylinder attains a temperature of 87°C.

EXAMPLE 5.14 Calculate the energy transferred from the cylinder of Example 5.13.

SOLUTION We know that the energy transferred from the 2-dimensional body is given by

$$\left(\frac{Q}{Q_i}\right)_{\text{2-D body}} = \left(\frac{Q}{Q_i}\right)_1 + \left(\frac{Q}{Q_i}\right)_2 \left[1 - \left(\frac{Q}{Q_i}\right)_1\right]$$

From Example 5.13, we have

$$(Bi)_{\text{plate}} = (Bi)_1 = 3.95$$

$$(Fo)_{\text{plate}} = (Fo)_1 = 0.14$$

$$\left[(Fo)(Bi)^2\right]_1 = 0.14(3.95)^2 = 2.18$$

$$(Bi)_{\text{cylinder}} = (Bi)_2 = 1.32$$

$$(Fo)_{\text{cylinder}} = (Fo)_2 = 1.26$$

$$\left[(Fo)(Bi)^2\right]_2 = 1.26(1.32)^2 = 2.2$$

Then we read from the Heisler chart for a flat plate

$$\left(\frac{Q}{Q_i}\right)_1 = 0.25 \quad \text{at} \quad (Fo)(Bi)^2 = 2.18 \quad \text{and} \quad Bi = 3.95$$

Then we read from the Heisler chart for an infinite cylinder

$$\left(\frac{Q}{Q_i}\right)_2 = 0.9 \quad \text{at} \quad (Fo)(Bi)^2 = 2.2 \quad \text{and} \quad Bi = 1.32$$

Therefore,

$$\left(\frac{Q}{Q_i}\right)_{\text{finite cylinder}} = 0.25 + 0.9(1 - 0.25) = 0.925$$

$$Q_i = \rho c V(T_i - T_\infty) = \rho c \pi r_0^2 (2\ell)(T_i - T_\infty)$$

For nickel-steel,

$$\rho = 7933 \text{ kg/m}^3; \quad c = 0.46 \text{ kJ/kg K}$$

Therefore,

$$Q_i = 7933 \times 0.46 \times 10^3 \times \pi(0.05)^2 \times 0.3(530 - 30) = 4.299 \text{ MJ}$$

Hence, $Q = 0.925 \times 4.299 \times 10^6 = 3.977$ MJ

5.7 Transient Heat Flow in a Semi-infinite Solid

Consider the semi-infinite solid shown in Fig. 5.23, which is initially at a temperature T_i. At time $t = 0$, the surface of the solid is suddenly changed to T_∞ and held at that temperature. Now, we would like to determine the temperature distribution in the solid as a function of time. The differential equation which governs the temperature distribution is given by

$$\frac{\partial T}{\partial t} = \alpha \frac{\partial^2 T}{\partial x^2} \tag{5.96}$$

The initial and boundary conditions are given by

$$T(x,0) = T_i \tag{5.97}$$

$$T(0,t) = T_\infty \quad \text{for} \quad t > 0 \tag{5.98}$$

The solution of Eq. (5.96) is given by

$$\frac{T - T_\infty}{T_i - T_\infty} = \text{erf}\left(\frac{x}{2\sqrt{\alpha t}}\right) \tag{5.99}$$

where the Gauss error function is defined as

$$\text{erf}\left(\frac{x}{2\sqrt{\alpha t}}\right) = \frac{2}{\sqrt{\pi}} \int_0^{\frac{x}{2\sqrt{\alpha t}}} e^{-\beta^2}\, d\beta \tag{5.100}$$

where β is a dummy variable. The values of the error function are given in Table 5.1.

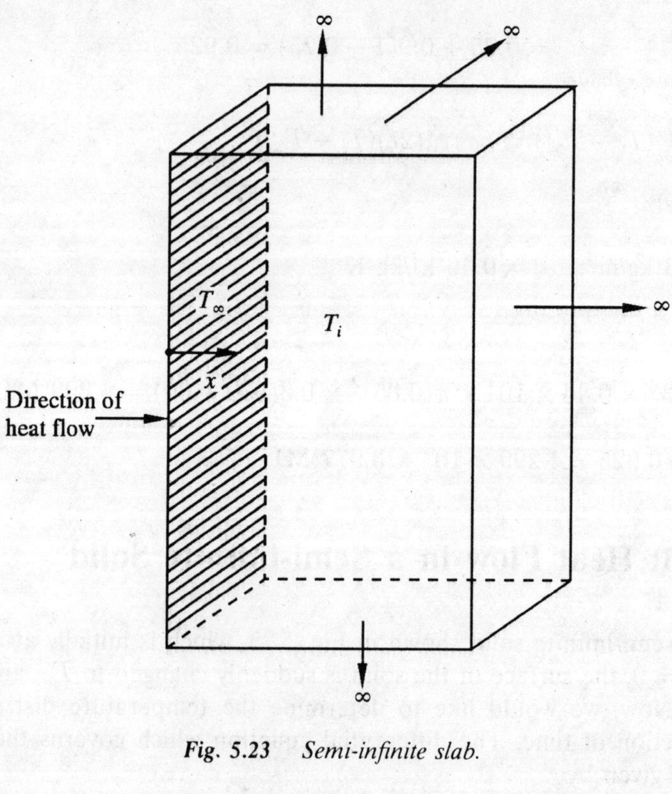

Fig. 5.23 Semi-infinite slab.

The rate of heat transfer at the surface can be obtained from

$$(q)_{x=0} = -kA\left(\frac{\partial T}{\partial x}\right)_{x=0} \tag{5.101}$$

Differentiating Eq. (5.99), we get

$$\frac{\partial T}{\partial x} = (T_i - T_\infty)\frac{2}{\sqrt{\pi}}e^{-x^2/4\alpha t}\frac{\partial}{\partial x}\left(\frac{x}{2\sqrt{\alpha t}}\right) = \frac{T_i - T_\infty}{\sqrt{\alpha t}}e^{-x^2/4\alpha t} \tag{5.102}$$

Therefore,

$$(q)_{x=0} = \frac{kA(T_\infty - T_i)}{\sqrt{\pi\alpha t}} \tag{5.103}$$

The total change in the energy of the solid during the time $t = 0$ to $t = t$ is given by

$$Q = \int_0^t q dt = \int_0^t kA\frac{T_\infty - T_i}{\sqrt{\pi\alpha}}t^{-1/2}dt = 2kA(T_\infty - T_i)\sqrt{\frac{t}{\pi\alpha}} \tag{5.104}$$

EXAMPLE 5.15 One surface of a large block of nickel-steel which is initially at 30°C is suddenly raised to a temperature of 530°C. Calculate the temperature at a depth of 5 cm after a time of 50 s.

SOLUTION For nickel-steel, k = 19 W/m K and $\alpha = 0.526 \times 10^{-5}$ m²/s. The temperature distribution in the solid is given by

$$\frac{T - T_\infty}{T_i - T_\infty} = \text{erf}\left(\frac{x}{2\sqrt{\alpha t}}\right)$$

$$\frac{x}{2\sqrt{\alpha t}} = \frac{0.05}{2\sqrt{0.526 \times 10^{-5} \times 50}} = 1.54$$

From Table 5.1, we read erf(1.54) = 0.9706. Therefore,

$$\frac{T - T_\infty}{T_i - T_\infty} = 0.9706$$

or

$$\frac{T - 530}{30 - 530} = 0.9706$$

or $\quad T = 44.7°\text{C}$

Table 5.1 The error function values

z	erf(z)	z	erf(z)	z	erf(z)
0.00	0.0	0.72	0.6914	1.44	0.9583
0.02	0.0226	0.74	0.7047	1.46	0.9611
0.04	0.0451	0.76	0.7175	1.48	0.9637
0.06	0.0676	0.78	0.7300	1.50	0.9661
0.08	0.0901	0.80	0.7421	1.52	0.9684
0.10	0.1125	0.82	0.7538	1.54	0.9706
0.12	0.1348	0.84	0.7651	1.56	0.9726
0.14	0.1570	0.86	0.7761	1.58	0.9746
0.16	0.1790	0.88	0.7867	1.60	0.9764
0.18	0.2009	0.90	0.7969	1.62	0.9780
0.20	0.2227	0.92	0.8068	1.64	0.9796
0.22	0.2443	0.94	0.8163	1.66	0.9811
0.24	0.2657	0.96	0.8254	1.68	0.9825
0.26	0.2869	0.98	0.8342	1.70	0.9838
0.28	0.3079	1.00	0.8427	1.72	0.9850
0.30	0.3286	1.02	0.8508	1.74	0.9861
0.32	0.3491	1.04	0.8587	1.76	0.9872
0.34	0.3694	1.06	0.8661	1.78	0.9882
0.36	0.3893	1.08	0.8733	1.80	0.9891
0.38	0.4090	1.10	0.8802	1.82	0.9899
0.40	0.4284	1.12	0.8808	1.84	0.9907
0.42	0.4475	1.14	0.8931	1.86	0.9915
0.44	0.4662	1.16	0.8991	1.88	0.9922
0.46	0.4847	1.18	0.9048	1.90	0.9928
0.48	0.5028	1.20	0.9103	1.92	0.9934
0.50	0.5205	1.22	0.9155	1.94	0.9939
0.52	0.5379	1.24	0.9205	1.96	0.9944
0.54	0.5549	1.26	0.9252	1.98	0.9949
0.56	0.5716	1.28	0.9297	2.00	0.9953
0.58	0.5897	1.30	0.9340	2.10	0.9970
0.60	0.6039	1.32	0.9381	2.20	0.9981
0.62	0.6194	1.34	0.9419	2.30	0.9989
0.64	0.6346	1.36	0.9456	2.40	0.9993
0.66	0.6494	1.38	0.9490	2.50	0.9996
0.68	0.6628	1.40	0.9523	2.60	0.9998
0.70	0.6778	1.42	0.9554	3.60	1.0000

EXAMPLE 5.16 A large slab of wrought-iron is at a uniform temperature of 375°C. The temperature of one surface of this slab is suddenly changed to 75°C. Calculate the time required for the temperature to reach 275°C at a depth of 5 cm and the quantity of energy transferred per unit area of the surface during this period.

SOLUTION For wrought-iron, k =59 W/m K and α = 1.626 × 10^{-5} m^2/s.

$$\frac{x}{2\sqrt{\alpha t}} = \frac{0.05}{2\sqrt{1.626 \times 10^{-5} t}} = \frac{6.2}{\sqrt{t}}$$

We know that the temperature distribution in the solid is given by

$$\frac{T - T_\infty}{T_i - T_\infty} = \text{erf}\left(\frac{x}{2\sqrt{\alpha t}}\right)$$

or $$\frac{275 - 75}{375 - 75} = 0.6667 = \text{erf}\left(\frac{x}{2\sqrt{\alpha t}}\right)$$

We read the value of $z = (x/2\sqrt{\alpha t}) = 0.685$ corresponding to $\text{erf}(z) = 0.6667$ (obtained by interpolation).

Therefore, $6.2/\sqrt{t} = 0.685$ or $t = 82$ s

Hence time required = 82 s.

The quantity of energy transferred per unit area of the surface is given by

$$\frac{Q}{A} = 2k(T_\infty - T_i)\sqrt{\frac{t}{\pi\alpha}} = 2 \times 59(75 - 375)\sqrt{\frac{82}{\pi \times 1.626 \times 10^{-5}}}$$

$$= -44.85 \text{ MJ/m}^2$$

5.7.1 Transient Heat Flow in a Semi-infinite Solid Exposed to a Constant Heat Flux

Now, let us consider the semi-infinite solid shown in Fig. 5.23. Initially the solid is at a uniform temperature T_i. At time t = 0, one surface of the solid is suddenly exposed to a constant heat flux q_0/A. Then the differential equation governing the temperature distribution is given by

$$\frac{\partial T}{\partial t} = \alpha\frac{\partial^2 T}{\partial x^2} \tag{5.105}$$

The initial and boundary conditions are:

$$T(x,0) = T_i \tag{5.106}$$

$$\frac{q_0}{A} = -k\left[\frac{\partial T}{\partial x}\right]_{x=0} \quad \text{for} \quad t > 0 \tag{5.107}$$

The solution of Eq. (5.105) subject to the boundary conditions [Eqs. (5.106) and (5.107)] is given by

$$T - T_i = \frac{2q_0\sqrt{\alpha t/\pi}}{kA}e^{-x^2/4\alpha t} - \frac{q_0 x}{kA}\left[1 - \text{erf}\left(\frac{x}{2\sqrt{\alpha t}}\right)\right] \tag{5.108}$$

EXAMPLE 5.17 A large block of wrought-iron is initially at a uniform temperature of 30°C. One surface of the block is suddenly exposed to a constant surface heat flux of 2 MW/m^2. Calculate the temperature at a depth of 5 cm after a time of 50 seconds.

SOLUTION We know that for wrought-iron, k = 59 W/m K and α =1.626 × 10^{-5} m^2/s. It is given that T_i = 30°C; q_0/A = 2 MW/m^2; x = 5 cm and t = 50 s.

$$\frac{x}{2\sqrt{\alpha t}} = \frac{0.05}{2\sqrt{1.626 \times 10^{-5} \times 50}} = 0.88$$

$$\frac{x^2}{4\alpha t} = 0.7744$$

From Table 5.1, we read erf(0.88) = 0.7867.

The temperature distribution in the solid is given by

$$T - T_i = \frac{2q_0\sqrt{\alpha t/\pi}}{kA}e^{-x^2/4\alpha t} - \frac{q_0 x}{kA}\left[1 - \text{erf}\left(\frac{x}{2\sqrt{\alpha t}}\right)\right]$$

or

$$T - 30 = \frac{2 \times 2 \times 10^6\sqrt{(1.626 \times 10^{-5} \times 50)/\pi}}{59}e^{-0.7744}$$

$$- \frac{2 \times 10^6 \times 0.05}{59}(1 - 0.7867)$$

$$= 141.2$$

or $T = 171.2$°C

5.7.2 Transient Heat Flow in a Semi-infinite Solid with Convection at the Surface

In many practical situations, the temperature of the surface of a semi-infinite body is not changed suddenly to a different value, but its surface is suddenly exposed to an environment at a different temperature. In such cases, the surface temperature changes due to convection heat transfer. To obtain the temperature distribution in a solid, Eq. (5.105) is to be solved with the boundary condition.

$$hA(T_\infty - T)_{x=0} = -kA\left(\frac{\partial T}{\partial x}\right)_{x=0} \tag{5.109}$$

which states that the heat transferred by convection at the surface is equal to the heat transferred by conduction into the solid from its surface. The solution of this problem is given by

$$\frac{T-T_i}{T_\infty - T_i} = 1 - \text{erf}\left(\frac{x}{2\sqrt{\alpha t}}\right) - \left[1 - \text{erf}\left(\frac{x}{2\sqrt{\alpha t}} + \frac{h\sqrt{\alpha t}}{k}\right)\right] e^{\left(\frac{hx}{k}+\frac{h^2\alpha t}{k^2}\right)} \tag{5.110}$$

The solution of this equation is presented in graphical form in Fig. 5.24.

EXAMPLE 5.18 A large block of wrought-iron is initially at a uniform temperature of 30°C. One surface of the block is suddenly exposed to an environment at 530°C, with a convection heat transfer coefficient of 400 W/m^2 K. Calculate the temperature at a depth of 5 cm after a time of 60 seconds.

SOLUTION We can use either Eq. (5.110) or Fig. 5.24 to obtain the solution of the problem. We know that for wrought-iron, k = 59 W/m K and $\alpha = 1.626 \times 10^{-5}$ m^2/s. It is given that x= 5 cm; t = 60 s; T_i = 30°C; T_∞ = 530°C and h = 400 W/m^{2}K.

$$\frac{x}{2\sqrt{\alpha t}} = \frac{0.05}{2\sqrt{1.626 \times 10^{-5} \times 60}} = 0.80$$

$$\frac{h\sqrt{\alpha t}}{k} = \frac{400 \times \sqrt{1.626 \times 10^{-5} \times 60}}{59} = 0.212$$

The temperature distribution in the block is given by

$$\frac{T-T_i}{T_\infty - T_i} = 1 - \text{erf}\left(\frac{x}{2\sqrt{\alpha t}}\right) - \left[1 - \text{erf}\left(\frac{x}{2\sqrt{\alpha t}} + \frac{h\sqrt{\alpha t}}{k}\right)\right] e^{\left(\frac{hx}{k}+\frac{h^2\alpha t}{k^2}\right)}$$

$$\text{or}\quad \frac{T-30}{530-30} = 1 - \text{erf}\,(0.80) - [1 - \text{erf}\,(0.80 + 0.212)]\,\exp\left[\frac{400 \times 0.05}{59} + (0.212)^2\right]$$

$$\text{or}\quad \frac{T-30}{500} = 1 - 0.7421 - (1 - 0.8476)1.468 = 0.034 \quad \text{or} \quad T = 47°\text{C}$$

where erf(0.80) = 0.7421 and erf(1.012)=0.8476. We can also obtain the solution from Fig. 5.24. We read

$$\frac{T - T_i}{T_\infty - T_i} = 0.034 \quad \text{at} \quad \frac{x}{2\sqrt{\alpha t}} = 0.80 \quad \text{and} \quad \frac{h\sqrt{\alpha t}}{k} = 0.212$$

$$\text{or}\quad \frac{T-30}{530-30} = 0.034 \quad \text{or} \quad T = 47°\text{C}$$

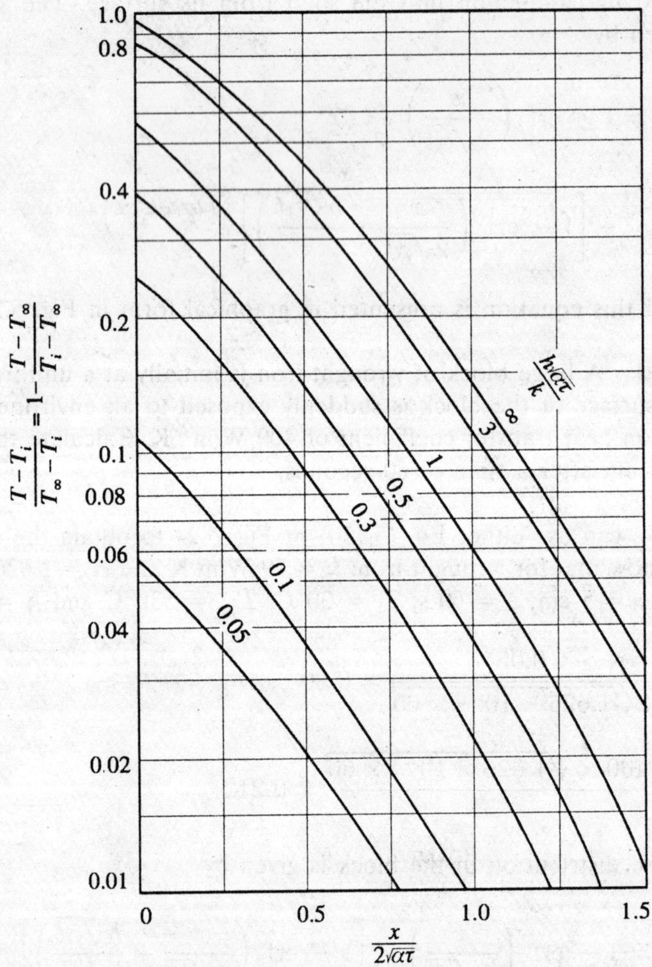

Fig. 5.24 Temperature distribution in a semi-infinite solid with convection at the surface.

EXAMPLE 5.19 For the wrought-iron block of Example 5.18, determine the time required for the temperature to reach 130°C at a depth of 5 cm.

SOLUTION The problem can be solved either by using Eq. (5.110) or Fig. 5.24. However, the use of Fig. 5.24 is easier. In Fig. 5.24, $(T - T_i)/(T_\infty - T_i)$ is given as a function of $(x/(2\sqrt{\alpha t}))$ and $(h\sqrt{\alpha t}/k)$. Hence a trial and error procedure is required to determine the time. We need the value of t such that

$$\frac{T - T_i}{T_\infty - T_i} = \frac{130 - 30}{530 - 30} = 0.2$$

Therefore, we assume a value of t and calculate $(x/2\sqrt{\alpha t})$ and $(h\sqrt{\alpha t}/k)$ and read the value of $(T - T_i)/(T_\infty - T_i)$ from Fig. 5.24 corresponding to the calculated values of $[x/(2\sqrt{\alpha t})]$ and $(h\sqrt{\alpha t}/k)$. If the value of $(T - T_i)/(T_\infty - T_i)$ thus read from the figure is equal to 0.2, then the assumed value of t is correct. Or else, we repeat the procedure till agreement is reached. The detailed calculations are presented below.

Assume $t = 200$ s. Then

$$\frac{x}{2\sqrt{\alpha t}} = \frac{0.05}{2\sqrt{1.626 \times 10^{-5} \times 200}} = 0.44$$

$$\frac{h\sqrt{\alpha t}}{k} = \frac{400\sqrt{1.626 \times 10^{-5} \times 200}}{59} = 0.39$$

Then read

$$\frac{T - T_i}{T_\infty - T_i} = 0.12 \text{ from Fig. 5.24 at } \frac{x}{2\sqrt{\alpha t}} = 0.44 \text{ and } \frac{h\sqrt{\alpha t}}{k} = 0.39.$$

Since the value of $[(T - T_i)/(T_\infty - T_i)] = 0.12$ is less than 0.2, the assumed value of t is low. Hence let us assume $t = 400$ s. Then

$$\frac{x}{2\sqrt{\alpha t}} = \frac{0.05}{2\sqrt{1.626 \times 10^{-5} \times 400}} = 0.31$$

$$\frac{h\sqrt{\alpha t}}{k} = \frac{400\sqrt{1.626 \times 10^{-5} \times 400}}{59} = 0.55.$$

Then read

$$\frac{T - T_i}{T_\infty - T_i} = 0.25 \text{ at } \frac{x}{2\sqrt{\alpha t}} = 0.31 \text{ and } \frac{h\sqrt{\alpha t}}{k} = 0.55.$$

The value of $[(T - T_i)/(T_\infty - T_i)] = 0.25$ is greater than 0.2. Hence, the assumed value of t is large. Now, let us revise the guess value of t as 290 seconds. Then

$$\frac{x}{2\sqrt{\alpha t}} = \frac{0.05}{2\sqrt{1.626 \times 10^{-5} \times 290}} = 0.36$$

$$\frac{h\sqrt{\alpha t}}{k} = \frac{400\sqrt{1.626 \times 10^{-5} \times 290}}{59} = 0.47$$

Then we read $\dfrac{T - T_i}{T_\infty - T_i} = 0.195 \simeq 0.2.$

Hence, the time required is approximately 290 s.

5.8 Summary

Analytical methods can be used to analyze transient heat transfer problems. The simplest problems in transient heat transfer are those in which the temperature is a function of time only and it is uniform throughout the system. In several situations, the internal resistance for conduction is very small compared to the external resistance for convection. In such cases the lumped heat capacity analysis can be employed. A measure of the relative importance of the internal conduction resistance is given by the Biot modulus which is defined as

$$Bi = \frac{\text{Internal conduction resistance}}{\text{External convection resistance}} = \frac{\ell/kA}{1/hA} = \frac{h\ell}{k}$$

If a body having density ρ, specific heat c and volume V, which is initially at temperature T_0, is suddenly exposed to an environment at T_∞, the temperature–time history of the body is given by

$$\frac{T - T_\infty}{T_0 - T_\infty} = \exp\left(\frac{-hAt}{\rho Vc}\right) = e^{-t/\tau}$$

where

$$\tau = \frac{\rho Vc}{hA} = R_{th}\,C_{th} = \text{time constant}$$

$$R_{th} = 1/hA = \text{thermal resistance}$$

$$C_{th} = \rho Vc = \text{thermal capacitance.}$$

The above relation is analogous to the equation describing the voltage-time history in an electrical system consisting of a capacitor C_e and a resistance R_e in series.

$$\frac{E - E_\infty}{E_0 - E_\infty} = \exp\left(-\frac{t}{R_e C_e}\right) = \exp(-t/\tau)$$

$$\frac{hAt}{\rho Vc} = \frac{ht}{\rho c\ell}\left(\frac{\ell k}{\ell k}\right) = \left(\frac{h\ell}{k}\right)\left(\frac{k}{\rho c}\right)\left(\frac{t}{\ell^2}\right) = \left(\frac{h\ell}{k}\right)\left(\frac{\alpha t}{\ell^2}\right) = (Bi)(Fo)$$

where $Fo = \dfrac{\alpha t}{\ell^2}$ = Fourier modulus.

The temperature–time history of the body can be expressed as

$$\frac{T - T_\infty}{T_0 - T_\infty} = \exp\left(\frac{-hAt}{\rho Vc}\right) = e^{-(Bi)(Fo)}$$

The quantity of energy transferred to the body in time t is given by

$$Q = \rho VchA\left[1 - e^{-(Bi)(Fo)}\right]$$

The lumped heat capacity analysis can also be extended to composite systems.

If a large flat plate of thickness 2ℓ, which is originally at temperature T_i. is suddenly exposed to an environment at T_∞ with convection heat transfer coefficient h, the temperature distribution in the plate, in the absence of internal heat generation, is governed by the differential equation

$$\frac{\partial T}{\partial t} = \alpha \frac{\partial^2 T}{\partial x^2}$$

The initial and boundary conditions are given by

$$T = T_i \quad \text{at} \quad t = 0$$

$$hA(T_{x=\ell} - T_\infty) = -kA\frac{\partial T}{\partial x} \quad \text{at} \quad x = \ell$$

(The coordinate axis passes through the centre-line of the plate.)

The solution of the above differential equation is given by

$$\frac{T - T_\infty}{T_i - T_\infty} = 2\sum_{n=1}^{\infty} \frac{\sin(\lambda_n \ell)\cos(\lambda_n x)e^{-\alpha\lambda_n^2 t}}{\ell\lambda_n + \sin(\lambda_n \ell)\cos(\lambda_n \ell)}$$

$$= \text{function of } Bi, Fo \text{ and } x/\ell$$

where λ_n is a characteristic value which satisfies the relation

$$\cot(\lambda\ell) = \lambda\ell/Bi$$

The above solution is in the form of an infinite series. Graphical solutions of the above relation(temperature distribution) are available in the form of charts, known as Heisler charts.

The analytical solutions of one-dimensional transient heat flow problems with boundary conditions of practical importance have been obtained by Heisler for the following geometries.

- Infinite flat plates whose thickness is small compared to the other dimensions.
- Infinite cylinders or cylinders whose diameter is small compared to the length.
- Spheres.

The solutions to the transient heat flow problems for the above geometries are available in the form of Heisler charts. These charts can be used if $Fo > 0.2$ and $0 < 1/Bi < 100$.

The Heisler charts provide the temperature distribution in the body and the energy transferred through the following three different types of plots.

1. Plots of dimensionless temperature ratio $(T_0 - T_\infty)/(T_i - T_\infty)$ at the centre versus Fo, with $1/Bi$ as a parameter. T_0 denotes temperature at $x = 0$ or $r = 0$. These plots can be used to determine the centre temperature from a knowledge of Bi and Fo.

2. Plots of dimensionless temperature ratio $(T - T_\infty)/(T_0 - T_\infty)$ versus $(1/Bi)$, with (x/ℓ) or (r/r_0) as a parameter. These plots along with plots of type 1 can be used to determine the temperature at any location from a knowledge of Bi, Fo and (x/ℓ) or (r/r_0). The temperature at any location can be calculated as

$$\frac{T - T_\infty}{T_i - T_\infty} = \left(\frac{T_0 - T_\infty}{T_i - T_\infty}\right)\left(\frac{T - T_\infty}{T_0 - T_\infty}\right)$$

3. Plots of dimensionless heat transfer Q/Q_i versus $(Fo)(Bi)^2$, with Bi as a parameter. These plots can be used to estimate the energy transferred during the time interval $t = 0$ to $t = t$ from a knowledge of Bi and Fo. Here Q_i denotes the initial energy of the body with reference to the base temperature T_∞, and is given by

$$Q_i = \rho c V(T_i - T_\infty)$$

Transient heat flow in two- and three-dimensional bodies can also be analyzed by using the Heisler charts. The solution for a two-dimensional $2\ell_1 \times 2\ell_2$ rectangular bar can be obtained as a product of the solutions of two infinite flat plates — plate 1 with thickness $2\ell_1$ and plate 2 with thickness $2\ell_2$. Similarly, the solution for a three-dimensional body of size $2\ell_1 \times 2\ell_2 \times 2\ell_3$ can be obtained as a product of the solutions of three infinite plates 1, 2 and 3 with thickness $2\ell_1, 2\ell_2$ and $2\ell_3$, respectively. The solution for a cylinder of finite length can be expressed as a product of the solutions of an infinitely long cylinder and an infinite flat plate having a thickness equal to the length of the cylinder. While using the product solutions, care should be taken to satisfy the boundary conditions of the one-dimensional heat flow problems to which the Heisler charts apply. This requirement can be easily satisfied by choosing the coordinate axes of the two-and three-dimensional bodies such that they lie along the axes of symmetry and intersect at the centre of the body. The heat transfer from a two-and three-dimensional bodies can be obtained by superimposing the heat transfer solutions from a one-dimensional body as

$$\left(\frac{Q}{Q_i}\right)_{\text{2-D body}} = \left(\frac{Q}{Q_i}\right)_1 + \left(\frac{Q}{Q_i}\right)_2 \left[1 - \left(\frac{Q}{Q_i}\right)_1\right]$$

$$\left(\frac{Q}{Q_i}\right)_{\text{3-D body}} = \left(\frac{Q}{Q_i}\right)_1 + \left(\frac{Q}{Q_i}\right)_2 \left[1 - \left(\frac{Q}{Q_i}\right)_1\right] + \left(\frac{Q}{Q_i}\right)_3 \left[1 - \left(\frac{Q}{Q_i}\right)_1\right]\left[1 - \left(\frac{Q}{Q_i}\right)_2\right]$$

where the subscripts 1, 2 and 3 refer to the intersecting one-dimensional bodies 1, 2 and 3, respectively.

If one surface of a semi-infinite solid originally at a uniform temperature T_i is suddenly changed to temperature T_∞ at time t = 0, the temperature distribution in the solid is governed by the differential equation

$$\frac{\partial T}{\partial t} = \alpha \frac{\partial^2 T}{\partial x^2}$$

subject to the following initial and boundary conditions:

$$T(x, 0) = T_i$$

$$T(0, t) = T_\infty \quad \text{for} \quad t > 0$$

The solution of this equation is given by

$$\frac{T - T_\infty}{T_i - T_\infty} = \text{erf}\left(\frac{x}{2\sqrt{\alpha t}}\right) = \frac{2}{\sqrt{\pi}} \int_0^{x/2\sqrt{\alpha t}} e^{-\beta^2} d\beta$$

where β is a dummy variable.

The change in the energy of the body during the time $t = 0$ to $t = t$ is given by

$$Q = 2kA(T_\infty - T_i)\sqrt{\frac{t}{\pi\alpha}}$$

If one surface of the semi-infinite body is exposed to a constant heat flux (q_0/A), the boundary condition has to be modified as

$$\frac{q_0}{A} = -k\left(\frac{\partial T}{\partial x}\right)_{x=0} \quad \text{for} \quad t > 0$$

Then the temperature distribution in the body is given by

$$T - T_i = \frac{2q_0\sqrt{\alpha t/\pi}}{kA}e^{-x^2/4\alpha t} - \frac{q_0 x}{kA}\left[1 - \text{erf}\left(\frac{x}{2\sqrt{\alpha t}}\right)\right]$$

If the surface of the semi-infinite body is suddenly exposed to an environment at T_∞ with convection heat transfer, the boundary condition is to be modified as

$$hA(T_\infty - T)_{x=0} = -kA\left(\frac{\partial T}{\partial x}\right)_{x=0}$$

Then the temperature distribution in the body is given by

$$\frac{T - T_i}{T_\infty - T_i} = 1 - \text{erf}\left(\frac{x}{2\sqrt{\alpha t}}\right) - \left[1 - \text{erf}\left(\frac{x}{2\sqrt{\alpha t}} + \frac{h\sqrt{\alpha t}}{k}\right)\right].e^{\left(\frac{hx}{k} + \frac{h^2\alpha t}{k^2}\right)}$$

Review Questions

1. What is meant by transient heat transfer?
2. Which is the simplest problem in transient heat transfer for which an analytical solution can be obtained?
3. What is meant by internal resistance and external resistance for heat transfer?
4. What is meant by lumped heat capacity analysis?
5. What assumptions are made in applying the lumped heat capacity analysis to solve transient heat flow problems?
6. What is the physical significance of Biot modulus?
7. What is the limiting value of Biot number to assume that a body is at uniform temperature?
8. What is meant by time constant in the anlaysis of transient heat transfer by lumped capacity method?

9. Describe a simple electrical circuit which is analogous to transient heat transfer from a body at uniform temperature.
10. Suppose a body (with negligible internal resistance for convection) originally at T_0 is allowed to cool by exposing it to an environment at T_∞. After a time $t = \tau$ (time constant), what will be the ratio of the temperature difference $(T - T_\infty)$ to the initial temperature difference $(T_0 - T_\infty)$?
11. The temperature–time history in the lumped heat capacity analysis is given by $[(T - T_\infty)/(T_0 - T_\infty)] = e^{-hAt/\rho Vc}$. Rewrite the argument $(hAt/\rho Vc)$ of the exponential term in terms of Biot and Fourier numbers.
12. Name a practical application of transient heat flow in an infinite plate.
13. What initial and boundary conditions are imposed in obtaining the solutions, which are presented in the form of Heisler charts, for transient heat transfer problems?
14. What is meant by an infinite cylinder?
15. Describe how one-dimensional transient heat transfer solutions can be used to obtain solutions for two- and three-dimensional bodies.
16. What is meant by a semi-infinite body?
17. Define error function.

Problems

5.1 An aluminium sphere of 30 cm diameter at a uniform temperature of 425°C is suddenly quenched in a large quantity of oil at 25°C. If the convection heat transfer coefficient between the aluminium sphere and the oil is 20 W/m^2 K, determine the time required for the sphere to reach a temperature of 225°C.

5.2 A cylindrical nickel-steel billet of 10 cm diameter and 30 cm length, initially at 825°C, is suddenly dropped in a large vessel containing oil at 25°C. The convection heat transfer coefficient between the billet and the oil is 15 W/m^2 K. Calculate the time required for the billet to reach a temperature of 225°C.

5.3 An aluminium sphere of 10 cm diameter at a uniform temperature of 525°C is suddenly exposed to an environment at 25°C, with convection heat transfer coefficient 30 W/m^{2}K. Calculate the temperature of the sphere (a) 100 s (b) 300 s and (c) 500 s after it is exposed to the environment.

5.4 A cylindrical carbon-steel billet of 20 cm diameter and 50 cm length, initially at 630°C, is suddenly dropped in a large quantity of oil at 30°C. The convection heat transfer coefficient between the oil and the steel billet is 50 W/m^2 K. Calculate the temperature of the billet (a) 200 s and (b) one hour after it is dropped in the oil.

5.5 A chrome-nickel wire of 2 mm diameter, initially at 25°C, is suddenly exposed to hot gases at 725°C. If the convection heat transfer coefficient is 10 W/m^2 K, determine the time required for the wire to reach a temperature of 625°C.

5.6 Calculate the time constant for the chrome-nickel wire of Problem 5.5.

5.7 Calculate the amount of energy transferred from the carbon-steel billet of Problem 5.4 during the time period (a) t = 50 s to t = 150 s and (b) t = 200 s to t = 300 s after it is dropped in oil.

5.8 A copper sphere of 5 cm diameter is initially at a uniform temperature of 250°C. It is suddenly exposed to ambient air at 25°C, with h = 10 W/m^2 K. How long does it take for the sphere to reach a temperature of 100°C and what is the time constant for the sphere ?

5.9 To determine the convection heat transfer coefficient between a heated copper sphere and air, a copper sphere of 10 cm diameter is heated to a uniform temperature of 325°C. Then the copper sphere is held in air at 25°C and it is found that the sphere has reached a temperature of 125°C in 12 minutes. Estimate the convection heat transfer coefficient.

5.10 A thin cylindrical aluminium can of 8 cm diameter and 8 cm length contains a soft drink at 5°C. If the can is exposed to ambient air at 40°C, with convection heat transfer coefficient of 10 W/m^2 K, estimate the time required for the soft drink to reach a temperature of 25°C. The properties of the soft drink are: ρ= 995 kg/m^3; c = 4 kJ/kg K and k = 0.6 W/m K. Use lumped heat capacity analysis to solve the problem.

5.11 A large slab of carbon-steel having a thickness of 30 cm is initially at a temperature of 630°C. It is then suddenly exposed to a convection environment at 30°C, with convection heat transfer coefficient of 10 W/m^2 K. Determine the time required for the centre-line of the slab to reach a temperature of 180°C.

5.12 A large plate of carbon-steel having a thickness of 10 cm is initially at 325°C. The plate is suddenly exposed to a convection environment at 25°C, with h = 150 W/m^2 K. Calculate the centre-line temperature of the plate (a) 10 minutes after it is exposed to the environment and (b) 50 minutes after it is exposed to the environment.

5.13 Calculate the time required for the surface of the plate of Problem 5.12 to reach a temperature of 125°C.

5.14 Calculate the temperature at a depth of 3 cm from the surface for the plate of Problem 5.12, 10 minutes after it is exposed to the environment.

5.15 A large aluminium plate of 12 cm thickness is originally at 225°C. It is suddenly exposed to an environment at 25°C, with convection heat transfer coefficient of 50 W/m^2 K. Calculate the temperature at a depth of 3 cm from one of the faces 5 minutes after the plate is exposed to the environment. How much energy is transferred per unit area of the plate during this time?

5.16 A large plate of aluminium of thickness 20 cm is originally at 525°C. It is suddenly exposed to a convection environment at 25°C, with convection heat transfer coefficient of 500 W/m^2 K. Calculate

(a) the time required for the centre of the plate to reach a temperature of 225°C.

(b) the surface temperature of the plate 2 minutes after it is exposed to the environment.

(c) the temperature at a depth of 5 cm from one of the surfaces 5 minutes after the plate is exposed to the environment.

5.17 Calculate the energy transferred per unit area of the plate during the first 3 minutes after the plate of Problem 5.16 is exposed to the environment.

5.18 A copper plate of 10 cm thickness is initially at a uniform temperature of 330°C. One surface of the copper plate is insulated and the second surface is suddenly exposed to an environment at 30°C. After 10 minutes, the surface temperature of the plate is found to be 180°C. Calculate the convection heat transfer coefficient.

5.19 A long aluminium rod of 30 cm diameter is originally at a uniform temperature of 325°C. The rod is suddenly exposed to an environment at 25°C, with convection heat transfer coefficient of 100 W/m^2 K. How long does it take for the centre to reach a temperature of 100°C?

5.20 Calculate the heat transferred from the aluminium rod of Problem 5.19.

5.21 For the aluminium rod of Problem 5.19, how long would it take for the surface to reach a temperature of 100°C?

5.22 A large carbon-steel cylinder of diameter 30 cm is originally at a uniform temperature of 330°C. It is suddenly exposed to an environment at 30°C, with convection heat transfer coefficient of 500 W/m^2 K. Calculate

(a) the centre temperature

(b) the surface temperature

(c) the temperature at a radius of 5 cm and

(d) the heat transferred per unit length

4 minutes after it is exposed to the environment.

5.23 A long brass rod of 20 cm diameter is heated to a uniform temperature of 225°C. Then it is suddenly exposed to an environment at 25°C. If the surface of the rod reaches a temperature of 125°C in 5 minutes, determine the convection heat transfer coefficient.

5.24 A chrome-steel sphere of radius 10 cm is initially at a uniform temperature of 300°C. The sphere is suddenly exposed to an environment at 30°C, with convection heat transfer coefficient of 600 W/m^2 K. How long does it take for the centre of the sphere to reach a temperature of 90°C?

5.25 For the sphere of Problem 5.24, what will be the centre temperature 2 minutes after it is exposed to the environment?

5.26 For the sphere of Problem 5.24, determine the temperatures at radii 4 cm, 6 cm and 8 cm, one minute after it is exposed to the environment.

5.27 For the sphere of Problem 5.24, estimate the surface temperature 60 s after it is exposed to the environment.

5.28 A lead sphere of diameter 5 mm at a uniform temperature of 200°C is suddenly exposed to a convection environment at 20°C, with h = 200 W/m^2 K. How much time does it take for the centre to reach a temperature of 110°C? Also calculate the amount of energy transferred.

5.29 For the lead sphere of Problem 5.28, calculate the temperature at a radius of 2 mm, 10 seconds after it is exposed to the environment.

5.30 A large carbon-steel bar of rectangular cross-section 30 cm × 20 cm is initially at a temperature of 525°C. Then it is suddenly exposed to a convection environment at 25°C, with h = 1000 W/m^2 K. Calculate the temperature at the centre of the bar 3 minutes after it is exposed to the environment.

5.31 For the carbon-steel bar of Problem 5.30, estimate the amount of energy transferred per unit length.

5.32 In a building, a long wooden beam of cross-section 30 cm × 20 cm is initially at a temperature of 30°C. Due to an accidental fire the beam is suddenly exposed to hot gases at 730°C, with h = 50 W/m^2 K. Estimate the time required for the wood to reach the ignition temperature of 500°C.

5.33 A large iron beam of cross-section 40 cm × 30 cm is originally at 25°C, with h = 500 W/m^2 K. Estimate the temperature at a point which is at a depth of 5 cm from each surface, 60 minutes after it is exposed to the environment at 725°C.

5.34 A cube of aluminium of 30 cm side length is initially at a temperature of 430°C. It is suddenly immersed in an oil at 30°C. The convection heat transfer coefficient is 600 W/m^2 K. Calculate the temperature at the centre of the cube one minute after it is immersed in the oil.

5.35 It is desired to bake a cake of size 30 cm × 20 cm × 10 cm. The dough is initially at a temperature of 25°C. The dough is suddenly placed in an oven which is at 175°C, and the convection heat transfer coefficient is 50 W/m^2 K. The cake is treated as completely baked if the centre reaches a temperature of 95°C. Determine the time required to bake the cake. The properties of dough are: ρ = 700 kg/m^3; c = 2 kJ/kg K and k =0.4 W/m K.

5.36 An aluminium cylinder of 30 cm diameter and 50 cm length is initially at a temperature of 30°C. It is placed inside the deep freezer chest of a refrigerator, where the temperature is −20°C, with a convection heat transfer coefficient of 20 W/m^2 K. Calculate the time required for the centre of the cylinder to reach a temperature of 0°C.

5.37 An iron cylinder of 20 cm diameter and 40 cm length is initially at a temperature of 25°C. It is suddenly dropped in boiling water at 100°C, with the convection heat transfer coefficient of 800 W/m^2 K. Determine the temperature at the centre of the cylinder 2 minutes after it is dropped in water.

5.38 Calculate the amount of energy transferred for the iron cylinder of Problem 5.37.

5.39 It is desired to anneal a glass cylinder of diameter 20 cm and length 40 cm to remove stresses before it is subjected to further processing. The glass cylinder is initially at 25°C. How long should the glass cylinder be kept in an oven so that every part of the glass attains at least a temperature of 425°C? The air in the oven is at 450°C, and h = 10 W/m^2 K.

5.40 A large thick layer of ice is initially at a temperature of −20°C. If the surface temperature is suddenly changed to −5°C, calculate the time required for the temperature at a depth of 2 cm to reach −10°C. The properties of ice are: ρ = 900 kg/m^3; c = 2 kJ/kg K and k = 2.2 W/m K.

5.41 A large slab of concrete (ρ = 2000 kg/m^3; c = 0.88 kJ/kg K and k = 1.37 W/m K) is initially at a temperature of 25°C. One surface of the slab is suddenly exposed to hot gases at 525°C, with convection heat transfer coefficient of 50 W/m^2 K. Calculate the temperature at a depth of 10 cm in the slab after a time of 2 hours.

5.42 A thick masonry brick wall is initially at 40°C. Suddenly one surface of the wall is exposed to a convection environment at 0°C, with h = 10 W/m^2 K. Calculate the time required for the temperature to reach 10°C at a depth of 5 cm from the surface.

5.43 In the installation of underground water pipes, it is necessary to estimate the depth to which a change in surface temperature is felt during a 12 hour period. If the ground is originally at 25°C and the surface temperature suddenly drops to −5°C, determine the depth at which the temperature reaches 0°C. For earth, ρ = 320 kg/m^3, c = 2 kJ/kg K and k =0.061 W/m K.

5.44 A long concrete (ρ = 2000 kg/m^3 and α = 7.784 × 10^{-7} m^2 /s) beam of 16 cm × 30 cm cross-section is initially at 30°C. If it is suddenly exposed to hot gases at 800°C because of fire, estimate the temperature at the surface of the beam after 12 hours. The convection heat transfer coefficient between concrete and hot gases is 5 W/m^2 K.

5.45 Estimate the time required to cook a chicken of 1.5 kg in an oven at 200°C. The chicken is considered to be cooked if the centre reaches a temperature of 110°C. The initial temperature of the chicken is 30°C; h = 50 W/m^2 K. For chicken, ρ = 700 kg/m^3; k = 0.4 W/m K and c = 1 kJ/kg K.

5.46 A thermoplastic material must be brought to about 90°C and its temperature should not exceed 110°C at any point before moulding. To prevent the material from damage it is kept in an oven at 110°C. If a sheet of 2 cm thickness at 30°C is kept in an oven, how long should it be kept before moulding? For plastic, k = 5 W/m K and α = 4 × 10^{-7} m^2/s. The convection heat transfer coefficient between the air in the oven and the plastic is 100 W/m^2 K.

CHAPTER SIX

TRANSIENT HEAT TRANSFER—NUMERICAL AND GRAPHICAL METHODS OF ANALYSIS

Learning Objectives

This chapter will enable the readers to

- realize the advantages of numerical and graphical methods over the analytical method in analyzing transient heat transfer problems.
- understand the forward difference or explicit and the backward difference or implicit formulation of finite difference equations.
- solve one-, two- and three-dimensional transient heat transfer problems using the forward difference technique.
- develop nodal equations for the forward difference technique for one-, two- and three-dimensional transient heat transfer.
- understand the thermal resistance–capacitance formulation for the solution of transient heat flow problems.
- solve the transient heat flow problems by applying the resistance–capacitance formulation technique.
- understand the graphical or Schmidt method of analyzing transient heat flow problems.

6.1 Introduction

The analytical method of analysis for transient heat transfer was discussed in Chapter 5. We found that the analytical solution, even in the case of a simple system—an infinite flat plate—is in the form of an infinite series. The results of the analytical solutions for some selected geometries are presented in Heisler charts. The Heisler charts are very useful in analyzing transient heat transfer in certain regular shaped solids. Unfortunately, quite a few practical situations are not covered by these regular shaped solids. In addition to this, an engineer may encounter situations in which the boundary conditions also vary with time. An analytical solution of a transient heat transfer problem with transient boundary conditions will be quite complicated. At times it may not be possible to obtain an analytical solution for a transient heat transfer problem for complicated geometries. In order to circumvent the mathematical difficulty, numerical and graphical methods have been developed which can be used to obtain approximate solutions.

The numerical methods for analyzing transient heat transfer are quite different from the methods used for analyzing steady-state heat transfer. In the steady-state analysis, the temperature distribution is determined for a network of points by using residual equations. In transient heat transfer problems, the objective is to determine the temperature distribution as a function of time, starting from the given initial temperature distribution. That is, starting from the given initial temperature distribution we should find the temperature distribution at a subsequent time. Therefore, we need to develop equations which guide us to determine the temperature distribution at time $(t + dt)$ from a knowledge of the temperature distribution at time t. The relevant expressions are presented in the following section.

6.2 Numerical Methods of Analysis—Finite Difference Equation

The differential equation governing the temperature distribution in a two-dimensional body is given by

$$\frac{\partial T}{\partial t} = \alpha \left(\frac{\partial^2 T}{\partial x^2} + \frac{\partial^2 T}{\partial y^2} \right) \tag{6.1}$$

If the differential increments in the temperature and space coordinates are approximated by finite differences in Eq. (6.1), we obtain the finite difference equation. In general, the smaller the intervals or increments in space coordinates and time, the closer will be the predicted results to the true values. Consider a two-dimensional body divided into a network of elements as shown in Fig. 6.1. The subscript m refers to the x coordinate position and n refers to the y coordinate position. Consider $T = T(t)$ at a specified location (m, n) as shown in Fig. 6.2.

The superscript p refers to the time increment. The first derivative of T with respect to t can be estimated in terms of (a) central, (b) forward or (c) backward differences as

$$\frac{\partial T}{\partial t} = \frac{T_{m,n}^{p+1} - T_{m,n}^{p-1}}{2\Delta t} \quad \text{central estimate} \tag{6.2}$$

$$\frac{\partial T}{\partial t} = \frac{T_{m,n}^{p+1} - T_{m,n}^{p}}{\Delta t} \quad \text{forward estimate} \tag{6.3}$$

$$\frac{\partial T}{\partial t} = \frac{T_{m,n}^{p} - T_{m,n}^{p-1}}{\Delta t} \quad \text{backward estimate} \tag{6.4}$$

In these estimates three temperatures T^{p-1}, T^p and T^{p+1} are involved. These three estimates will have different error levels. The forward difference estimate is in terms of the present value (T^p) and the value at the subsequent time increment (T^{p+1}).

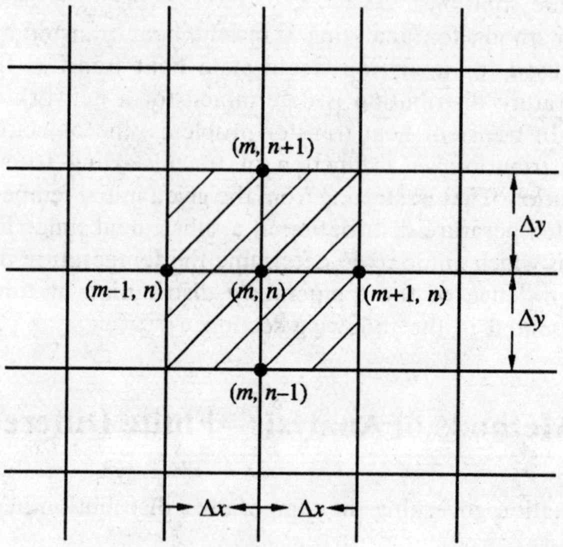

Fig. 6.1 Nomenclature for the network of points in a two-dimensional body.

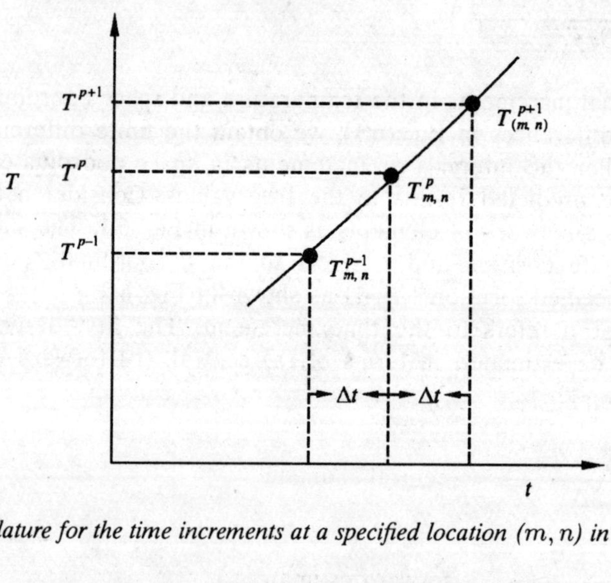

Fig. 6.2 Nomenclature for the time increments at a specified location (m, n) in a two-dimensional body.

The second partial derivatives of T with respect to the space coordinates x and y, at time increment p, can be written as

$$\frac{\partial^2 T}{\partial x^2} = \frac{T^p_{m+1,n} + T^p_{m-1,n} - 2T^p_{m,n}}{(\Delta x)^2} \tag{6.5}$$

$$\frac{\partial^2 T}{\partial y^2} = \frac{T^p_{m,n+1} + T^p_{m,n-1} - 2T^p_{m,n}}{(\Delta y)^2} \tag{6.6}$$

Substituting Eqs. (6.3),(6.5) and (6.6) in Eq. (6.1), we get

$$\frac{T^{p+1}_{m,n} - T^p_{m,n}}{\Delta t} = \alpha \left[\frac{T^p_{m+1,n} + T^p_{m-1,n} - 2T^p_{m,n}}{(\Delta x)^2} + \frac{T^p_{m,n+1} + T^p_{m,n-1} - 2T^p_{m,n}}{(\Delta y)^2} \right] \tag{6.7}$$

Therefore, knowing the temperatures at nodes $(m, n), (m+1, n), (m-1, n), (m, n+1)$ and $(m, n-1)$ at time increment p, we can calculate the temperature at node (m, n) at time increment $(p+1)$. This procedure can be repeated to obtain the temperature distribution, in a two-dimensional body, after any desired number of time increments. If the increments in the space coordinates are chosen such that

$$\Delta x = \Delta y \tag{6.8}$$

Equation (6.7) can be rewritten as

$$T^{p+1}_{m,n} = \frac{\alpha \Delta t}{(\Delta x)^2}(T^p_{m+1,n} + T^p_{m-1,n} + T^p_{m,n+1} + T^p_{m,n-1}) + \left[1 - \frac{4\alpha\Delta t}{(\Delta x)^2}\right] T^p_{m,n}$$

or

$$T^{p+1}_{m,n} = (Fo)_\Delta (T^p_{m+1,n} + T^p_{m-1,n} + T^p_{m,n+1} + T^p_{m,n-1}) + (1 - 4(Fo)_\Delta)\, T^p_{m,n} \tag{6.9}$$

where

$$(Fo)_\Delta = \frac{\alpha \Delta t}{(\Delta x)^2} = \text{Network Fourier number} \tag{6.10}$$

Equation (6.9) based on the forward difference estimate of $\partial T/\partial t$ permits the explicit calculation of the unknown $T^{p+1}_{m,n}$ from the known values of $T^p_{m+1,n}$; $T^p_{m-1,n}$; $T^p_{m,n+1}$ and $T^p_{m,n-1}$. This formulation is known as the *explicit method or forward difference technique* of transient numerical analysis. In this method it is possible to estimate the temperatures at the nodal points at a future time increment in terms of the temperatures of the surrounding nodal points at the beginning of the time increment. The size of the network chosen for a particular problem affects the accuracy of the estimated result. In Eq. (6.9) the coefficients of temperatures $(Fo)_\Delta$ and $[1-4(Fo)_\Delta]$ indicate the magnitude of the effect of the various temperatures on $T^{p+1}_{m,n}$. Therefore, none of these coefficients can have negative values. A negative value of the coefficient indicates that a high temperature at node (m, n) or at the surrounding nodes leads to a low value of temperature, at subsequent time increment, at node (m, n), which is not an acceptable result. Hence, the minimum acceptable value for a coefficient is zero. In order to make the coefficient of $T^p_{m,n}$ equal to zero or greater than zero, we should have $(Fo)_\Delta \leq 1/4$ for a two-dimensional body. Hence a choice of the network spacing (Δx) fixes the value of the time increment Δt. The larger the value of Δx or Δt, the more rapidly the solution converges. On the other hand, smaller values of the increments leads to more accuracy. The finite difference solutions will not converge unless the above conditions on (Fo) are satisfied.

We can write down explicit expressions similar to Eq. (6.9) for one- and three-dimensional bodies as

$$T^{p+1}_m = (Fo)_\Delta(T^p_{m+1} + T^p_{m-1}) + [1 - 2(Fo)_\Delta]T^p_m \tag{6.11}$$

$$T^{p+1}_{m,n,\ell} = (Fo)_\Delta \Big[T^p_{m+1,n,\ell} + T^p_{m-1,n,\ell} + T^p_{m,n+1,\ell} + T^p_{m,n-1,\ell} + T^p_{m,n,\ell+1} + T^p_{m,n,\ell-1}\Big] + [1 - 6(Fo)_\Delta]\, T^p_{m,n,\ell} \tag{6.12}$$

The values of $(Fo)_\Delta$ should be chosen such that

$$(Fo)_\Delta \leq \frac{1}{2} \text{ for a one-dimensional body} \tag{6.13}$$

$$(Fo)_\Delta \leq \frac{1}{4} \text{ for a two-dimensional body} \tag{6.14}$$

$$(Fo)_\Delta \leq \frac{1}{6} \text{ for a three-dimensional body} \tag{6.15}$$

If we choose $(Fo)_\Delta = 1/4$ for a two-dimensional body, Eq. (6.9) reduces to

$$T^{p+1}_{m,n} = \frac{T^p_{m+1,n} + T^p_{m-1,n} + T^p_{m,n+1} + T^p_{m,n-1}}{4} \tag{6.16}$$

$$(Fo)_\Delta = \frac{1}{4} = \frac{\alpha \Delta t}{(\Delta x)^2}$$

or

$$\Delta t = \frac{(\Delta x)^2}{4\alpha} \tag{6.17}$$

The application of the forward difference method to analyze a transient heat transfer problem in a one-dimensional case is illustrated through Example 6.1.

EXAMPLE 6.1 A nickel-steel plate ($\alpha = 0.5 \times 10^{-5}$ m²/s) of 30 cm thickness is initially at 30°C. Suddenly one face of the wall is raised to 670°C while the other face is insulated. Estimate the temperature distribution that exists in the plate 25 minutes after the surface's temperature is raised.

SOLUTION Let us choose Δx = 5 cm. Then

$$(Fo)_\Delta = \frac{1}{2} = \frac{\alpha \Delta t}{(\Delta x)^2} = \frac{0.5 \times 10^{-5} \Delta t}{(0.05)^2}$$

or $\quad \Delta t = 250$ s.

$$\text{Therefore number of time increments} = \frac{25 \times 60}{250} = 6.$$

Since one face of the wall is insulated, we can consider the insulated surface as the centre of a plate of thickness 60 cm, both the surfaces of which are maintained at 670°C as shown in Fig. 6.3.

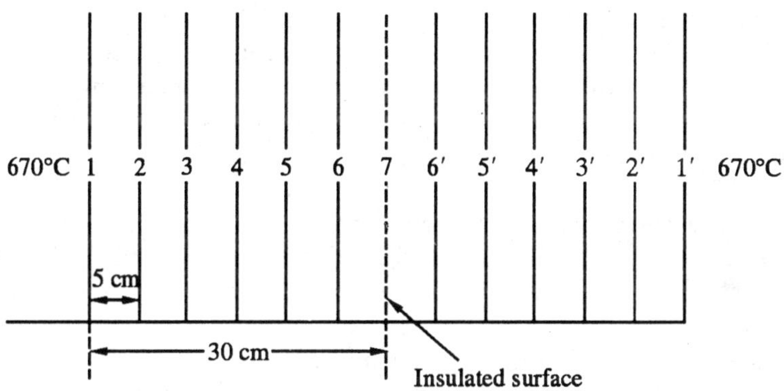

Fig. 6.3 Equivalent network for the plate of Example 6.1.

For a one-dimensional transient heat flow problem, we have

$$T_m^{p+1} = (Fo)_\Delta (T_{m+1}^p + T_{m-1}^p) + [1 - 2(Fo)_\Delta]\, T_m^p$$

We have chosen $(Fo)_\Delta = \frac{1}{2}$. Hence the above equation reduces to

$$T_m^{p+1} = \frac{T_{m+1}^p + T_{m-1}^p}{2}$$

At time t = 0, the surface 1 is at 670°C and the rest of the plate (that is, all the nodes from 2 to 7) is at 30°C. The calculations are summarized in the following table.

Time	Temperature in °C at nodes							
increment	1	2	3	4	5	6	7	6′
0	670	30	30	30	30	30	30	30
1	670	350	30	30	30	30	30	30
2	670	350	190	30	30	30	30	30
3	670	430	190	110	30	30	30	30
4	670	430	270	110	70	30	30	30
5	670	470	270	170	70	50	30	50
6	670	470	320	170	110	50	50	50

The final temperature distribution in the plate is shown below.

Distance (cm) from the hot surface	0	5	10	15	20	25	30
Temperature (°C)	670	470	320	170	110	50	50

EXAMPLE 6.2 A large iron safe is to be designed so that the material inside the safe can be protected from fire. For this purpose it was decided to place asbestos in between two steel plates. Initially the safe is at 30°C. If a fire occurs, the outer surface of the safe is expected to suddenly reach a temperature of 970°C. It is required to determine the thickness of asbestos packing such that the inner surface temperature remains at 30°C for 1 hour after the external surface is raised to 970°C. Use the transient numerical method to solve the problem.

SOLUTION For asbestos, $\alpha = 2.88 \times 10^{-7}\ \text{m}^2/\text{s}$.

Let us arbitrarily divide the thickness of asbestos into six equal layers. (The number of layers is chosen arbitrarily). Then we have seven nodes — the first node being the external surface and the seventh node the interior surface. At time $t = 0$, the first node attains a temperature of 970°C while the other nodes are at 30°C. First, we determine the number of time increments at which the inner surface temperature starts changing. Then, the value of each time increment is calculated. Later, we calculate the thickness of each layer and hence the total thickness of asbestos packing, from $(Fo)_\Delta = 1/2$. The calculations are presented in a table on the next page.

In this table, the eighth node is added to denote the inside of the iron safe and it is assumed that the temperature of the inside material will be the same as the temperature at node 7.

From the table we also find that after six time intervals, the inside surface temperature changes to 44.7°C.

$$\text{Hence, time increment} = \frac{60}{6} = 10\,\text{min} = 600\ \text{s}.$$

$$(Fo)_\Delta = \frac{1}{2} = \frac{\alpha \Delta t}{(\Delta x)^2}$$

or $\quad \dfrac{1}{2} = \dfrac{2.88 \times 10^{-7} \times 600}{(\Delta x)^2} \quad$ or $\quad \Delta x = 0.0186$ m

Total thickness of asbestos required = $6 \times 0.0186 = 0.1116$ m = 11.16 cm.

Time increment	Temperature in °C at nodes							
	1	2	3	4	5	6	7	8
0	970	30	30	30	30	30	30	
1	970	500	30	30	30	30	30	
2	970	500	265	30	30	30	30	
3	970	617.5	265	147.5	30	30	30	
4	970	617.5	382.5	147.5	88.75	30	30	
5	970	676.25	382.5	235.6	88.75	59.4	30	
6	970	676.25	456	235.6	147.5	59.4	44.7	

The difference equations (6.9), (6.11) and (6.12) can be used for estimating the internal temperature in a solid as a function of space and time. When a solid is exposed to a convection environment, a convectional resistance at the boundary of the solid exists, and these equations [(6.9), (6.11) and (6.12)] are no longer applicable. Depending on the geometry of the boundary, the convection boundary condition has to be applied. If the surface of a flat plate is exposed to a convection environment, the following boundary condition must be satisfied:

$$hA(T_o - T_\infty) = -kA\left(\frac{\partial T}{\partial x}\right)_{x=0} \tag{6.18}$$

where

T_o = temperature of surface exposed to a fluid (at $x = 0$)

T_∞ = temperature of the fluid.

Consider the two-dimensional wall shown in Fig. 6.4.

Let us make an energy balance on node (m, n). The change in the energy of the node is equal to the energy conducted and convected into the node. That is,

$$\rho c \frac{\Delta x}{2} \Delta y \frac{(T^{p+1}_{m,n} - T^p_{m,n})}{\Delta t} = k\frac{\Delta x}{2}\frac{(T^p_{m,n+1} - T^p_{m,n})}{\Delta y} + k\frac{\Delta x}{2}\frac{(T^p_{m,n-1} - T^p_{m,n})}{\Delta y}$$

$$+ k\Delta y \frac{(T^p_{m-1,n} - T^p_{m,n})}{\Delta x} + h\Delta y(T_\infty - T^p_{m,n}) \tag{6.19}$$

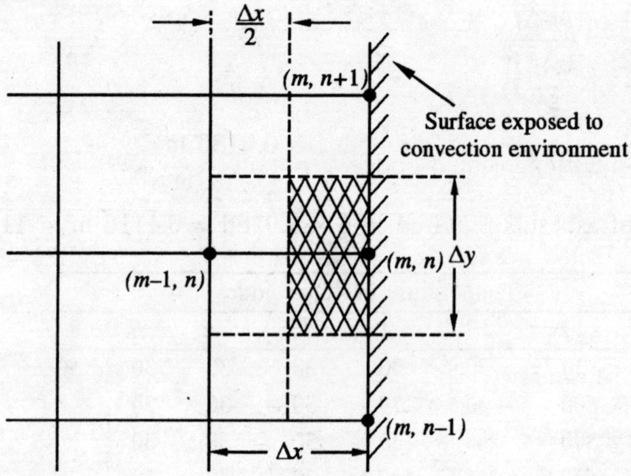

Fig. 6.4 A segment of a two-dimensional body exposed to a convection environment at T_∞.

If we choose $\Delta x = \Delta y$, then Eq. (6.19) reduces to

$$T^{p+1}_{m,n} = \frac{\alpha \Delta t}{(\Delta x)^2}\left[\frac{2h\Delta x}{k}T_\infty + 2T^p_{m-1,n} + T^p_{m,n+1} + T^p_{m,n-1}\right]$$

$$+\frac{\alpha \Delta t}{(\Delta x)^2}\left[\frac{(\Delta x)^2}{\alpha \Delta t} - \frac{2h\Delta x}{k} - 4\right]T^p_{m,n} \tag{6.20}$$

One can define the Biot modulus for the element as

$$(Bi)_\Delta = \frac{h\Delta x}{k} \tag{6.21}$$

Then Eq. (6.20) can be rewritten as

$$T^{p+1}_{m,n} = (Fo)_\Delta\left[2(Bi)_\Delta T_\infty + 2T^p_{m-1,n} + T^p_{m,n+1} + T^p_{m,n-1}\right]$$

$$+ (Fo)_\Delta\left(\frac{1}{(Fo)_\Delta} - 2(Bi)_\Delta - 4\right)T^p_{m,n} \tag{6.22}$$

To make the coefficient of $T^p_{m,n}$ in Eq. (6.22) as zero, we choose

$$(Fo)_\Delta = \frac{1}{2(2+(Bi)_\Delta)} \tag{6.23}$$

With this choice of $(Fo)_\Delta$, Eq. (6.22) reduces to

$$T_{m,n}^{p+1} = (Fo)_\Delta \left[2(Bi)_\Delta T_\infty + 2T_{m-1,n}^p + T_{m,n+1}^p + T_{m,n-1}^p\right] \tag{6.24}$$

Similarly, for a one-dimensional body we obtain

$$T_m^{p+1} = (Fo)_\Delta \left[2(Bi)_\Delta T_\infty + 2T_{m-1}^p\right] \tag{6.25}$$

when we choose

$$(Fo)_\Delta = \frac{1}{2[1+(Bi)_\Delta]} \tag{6.26}$$

Thus, we find the choice of Δt and Δx are governed by

$$(Fo)_\Delta = \frac{\alpha \Delta t}{(\Delta x)^2} = \frac{1}{2[2+(Bi)_\Delta]} \quad \text{for a two-dimensional body} \tag{6.23}$$

$$(Fo)_\Delta = \frac{\alpha \Delta t}{(\Delta x)^2} = \frac{1}{2[1+(Bi)_\Delta]} \quad \text{for a one-dimensional body} \tag{6.26}$$

Some typical nodal equations for a two-dimensional body are given in Table 6.1.

Table 6.1 *Nodal equations for forward difference estimation for two-dimensional transient heat flow.*

Physical situation	Nodal equation for $\Delta x = \Delta y$	Stability requirement
(a) Interior node 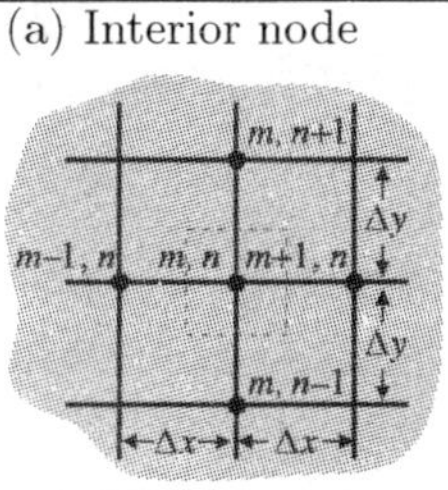	$T_{m,n}^{p+1} = Fo[T_{m-1,n}^p + T_{m,n+1}^p + T_{m+1,n}^p + T_{m,n-1}^p] + [1-4(Fo)]T_{m,n}^p$	$(Fo) \le \frac{1}{4}$
(b) Convection boundary node 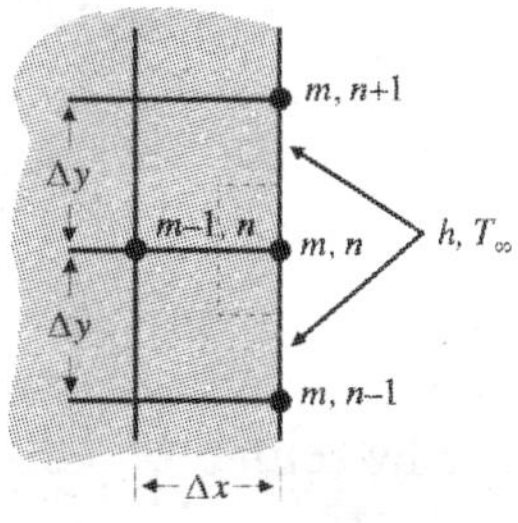	$T_{m,n}^{p+1} = Fo[2T_{m-1,n}^p + T_{m,n+1}^p + T_{m,n-1}^p + 2(Bi)T_\infty^p] + [1-4(Fo)-2(Fo)(Bi)]T_{m,n}^p$	$Fo(2+Bi) \le \frac{1}{2}$

Continues

Table 6.1 Continued

Physical situation	Nodal equation for $\Delta x = \Delta y$	Stability requirement
(c) Exterior corner with convection boundary 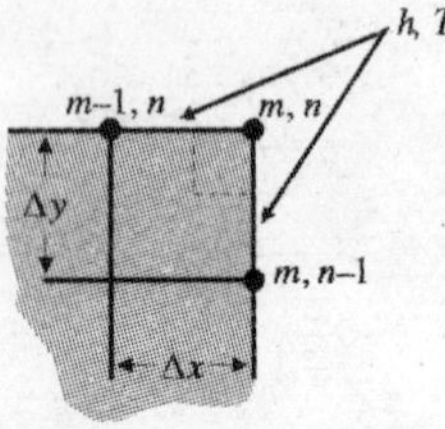	$T^{p+1}_{m,n} = 2(Fo)[T^p_{m-1,n} + T^p_{m,n-1} + 2(Bi)T^p_\infty] + [1 - 4(Fo) - 4(Fo)(Bi)]T^p_{m,n}$	$Fo(1+Bi) \le \frac{1}{4}$
(d) Interior corner with convection boundary 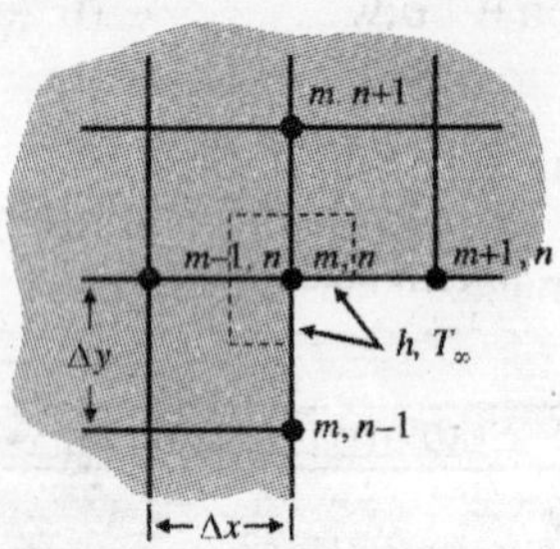	$T^{p+1}_{m,n} = \frac{2}{3}(Fo)[2T^p_{m,n+1} + 2T^p_{m+1,n} + 2T^p_{m-1,n} + T^p_{m,n-1} + 2(Bi)T^p_\infty] + [1 - 4(Fo) - \frac{4}{3}(Fo)(Bi)]T^p_{m,n}$	$Fo(3+Bi) \le \frac{3}{4}$
(e) Insulated boundary 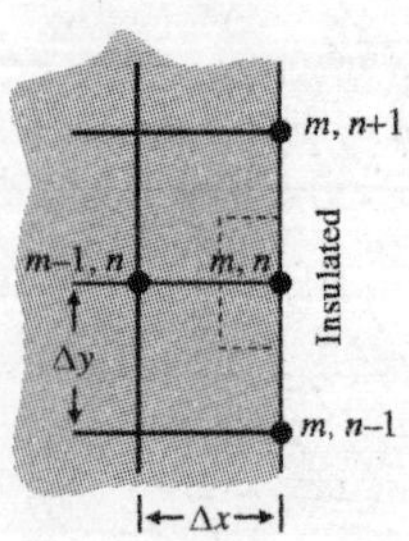	$T^{p+1}_{m,n} = (Fo)[2T^p_{m-1,n} + T^p_{m,n+1} + T^p_{m,n-1}] + [1 - 4(Fo)]T^p_{m,n}$	$Fo \le \frac{1}{4}$

The convection surface can be changed to an insulated surface by setting $h = 0$ or $Bi = 0$ in the above nodal equations.

The nodal equations for forward difference estimation (explicit equation) for one-dimensional transient heat flow are given in Table 6.2.

Table 6.2 Nodal equations (explicit equations) for finite difference approximation for one-dimensional transient heat flow.

Physical situation	Node equation	Stability requirement
(a) Interior node 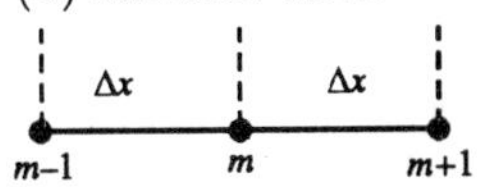	$T_m^{p+1} = (Fo)\left(T_{m-1}^p + T_{m+1}^p\right) + (1-2Fo)T_m^p$	$Fo \le \frac{1}{2}$
(b) Convection at surface 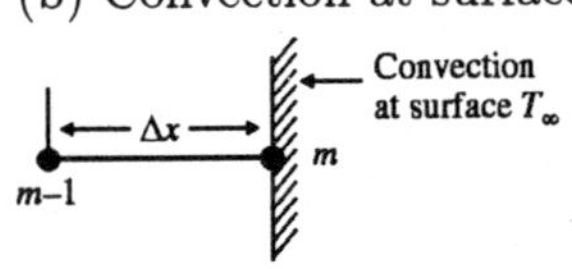	$T_m^{p+1} = 2(Fo)\left(T_{m-1}^p + BiT_\infty^p\right) + (1-2Fo-2FoBi)\,T_m^p$	$Fo \le \frac{1}{2(1+Bi)}$
(c) Adiabatic surface 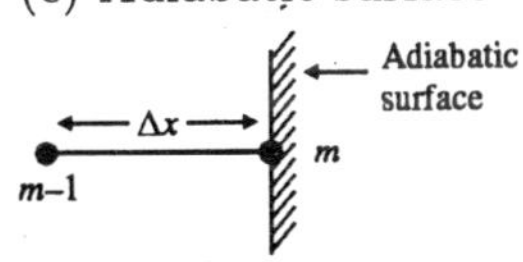	$T_m^{p+1} = 2(Fo)T_{m-1}^p + (1-2Fo)T_m^p$	$Fo \le \frac{1}{2}$

EXAMPLE 6.3 A long concrete structure 15 cm $\times$ 20 cm in cross-section is initially at a uniform temperature of 250°C. Two adjacent faces of the concrete are suddenly raised to 200°C while the other two adjacent faces are insulated. Determine the temperature distribution in the concrete 1 hour after its temperature is raised. For concrete $\alpha = 8 \times 10^{-7}$ m²/s.

SOLUTION Let us choose $\Delta x = \Delta y = 5$ cm and construct a grid of nodes as shown in Fig. 6.5

$$Fo = \frac{\alpha \Delta t}{(\Delta x)^2} = \frac{1}{4}$$

or
$$\frac{8 \times 10^{-7} \Delta t}{(0.05)^2} = \frac{1}{4}$$

or $t = 781 \text{ s} = 0.217 \text{ h}$

$$\text{Number of time increments} = \frac{1}{0.217} = 4.6.$$

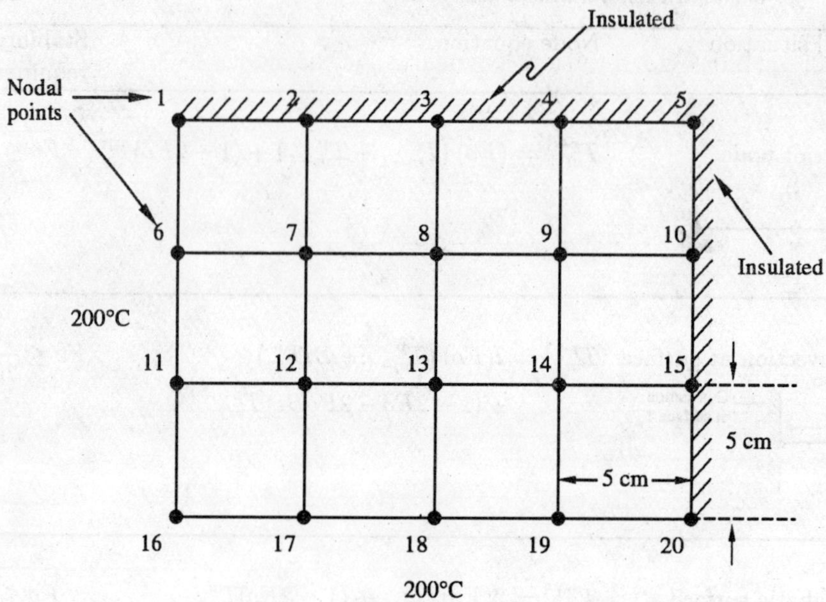

Fig. 6.5 Physical situation showing the nodal points.

Therefore, let us calculate for 5 time increments. The nodes 1, 6, 11, 16, 17, 18, 19 and 20 are raised to 200°C at time $t = 0$ and held at that temperature. All other nodes are at (at the same time) $T = 25°C$. The interior nodes are 7, 8, 9, 12, 13 and 14. The nodal equations [see Table 6.1] for these nodes are given by

$$T_7^{p+1} = \frac{T_6^p + T_8^p + T_{12}^p + T_2^p}{4} \tag{A}$$

$$T_8^{p+1} = \frac{T_7^p + T_9^p + T_{13}^p + T_3^p}{4} \tag{B}$$

$$T_9^{p+1} = \frac{T_8^p + T_{10}^p + T_{14}^p + T_4^p}{4} \tag{C}$$

$$T_{12}^{p+1} = \frac{T_{11}^p + T_{13}^p + T_{17}^p + T_7^p}{4} \tag{D}$$

$$T_{13}^{p+1} = \frac{T_{12}^p + T_{14}^p + T_8^p + T_{18}^p}{4} \tag{E}$$

$$T_{14}^{p+1} = \frac{T_{13}^p + T_{15}^p + T_9^p + T_{19}^p}{4} \tag{F}$$

The nodal equations (for insulated boundary) for the nodes 2, 3, 4, 10 and 15 are given by

$$T_2^{p+1} = \frac{T_1^p + T_3^p + 2T_7^p}{4} \tag{G}$$

$$T_3^{p+1} = \frac{T_2^p + T_4^p + 2T_8^p}{4} \tag{H}$$

$$T_4^{p+1} = \frac{T_3^p + T_5^p + 2T_9^p}{4} \tag{I}$$

$$T_{10}^{p+1} = \frac{T_5^p + T_{15}^p + 2T_9^p}{4} \tag{J}$$

$$T_{15}^{p+1} = \frac{T_{10}^p + T_{20}^p + 2T_{14}^p}{4} \tag{K}$$

The nodal equation for node 5 (exterior corner node) is given by

$$T_5^{p+1} = \frac{T_4^p + T_{10}^P}{2} \tag{L}$$

Substituting the values in equations (A)–(L), we can calculate the temperatures at these nodes for several time increments. These calculations are summarized in the following table.

Summary of calculated results

	Temperature (°C) at nodes									
p	1	2	3	4	5	6	7	8	9	10
0	200	25	25	25	25	200	25	25	25	25
1	200	68.75	25	25	25	200	68.75	25	25	25
2	200	90.63	35.94	25	25	200	101.56	31.38	35.94	41.41
3	200	109.77	44.6	33.21	33.21	200	114.1	68.75	47.11	41.41
4	200	118.2	70.12	43.01	37.31	200	132.33	82.31	61.23	55.54
5	200	133.7	81.46	57.47	49.28	200	139.98	98.49	74.3	67.73

p	11	12	13	14	15	16	17	18	19	20
0	200	25	25	25	25	200	200	200	200	200
1	200	112.5	68.75	68.75	68.75	200	200	200	200	200
2	200	134.38	101.56	90.63	68.75	200	200	200	200	200
3	200	150.78	123.44	101.56	94.73	200	200	200	200	200
4	200	159.39	130.27	116.32	111.13	200	200	200	200	200
5	200	165.65	139.51	125.66	122.05	200	200	200	200	200

The temperature distribution in the concrete 1 hour after the surfaces are raised to a temperature of 200°C is shown in Fig. 6.6.

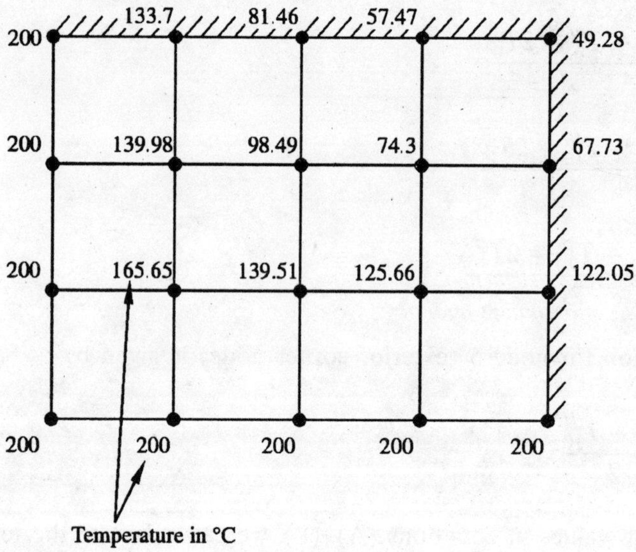

Fig. 6.6 Temperature distribution in the concrete slab.

6.3 Numerical Method of Analysis—Resistance–Capacitance Formulation

So far we have discussed the method of solving transient heat flow problems by finite difference approximation to the differential equation. In this method, an equation is formulated for each node and the set of resulting equations are solved to obtain the temperature distribution in the body. A useful alternative to the formulation for the nodal equations is to use the concept of thermal resistance between the nodes. Consider the node m, shown in Fig. 6.7, surrounded by a volume ΔV_m of thermal capacitance

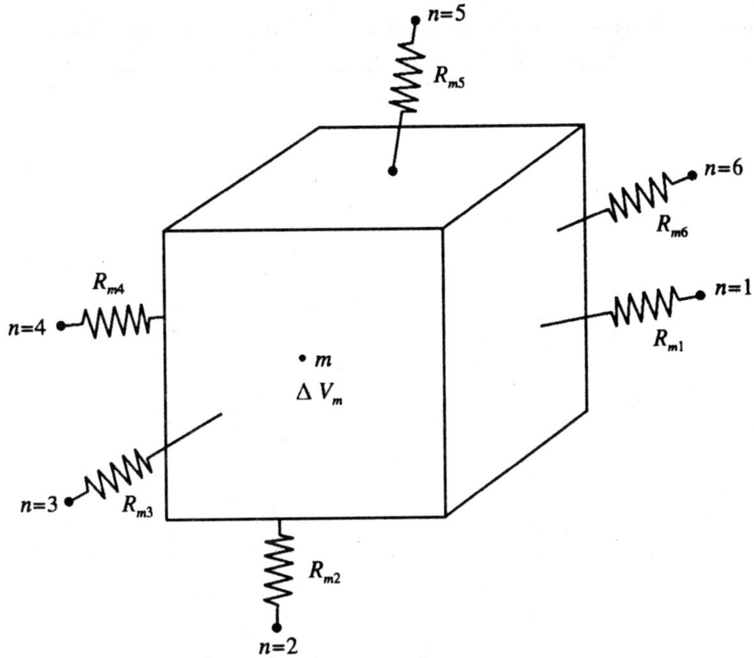

Fig. 6.7 A small volume element ΔV_m surrounding the node m. The thermal resistance between the node m and the surrounding node n is denoted by R_{mn}.

$$C_m = \rho_m c_m \Delta V_m \tag{6.27}$$

The surrounding nodes are denoted by n. The conduction from the volume ΔV_n of the node n into the volume ΔV_m is expressed in terms of thermal resistance R_{mn}. Each volume element behaves like a small lumped capacity and the interaction of all the volume elements determines the behaviour of the solid during a transient process. Then the rate of change of the energy of the volume element can be approximated as

$$\frac{\Delta E_m}{\Delta t} = \rho_m C_m \Delta V_m \frac{T_m^{p+1} - T_m^p}{\Delta t}$$

or

$$\frac{\Delta E_m}{\Delta t} = C_m \frac{T_m^{p+1} - T_m^p}{\Delta t} \tag{6.28}$$

An energy balance on volume ΔV_m over a time interval Δt gives

$$C_m \frac{T_m^{p+1} - T_m^p}{\Delta t} = \sum_n \frac{T_n^p - T_m^p}{R_{mn}} \tag{6.29}$$

Equation (6.29) is an explicit formulation and it is applicable for an interior node. A similar expression can also be written down for a boundary node in which the resistances R_{mn} can be modified. Equation (6.29) can be rewritten as

$$T_m^{p+1} = \left(1 - \frac{\Delta t}{C_m} \sum_n \frac{1}{R_{mn}}\right) T_m^p + \left(\sum_n \frac{T_n^p}{R_{mn}}\right) \frac{\Delta t}{C_m} \tag{6.30}$$

The stability condition requires that the coefficient of T_m^p should not be negative. Hence, the minimum stability requirement is given by

$$1 - \frac{\Delta t}{C_m} \sum_n \frac{1}{R_{mn}} \geq 0 \tag{6.31}$$

The resistance–capacitance formulation is usually applied to problems dealing with complicated geometries and boundary conditions. This formulation is quite useful if the thermal properties of the material are temperature dependent. In that case, the capacitance C_m and resistances R_{mn} are evaluated at time step p in the forward difference equation (or explicit formulation). The time step is to be selected to ensure stability and hence Δt should be equal to or less than a value obtained from the most restrictive nodal relation given by Eq. (6.31). From Eq. (6.31), we get

$$\Delta t \leq \left[\frac{C_m}{\sum_n \left(\frac{1}{R_{mn}}\right)}\right]_{\min} \tag{6.32}$$

For convenience of calculation, Eq. (6.30) can be rearranged as

$$T_m^{p+1} = \frac{\Delta t}{C_m} \left(\sum_n \frac{T_n^p - T_m^p}{R_{mn}}\right) + T_m^p \tag{6.33}$$

The volume elements for the three common coordinate systems are shown in Fig. 6.8.

The volume elements and internal nodal resistances for three common coordinate systems are presented in Table 6.3.

The plus or minus on the resistance subscripts denotes the resistance in a positive or negative direction from the central node (m, n, k). In a practical problem, usually the coordinate increments are frequently chosen so that $\Delta x = \Delta y = \Delta z$, so that the resistances are simplified. Table 6.4 presents the resistances for two-dimensional cartesian coordinates, $\Delta x = \Delta y$.

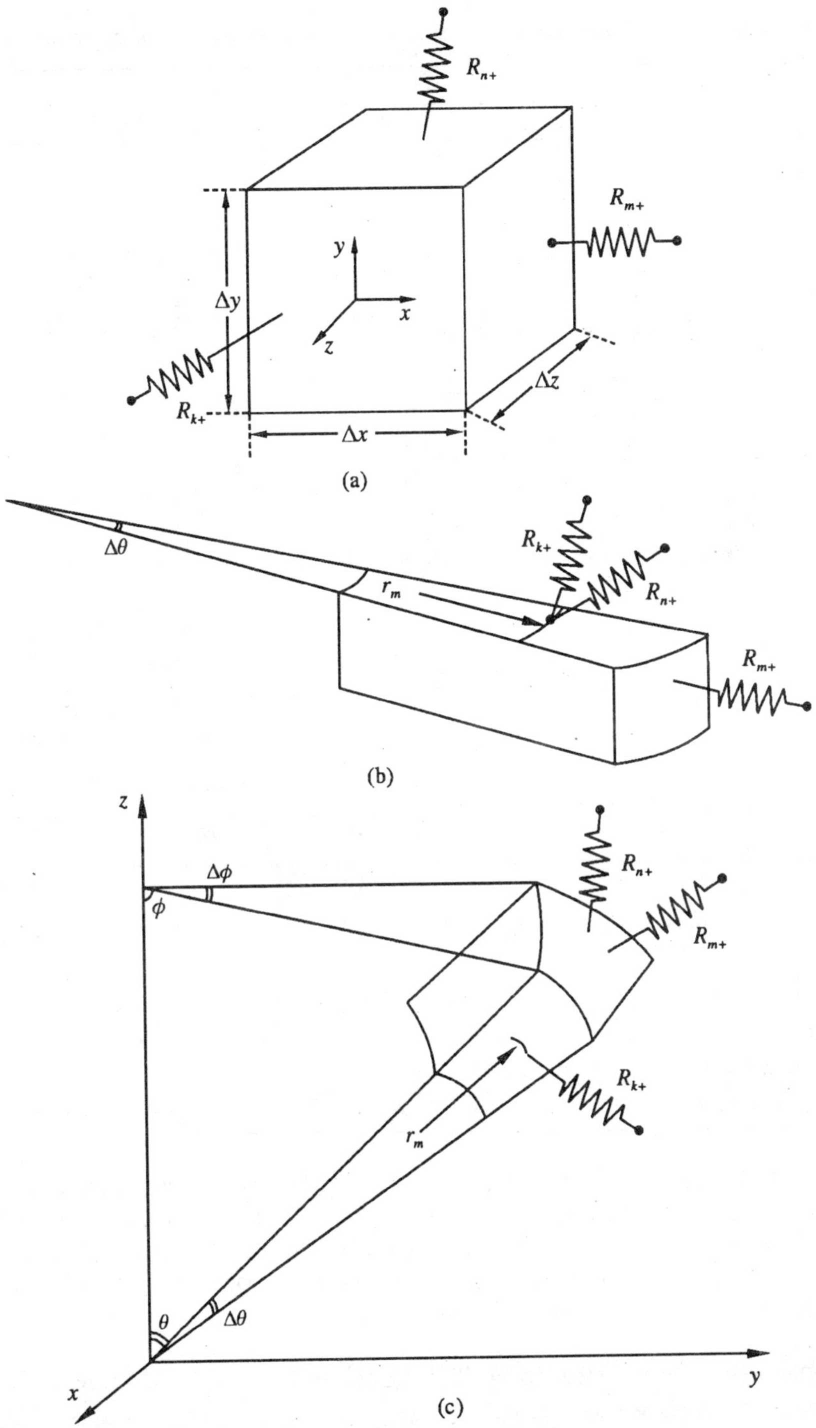

Fig. 6.8 *Volume of resistance elements (a) cartesian (b) cylindrical and (c) spherical coordinate systems.*

Table 6.3 Internal nodal resistances in cartesian, cylindrical and spherical coordinate systems

	Coordinate system		
	Cartesian	Cylindrical	Spherical
Coordinates	x, y, z	r, θ, z	r, ϕ, θ
Nomenclature	m, n, k	m, n, k	m, n, k
Volume element	$\Delta x \Delta y \Delta z$	$r_m \Delta r \Delta\theta \Delta z$	$r_m^2 \sin\theta \Delta r \Delta\phi \Delta\theta$
R_{m+}	$\dfrac{\Delta x}{\Delta y \Delta z k}$	$\dfrac{\Delta r}{(r_m + \frac{\Delta r}{2})\Delta\theta \Delta z k}$	$\dfrac{\Delta r}{(r_m + \frac{\Delta r}{2})^2 \sin\theta \Delta\phi \Delta\theta k}$
R_{m-}	$\dfrac{\Delta x}{\Delta y \Delta z k}$	$\dfrac{\Delta r}{(r_m - \frac{\Delta r}{2})\Delta\theta \Delta z k}$	$\dfrac{\Delta r}{(r_m - \frac{\Delta r}{2})^2 \sin\theta \Delta\phi \Delta\theta k}$
R_{n+}	$\dfrac{\Delta y}{\Delta x \Delta z k}$	$\dfrac{r_m \Delta\theta}{\Delta r \Delta z k}$	$\dfrac{\Delta\phi \sin\theta}{\Delta r \Delta\theta k}$
R_{n-}	$\dfrac{\Delta y}{\Delta x \Delta z k}$	$\dfrac{r_m \Delta\theta}{\Delta r \Delta z k}$	$\dfrac{\Delta\phi \sin\theta}{\Delta r \Delta\theta k}$
R_{k+}	$\dfrac{\Delta z}{\Delta x \Delta y k}$	$\dfrac{\Delta z}{r_m \Delta\theta \Delta r k}$	$\dfrac{\Delta\theta}{\sin(\theta + \frac{\Delta\theta}{2})\Delta r \Delta\phi k}$
R_{k-}	$\dfrac{\Delta z}{\Delta x \Delta y k}$	$\dfrac{\Delta z}{r_m \Delta\theta \Delta r k}$	$\dfrac{\Delta\theta}{\sin(\theta - \frac{\Delta\theta}{2})\Delta r \Delta\phi k}$

Table 6.4 Volumes and resistances for two-dimensional cartesian coordinates, $\Delta x = \Delta y$

Physical Situation	R_{01}	R_{02}	R_{03}	R_{04} or $R_{0\infty}$	ΔV
Interior node	$1/k$	$1/k$	$1/k$	$1/k$	$(\Delta x)^2$
Convection boundary node	$1/k$	$2/k$	$2/k$	$1/(h\Delta x)$	$(\Delta x)^2/2$
Exterior corner (convection boundary)	$2/k$	$2/k$	-	$1/(h\Delta x)$	$(\Delta x)^2/4$
Interior corner (Convection boundary)	$2/k$ $R_\infty = 1/(h\Delta x)$	$1/k$	$1/k$	$2/k$	$3(\Delta x)^2/4$
Insulated boundary	$1/k$	$2/k$	$2/k$	∞	$(\Delta x)^2/2$

EXAMPLE 6.4 A chrome-steel rod (k = 40 W/m K) of 10 mm diameter and 50 cm length is initially at a uniform temperature of 500°C. This rod is suddenly exposed to a convection environment at 30°C with h = 40 W/m^2 K, while one end of the rod is maintained at 500°C. Determine the temperature distribution in the rod 6 minutes after it is exposed to the convection environment.

SOLUTION For chrome-steel, ρ = 7833 kg/m^3 and c = 0.46 kJ/kg K. A schematic of the steel rod is shown in Fig. 6.9.

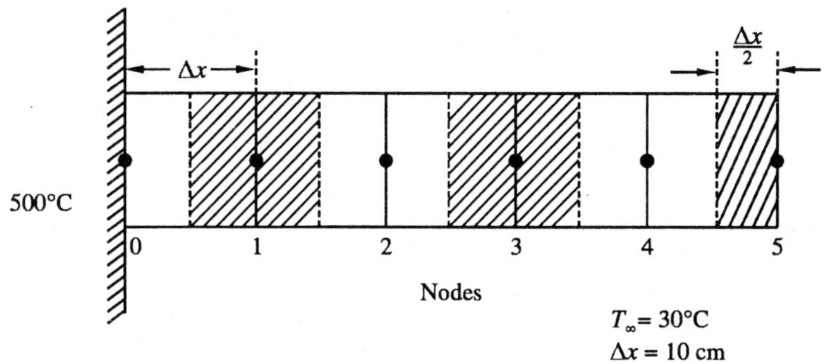

Fig. 6.9 Schematic of steel rod.

Cross-sectional area of rod,

$$A = \frac{\pi d^2}{4} = \frac{\pi (0.01)^2}{4} = 7.854 \times 10^{-5} \text{ m}^2$$

$$\Delta V = A\Delta x = 7.854 \times 10^{-5} \times 0.1 = 7.854 \times 10^{-6} \text{ m}^3$$

For nodes 1,2,3 and 4, $\Delta V = 7.854 \times 10^{-6}$ m^3.

Node 5 has a volume of $\dfrac{\Delta V}{2} = 3.927 \times 10^{-6}$ m^3.

For nodes 1,2,3 and 4, we get

$$C = \rho c \Delta V = 7833 \times 0.46 \times 10^3 \times 7.854 \times 10^{-6} = 28.3 \text{ J/K}$$

$$R_{m+} = R_{m-} = \frac{\Delta x}{kA} = \frac{0.1}{40 \times 7.854 \times 10^{-5}} = 31.83 \text{ K/W}$$

$$R_{\infty} = \frac{1}{h(\pi d \Delta x)} = \frac{1}{40\pi(0.01 \times 0.1)} = 7.96 \text{ K/W}$$

For node 5, we get

$$C = \rho c \Delta V = 7833 \times 0.46 \times 10^3 \times 3.927 \times 10^{-6} = 14.15 \text{ J/K}$$

$$R_{m+} = \frac{1}{hA} = \frac{1}{40 \times 7.854 \times 10^{-5}} = 318.3 \text{ K/W}$$

$$R_{m-} = \frac{\Delta x}{kA} = \frac{0.1}{40 \times 7.854 \times 10^{-5}} = 31.83 \text{ K/W}$$

$$R_\infty = \frac{2}{h\pi d\Delta x} = 15.92 \text{ K/W}$$

To determine the stability requirement, we have to determine the most restrictive $C_m \Big/ \sum \frac{1}{R_{mn}}$

Node	$\sum \frac{1}{R_{mn}}$	C_m	$C_m \Big/ \sum \frac{1}{R_{mn}}$
1	0.18846	28.3	150.2
2	0.18846	28.3	150.2
3	0.18846	28.3	150.2
4	0.18846	28.3	150.2
5	0.09737	14.15	145.3

Hence $\Delta t \leq 145.3$ s.

Let us choose $\Delta t = 120$ s. Then

$$\text{Number of time increments} = \frac{6 \times 60}{120} = 3.$$

Then we calculate the temperatures at the nodes using the relation

$$T_m^{p+1} = \frac{\Delta t}{C_m}\left(\sum_n \frac{T_n^p - T_m^p}{R_{mn}}\right) + T_m^p$$

$$T_1^1 = T_2^1 = T_3^1 = T_4^1 = \frac{\Delta t}{C_m}\left(\frac{T_\infty - T_m^p}{R_\infty}\right) + T_m^p$$

$$= \frac{120}{28.3}\left(\frac{30-500}{7.96}\right) + 500 = 249.6°\text{C}$$

$$T_5^1 = \frac{\Delta t}{C_5}\left(\frac{T_\infty - T_m^p}{R_{m+}} + \frac{T_\infty - T_m^p}{R_\infty}\right) + T_m^p$$

$$= \frac{120}{14.15}\left[\frac{30-500}{15.92} + \frac{30-500}{318.3}\right] + 500 = 237.1°\text{C}$$

$$T_1^2 = \frac{\Delta t}{C_1}\left(\frac{T_o - T_1^1}{R_{1-}} + \frac{T_2^1 - T_1^1}{R_{1+}} + \frac{T_\infty - T_1^1}{R_\infty}\right) + T_1^1$$

$$= \frac{120}{28.3}\left(\frac{500 - 249.6}{31.83} + \frac{30 - 249.6}{7.96}\right) + 249.6 = 166°\text{C}$$

$$T_2^2 = \frac{\Delta t}{C_2}\left(\frac{T_1^1 - T_2^1}{R_{2-}} + \frac{T_3^1 - T_2^1}{R_{2+}} + \frac{T_\infty - T_2^1}{R_\infty}\right) + T_2^1$$

$$= \frac{120}{28.3}\left(\frac{30 - 249.6}{7.96}\right) + 249.6 = 132.6°\text{C}$$

$$T_3^2 = \frac{\Delta t}{C_3}\left(\frac{T_2^1 - T_3^1}{R_{3-}} + \frac{T_4^1 - T_3^1}{R_{3+}} + \frac{T_\infty - T_3^1}{R_\infty}\right) + T_3^1$$

$$= \frac{120}{28.3}\left(\frac{30 - 249.6}{7.96}\right) + 249.6 = 132.6°\text{C}$$

$$T_4^2 = \frac{\Delta t}{C_4}\left(\frac{T_3^1 - T_4^1}{R_{4-}} + \frac{T_5^1 - T_4^1}{R_{4+}} + \frac{T_\infty - T_4^1}{R_\infty}\right) + T_4^1$$

$$= \frac{120}{28.3}\left(\frac{237.1 - 249.6}{31.83} + \frac{30 - 249.6}{7.96}\right) + 249.6 = 131°\text{C}$$

$$T_5^2 = \frac{\Delta t}{C_5}\left(\frac{T_4^1 - T_5^1}{R_{5-}} + \frac{T_\infty - T_5^1}{R_{5+}} + \frac{T_\infty - T_5^1}{R_\infty}\right) + T_5^1$$

$$= \frac{120}{14.15}\left(\frac{249.6 - 237.1}{31.83} + \frac{30 - 237.1}{318.3} + \frac{30 - 237.1}{15.92}\right) + 237.1 = 124.6°\text{C}$$

Similarly, the temperature at all the nodes are calculated for 3 time increments. The results of the calculations are shown in the following table.

Time increment	Node temperature (°C)				
	T_1	T_2	T_3	T_4	T_5
0	500	500	500	500	500
1	249.6	249.6	249.6	249.6	237.1
2	166	132.6	132.6	131	124.6
3	133.6	82.4	77.7	76.6	73.4

Thus the temperature distribution in the steel rod 6 minutes after it is exposed to the convection environment is as follows.

Location	Temperature (°C)
Base	500
10 cm from base	133.6
20 cm from base	82.4
30 cm from base	77.7
40 cm from base	76.6
End	73.4

EXAMPLE 6.5 Calculate the rate of heat transfer from the steel rod of Example 6.4 at the end of 6 minutes after the rod is exposed to the convection environment.

SOLUTION The rate of heat transfer from the rod at the end of 6 minutes after it is exposed to the convection environment can be calculated by summing the heat losses by convection from the surface of the rod. Thus

$$q = \sum_n \frac{T_n - T_\infty}{R_{n\infty}}$$

or

$$q = \frac{1}{2}\left(\frac{500 - 30}{7.96}\right) + \frac{(133.6 - 30) + (82.4 - 30) + (77.7 - 30) + (76.6 - 30)}{7.96}$$

$$+ \frac{(73.4 - 30)}{318.3} + \frac{(73.4 - 30)}{15.92} = 63.829 \text{ W}$$

EXAMPLE 6.6 A rectangular ceramic plate of size 10 cm × 5 cm is initially at a uniform temperature of 400°C. The two sides of 5 cm length are maintained at a constant temperature of 400°C, and one side of length 10 cm is insulated while the other side of 10 cm length is suddenly exposed to a convection environment at 25°C with h = 250 W/m^2 K. Determine the temperature distribution in the plate at the end of 2 minutes after it is exposed to the convection environment. Solve the problem using resistance–capacitance formulation and use $\Delta x = \Delta y = 2.5$ cm.

SOLUTION A schematic of the ceramic plate showing the nodes is presented in Fig. 6.10.

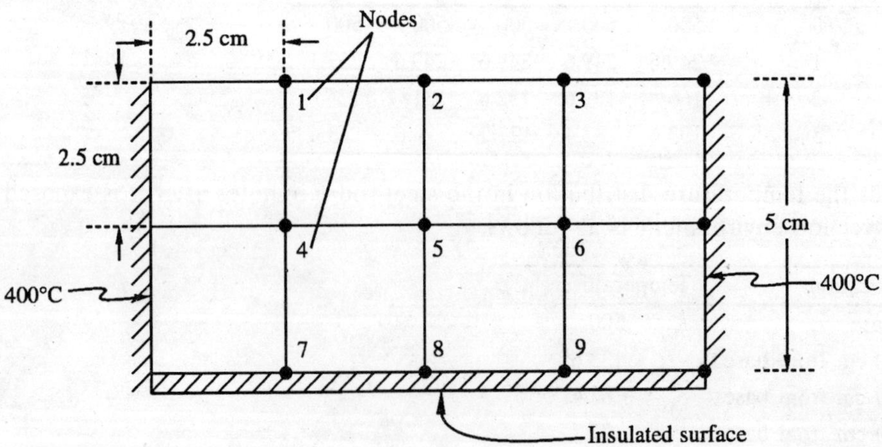

Fig. 6.10 Schematic of the nickel plate.

For ceramic, ρ= 1600 kg/m^3; c = 0.8 kJ/kg K and k = 3 W/m K.

Based on the symmetry of the nodes shown in Fig. 6.10, we find

$$T_1 = T_3; \quad T_4 = T_6; \quad T_7 = T_9$$

Therefore, we have to estimate the temperatures at nodes 1, 2, 4, 5, 7 and 8 only. Nodes 1 and 2 are convection boundary nodes.

$$R_{m+} = R_{03} = \frac{2}{k} = \frac{2}{3} = 0.6667 \text{ K/W}$$

$$R_{m-} = R_{02} = \frac{2}{k} = \frac{2}{3} = 0.6667 \text{ K/W}$$

$$R_{n+} = R_{0\infty} = \frac{1}{h\Delta x} = \frac{1}{250 \times 0.025} = 0.16 \text{ K/W}$$

$$R_{n-} = R_{01} = \frac{1}{k} = \frac{1}{3} = 0.3333 \text{ K/W}$$

$$\Delta V = \frac{(\Delta x)^2}{2} = \frac{(0.025)^2}{2} = 3.125 \times 10^{-4}$$

$$C_m = \rho c \Delta V = 1600 \times 0.8 \times 10^3 \times 3.125 \times 10^{-4} = 400 \text{ J/K}$$

Nodes 4 and 5 are interior nodes.

$$R_{m+} = R_{m-} = R_{n+} = R_{n-} = \frac{1}{k} = \frac{1}{3} = 0.3333 \text{ K/W}$$

$$\Delta V = (\Delta x)^2 = (0.025)^2 = 6.25 \times 10^{-4}$$

$$C_m = \rho c \Delta V = 1600 \times 0.8 \times 10^3 \times 6.25 \times 10^{-4} = 800 \text{ J/K}$$

Nodes 7 and 8 are insulated boundary nodes.

$$R_{m+} = R_{03} = \frac{2}{k} = \frac{2}{3} = 0.6667 \text{ K/W}$$

$$R_{m-} = R_{02} = \frac{2}{k} = \frac{2}{3} = 0.6667 \text{ K/W}$$

$$R_{n+} = R_{01} = \frac{1}{k} = \frac{1}{3} = 0.3333 \text{ K/W}$$

$$R_{n-} = \infty$$

$$\Delta V = \frac{(\Delta x)^2}{2} = \frac{(0.025)^2}{2} = 3.125 \times 10^{-4}$$

$$C_m = \rho c \Delta V = 400 \text{ J/K}$$

Node	$\sum_n \frac{1}{R_{mn}}$	C_m	$\frac{C_m}{\sum_n \frac{1}{R_{mn}}}$
1	12.25	400	32.65
2	12.25	400	32.65
4	12	800	66.67
5	12	800	66.67
7	6	400	66.67
8	6	400	66.67

The two nodes 1 and 2 are most restrictive. Therefore, we should choose Δt such that $\Delta t \leq 32.65$ s.

Let us choose $\Delta t = 30$ s.

Number of time increments $= \dfrac{120}{30} = 4$.

Then we calculate the temperatures at the nodes 1, 2, 4, 5, 7 and 8 using the relation

$$T_m^{p+1} = \frac{\Delta t}{C_m}\left(\sum \frac{T_n^p - T_m^p}{R_{mn}}\right) + T_m^p$$

$T^\circ = 400°$C for all the nodes.

$$T_1^1 = T_3^1 = \frac{30}{400}\left(\frac{25 - 400}{0.16}\right) + 400 = 224.22°\text{C}$$

$$T_2^1 = \frac{30}{400}\left(\frac{25 - 400}{0.16}\right) + 400 = 224.22°\text{C}$$

$$T_1^2 = T_3^2 = \frac{30}{400}\left(\frac{400 - 224.22}{0.6667} + \frac{400 - 224.22}{0.3333} + \frac{25 - 224.22}{0.16}\right) + 224.22$$

$$= 190.16°\text{C}$$

$$T_2^2 = \frac{30}{400}\left(\frac{400 - 224.22}{0.3333} + \frac{25 - 224.22}{0.16}\right) + 224.22 = 170.39°\text{C}$$

$$T_4^2 = T_6^2 = \frac{30}{800}\left(\frac{224.22 - 400}{0.3333}\right) + 400 = 380.22°\text{C}$$

$$T_5^2 = \frac{30}{800}\left(\frac{224.22 - 400}{0.3333}\right) + 400 = 380.22°\text{C}$$

Similar calculations are carried out for four time increments and the results are presented in the following table.

Time increment	Node temperature °C					
	1	2	4	5	7	8
0	400	400	400	400	400	400
1	224.22	224.22	400	400	400	400
2	190.16	170.39	380.22	380.22	400	400
3	176.89	153.90	363.28	358.84	391.74	391.74
4	170.15	144.77	350.07	341.41	386.27	384.34

The temperature distribution in the ceramic plate at the end of four time periods (2 minutes) after it is exposed to the convection environment is given below.

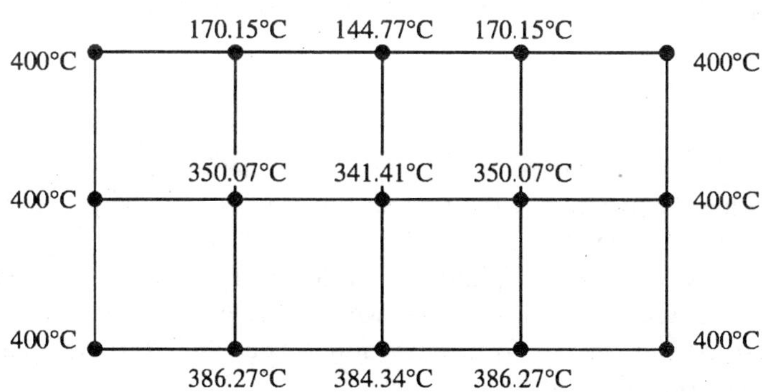

6.4 Graphical Method of Analysis

A graphical method of analysis for solving transient heat transfer problems is discussed in this section. This technique, known as *Schmidt method*, is flexible and yields approximate results for problems which cannot be handled conveniently by analytical methods.

6.4.1 Plane Wall with Constant Surface Temperature

Consider a thick plane wall initially at a uniform temperature T_i. Suppose one of the surfaces of the wall is suddenly changed to a temperature T_∞ and we are interested in determining the temperature distribution as a function of time. The differential equation which governs the transient heat conduction in a one-dimensional body is given by

$$\frac{\partial T}{\partial t} = \alpha \frac{\partial^2 T}{\partial x^2} \tag{6.34}$$

Equation (6.34) can be expressed in the finite difference form as

$$\frac{T_m^{p+1} - T_m^p}{\Delta t} = \alpha \left(\frac{T_{m+1}^p + T_{m-1}^p - 2T_m^p}{(\Delta x)^2} \right)$$

or

$$T_m^{p+1} = \frac{\alpha \Delta t}{(\Delta x)^2} \left(T_{m+1}^p + T_{m-1}^p \right) + \left[1 - \frac{2\alpha \Delta t}{(\Delta x)^2} \right] T_m^p$$

or

$$T_m^{p+1} = (Fo)_\Delta \left(T_{m+1}^p + T_{m-1}^p \right) + \left(1 - 2(Fo)_\Delta \right) T_m^p \tag{6.35}$$

If we choose $(Fo)_\Delta = 1/2$ for a one-dimensional situation, Eq. (6.35) reduces to

$$T_m^{p+1} = \frac{T_{m+1}^p + T_{m-1}^p}{2} \tag{6.36}$$

Equation (6.36) shows that the temperature at time increment $(p+1)$ for node m is given by the arithmetic average of the temperatures of the two adjacent nodes ($m+1$ and $m-1$) at time increment p.

In this method, we divide the plane wall into layers, each of thickness Δx, and label the planes between the layers as shown in Fig. 6.11. Then we plot the initial temperature distribution (at $t = 0$) at each node. The time increment Δt is fixed by the choice of Δx according to the criterion

$$(Fo)_\Delta = \frac{\alpha \Delta t}{(\Delta x)^2} = \frac{1}{2} \tag{6.37}$$

Then, if we draw a straight line connecting the temperatures at nodes $(m+1)$ and $(m-1)$ at time increment p, the point of intersection of this line with the plane gives the temperature at node m at time increment $(p+1)$. The graphical construction is illustrated in Fig. 6.11.

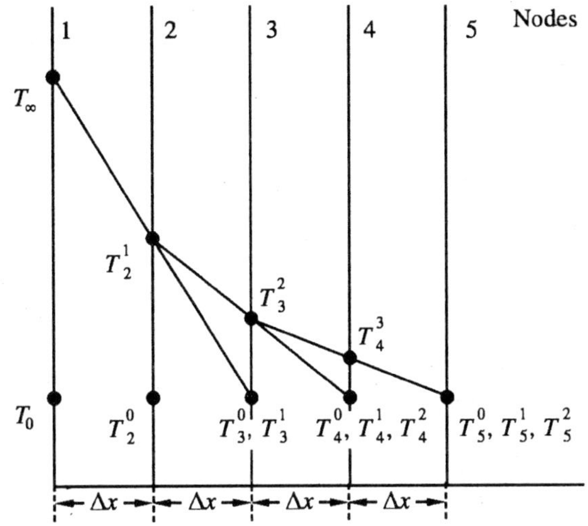

Fig. 6.11 Schmidt plot for transient heat conduction in a plane wall.

EXAMPLE 6.7 Rework Example 6.1 using the Schmidt method.

SOLUTION Let us choose Δx = 5 cm. Then the total number of layers is equal to 30/5 = 6. Let us denote the nodal planes as 1, 2, ..., 7. Plane 1 refers to the surface which is suddenly raised to 670°C and plane 7 refers to the insulated surface. For a one-dimensional problem

$$(Fo)_\Delta = \frac{1}{2} = \frac{\alpha \Delta t}{(\Delta x)^2} = \frac{0.5 \times 10^{-5} \Delta t}{(0.05)^2} \quad \text{or} \quad \Delta t = 250 \text{ s}$$

Therefore, the number of time increments $= \dfrac{25 \times 60}{250} = 6.$

The Schmidt plot, showing the details of the solution, is shown in Fig. 6.12.

It may be observed from Fig. 6.12 that the temperature at any nodal plane, separating the two adjacent layers, changes during alternate periods only. In this problem the temperature at surface 7 does not change for 5 time increments and starts changing in the sixth time increment only. Surface 7 is insulated and hence the boundary condition demands that the temperature gradient at surface 7 should be zero. Therefore, at the end of the sixth time interval (when the insulated surface temperature changes), the temperature of the insulated surface can be obtained by drawing a horizontal line from T_6^6 to surface 7.

The temperature distribution at the end of 6 time increments is given below.

Nodal point	1	2	3	4	5	6	7
Temperature (°C)	670	470	320	170	110	50	50

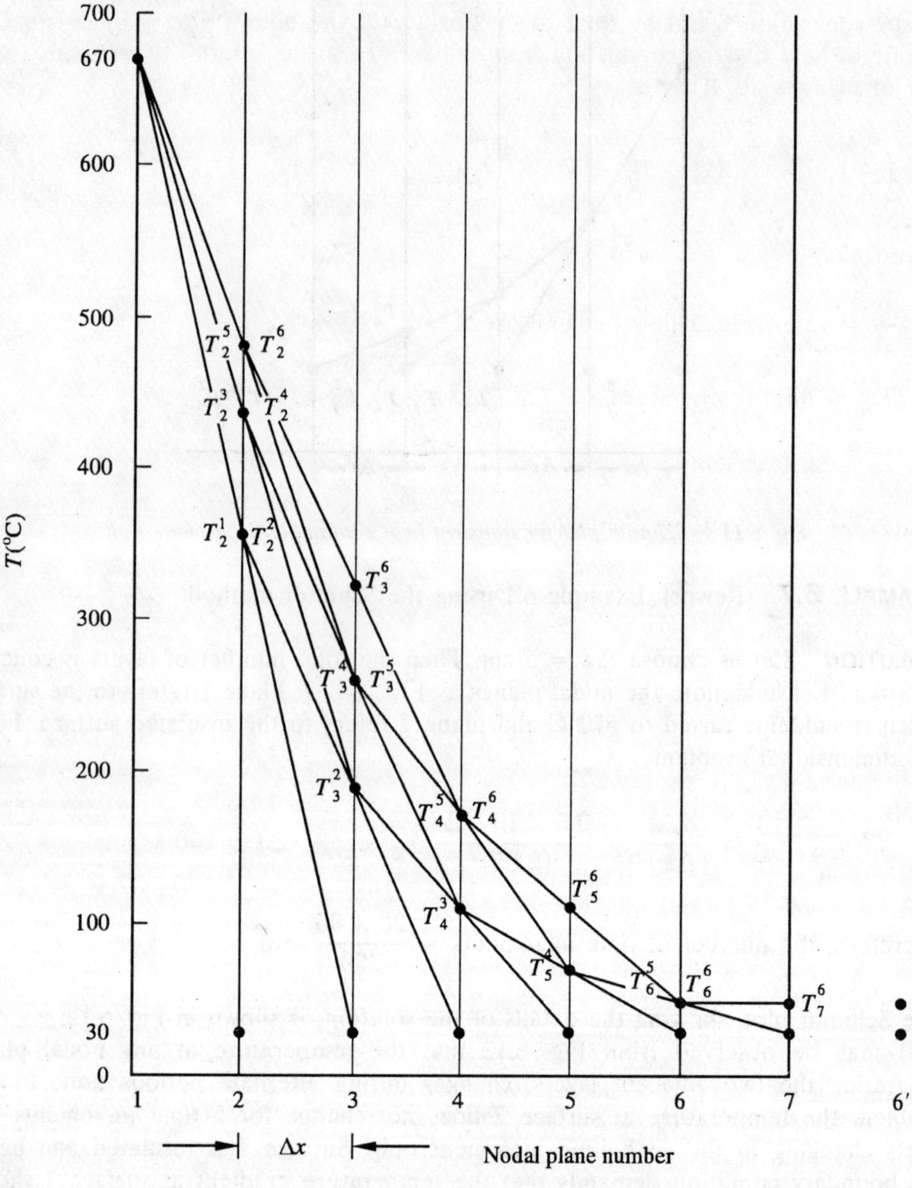

Fig. 6.12 Sketch for Example 6.7.

6.4.2 Plane Wall with Convection Boundary Condition

We have discussed the Schmidt method of analyzing transient heat conduction through a plane wall when the surface is suddenly raised to a different temperature. In practice, the surfaces are exposed to a convection environment and the wall will be in contact with a fluid. Then the boundary condition states that at any instance, the rate of heat flow by convection from the fluid to the surface of the body (at $x = 0$) is equal to the rate of heat flow by conduction from the surface to the interior of the body (next layer of plane wall). That is,

$$\left(\frac{q}{A}\right)_{x=0} = h(T_\infty - T_0) = -k\left(\frac{\partial T}{\partial x}\right)_{x=0} \tag{6.38}$$

where

$\frac{q}{A}$ = rate of heat flow per unit area

T_∞ = fluid temperature

T_0 = surface temperature of the plane wall (at $x = 0$).

Equation (6.38) can be rewritten in the finite difference form as

$$\left(\frac{\partial T}{\partial x}\right)_0^p = \frac{T_0^p - T_\infty}{(k/h)} \tag{6.39}$$

Equation (6.39) states that at any time $(p\Delta t)$, the temperature gradient at the surface $\left(\frac{\partial T}{\partial x}\right)_{x=0}^p$ is equal to the temperature difference between the surface and the fluid $(T_0^p - T_\infty)$ divided by (k/h). The quantity (k/h) has the dimensions of length. Therefore, in the Schmidt plot we can extend the solid by the distance (k/h) and represent the temperature variation by a continuous straight line between the temperature at plane 1, through the surface of the solid, to the temperature of the fluid T_∞ which is located at a distance (k/h) from the surface. The extended wall (which is fictitious and has length k/h) represents the thermal resistance between the surface and the fluid. The fictitious length k/h is not to be subdivided into layers or segments while constructing the Schmidt plot. The method of solution is illustrated in the following example.

EXAMPLE 6.8 A large concrete slab of thickness 30 cm is initially at 25°C. The slab is suddenly exposed to fire and the hot gases are at 900°C. If h = 13.7 W/m^2 K between the hot gases and the concrete surface, how long does it take for the centre of the slab to reach a temperature of 110°C? Use the Schmidt method to solve the problem. For concrete, $\alpha = 8 \times 10^{-7}$ m^2/s and k = 1.37 W/m K.

SOLUTION Let us divide the slab into 10 layers, each having a thickness of 3 cm. Because of the symmetry, only one half of the slab need be considered. As there is no heat transfer across the central line of the slab, it can be treated as an insulated surface ($\partial T/\partial x = 0$ at the centre). The thermal resistance at the surface is estimated by extending the slab thickness by a distance of k/h.

$$\frac{k}{h} = \frac{1.37}{13.7} = 0.1 \text{ m}$$

The steps of the graphical solution are shown in Fig. 6.13. Only one half of the slab with 5 layers is considered. The layer labelled 0 refers to the surface of the slab and the layer labelled 5 refers to the centre of the slab. To the left of the surface at a distance of 10 cm, the fluid temperature T_∞ is shown. First draw a straight line

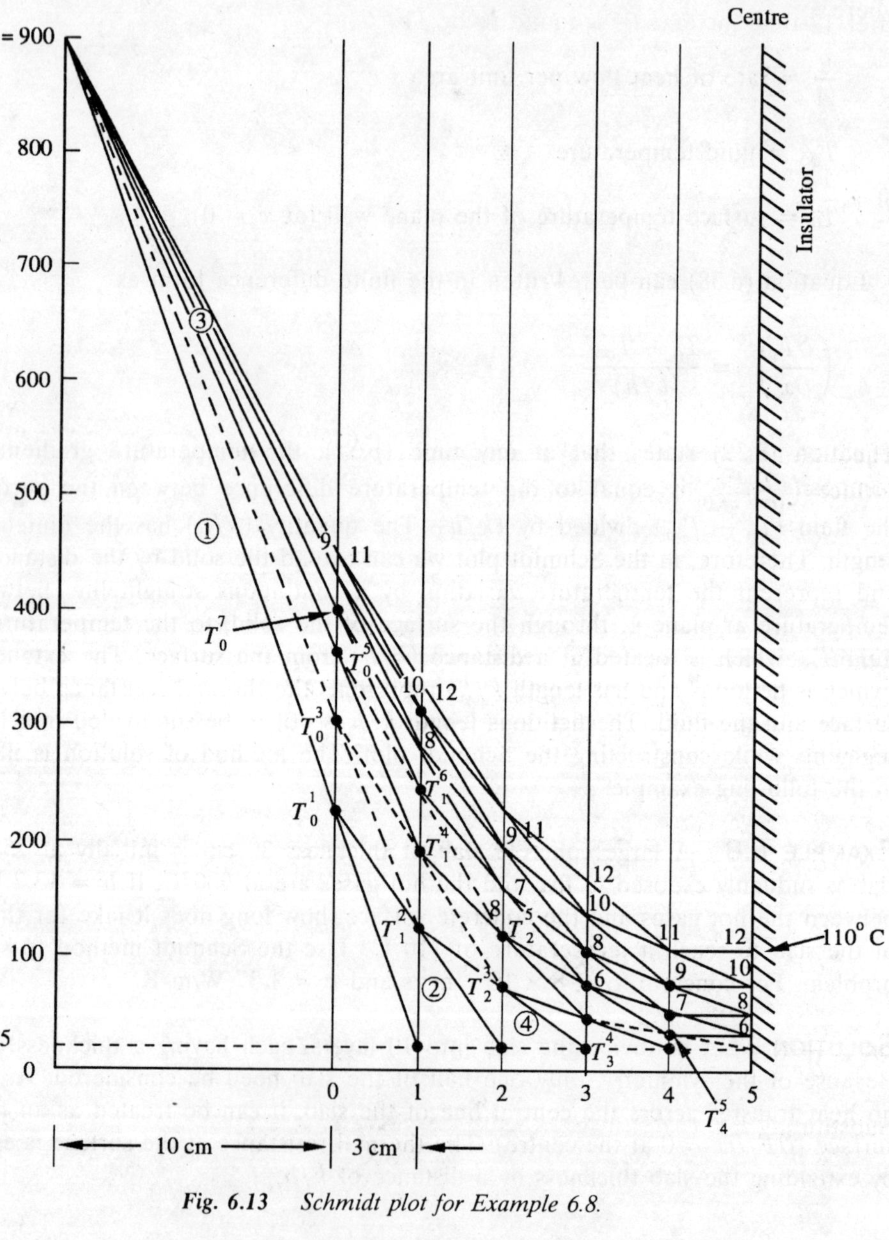

Fig. 6.13 Schmidt plot for Example 6.8.

connecting T_∞ (=900°C) and the initial temperature T_0^0 (=25°C) of the layer labelled 1. The intersection of this line with the layer 0 gives the temperature (225°C) of the surface after one time increment. The temperatures of all the other planes 1 to 5 remain at 25°C after one time increment. The temperature of plane 1 after 2 time increments is obtained by drawing a straight line connecting T_0^1 and T_2^1 and by determining the intersection of this line with plane 1. This gives T_1^2 = 125°C. This procedure is continued for successive time increments. The temperature at the centre after 12 time increments is found to be 110°C.

$$(Fo)_\Delta = \frac{1}{2} = \frac{\alpha \Delta t}{(\Delta x)^2}$$

or $$\frac{1}{2} = \frac{8 \times 10^{-7} \Delta t}{(0.03)^2} \quad \text{or} \quad \Delta t = 562.5 \text{ s}$$

Therefore, the time required for the centre to reach 110°C = $12 \times \Delta t = 12 \times 562.5 =$ 6750 s = 1.875 hours

6.4.3 Composite Wall

The Schmidt method can also be extended to analyze transient heat flow through composite walls. The differential equation for one-dimensional heat flow is given by

$$\frac{\partial T}{\partial t} = \alpha \frac{\partial^2 T}{\partial x^2} = \frac{k}{\rho c} \frac{\partial^2 T}{\partial x^2} \tag{6.40}$$

Let us define $\eta = \dfrac{x}{k}$ (6.41)

Then $d\eta = \dfrac{dx}{k}$

$$\frac{\partial T}{\partial x} = \frac{\partial T}{\partial \eta}\frac{\partial \eta}{\partial x} = \frac{1}{k}\frac{\partial T}{\partial \eta}$$

$$\frac{\partial^2 T}{\partial x^2} = \frac{\partial}{\partial x}\left(\frac{\partial T}{\partial x}\right) = \frac{\partial}{\partial \eta}\left(\frac{\partial T}{\partial x}\right)\frac{\partial \eta}{\partial x} = \frac{\partial}{\partial \eta}\left(\frac{1}{k}\frac{\partial T}{\partial \eta}\right)\frac{1}{k}$$

or $$\frac{\partial^2 T}{\partial x^2} = \frac{1}{k^2}\frac{\partial^2 T}{\partial \eta^2} \tag{6.42}$$

Substituting Eq. (6.42) in Eq. (6.40), we get

$$\frac{\partial T}{\partial t} = \frac{1}{k\rho c}\frac{\partial^2 T}{\partial \eta^2} \tag{6.43}$$

In the finite difference form, Eq. (6.43) can be rewritten as

$$\frac{T_m^{p+1} - T_m^p}{\Delta t} = \frac{1}{k\rho c}\left[\frac{T_{m+1}^p + T_{m-1}^p - 2T_m^p}{(\Delta\eta)^2}\right]$$

or

$$T_m^{p+1} = \frac{\Delta t}{k\rho c(\Delta\eta)^2}(T_{m+1}^p + T_{m-1}^p) + \left(1 - \frac{2\Delta t}{k\rho c(\Delta\eta)^2}\right)T_m^p$$

$$= \frac{\alpha\Delta t}{k^2(\Delta\eta)^2}(T_{m+1}^p + T_{m-1}^p) + \left(1 - \frac{2\alpha\Delta t}{k^2(\Delta\eta)^2}\right)T_m^p$$

or

$$T_m^{p+1} = (Fo)_\Delta\left(T_{m+1}^p + T_{m-1}^p\right) + (1 - 2(Fo)_\Delta)T_m^p \tag{6.44}$$

where

$$(Fo)_\Delta = \frac{\alpha\Delta t}{k^2(\Delta\eta)^2} = \frac{\alpha\Delta t}{(\Delta x)^2} \tag{6.45}$$

Now, let us consider a composite wall consisting of slab I with thermal conductivity k_1 and thickness x_1 and slab II of thickness x_2 with thermal conductivity k_2. At the interface of the two slabs, the following condition must be satisfied:

$$-k_1\frac{\partial T}{\partial x_1} = -k_2\frac{\partial T}{\partial x_2} \tag{6.46}$$

If $k_1 \neq k_2$, there will be a discontinuity in the temperature gradient. In the graphical method, we should not have any discontinuity at the interface. This can be avoided by changing the variable from x to η. Equation (6.46) can be rewritten as

$$\frac{\partial T}{(\partial x_1/k_1)} = \frac{\partial T}{(\partial x_2/k_2)} \quad \text{or} \quad \frac{\partial T}{\partial\eta_1} = \frac{\partial T}{\partial\eta_2} \tag{6.47}$$

or

$$\left(\frac{\Delta T}{\Delta\eta}\right)_I = \left(\frac{\Delta T}{\Delta\eta}\right)_{II} \tag{6.48}$$

Therefore, to apply the Schmidt method for a composite wall, we divide each slab into equal increments of $\Delta\eta$. Then we choose Δt such that

$$(Fo)_\Delta = \frac{1}{2} \quad \text{or} \quad \Delta t = \frac{k^2(\Delta\eta)^2}{2\alpha} = \frac{(\Delta x)^2}{2\alpha} \tag{6.49}$$

With this choice of $(Fo)_\Delta$, Eq. (6.44) reduces to

$$T_m^{p+1} = \frac{T_{m+1}^p + T_{m-1}^p}{2} \tag{6.50}$$

which is identical to the relation for a plane wall of homogeneous material.

In order that the time increments (Δt) are the same in both the materials, the condition to be satisfied is

$$(\Delta t)_I = (\Delta t)_{II} \quad \text{or} \quad \frac{k_1^2(\Delta\eta_1)^2}{\alpha_1} = \frac{k_2^2(\Delta\eta_2)^2}{\alpha_2}$$

or

$$\frac{k_1\Delta\eta_1}{k_2\Delta\eta_2} = \frac{\Delta x_1}{\Delta x_2} = \sqrt{\frac{\alpha_1}{\alpha_2}} \tag{6.51}$$

In applying the Schmidt method to a composite wall, we change the variable from x to η, and in the plot η will be taken as the length variable. Then we decide on the widths of the layers such that Eq. (6.51) is satisfied. The method of construction of the plot is illustrated in the following example.

EXAMPLE 6.9 A composite wall for a furnace is made of two different materials. The inside wall of thickness 25 cm is made of fireclay brick (k = 1 W/m K and $\alpha = 5 \times 10^{-7}$ m^2/s) while the outer wall of thickness 18.75 cm is made of concrete (k = 1.56 W/m K and $\alpha = 7.81 \times 10^{-7}$m^2/s). The composite wall is initially at 25°C. The inner surface of the fireclay brick is suddenly raised to a temperature of 900°C, while the outer surface, of the concrete is insulated. How long can the furnace be operated before the outer concrete wall reaches a temperature of 45°C ? Determine the temperature distribution in the wall at that time. Use the Schmidt method to solve the problem.

SOLUTION Let us divide the fireclay brick wall into 5 equal layers. Then Δx_2 = 25/5 = 5 cm = thickness of each fireclay brick layer. We know that

$$\frac{\Delta x_1}{\Delta x_2} = \sqrt{\frac{\alpha_1}{\alpha_2}} = \sqrt{\frac{7.81 \times 10^{-7}}{5 \times 10^{-7}}} = 1.25$$

Then $\Delta x_1 = 1.25, \Delta x_2 = 1.25 \times 5 = 6.25$ cm = thickness of each concrete layer.

$$\text{Number of concrete layers} = \frac{18.75}{6.25} = 3.$$

$$\text{Actual thickness of fireclay brick lining in the plot} = \frac{x_2}{k_2} = \frac{0.25}{1} = 0.25 \text{ m} = 25 \text{ cm}$$

Actual thickness of concrete lining in the plot $= \dfrac{x_1}{k_1}$

$$= \frac{18.75 \times 10^{-2}}{1.56} = 0.12 \text{ m} = 12 \text{ cm}$$

Therefore, we choose 25 units thick fireclay and divide it into 5 equal parts and 12 units thick concrete and divide it into 3 equal parts in the Schmidt plot as shown in Fig. 6.14. We calculate Δt from the relation

$$(Fo)_\Delta = \frac{1}{2} = \frac{\alpha \Delta t}{(\Delta x)^2} \quad \text{or} \quad \Delta t = \frac{(\Delta x)^2}{2\alpha} = \frac{(0.05)^2}{2 \times 5 \times 10^{-7}} = 2500 \text{ s}$$

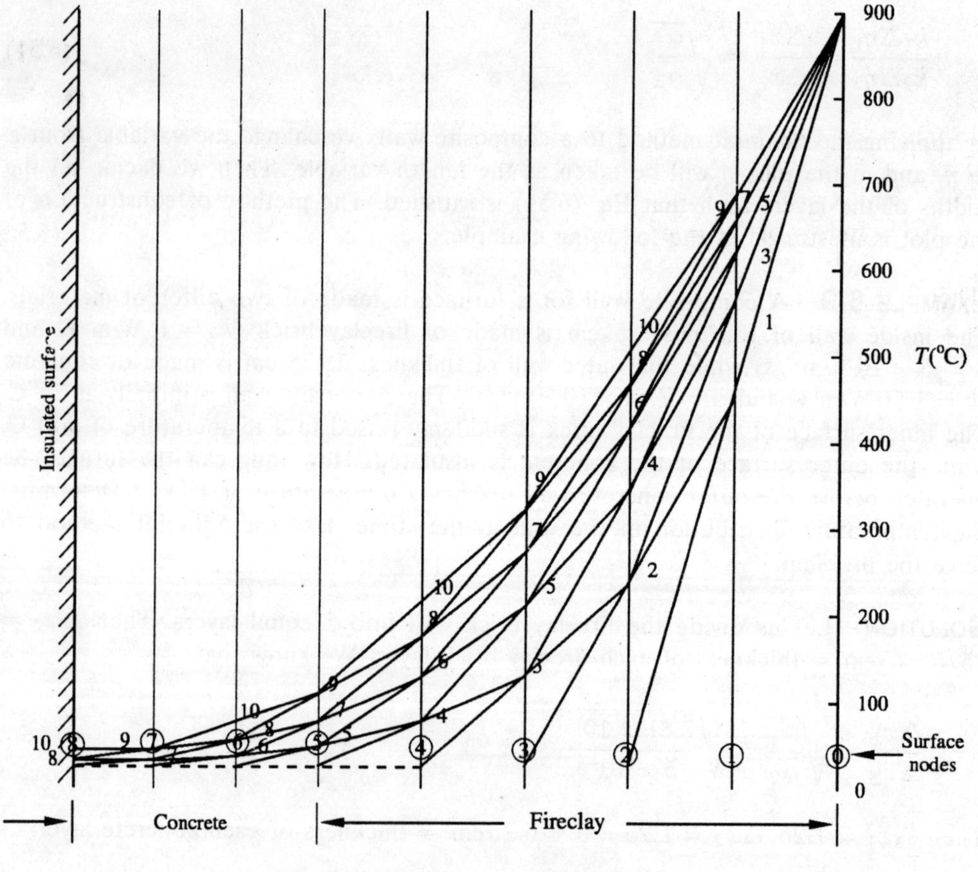

Fig. 6.14 Schmidt plot for Example 6.9.

From Fig. 6.14 we find that the outer surface of concrete lining (nodal surface no.8) reaches a temperature of 45°C after 10 time increments. Therefore, total time of operation is 10 × 2500 = 25000 s = 6.944 hours.

The temperature distribution in the composite wall at that time is as follows:

Surface node	0	1	2	3	4	5	6	7	8
Distance from inner surface(cm)	0	5	10	15	20	25	31.25	37.5	43.75
Temperature (°C)	900	685	505	325	215	110	80	45	45

6.4.4 Long Cylinder

The Schmidt method can also be extended to long solid or hollow cylinders. The differential equation describing the temperature distribution in a long cylinder is given by

$$\frac{\partial T}{\partial t} = \alpha\left(\frac{\partial^2 T}{\partial r^2} + \frac{1}{r}\frac{\partial T}{\partial r}\right) \tag{6.52}$$

To transform Eq. (6.52) into a finite difference equation, let us define a new variable β as

$$\beta = \ln r \quad \text{or} \quad r = e^{\beta} \tag{6.53}$$

Then

$$\frac{d\beta}{dr} = \frac{1}{r} = e^{-\beta} \tag{6.54}$$

$$\frac{\partial T}{\partial r} = \frac{\partial T}{\partial \beta}\frac{d\beta}{dr} = \frac{1}{r}\frac{\partial T}{\partial \beta} \tag{6.55}$$

$$\frac{\partial^2 T}{\partial r^2} = \frac{\partial}{\partial r}\left(\frac{\partial T}{\partial r}\right) = \frac{\partial}{\partial r}\left(\frac{1}{r}\frac{\partial T}{\partial \beta}\right) = \frac{\partial}{\partial \beta}\left(\frac{1}{r}\frac{\partial T}{\partial \beta}\right)\frac{d\beta}{dr}$$

or

$$\frac{\partial^2 T}{\partial r^2} = \left[\frac{1}{r}\frac{\partial^2 T}{\partial \beta^2} + \frac{\partial T}{\partial \beta}\frac{\partial(\frac{1}{r})}{\partial \beta}\right]\frac{1}{r} \tag{6.56}$$

$$\frac{\partial(\frac{1}{r})}{\partial \beta} = \frac{\partial(e^{-\beta})}{\partial \beta} = -e^{-\beta} = -\frac{1}{r} \tag{6.57}$$

Substituting Eq. (6.57) in Eq. (6.56), we get

$$\frac{\partial^2 T}{\partial r^2} = \frac{1}{r^2}\frac{\partial^2 T}{\partial \beta^2} - \frac{1}{r^2}\frac{\partial T}{\partial \beta} \tag{6.58}$$

Substituting Eqs. (6.55) and (6.58) in Eq. (6.52) we obtain

$$\frac{\partial T}{\partial t} = \frac{\alpha}{r^2}\frac{\partial^2 T}{\partial \beta^2} \tag{6.59}$$

Equation (6.59) can be written in the finite difference form as

$$\frac{T_m^{p+1} - T_m^p}{\Delta t} = \frac{\alpha}{r^2}\left(\frac{T_{m+1}^p + T_{m-1}^p - 2T_m^p}{(\Delta\beta)^2}\right)$$

or

$$T_m^{p+1} = \frac{\alpha\Delta t}{(r\Delta\beta)^2}\left(T_{m+1}^p + T_{m-1}^p\right) + \left(1 - \frac{2\alpha\Delta t}{(r\Delta\beta)^2}\right)T_m^p$$

or

$$T_m^{p+1} = (Fo)_\Delta\left(T_{m+1}^p + T_{m-1}^p\right) + (1 - 2(Fo)_\Delta)T_m^p \tag{6.60}$$

where

$$(Fo)_\Delta = \frac{\alpha\Delta t}{(r\Delta\beta)^2} \tag{6.61}$$

If we choose

$$(Fo)_\Delta = \frac{1}{2} \tag{6.62}$$

Eq. (6.44) reduces to

$$T_m^{p+1} = \frac{T_{m+1}^p + T_{m-1}^p}{2} \tag{6.63}$$

which is identical to the expression used for a flat plate. The value of Δt and $\Delta\beta$ are to be chosen such that

$$(Fo)_\Delta = \frac{\alpha\Delta t}{(r\Delta\beta)^2} = \frac{1}{2} \tag{6.64}$$

In applying the Schmidt method for a slab, we divided the slab into layers of equal thickness Δx. In the case of a cylinder, we divide the thickness into equal increments or $r\Delta\beta$. From Eq. (6.54), we find $\Delta\beta = \Delta r/r$. Hence, the cylinder is to be partitioned into several layers, and the thickness of a layer is $\Delta r/r$. If the cylinder is not at uniform temperature initially, then the temperature distribution must be plotted on a scale corresponding to $\Delta\beta$ increment in accordance with Eq. (6.53). The application of the Schmidt method to a cylinder is illustrated in the following example.

EXAMPLE 6.10 A fluid at 300°C is flowing through a thick pipe of 6 cm ID and 16 cm OD. The outer surface of the pipe is at 100°C under steady-state conditions. Suddenly the fluid is heated to 900°C. Determine the temperature distribution in the pipe 30 seconds after the fluid is heated, if the convection heat transfer coefficient between the fluid and pipe is 500 W/m² K, and for the pipe material k = 30 W/m K and $\alpha = 1 \times 10^{-5}\,\text{m}^2/\text{s}$.

SOLUTION Let us divide the pipe into 5 annular sections of $\Delta r = 1$ cm. Then the corresponding values of $\Delta\beta$ are calculated and presented below.

Section	r_i(cm)	r_0(cm)	r_{mean}(cm)	$\Delta\beta = (\Delta r)/r$
1	3	4	3.5	0.2857
2	4	5	4.5	0.2222
3	5	6	5.5	0.1818
4	6	7	6.5	0.1538
5	7	8	7.5	0.1333
				$\sum \Delta\beta = 0.9768$

The boundary condition at the inner surface is given by

$$q = h(2\pi r_i)(T_\infty - T_i) = 2\pi r_i k \left(\frac{\partial T}{\partial r}\right)_{r=r_i}$$

or $$h(T_\infty - T_i) = k\left(\frac{\partial T}{\partial \beta}\frac{d\beta}{dr}\right)_{r=r_i} = \frac{k}{r_i}\left(\frac{\partial T}{\partial \beta}\right)$$

or $$\frac{\partial T}{\partial \beta} = \frac{T_\infty - T_i}{(k/hr_i)}$$

$\dfrac{k}{hr_i}$ = fictitious distance of the pipe wall to take into account the convection boundary condition.

$$\frac{k}{hr_i} = \frac{30}{500 \times 0.03} = 2$$

The time increment is calculated from the condition

$$(Fo)_\Delta = \frac{\alpha\Delta t}{(r\Delta\beta)^2} = \frac{\alpha\Delta t}{(\Delta r)^2} = \frac{1}{2} \quad \text{or} \quad \frac{1 \times 10^{-5}\Delta t}{(0.01)^2} = \frac{1}{2} \quad \text{or} \quad \Delta t = 5 \text{ s}$$

The Schmidt plot for this example is shown in Fig. 6.15. The temperature distribution in the pipe wall is given below:

r (cm)	3	4	5	6	7	8
T (°C)	290	215	160	130	110	100

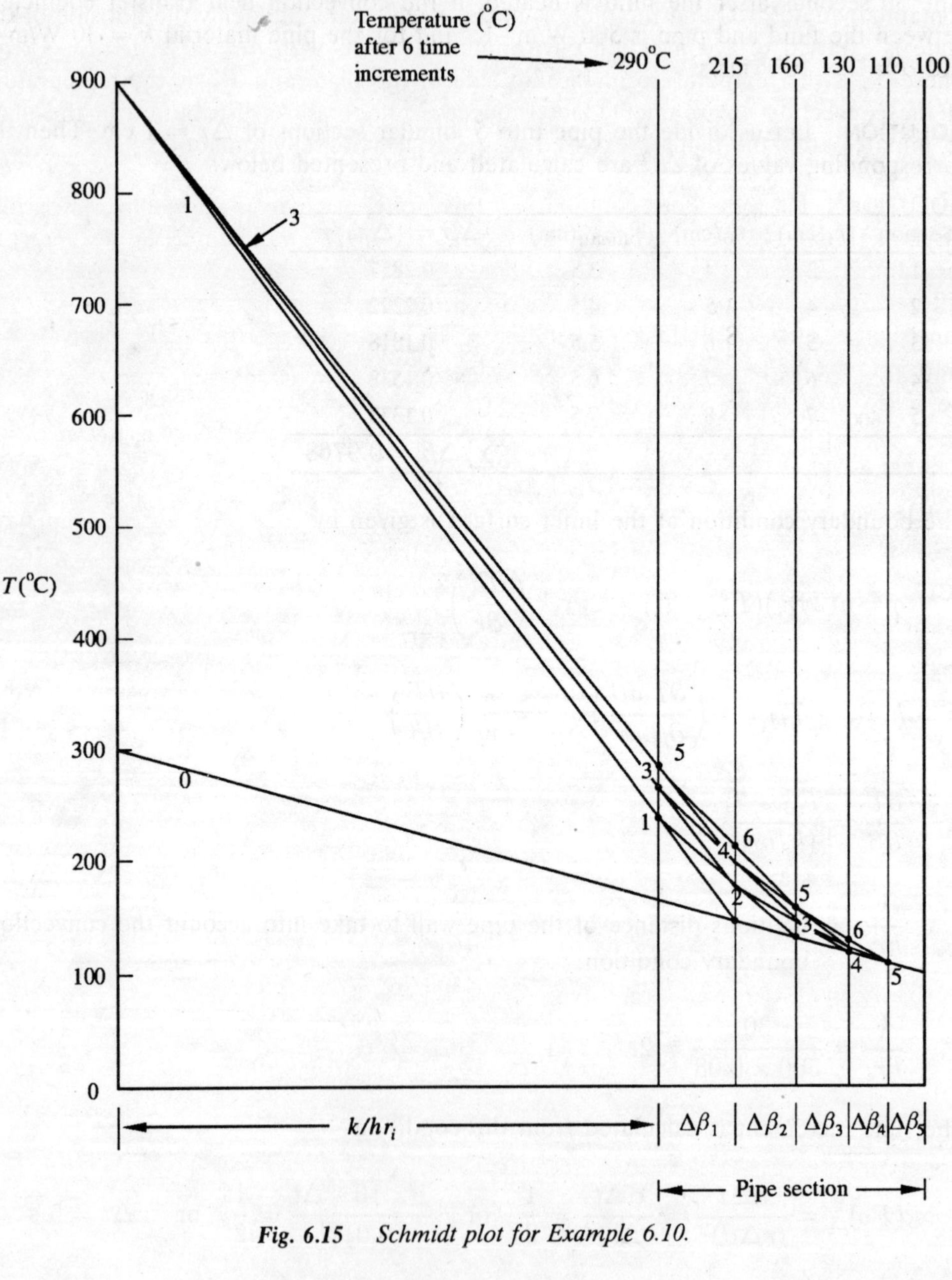

Fig. 6.15 Schmidt plot for Example 6.10.

6.4.5 Time Dependent Boundary Conditions

The Schmidt method can also be applied if the boundary condition is time dependent. The method of application is illustrated through the following example.

EXAMPLE 6.11 One surface of a large steel plate ($k = 20$ W/m K and $\alpha = 0.5 \times 10^{-5}$ m^2/s) of 30 cm thickness is maintained at 600°C, while the second surface is maintained at 25°C and the plate is in the steady-state. Suddenly the hot side of the plate is exposed to a convection environment at 25°C with $h = 50 + 0.1(T_0 - T_\infty)$, where T_0 is the temperature of the hot side of the plate and T_∞ is the temperature of the convection environment. Determine the temperature distribution in the plate 25 minutes after the plate is exposed to the convection environment.

SOLUTION The boundary condition in this problem changes with time, since h is varying. Hence we should change the thermal resistance with time. We shall record the temperature of the hot surface, the value of h and the surface resistance for different time increments in a tabular form (as shown on the next page) while the construction of the Schmidt plot is in progress. Let us arbitrarily decide to divide the plate into six layers of $\Delta x = 5$ cm each.

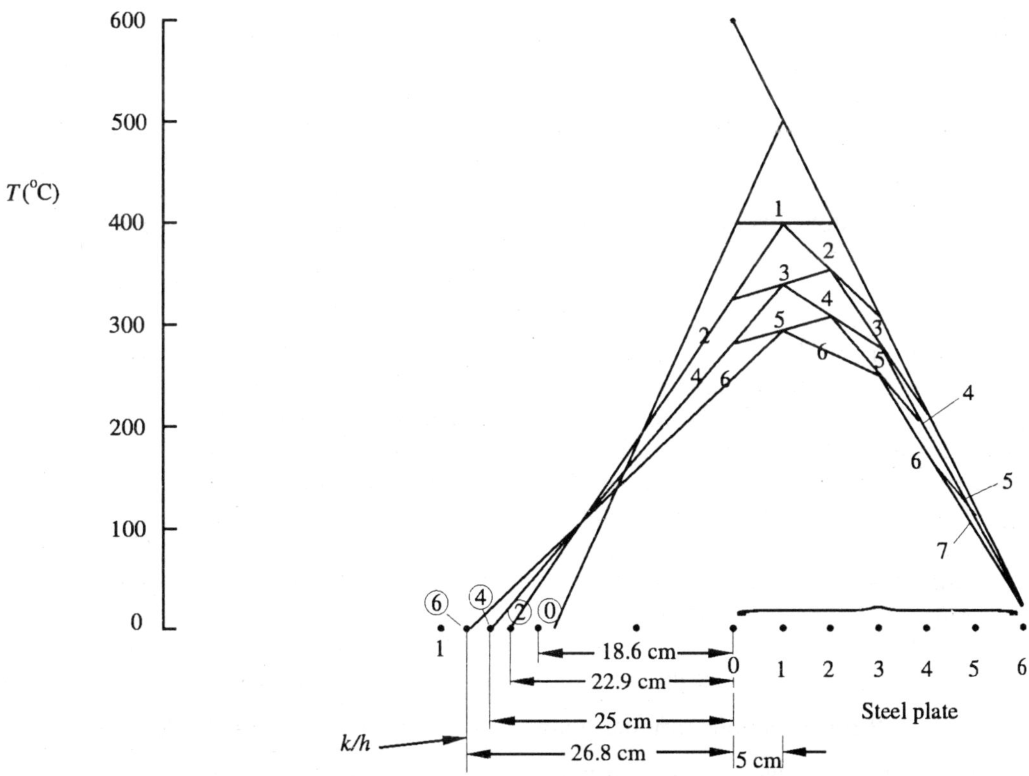

Fig. 6.16 Schmidt plot for Example 6.11.

Then $(Fo)_{\Delta} = \dfrac{\alpha \Delta t}{(\Delta x)^2} = \dfrac{1}{2}$ or $\dfrac{0.5 \times 10^{-5} \Delta t}{(0.05)^2} = \dfrac{1}{2}$ or $\Delta t = 250$ s

Δt	T_0	$T_0 - T_\infty$	h	k/h(cm)
0	600	575	107.5	18.6
2	400	375	87.5	22.9
4	325	300	80	25
6	270	245	74.5	26.8

The temperature distribution in the plate after 6 time intervals (25 minutes) as read from the Schmidt plot shown in Fig. 6.16 is given below.

x (cm)	0	5	10	15	20	25	30
T(°C)	250	295	275	250	175	115	25

6.5 Summary

Transient heat flow problems, in some simple cases, can be solved analytically. The analytical solutions for a majority of problems deal with infinite series. The results of analytical solutions for simple geometries are available in the form of Heisler charts. Several practical problems dealing with irregular and complex geometries cannot be analyzed using Heisler charts. At times it may not be possible to find analytical solutions for transient heat flow problems in complex geometries. In order to circumvent the mathematical difficulty, numerical and graphical methods have been developed to obtain approximate solutions.

The widely used numerical methods of analysis replace the differential equation by a finite difference equation. If the differential increments in the space coordinates and time, in the differential equation which governs the temperature distribution in a body, are replaced by finite differences, the differential equation reduces to a finite difference equation. The finite difference equation can be written by using either the forward difference estimate or the backward difference estimate. The forward difference estimate results in an explicit equation or explicit method. The explicit method provides a means of estimating the temperature at a node at a later time increment from a knowledge of the temperatures of the surrounding nodes at the present time. The size of the network chosen for a particular problem affects the accuracy of the estimated result.

The choice of the time increment is governed by the stability criteria. The stability criteria for conduction is given by

$Fo \leq 1/2$ for a one-dimensional body

$Fo \leq 1/4$ for a two-dimensional body

$Fo \leq 1/6$ for a three-dimensional body

If the body is exposed to a convection environment, the stability criteria is given by

$$Fo = \frac{1}{2(1+Bi)} \quad \text{for a one-dimensional body}$$

$$Fo = \frac{1}{2(2+Bi)} \quad \text{for a two-dimensional body}$$

An useful alternative to the formulation for the nodal equations is to use the concepts of thermal resistance and thermal capacitance. The thermal capacitance of a node m is given by

$$C_m = \rho_m c_m \Delta V_m$$

where

C_m = thermal capacitance of node m

ρ_m = density of the material

c_m = heat capacity of the material

ΔV_m = volume surrounding the node m.

The stability criteria demands that

$$1 - \frac{\Delta t}{C_m} \sum_n \frac{1}{R_{mn}} \geq 0$$

where R_{mn} is the thermal resistance between the nodes m and n. The resistance–capacitance formulation is usually applied to problems dealing with complex geometries and boundary conditions. This formulation is quite useful if the thermal properties of the material are temperature dependent. The time increment Δt is chosen such that its value is less than the value obtained from the most restrictive nodal relation. That is,

$$\Delta t \leq \left[\frac{C_m}{\sum \frac{1}{R_{mn}}} \right]_{\min}$$

The Schmidt method or the graphical method of analysis for solving transient heat transfer problems is flexible and yields approximate values for problems which cannot be conveniently handled by analytical methods. The Schmidt plots for a plane wall with constant surface temperature, with convection boundary conditions, composite walls, long cylinders and for time dependent boundary conditions are discussed.

Review Questions

1. What are the drawbacks associated with the analytical solutions of transient heat flow problems?
2. What are the alternative methods of analyzing transient heat flow problems?
3. What is the difference between forward difference and backward difference estimates ?
4. Define the network Fourier modulus.
5. What is the difference between explicit and implicit formulations ?
6. What is the advantage in using explicit formulation?
7. How does the choice of the space coordinate increment affect the time increment Δt in using the explicit nodal equation?
8. What is meant by stability criteria?
9. Define the network Biot modulus.
10. What are the stability criteria for one-, two- and three-dimensional heat flow by conduction?
11. What are the stability criteria for one- and two-dimensional heat flow with convection boundary conditions?
12. How do you modify a nodal equation which is valid for a convection boundary condition, so that it is applicable for an insulated boundary?
13. What is meant by thermal resistance–capacitance formulation?
14. Define the thermal capacitance of a node.
15. Name the stability criteria used in resistance–capacitance formulation.
16. What are the advantages of the Schmidt method over the analytical solution in solving transient heat flow problems?
17. Describe the Schmidt method for analyzing transient heat flow through a plane, one surface of which is maintained at a constant temperature.
18. How do you modify the Schmidt plot if the surface of a plane wall is exposed to a convection environment?
19. Describe the Schmidt method for analyzing transient heat conduction through a composite wall.
20. How do you modify the space coordinates in the Schmidt plot for a composite wall to avoid discontinuity in the temperature gradient at the interface of two different materials?
21. Describe the Schmidt plot for analyzing transient heat transfer through the wall of a long cylinder.
22. In a Schmidt plot, how do you take into account the time dependent boundary condition?

Problems

6.1 A large iron (α = 2.034 × 10^{-5} m^2/s) slab of 30 cm thickness is initially at 30°C. Suddenly one face of the slab is raised to a temperature of 200°C and held at that temperature, while the other face is insulated. Using the forward difference technique, estimate the temperature distribution in the iron slab ten minutes after its surface temperature is raised.

6.2 A large oven in a bakery is to be insulated with asbestos. Initially the oven is at a temperature of 30°C. At time t = 0, the inside surface of the oven wall is suddenly raised to a temperature of 200°C and maintained at that temperature. It is required to determine the thickness of insulation so that the outer surface of the wall does not reach 45°C for 2 hours after the inside surface of the oven wall is raised to 200°C. Assume that the oven wall is an infinitely long flat plate and solve the problem using the transient numerical method.

6.3 A large chrome-steel slab of 30 cm thickness is initially at 30°C. Suddenly both the surfaces of the slab are raised to 800°C and subsequently maintained at that temperature. Determine the temperature distribution in the slab 4 minutes after the surfaces are raised to 800°C. Use the forward difference technique to solve the problem.

6.4 Consider an interior node in a two-dimensional body and develop the backward difference equation which can be used to solve the transient heat conduction problem for the plate.

6.5 Develop the explicit nodal equation for a node on the boundary of a rectangular plate which is exposed to a convection environment. Also obtain the stability criterion for the nodal equation.

6.6 Consider a node at the exterior corner of a two-dimensional body. If the surfaces of the body are exposed to a convection environment, derive the explicit nodal equation and the stability criterion for the nodal equation.

6.7 Consider the nodes 1,2,3 and 4 shown in Fig. 6.17. Initially the fireclay brick is at 25°C. Suddenly the surfaces of the brick are exposed to a convection environment at 300°C with h = 50 W/m^2 K. Estimate the maximum permissible Δt which can be used in a transient numerical analysis.

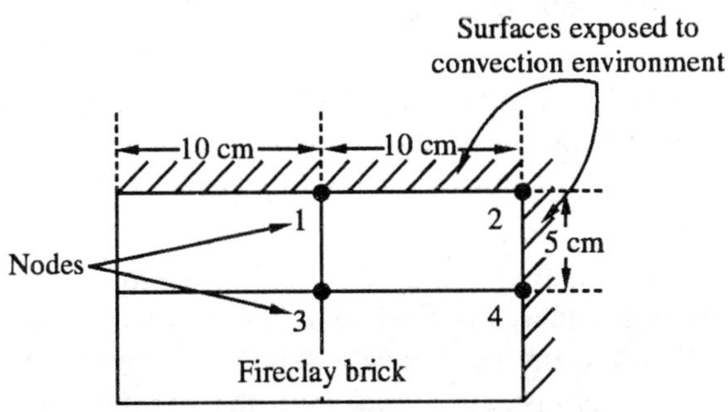

Fig. 6.17 Schematic of a fireclay brick.

6.8 Develop the explicit nodal equation for the transient numerical analysis for the node (m, n) shown in Fig. 6.18.

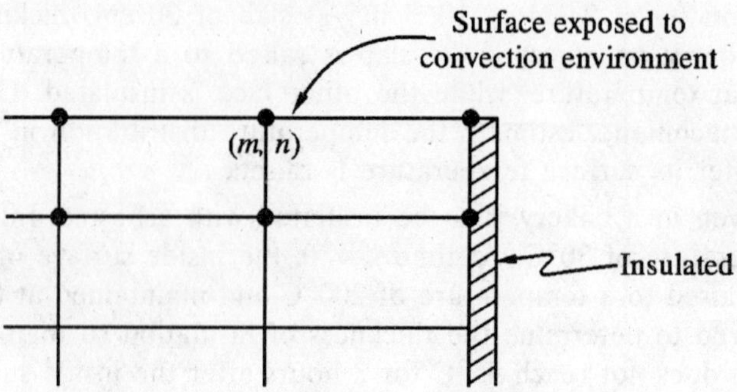

Fig. 6.18 Sketch for Problem 6.8.

6.9 Consider a one-dimensional transient heat conduction through an infinitely long plate which is exposed to a convection environment. Consider a node on the surface and develop the explicit nodal equation for the transient numerical analysis. Also determine the stability criterion.

6.10 If the two surfaces of the chrome-steel slab of Problem 6.3 are suddenly exposed to a convection environment at 800°C with h = 200 W/m^2 K, determine the temperature distribution in the slab 4 minutes after the surfaces are exposed to the convection environment.

6.11 A 10 cm thick resin (ρ = 2500 kg/m^3; c = 0.8 kJ/kg K and k = 1 W/m K) slab is to be cured by exposing its top surface to a stream of hot air at 100°C with h = 40 W/m^2 K, while the bottom surface is insulated. If the resin slab is initially at 30°C, determine the temperature distribution in the resin slab 1 hour after it is exposed to the stream of hot air. Solve the problem using the forward difference technique of numerical analysis.

6.12 A large iron plate 30 cm × 30 cm in cross-section is initially at a uniform temperature of 30°C. Two adjacent faces of the plate are insulated. The other two adjacent faces are suddenly raised to a temperature of 350°C. Determine the temperature distribution in the plate 15 minutes after the surfaces are raised to 350°C. Solve the problem using explicit nodal equations.

6.13 A nickel-steel plate shown in Fig. 6.19 is initially at a uniform temperature of 800°C. At time t = 0, the surfaces are changed to the temperatures as shown in the figure and held constant at those values. Calculate the temperatures at the nodes 1, 2, 3 and 4, eight minutes after the surfaces are changed to new values of temperatures.

6.14 A chrome-steel plate shown in Fig. 6.20 is initially at a uniform temperature of 30°C. The plate is suddenly exposed to a convection environment at 800°C with h = 50 W/m^2 K, while the right face is held at a constant temperature of 30°C and the bottom face is insulated. Calculate the temperatures at all the nodes 3 minutes after the plate is exposed to the convection environment.

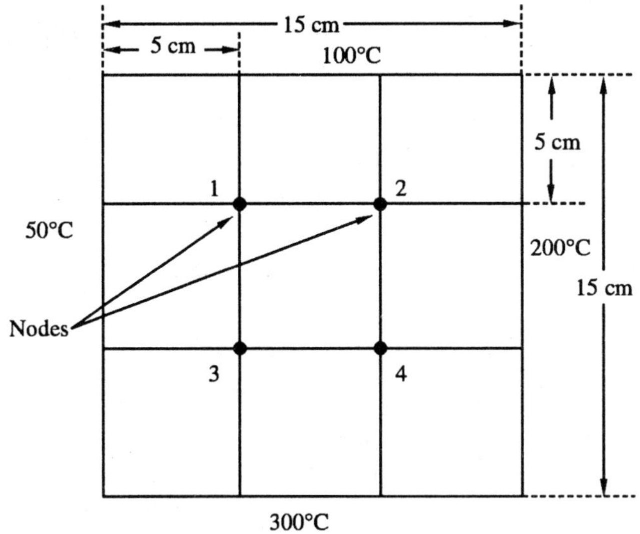

Fig. 6.19 Sketch for Problem 6.13.

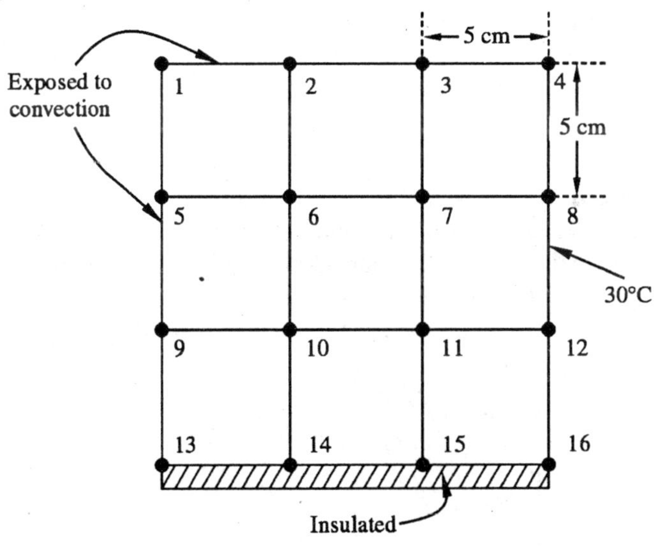

Fig. 6.20 Sketch for Problem 6.14.

6.15 An aluminium rod of 6 mm diameter and 15 cm length is used as a pin fin. The base of the fin is at 350°C. Initially the entire fin is at a uniform temperature of 350°C. If the fin is suddenly exposed to a convection environment at 30°C with h = 20 W/m^2 K, while the base of the fin is held at a constant temperature of 350°C, determine the temperature distribution in the fin 1 minute after it is exposed to the convection environment. Use the thermal resistance–capacitance formulation to solve the problem.

6.16 Calculate the rate of heat transfer from the fin of Problem 6.15 at the end of 1 minute.

6.17 A chrome-steel rod 25 mm in diameter and 30 cm in length has one end attached to a thermal reservoir at 300°C. The rod is initially at a uniform temperature of 300°C. It is then suddenly exposed to an air stream at 30°C with h = 20 W/m^2 K. Estimate the time required for the temperature of the rod at a distance of 15 cm from one end to attain a value of 200°C.

6.18 Rework Problem 6.1 using the Schmidt method.

6.19 Rework Problem 6.2 using the Schmidt method.

6.20 Rework Problem 6.3 using the graphical method.

6.21 Rework Problem 6.10 using the Schmidt method.

6.22 Rework Problem 6.11 using the graphical method.

6.23 A furnace wall is made in two layers. The inner layer of 10 cm thickness is of fireclay (k = 1 W/m K and $\alpha = 5 \times 10^{-7}$ m^2/s) and the outer layer of 10 cm thickness is of concrete (k= 1.4 W/m K and $\alpha = 8.8 \times 10^{-7}$ m^2/s). The wall is initially at 25°C. Suddenly the inner surface of the fireclay layer is raised to a temperature of 700°C and subsequently held at that temperature. How long can the furnace be operated before the outer surface of the concrete layer rises to a temperature of 45°C? Solve the problem using the graphical method. Divide the inner fireclay layer into four sections for graphical analysis.

6.24 Rework Problem 6.23 if the inner surface of the fireclay layer is suddenly exposed to a convection environment at 800°C with h = 200 W/m^2 K.

6.25 A large composite slab is made with two different materials. The inner layer of 27.5 cm thickness is of aluminium and the outer layer of 15 cm thickness is of chrome-steel. Initially the slab is at a uniform temperature of 25°C. Suddenly one side of the slab is exposed to a convection environment at 700°C with h= 200 W/m^2 K, while the second surface is insulated as shown in Fig. 6.21.

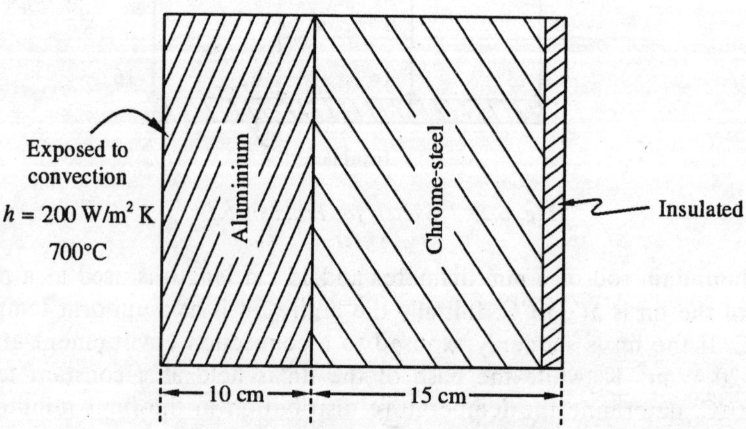

Fig. 6.21 Sketch for Problem 6.25.

Using the Schmidt method determine the temperature distribution in the slab 10 minutes after it is exposed to the convection environment. Divide the chrome-steel layer into three equal segments for the construction of a Schmidt plot.

6.26 A fluid at 100°C is flowing through a thick pipe of 5 cm ID and 15 cm OD. The outer surface of the pipe is at 20°C in steady-state. Suddenly the fluid is heated to 500°C. For the pipe material k = 20 W/m K and $\alpha = 1 \times 10^{-5}$ m^2/s. The convection heat transfer coefficient between the fluid and the pipe surface is 300 W/m^2 K. Using the Schmidt method, determine the temperature distribution in the pipe wall 40 seconds after the fluid is heated. Divide the pipe into 5 annular sections for constructing the Schmidt plot.

6.27 A large door is made in two sections. The outer layer of 5.5 cm thickness is made of nickel-steel and the inner layer of 2 cm is made of concrete. Initially the door is at a uniform temperature of 30°C. Suddenly the outer surface of the nickel-steel layer is exposed to hot gases at 600°C with h = 50 W/m^2 K. How long does it take for the external surface of the concrete to rise to 50°C? Solve the problem using the Schmidt method and divide the concrete layer into two equal segments while constructing the Schmidt plot.

6.28 A thick pipe of 5 cm ID and 15 cm OD is initially at a uniform temperature of 300 K. (For the pipe material k = 10 W/m k and $\alpha = 5 \times 10^{-6}$ m^2/s). Suddenly liquid nitrogen at 77 K is admitted into the pipe. The convection heat transfer coefficient between the liquid nitrogen and pipe wall is 200 W/m^2 K. Using the Schmidt method, estimate the temperature distribution in the pipe 1 minute after liquid nitrogen is admitted into the pipe. Divide the pipe wall into five annular sections for drawing the Schmidt plot.

6.29 A large resin (k = 1 W/m K; ρ = 2500 kg/m^3; c = 0.8 kJ/kg K) slab of 10 cm thickness is initially at 25°C. One surface of the slab is suddenly exposed to a stream of hot air at 100°C and the heat transfer coefficient between the air and the resin varies as $h = 10+0.1\,(T_\infty - T)$, where T_∞ is the temperature of air and T is the temperature of the resin surface. Using the Schmidt method, determine the temperature distribution in the resin slab 40 minutes after it is exposed to hot air. Divide the resin into five equal layers for drawing the Schmidt plot.

CHAPTER SEVEN

RADIATION HEAT TRANSFER

Learning Objectives

This chapter will enable the readers to

- understand the meaning of thermal radiation.
- define the terms—emissive power, reflectivity, absorptivity and transmissivity.
- distinguish between regular and diffuse reflection.
- understand Kirchhoff's law, the Stefan–Boltzmann law, Planck's law and Wein's displacement law.
- use radiation functions to determine the energy radiated in a particular range of wavelengths.
- define radiation intensity, gray body and shape factor.
- estimate shape factors for various geometrical configurations.
- understand and apply the shape factor algebra to determine the shape factors for several geometries.
- understand Hottel's string rule.
- draw equivalent electrical networks for radiation heat transfer between black bodies and gray bodies.
- analyze radiation heat exchange between black bodies and gray bodies.
- define surface resistance to radiation and radiation transfer factor or gray body shape factor.
- analyze radiation heat exchange in the presence of radiation shields.

7.1 Thermal Radiation

In the earlier chapters we have studied the conduction heat transfer. In conduction, heat flows from a region of high temperature to a region of low temperature within a medium. In conduction heat transfer, the energy is transmitted by direct molecular communication. In contrast to this, radiation is a process in which heat flows from a high temperature body to a low temperature body when the two bodies are separated from each other in space, even when a vacuum exits between them. The term *radiation* is generally applied to all kinds of electromagnetic phenomena. *Thermal radiation* is that electromagnetic radiation emitted by a body as a result of its temperature. The radiation energy transfer is explained as the consequence of energy carrying electromagnetic waves. These waves or photons are emitted from the matter. The electromagnetic spectrum ranging from radio waves to gamma rays is shown in Fig. 7.1. The electromagnetic radiation propagates at the speed of light. The wavelength of radiation λ is defined as the ratio of the propagation velocity C to the frequency ν. That is,

$$\lambda = C/\nu \tag{7.1}$$

The units of wavelength which are in normal use are centimeter, angstrom (1 Å = 10^{-10} m) or micron (micrometer, 1 μm = 10^{-6} m). Thermal radiation wavelengths range from about 0.1 to 100 μm, while visible light wavelengths range from about 0.35 to 0.75 μm .

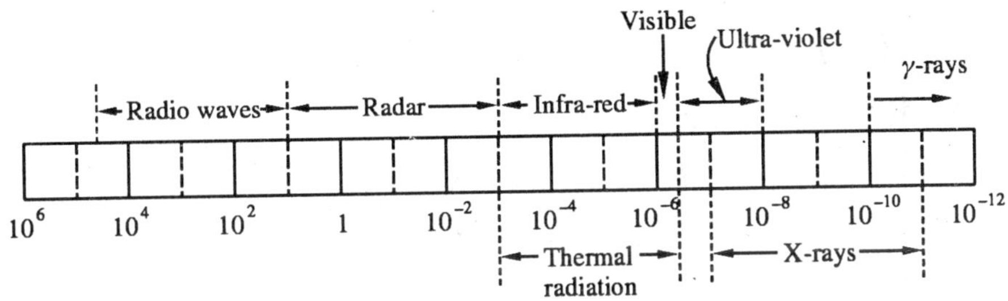

Fig. 7.1 *Electromagnetic spectrum.*

All bodies continuously emit radiation. Radiation in the wavelength region 0.1 to 100 μm, if received by a body, causes its temperature to rise. In this chapter we shall study thermal radiation only.

7.2 Radiation Properties

The total amount of radiation emitted by a body per unit area and time is called the total *emissive power E*. The emissive power of a body depends on its temperature and the characteristics of the surface of the body. At a given temperature the quantity of radiation emitted per unit wavelength is different at the various wavelengths. For an ideal radiator, the distribution of emissive power with wavelength is as shown in Fig. 7.2 for different temperatures. It can be observed from these curves that a major portion of radiation is emitted within a narrow band of wavelengths, and the emissive power reaches a peak value at a particular wavelength at a specified temperature. The peak of the curve shifts to shorter wavelengths at higher temperatures. One can characterize the quality of radiation in terms of the wavelength at which maximum emission occurs. For example, the Sun whose surface temperature is approximately

5800 K emits about 90 percent of its total radiation in the range of wavelengths 0.1 to 3 μm. A body about 1200°C emits most of its radiation in the range of wavelengths 1 to 20 μm. Ordinary window glass transmits radiation upto 2.5 μm but it is opaque to radiation emitted by bodies at 30°C. Hence solar radiation can pass through the window glass.

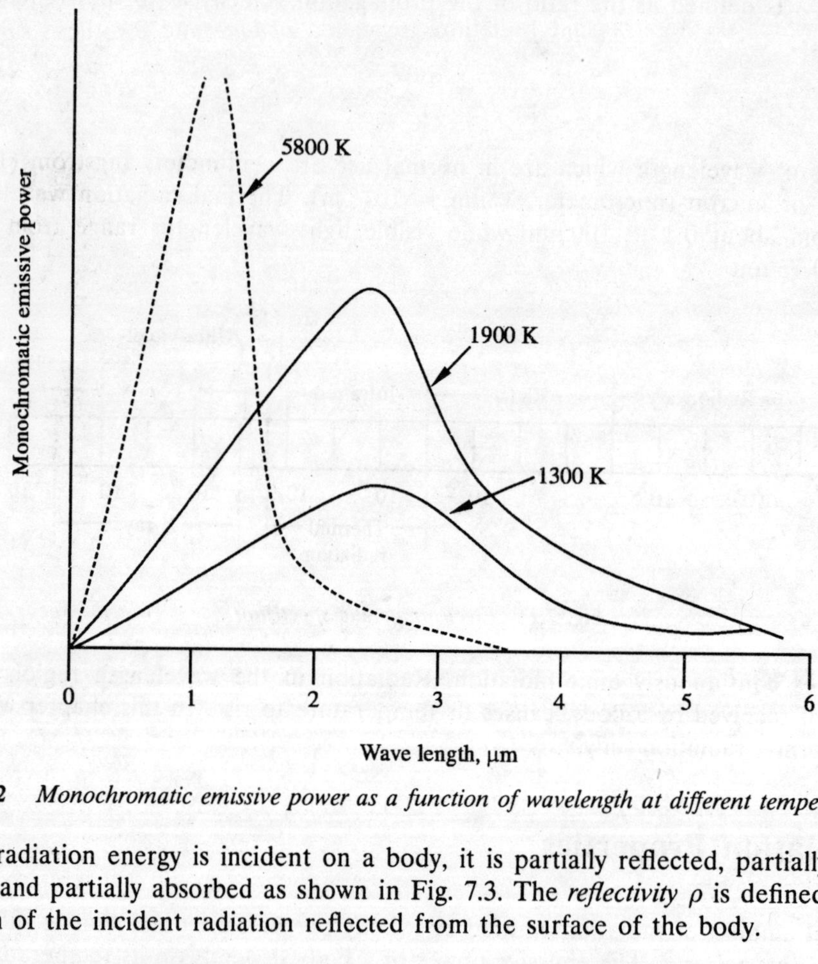

Fig. 7.2 Monochromatic emissive power as a function of wavelength at different temperatures.

When radiation energy is incident on a body, it is partially reflected, partially transmitted and partially absorbed as shown in Fig. 7.3. The *reflectivity* ρ is defined as the fraction of the incident radiation reflected from the surface of the body.

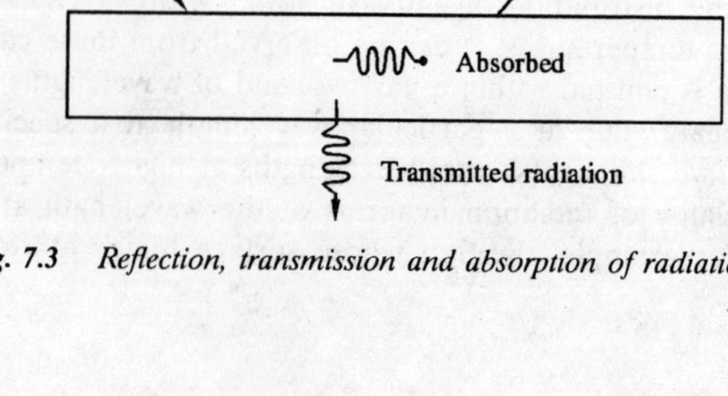

Fig. 7.3 Reflection, transmission and absorption of radiation.

The *transmissivity* τ is defined as the fraction of the incident radiation transmitted through the body and the *absorptivity* α is defined as the fraction of the incident radiation absorbed by the body. Therefore

$$\rho + \alpha + \tau = 1 \tag{7.2}$$

Bodies which do not transmit radiation are called *opaque* and for these substances Eq. (7.2) reduces to

$$\rho + \alpha = 1 \tag{7.3}$$

Most of the solid bodies are opaque to thermal radiation. Glass, rock salt and some inorganic crystals are exceptions among the solids. Many liquids and all gases are transparent to thermal radiation. The reflection of thermal radiation may be regular or diffuse. In *regular* or *specular reflection*, the angle of incidence is equal to the angle of reflection. Regular reflection can be observed in highly polished or smooth surfaces. Most of the materials used in industrial appliances do not have smooth surfaces; they have rough surfaces, in the sense that the surfaces have asperities which are larger than one wavelength. The reflection of radiation from a rough surface occurs in all directions and such a reflection is called *diffuse*. For most practical purposes one can treat the reflection from a surface as diffuse. The values of α, ρ and τ depend on the material, its surface conditions and the wavelength of radiation.

7.3 The Black Body and Kirchoff's Law

The black body is an idealization of a radiator. The *black body* or an *ideal radiator* is defined as a body which absorbs all the radiation incident upon it and does not reflect or transmit. As a radiator, the black body emits the maximum possible thermal radiation at all wavelengths. The black body is a theoretical concept and is used as a standard of perfection against which the radiation characteristics of real bodies are compared. A black body can be closely approximated in practice to a cavity, like a hollow sphere, the inner surface of which is maintained at a uniform temperature. If a small hole is made in the wall of the sphere, any radiation entering through the hole will get reflected several times at the inner surface as shown in Fig. 7.4. Each time it strikes the surface, a part of it is absorbed. Hence, the radiation beam which leaves the cavity, through the hole, carries negligible energy.

The radiation emitted by the interior surface of a cavity is absorbed and reflected several times and fills the cavity. Suppose such a perfect enclosure is available. Now, if we place a body in the enclosure and allow it to come to thermal equilibrium with the enclosure, then the energy absorbed by the body is equal to the energy emitted by the body. Let

A = surface area of the body

α = absorptivity of the body

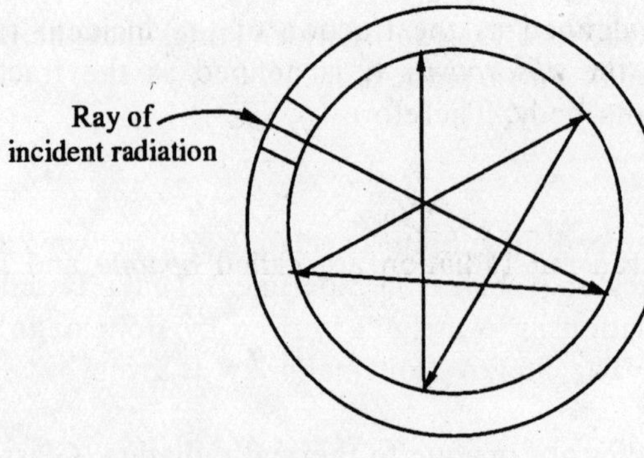

Fig. 7.4 Multiple reflections of radiation in a cavity.

E = emissive power (or energy emitted per unit area and time) of the body

q = radiant flux arriving at the area A in the enclosure.

Then, at equilibrium we have

$$EA = qA\alpha \tag{7.4}$$

Now, replace the body by a black body of the same surface area A and allow it to reach a state of thermal equilibrium with the enclosure. Then for the black body, we have

$$E_b A = qA \times 1 \tag{7.5}$$

where

E_b = emissive power of the black body

$\alpha = 1$ for a black body since it absorbs all the radiation incident upon it.

Dividing Eq. (7.4) with Eq. (7.5), we get

$$\frac{E}{E_b} = \alpha \tag{7.6}$$

That is, the absorptivity α of a body is the ratio of the emissive power of a body to the emissive power of a black body at the same temperature. This ratio is defined as the *emissivity* ε of the body. Therefore,

$$\varepsilon = \alpha \tag{7.7}$$

Equation (7.7) is called Kirchoff's identity which states that the emissivity and absorptivity of a body are equal. For a black body,

$$\varepsilon = \alpha = 1$$

Stefan, in 1879, deduced an empirical relationship between the temperature and emissive power of a black body, based on experimental data. Based on thermodynamical considerations, this relationship was later derived by Boltzmann and is known as the *Stefan–Boltzmann law*. The Stefan–Boltzmann law is given by

$$E_b = \sigma T^4 \tag{7.8}$$

where

E_b = energy radiated per unit area and per unit time by an ideal radiator (or black body)

σ = $5.669 \times 10^{-8}\ \text{W/m}^2\,\text{K}^4$ = Stefan–Boltzmann constant

T = temperature of the black body.

The Stefan–Boltzmann equation gives the total emissive power of a black body. It does not provide the distribution of energy among the different wavelengths. Based on quantum theory, Max Planck, in 1900, derived a relation which gives the emissive power of a black body as a function of wavelength and the relationship is known as *Planck's law*. Planck's law is given by

$$E_{b\lambda} = \frac{c_1 \lambda^{-5}}{\exp(c_2/\lambda T) - 1} \tag{7.9}$$

where

$E_{b\lambda}$ = monochromatic emissive power of a black body ($\text{W/m}^2\mu\text{m}$)

λ = wavelength (μm)

T = temperature (K)

c_1 = $3.743 \times 10^8\ \text{W}\mu\text{m}^4/\text{m}^2$

c_2 = $1.4387 \times 10^4\ \mu\text{m K}$.

The monochromatic emissive power of a black body as a function of wavelength and temperature is shown in Fig. 7.2. The wavelength at which the monochromatic emissive power is a maximum shifts to shorter wavelengths as the temperature of the black body increases. The relationship between the wavelength λ_{max} at which $E_b\lambda$ is a maximum and the temperature is given by *Wien's displacement law*

$$\lambda_{max}T = 2897.6\,\mu\text{m K} \tag{7.10}$$

At temperatures of engineering interest, the emissive power is appreciable in the range of wavelengths 0.3 to 50 μm. The visible range of wavelengths extends over a narrow range from about 0.4 to 0.7 μm. Only a small amount of the total energy falls in this range of wavelengths at temperatures below 600^{0}C. As the temperature of the black body increases, the amount of energy within the visible range increases and the human eye begins to detect the radiation. At about 750°C the amount of radiant energy in the wavelength region 0.6 to 0.7 μm is sufficient to be observed and the object appears to be dull red in colour. As the temperature of the object is further increased, the maximum intensity shifts to the shorter wavelengths and the colour of the object changes to bright red, then bright yellow and finally appears to be white at about 1350°C. As the temperature of the object is further increased, it appears much brighter because more of the total radiation falls within the visible range.

In many engineering applications, we are interested in determining the amount of energy radiated from a black body in a specified range of wavelengths. The fraction of the total energy radiated between 0 and λ is given by

$$\frac{E_{b0-\lambda}}{E_{b0-\infty}} = \frac{\int_0^{\lambda} E_{b\lambda}\,d\lambda}{\int_0^{\infty} E_{b\lambda}\,d\lambda} \tag{7.11}$$

From Eq. (7.9) we obtain

$$\frac{E_b\lambda}{T^5} = \frac{c_1}{(\lambda T)^5\,[\exp\,(c_2/\lambda T) - 1]} \tag{7.12}$$

Then, for any specified temperature, we find from Eqs. (7.11) and (7.12) $E_{b0-\lambda}/E_{b0-\infty}$ is a function of only one variable λT. The values of the ratio $E_{b0-\lambda}/E_{b0-\infty}$ are presented in Table 7.1 as a function of λT. If it is desired to determine the radiant energy emitted in the wavelength region λ_1 and λ_2 it can be obtained as

$$E_{b\lambda_1-\lambda_2} = E_{b0-\infty}\left(\frac{E_{b0-\lambda_2}}{E_{b0-\infty}} - \frac{E_{b0-\lambda_1}}{E_{b0-\infty}}\right) \tag{7.13}$$

where $E_{bo-\infty}$ is the total radiation emitted over all the wavelengths and is given by

$$E_{b0-\infty} = \sigma T^4 \tag{7.14}$$

which is obtained by integrating the Planck's distribution relation over all wavelengths.

Table 7.1 Radiation functions

$\lambda T(\mu\text{m K})$	$E_{b0-\lambda T}/\sigma T^4$	$\lambda T(\mu\text{m K})$	$E_{b0-\lambda T}/\sigma T^4$	$\lambda T(\mu\text{m K})$	$E_{b0-\lambda T}/\sigma T^4$
555.6	0.170×10^{-7}	4000.0	0.48085	7444.4	0.83166
666.7	0.756×10^{-6}	4111.1	0.50066	7555.6	0.83698
777.8	0.106×10^{-4}	4222.2	0.51974	7666.7	0.84209
888.9	0.738×10^{-4}	4333.3	0.53809	7777.8	0.84699
1000.0	0.321×10^{-3}	4444.4	0.55573	7888.9	0.85171
1111.1	0.00101	4555.6	0.57267	8000.0	0.85624
1222.2	0.00252	4666.7	0.58891	8111.1	0.86059
1333.3	0.00531	4777.8	0.60449	8222.2	0.86477
1444.4	0.00983	4888.9	0.61941	8333.3	0.86880
1555.6	0.01643	5000.0	0.63371	8888.9	0.88677
1666.7	0.02537	5111.1	0.64740	9444.4	0.90168
1777.8	0.03677	5222.2	0.66051	10000.0	0.91414
1888.9	0.05059	5333.3	0.67305	10555.6	0.92462
2000.0	0.06672	5444.4	0.68506	11111.1	0.93349
2111.1	0.08496	5555.6	0.69655	11666.7	0.94104
2222.2	0.10503	5666.7	0.70754	12222.2	0.94751
2333.3	0.12665	5777.8	0.71806	12777.8	0.95307
2444.4	0.14953	5888.9	0.72813	13333.3	0.95788
2555.6	0.17337	6000.0	0.73777	13888.9	0.96207
2666.7	0.19789	6111.1	0.74700	14444.4	0.96572
2777.8	0.22285	6222.2	0.75583	15000.0	0.96892
2888.9	0.24803	6333.3	0.76429	15555.6	0.97174
3000.0	0.27322	6444.4	0.77238	16111.1	0.97423
3111.1	0.29825	6555.6	0.78014	16666.7	0.97644
3222.2	0.32300	6666.7	0.78757	22222.2	0.98915
3333.3	0.34734	6777.8	0.79469	27777.8	0.99414
3444.4	0.37118	6888.9	0.80152	33333.3	0.99649
3555.6	0.39445	7000.0	0.80806	38888.9	0.99773
3666.7	0.41708	7111.1	0.81433	44444.4	0.99845
3777.8	0.43905	7222.2	0.82035	50000.0	0.99889
3888.9	0.46031	7333.3	0.82612	55555.6	0.99918

EXAMPLE 7.1 The window glass transmits 92 percent of the incident radiation in the wavelength range 0.35 and 2.7 μm and is opaque at shorter and longer wavelengths. Estimate the percent of solar radiation which the glass will transmit assuming that the Sun is a black body at 5800 K.

SOLUTION Given $T = 5800$ K; $\lambda_1 = 0.35\,\mu$m and $\lambda_2 = 2.7\,\mu$m

$$\lambda_1 T = 0.35 \times 5800 = 2030\ \mu\text{m K}; \qquad \lambda_2 T = 2.7 \times 5800 = 15660\ \mu\text{m K}$$

From Table 7.1 we read $E_{b0-\lambda T}/\sigma T^4$ values at $\lambda_1 T = 2030\ \mu$m K and $\lambda_2 T = 15660\ \mu$m K as 0.07165 and 0.97176 (obtained by interpolation).

$$\sigma T^4 = 5.669 \times 10^{-8} \times (5800)^4 = 64.153\ \text{MW/m}^2$$

The total incident radiation

$$E_{b\,\lambda_1-\lambda_2} = E_{b\,0-\infty}\left(\frac{E_{b0-\lambda_2}}{E_{b0-\infty}} - \frac{E_{b0-\lambda_1}}{E_{b0-\infty}}\right)$$

$$\text{Percent radiation in the range 0.35 to } 2.7\mu\text{m} = \frac{E_{b\lambda_1-\lambda_2}}{E_{b0-\infty}}$$

$$= (0.97176 - 0.07165)100 = 90.01$$

Percent radiation transmitted by the window glass = 0.92 × 90.01 = 82.81.

7.4 Radiation Intensity

So far we have discussed the total radiation from a black body in all directions. One can visualize this condition by placing a hemispherical surface over a surface element of area dA_1 as shown in Fig. 7.5. The hemispherical surface will intercept all the radiation emitted by the surface dA_1. When observed from a point on the hemisphere, displaced by an angle ϕ from the normal to the surface, the element dA_1 will appear as the projected area $dA_1 \cos\phi$.

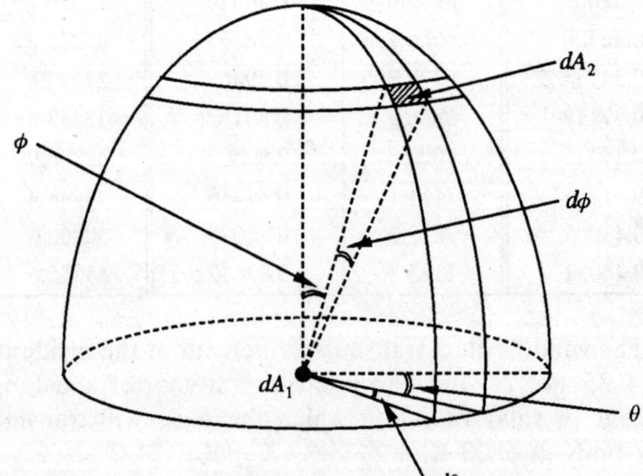

Fig. 7.5 Hemispherical surface enclosing the surface dA_1.

The intensity of radiation I from the surface dA_1 is defined as the radiant energy propagated in a particular direction per unit solid angle and per unit area of dA_1 as projected on a plane perpendicular to the direction of propagation. Then the rate of radiant heat flow from dA_1 to dA_2 is given by

$$dq_{1\to 2} = I\cos\phi dA_1 dA_2/r^2 \tag{7.15}$$

where r is the distance of dA_2 from dA_1.

The solid angle $d\omega_1$ subtended by dA_2 at dA_1 is given by

$$d\omega_1 = \frac{dA_2}{r^2} \tag{7.16}$$

Effective area seen from dA_2 or the projected area of dA_1 in the direction ϕ

$$= dA_1 \cos\phi$$

Therefore,

$$I = \frac{dq_{1\to 2}}{d\omega_1 dA_1 \cos\phi} \tag{7.17}$$

If the emitting surface is diffuse, the intensity does not change with the angle ϕ and remains constant. The emissive power of the body can be obtained by integrating the radiation intercepted by an element of area dA_2 over the surface of the hemisphere.

$$E = \int\int I\cos\phi d\omega_1 \tag{7.18}$$

The solid angle is defined as the area subtended on a sphere of unit radius [see Eq. (7.16)]. Therefore,

$$d\omega_1 = \frac{(r\sin\phi)d\theta(rd\phi)}{r^2} = \sin\phi d\phi d\theta \tag{7.19}$$

Substituting Eq. (7.19) in Eq. (7.18) we get

$$E = I\int_0^{2\pi} d\theta \int_0^{\pi/2} \sin\phi\cos\phi d\phi$$

or

$$E = \pi I \tag{7.20}$$

That is, the total emissive power of a diffuse surface is π times its intensity.

7.5 Gray Body

Radiation from real surfaces differs from black body radiation. A real surface always radiates less than a black body at the same temperature. If the monochromatic emissivity ε of a body is constant over the entire wavelength spectrum, then the body is said to be gray. Then the emissive power of a gray body (E_g) is given by

$$E_g = \varepsilon \sigma T^4 \tag{7.21}$$

7.6 Radiation Heat Exchange between Black Bodies

We have studied the emission and absorption of radiation by the bodies but did not discuss the net heat exchange between the bodies. To determine the net heat exchange by radiation between two bodies it is necessary to determine the fraction of the total radiation emitted from one body which reaches the other, and the fraction of the incident radiation absorbed by the other. If we consider only black bodies, all the incident radiation is absorbed by them. Let us consider two black surfaces A_1 and A_2 which are separated by a distance r in a medium which does not absorb any radiation, as shown in Fig. 7.6.

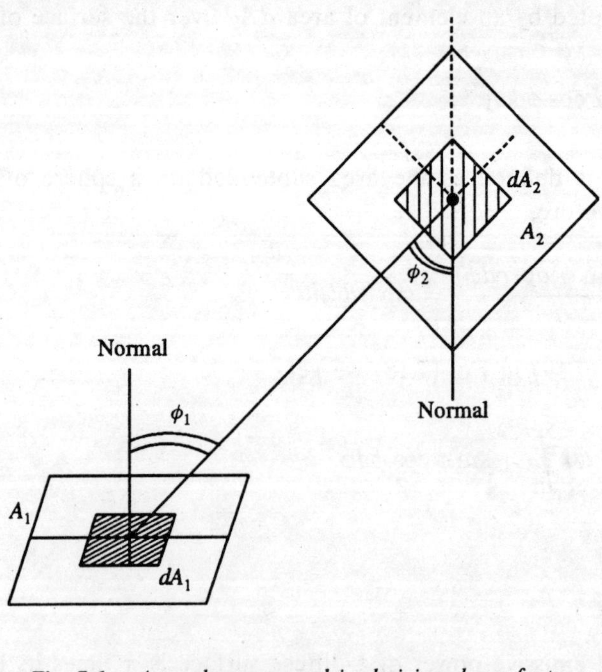

Fig. 7.6 Area elements used in deriving shape factor.

We would like to obtain a general expression to evaluate the energy exchange by radiation between the two surfaces when they are maintained at different temperatures. For this, we have to determine the fraction of the energy leaving surface A_1 that strikes the surface A_2. Let us consider two differential surfaces dA_1 and dA_2. Then the rate at which radiation from dA_1 strikes dA_2 is given by [see Eq. (7.15)]

$$dq_{1\to 2} = I_1 \cos\phi_1 dA_1 d\omega_{1-2} \tag{7.22}$$

where I_1 = intensity of radiation from dA_1

$dA_1 \cos\phi_1$ = projection of differential area dA_1 with respect to dA_2

$d\omega_{1\to 2}$ = solid angle subtended by area dA_2 with respect to dA_1.

The solid angle $d\omega_{1-2}$ subtended by dA_2 is equal to the projected area of the receiving surface (dA_2) in the direction of the incident radiation, divided by the square of the distance between the two surfaces. That is,

$$d\omega_{1-2} = \cos\phi_2 \frac{dA_2}{r^2} \tag{7.23}$$

Substituting for I_1 from Eq. (7.20) and for $d\omega_{1-2}$ from Eq. (7.23) in Eq. (7.22), we get

$$dq_{1\to 2} = E_{b1} dA_1 \frac{\cos\phi_1 \cos\phi_2 dA_2}{\pi r^2} \tag{7.24}$$

where

$\dfrac{\cos\phi_1 \cos\phi_2 dA_2}{\pi r^2}$ = fraction of the total radiation emitted from dA_1 that is intercepted by dA_2.

Similarly, the rate at which radiation from dA_2 that strikes dA_1 is given by

$$dq_{2\to 1} = E_{b2} dA_2 \frac{\cos\phi_2 \cos\phi_1 dA_1}{\pi r^2} \tag{7.25}$$

Therefore, the net rate of radiant heat transfer between dA_1 and dA_2 is given by

$$dq_{1\leftrightarrow 2} = (E_{b1} - E_{b2}) \frac{\cos\phi_1 \cos\phi_2 dA_1 dA_2}{\pi r^2} \tag{7.26}$$

To determine the net rate of radiant heat transfer between the surfaces A_1 and A_2, we integrate Eq. (7.26) over both surfaces

$$q_{net} = (E_{b1} - E_{b2}) \int_{A_1} \int_{A_2} \cos\phi_1 \cos\phi_2 \frac{dA_1 dA_2}{\pi r^2} \tag{7.27}$$

The integral in Eq. (7.27) is conveniently written as A_1F_{12} or A_2F_{21}, where F_{12} is the *shape factor* evaluated on the basis of A_1 and F_{21} is the shape factor evaluated on the basis of A_2. F_{12} represents the fraction of radiant energy leaving A_1 which reaches A_2. F_{21} represents the fraction of radiant energy leaving A_2 which reaches A_1. The shape factor is also called *view factor, angle factor* and *configuration factor.* In terms of F_{12} and F_{21}, Eq. (7.27) can be written as

$$q_{\text{net}} = (E_{b1} - E_{b2})A_1F_{12} = (E_{b1} - E_{b2})A_2F_{21} \tag{7.28}$$

Therefore,

$$A_1F_{12} = A_2F_{21} = \int_{A1}\int_{A2} \frac{\cos\phi_1 \cos\phi_2 dA_1 dA_2}{\pi r^2} \tag{7.29}$$

and $A_1F_{12} = A_2F_{21}$ (7.30)

The relation (7.30) is known as the *reciprocity theorem*. Equation (7.28) shows that the net rate of radiant heat flow between two black surfaces can be determined by evaluating either A_1F_{12} or A_2F_{21}. Since the choice of the terms F_{12} or F_{21} is not going to affect the end result, one selects that surface for which the shape factor can be determined easily. In general, the determination of shape factor is quite difficult, except for some simple geometries. The method of evaluation of shape factor in a simple case is illustrated in the following example.

EXAMPLE 7.2 Determine the shape factor for a very small disk A_1 and a large parallel disk A_2 located at a distance ℓ directly above the small disk.

SOLUTION The shape factor is defined as

$$A_1F_{12} = \int_{A1}\int_{A2} \frac{\cos\phi_1 \cos\phi_2 dA_1 dA_2}{\pi r^2} \tag{A}$$

To determine the shape factor, we should know the geometry of the surfaces A_1 and A_2. The geometry of the surfaces A_1 and A_2 are given as disks. Let us choose the surface A_1 as the small disk and the surface A_2 as the large disk located at a distance ℓ directly above the small disk as shown in Fig. 7.7.

Consider a differential surface dA_2 as the circular ring segment of radius x and width dx as shown in Fig. 7.7.

Let ϕ_1 = angle between the normal to the surface A_1 and the line joining the area elements dA_1 and dA_2

ϕ_2 = angle between the normal to the surface A_2 and the line joining the area elements dA_1 and dA_2

R = radius of the large disk (surface A_2).

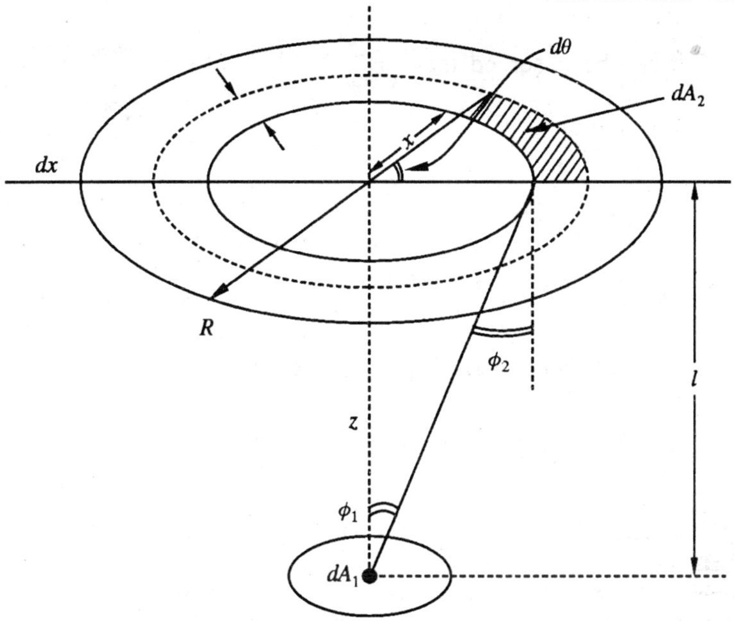

Fig. 7.7 *Sketch of a large disk located at a distance ℓ from a small disk.*

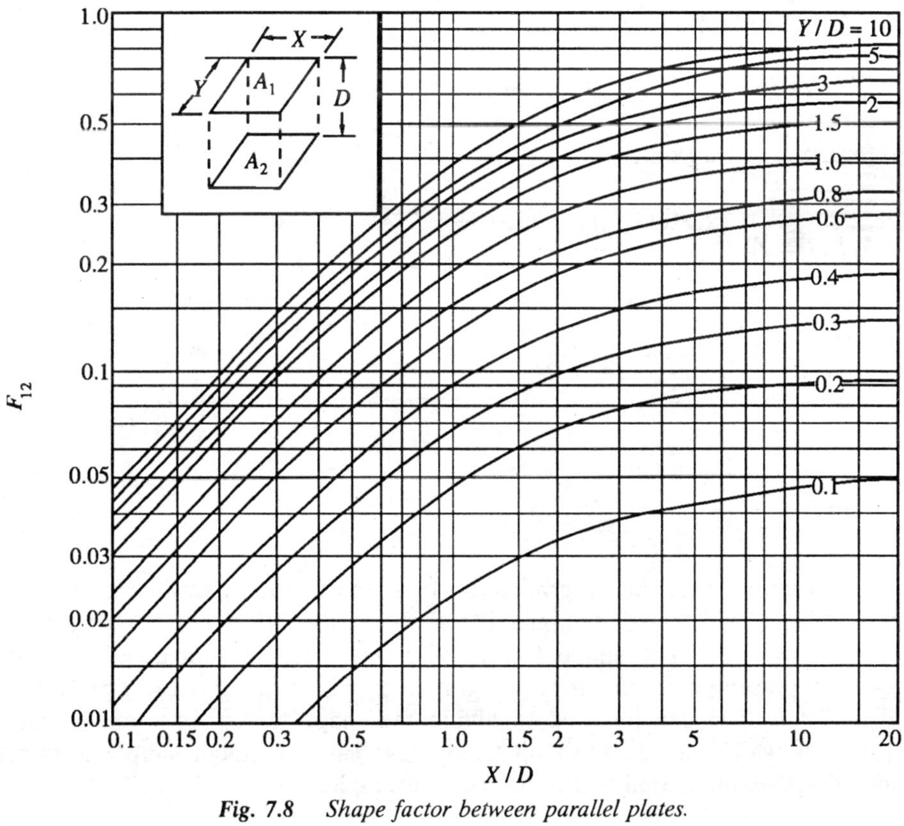

Fig. 7.8 *Shape factor between parallel plates.*

From Fig. 7.7 it can be observed that

$$\cos\phi_1 = \cos\phi_2 = \frac{\ell}{z} \tag{B}$$

$$z = \sqrt{\ell^2 + x^2} \tag{C}$$

$$dA_2 = x d\theta dx \tag{D}$$

Since the surface A_1 is very small, Eq. (A) becomes

$$A_1 F_{12} = \frac{A_1}{\pi} \int_{A2} \frac{\cos\phi_1 \cos\phi_2}{r^2} dA_2 \tag{E}$$

Substituting Eqs. (B), (C) and (D) in Eq. (E), we get

$$A_1 F_{12} = \frac{A_1}{\pi} \int_0^R \int_0^{2\pi} \frac{\ell^2}{(\ell^2 + x^2)^2} x dx d\theta = 2A_1 \ell^2 \int_0^R \frac{x dx}{(\ell^2 + x^2)^2} \tag{F}$$

From Eq. (C) we get

$$dz = 2x dx.$$

Therefore, Eq. (F) becomes

$$A_1 F_{12} = A_1 \ell^2 \int_{\ell^2}^{\ell_2 + R^2} \frac{dz}{z^2}$$

$$= A_1 \ell^2 \left[\frac{1}{\ell^2} - \frac{1}{\ell^2 + R^2}\right]$$

or

$$F_{12} = \frac{R_2}{\ell^2 + R^2}$$

Example 7.2 shows that the determination of shape factor by evaluating the double integral of Eq. (7.29) is very tedious. The shape factors for a number of geometries have been evaluated. The shape factors of a few selected geometries of practical interest are presented in the form of graphs in Figs. 7.8 to 7.13. The data from the Figs. 7.8 to 7.13 can be used to obtain the shape factors of some geometrical arrangements which can be built up from these cases, through simple arithmetical manipulations, as illustrated in the following example.

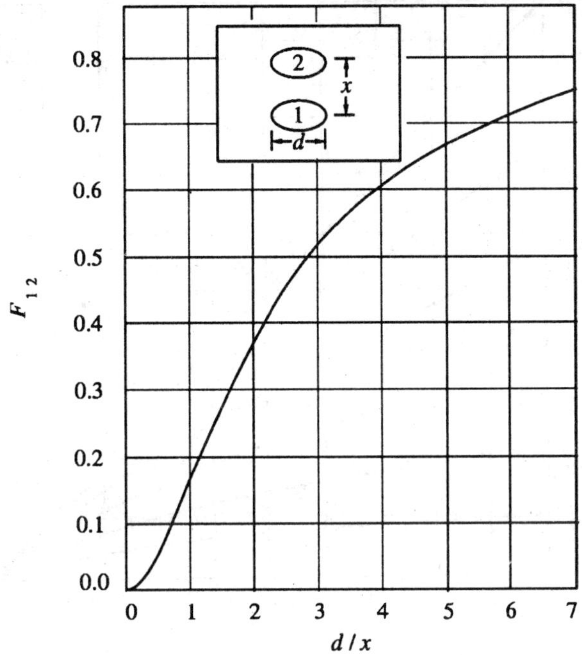

Fig. 7.9 Shape factor between parallel disks.

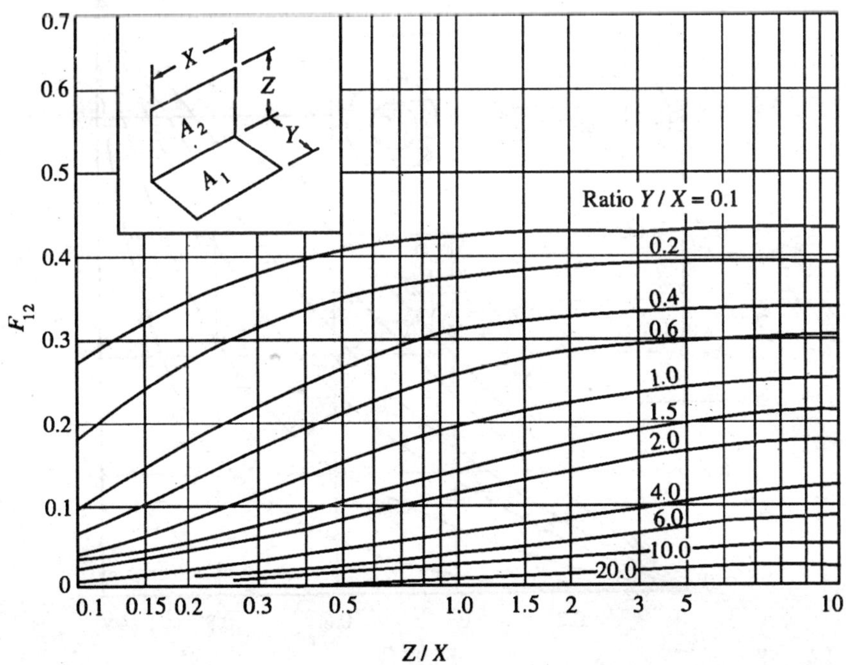

Fig. 7.10 Shape factor between perpendicular rectangles with a common edge.

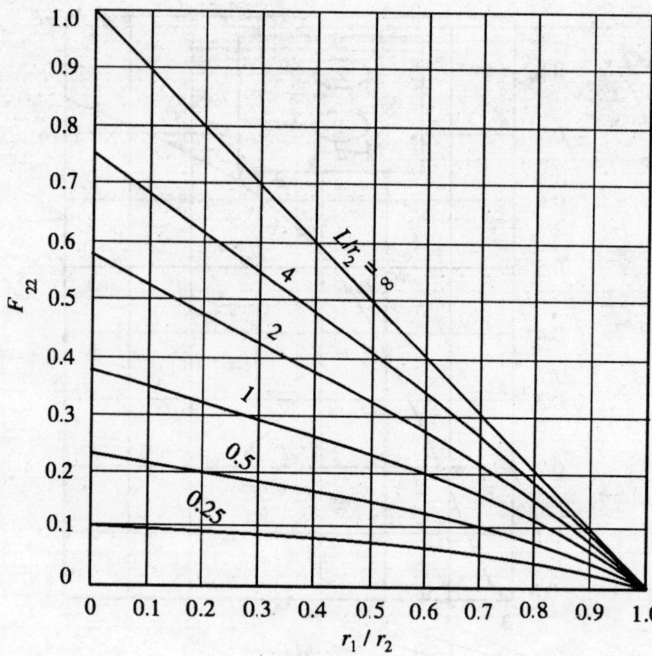

Fig. 7.11 *Shape factor for two concentric cylinders of finite length—outer cylinder to itself.*

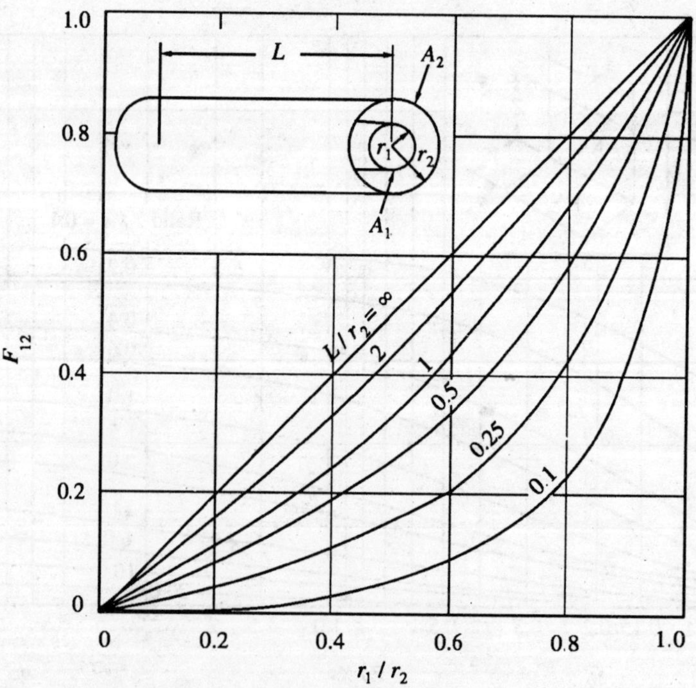

Fig. 7.12 *Shape factor for two concentric cylinders of finite length—outer cylinder to inner cylinder.*

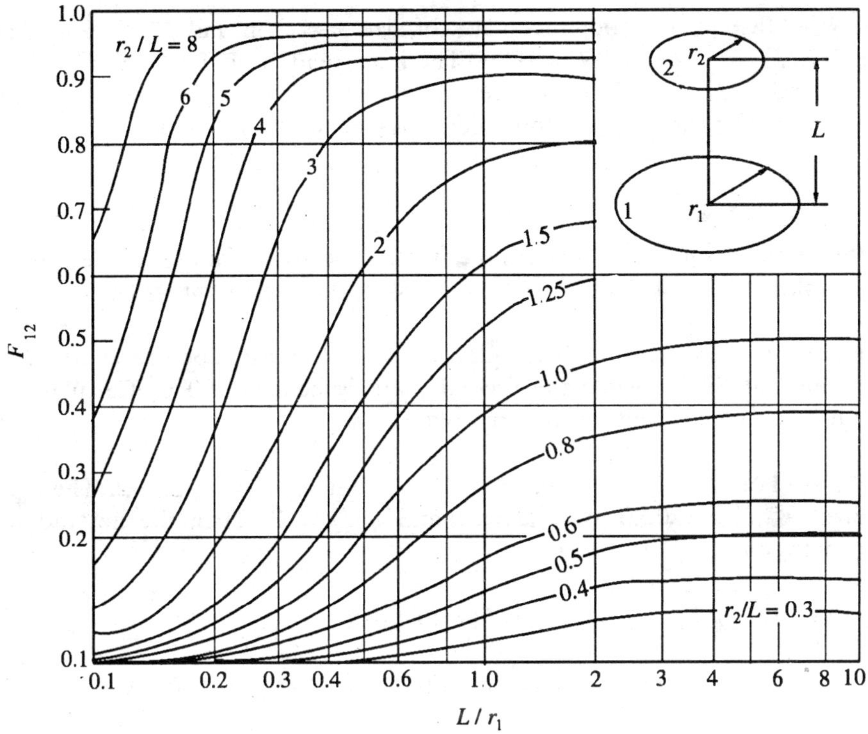

Fig. 7.13 Shape factor between two parallel concentric disks.

EXAMPLE 7.3 A room of floor size 3 m × 4 m has a ceiling height of 2 m. A small ventilator is located in the middle of the ceiling. Determine the fraction of energy leaving the floor which strikes the ventilator.

SOLUTION A schematic of the floor and ventilator is shown in Fig. 7.14

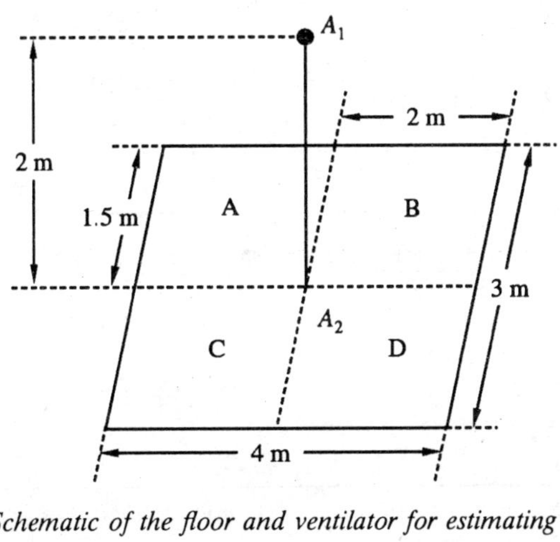

Fig. 7.14 Schematic of the floor and ventilator for estimating shape factor.

The floor is divided into four rectangles of size $x = 2$ m and $y = 1.5$ m. For each rectangle $x/D = 1$ and $y/D = 0.75$. Then we read from Fig. 7.8, $F_{12} = 0.15$ at $x/D = 1$ and $y/D = 0.75$. Therefore, the shape factor for the floor is obtained by adding the shape factors of the four rectangles in the floor. Thus

$$F_{12} = 4 \times 0.15 = 0.6$$

Since the shape factor F_{12} represents the fraction of the radiant energy leaving surface A_1 that strikes A_2, 60% of the radiation energy leaving the floor strikes the ventilator.

EXAMPLE 7.4 Two parallel black plates 1 m × 2 m are kept 1 m apart. One plate is maintained at 800^0C while the second plate is held at 200°C. Calculate the net radiant heat exchange between the two plates.

SOLUTION For the plates we have $x/D = 1$ and $y/D = 2/1 = 2$. Then we read from Fig. 7.8, $F_{12} = 0.29$ at $x/D = 1$ and $y/D = 2$. Then the net radiant heat transfer between the plates is given by

$$q_{\text{net}} = A_1 F_{12}(E_{b1} - E_{b2}) = A_1 F_{12}\sigma(T_1^4 - T_2^4)$$

$$= 1 \times 2 \times 0.29 \times 5.669 \times 10^{-8}(1073^4 - 473^4) = 41.94 \text{ kW}$$

The shape factors of some simple two- and three-dimensional configurations are presented in Table 7.2.

Table 7.2 Shape factors for two- and three-dimensional configurations

Configuration	Shape factors
Two-dimensional	
1. Concentric cylinders 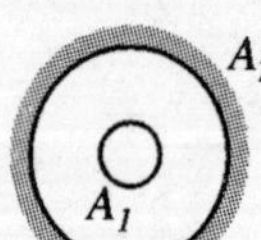	$F_{12} = 1;\ \ F_{21} = A_1/A_2$ $F_{22} = 1 - F_{21} = 1 - A_1/A_2$
2. Long duct with equilateral triangular cross-section: 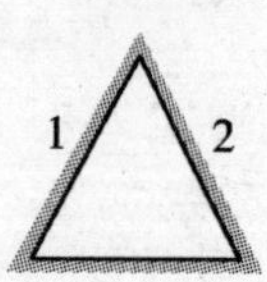	$F_{12} = F_{13} = 1/2$

Continues

Table 7.2 Continued

Configuration	Shape factors
3. Long parallel plates of equal width 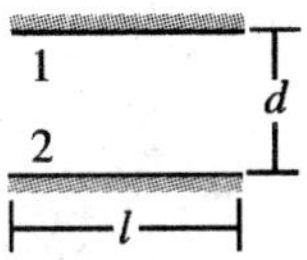	$F_{12} = F_{21} = \sqrt{1 + \left(\frac{d}{\ell}\right)^2} - \frac{d}{\ell}$
4. Long adjacent plates 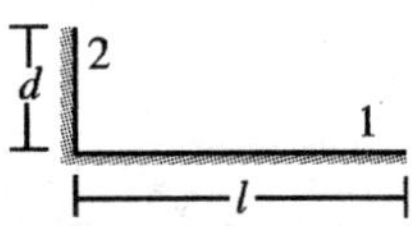	$F_{12} = \frac{1}{2}\left[1 + \frac{d}{\ell} - \sqrt{1 + \left(\frac{d}{\ell}\right)^2}\right]$
5. Long symmetrical wedge 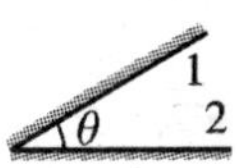	$F_{12} = F_{21} = 1 - \sin\frac{\theta}{2}$
6. Long cylinder parallel to a large plane area 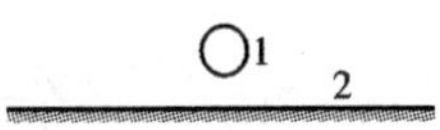	$F_{12} = \frac{1}{2}$
7. Long cylinder parallel to a plate 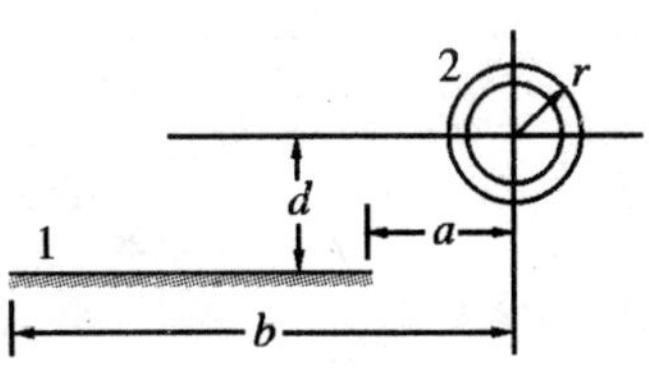	$F_{12} = \frac{r}{b-a}\left[\tan^{-1}\left(\frac{b}{d}\right) - \tan^{-1}\left(\frac{a}{d}\right)\right]$
8. Long adjacent parallel cylinders of equal diameters 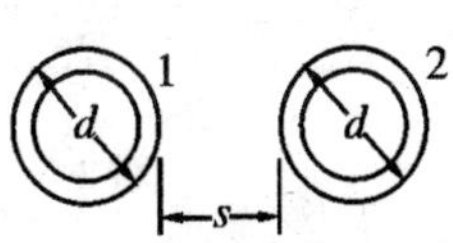	Let $X = 1 + \frac{s}{d}$, then $F_{12} = F_{21} = \frac{1}{\pi}\left[(x^2 - 1)^{1/2} + \sin^{-1}\frac{1}{x} - x\right]$

Continues

Table 7.2 Continued

Configuration	Shape factors
Three-dimensional configurations	
9. concentric spheres 	$F_{12} = 1;\ \ F_{21} = \dfrac{A_1}{A_2}$ $F_{22} = 1 - F_{21} = 1 - \dfrac{A_1}{A_2}$
10. Regular tetrahedron 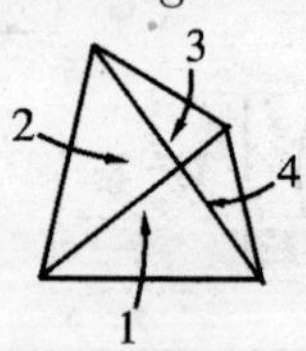	$F_{12} = F_{13} = F_{14} = \dfrac{1}{3}$
11. Sphere near a large plane area 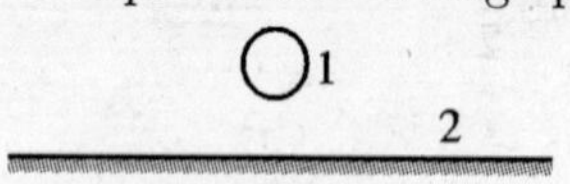	$F_{12} = \dfrac{1}{2}$
12. Small area perpendicular to the axis of a surface of revolution 	$F_{12} = \sin^2\theta$
13. Area on the inside of a sphere 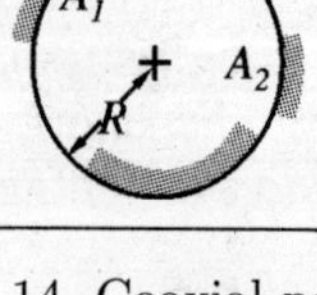	$F_{12} = \dfrac{A_2}{4\pi R^2}$
14. Coaxial parallel disks 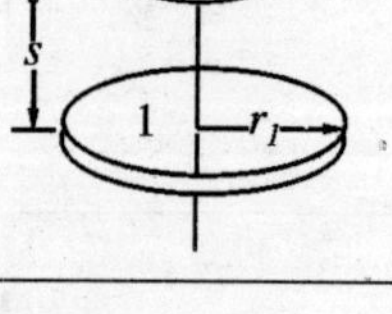	Let $R_1 = \dfrac{r_1}{s};\ R_2 = \dfrac{r_2}{s};\ X = 1 + \dfrac{1 + R_2^2}{R_1^2}$; then $F_{12} = \dfrac{1}{2}\left\{X - \left[X^2 - 4\left(\dfrac{R_2}{R_1}\right)^2\right]^{1/2}\right\}$

Continues

Table 7.2 Continued

Configuration	Shape factors
	Three-dimensional configurations
15. Opposite rectangles 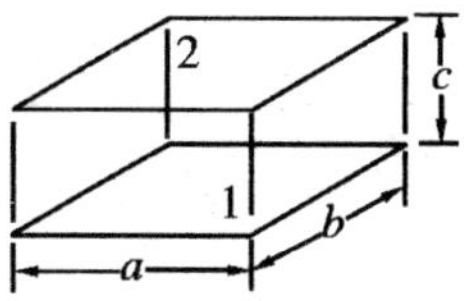	Let $X = \frac{b}{c}$, $Y = \frac{a}{c}$; then $F_{12} = \frac{2}{\pi XY}\left\{\ln\left[\frac{(1+X^2)(1+Y^2)}{1+X^2+Y^2}\right]^{1/2} - X\tan^{-1}X - Y\tan^{-1}Y + X(1+Y^2)^{1/2}\tan^{-1}\frac{X}{(1+Y^2)^{1/2}} + Y(1+X^2)^{1/2}\tan^{-1}\frac{Y}{(1+X^2)^{1/2}}\right\}$
16. Adjacent rectangles 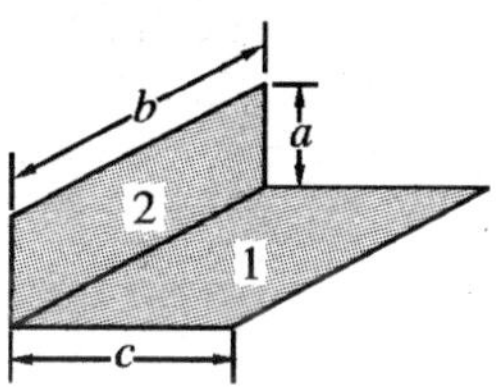	Let $L = c/b$; $N = a/b$; then $F_{12} = \frac{1}{\pi L}\left\{L\tan^{-1}\frac{1}{L} + N\tan^{-1}\frac{1}{N} - (N^2+L^2)^{1/2}\tan^{-1}(N^2+L^2)^{-1/2} + \frac{1}{4}\ln\left[\left(\frac{(1+L^2)(1+N^2)}{1+L^2+N^2}\right)\left(\frac{L^2(1+L^2+N^2)}{(1+L^2)(L^2+N^2)}\right)^{L^2}\left(\frac{N^2(1+L^2+N^2)}{(1+N^2)(L^2+N^2)}\right)^{N^2}\right]\right\}$

7.7 Shape Factor Algebra

The use of the shape factor formulae and graphs can be extended by the use of shape factor algebra and the reciprocity theorem. The shape factor algebra involves the judicious use of the reciprocity relation and the principle of energy conservation. Consider an enclosure consisting of n surfaces as shown in Fig. 7.15. The energy leaving the surface i must be intercepted by some surface of the enclosure, including i itself if it is concave. That is,

$$\sum_{j=1}^{n} F_{ij} = 1 \tag{7.31}$$

If surface j is subdivided into m subsurfaces as shown in Fig. 7.15, we can write

$$F_{ij} = \sum_{k=1}^{m} F_{ik} \tag{7.32}$$

F_{ii} is called the *self-viewing factor* and it is non-zero only for a concave surface. Now we can apply the relation (7.31) and the reciprocity relation to derive shape factors for simple geometries as illustrated in the following examples.

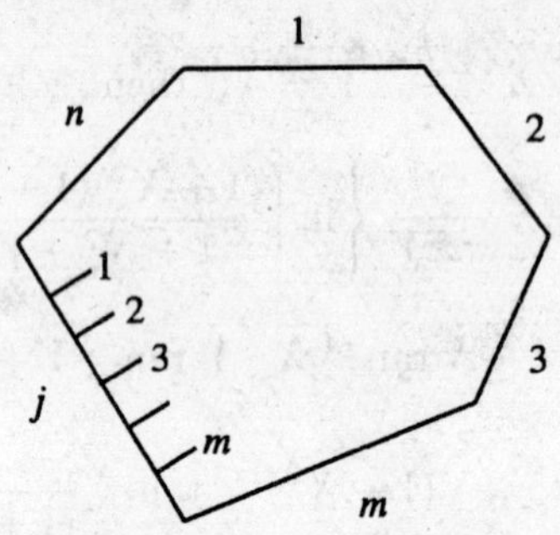

Fig. 7.15 An enclosure of n surfaces the surface j with subdivided into m surfaces.

EXAMPLE 7.5 Suppose a sphere of surface area A_1 is completely enclosed by an irregular surface of area A_2 as shown in Fig. 7.16. Determine the shape factors F_{12}, F_{21}, F_{11} and F_{22}.

SOLUTION The energy conservation principle gives

$$F_{11} + F_{12} = 1$$

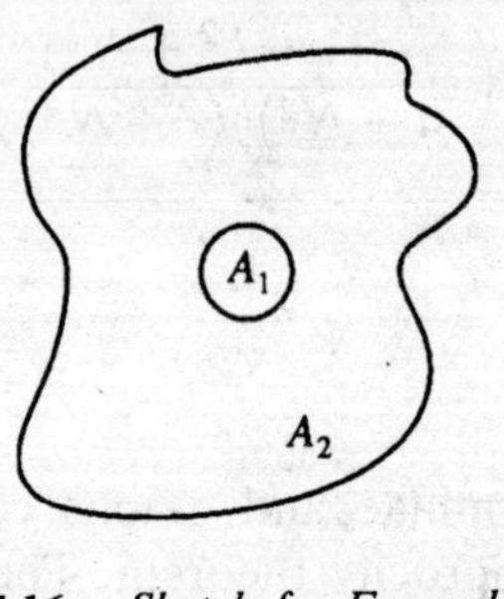

Fig. 7.16 Sketch for Example 7.5.

The spherical surface is convex and hence the self-viewing factor is given by $F_{11} = 0$. Hence $F_{12} = 1$.

The reciprocity relation gives $A_1 F_{12} = A_2 F_{21}$.

Therefore, $$F_{21} = \frac{A_1}{A_2} F_{12} = \frac{A_1}{A_2}$$

The energy conservation principle for surface A_2 gives

$$F_{21} + F_{22} = 1$$

or $\quad F_{22} = 1 - F_{21} = 1 - \dfrac{A_1}{A_2}$

EXAMPLE 7.6 Consider an enclosure of three non-concave surfaces shown in Fig. 7.17. Determine the shape factors F_{12}, F_{23} and F_{31}.

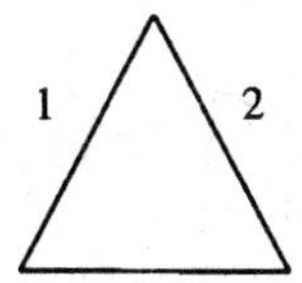

Fig. 7.17 An enclosure of three non-concave surfaces.

SOLUTION Application of the energy conservation principle to the surfaces 1, 2 and 3 gives

$$F_{12} + F_{13} = 1 \tag{A}$$

$$F_{21} + F_{23} = 1 \tag{B}$$

$$F_{31} + F_{32} = 1 \tag{C}$$

The reciprocity relations give

$$A_1F_{12} = A_2F_{21} \quad \text{or} \quad F_{21} = \frac{A_1}{A_2}F_{12} \tag{D}$$

$$A_1F_{13} = A_3F_{31} \quad \text{or} \quad F_{13} = \frac{A_3}{A_1}F_{31} \tag{E}$$

$$A_2F_{23} = A_3F_{32} \quad \text{or} \quad F_{32} = \frac{A_2}{A_3}F_{23} \tag{F}$$

Solving Eqs. (A) to (F), we get

$$F_{12} = \frac{A_1 + A_2 - A_3}{2A_1}$$

$$F_{23} = \frac{A_2 + A_3 - A_1}{2A_2}$$

$$F_{31} = \frac{A_3 + A_1 - A_2}{2A_3}$$

EXAMPLE 7.7 Consider the surfaces 1, 2 and 3 shown in Fig. 7.18. Determine the shape factor F_{13} in terms of the known shape factors for perpendicular rectangles with a common edge.

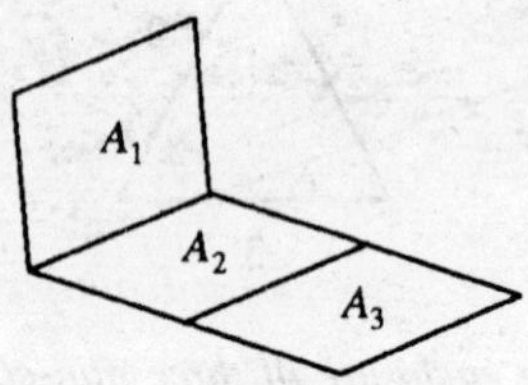

Fig. 7.18 Sketch for Example 7.7.

SOLUTION Applying Eq. (7.32) we can write

$$F_{1-2,3} = F_{12} + F_{13} \tag{A}$$

$F_{1-2,3}$ and F_{12} can be determined from Fig. 7.10 (shape factors for perpendicular rectangles with common edge). Hence Eq. (A) gives

$$F_{13} = F_{1-2,3} - F_{12}$$

Equation (A) can be rewritten as

$$A_1F_{1-2,3} = A_1F_{12} + A_1F_{13} \tag{B}$$

The reciprocity relation gives

$$A_1F_{1-2,3} = A_{2,3}F_{2,3-1} \tag{C}$$

$$A_1F_{12} = A_2F_{21} \tag{D}$$

$$A_1F_{13} = A_3F_{31} \tag{E}$$

Substituting Eqs. (C), (D) and (E) in Eq. (B), we obtain

$$A_{2,3}F_{2,3-1} = A_2F_{21} + A_3F_{31} \tag{F}$$

The right-hand side of Eq. (F) represents the sum of the radiations reaching surface A_1 from surfaces A_2 and A_3. Hence the total radiation reaching surface A_1 from the combined surface $A_{2,3}$ is the sum of the radiations reaching A_1 from the individual surfaces A_2 and A_3.

EXAMPLE 7.8 Determine the shape factor F_{14} for the surfaces shown in Fig. 7.19 in terms of the known shape factors for perpendicular rectangles with a common edge.

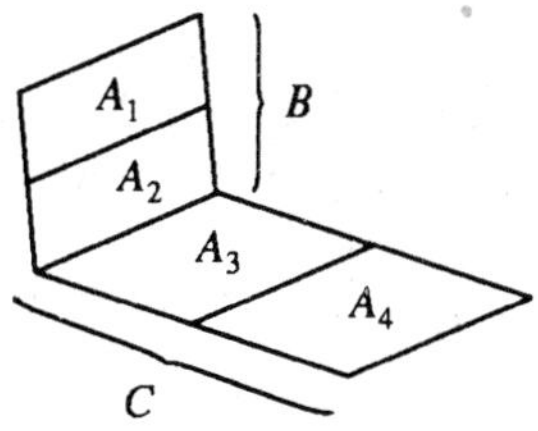

Fig. 7.19 Sketch for Example 7.8.

SOLUTION Let us call the combined surfaces A_1 and A_2 as B and the combined surfaces A_3 and A_4 as C. Then we can write

$$A_B F_{BC} = A_1 F_{1C} + A_2 F_{2c} \tag{A}$$

Since the radiation leaving the surface B is the sum of the radiations leaving the surfaces A_1 and A_2, $A_1 F_{1c}$ can be written as

$$A_1 F_{1C} = A_1 F_{13} + A_1 F_{14} \tag{B}$$

$$A_B F_{B3} = A_1 F_{13} + A_2 F_{23}$$

or

$$A_1 F_{13} = A_B F_{B3} - A_2 F_{23} = A_{1,2} F_{1,2-3} - A_2 F_{23} \tag{C}$$

Substituting Eq. (C) in Eq. (B) we get

$$A_1 F_{1C} = A_1 F_{1-3,4} = A_{1,2} F_{1,2-3} - A_2 F_{23} + A_1 F_{14} \tag{D}$$

Now, substitute Eq. (D) in Eq. (A) to obtain

$$A_B F_{BC} = A_{1,2} F_{1,2-3,4} = A_{1,2} F_{1,2-3} - A_2 F_{23} + A_1 F_{14} + A_2 F_{2-3,4}$$

or

$$F_{14} = \frac{A_{1,2} F_{1,2-34} + A_2 F_{23} - A_{1,2} F_{1,2-3} - A_2 F_{2-3,4}}{A_1} \tag{E}$$

In Eq. (E) all the shape factors are for perpendicular rectangles and hence they can be read from Fig. 7.10. Thus F_{14} can be determined.

The shape factors of two-dimensional configurations can be determined by another method also. This method is known as *Hottel's string rule*. This method is restricted to surfaces which are infinite in one direction, and there is no need to perform spatial integration. Hottel's string rule is based on the shape factor analysis for triangular enclosures. Consider the surfaces 1 and 2 shown in Fig. 7.20. The shape factor is given by

$$2\ell_1 F_{12} = \begin{pmatrix}\text{Sum of the lengths of}\\ \text{the two crossed strings}\end{pmatrix} - \begin{pmatrix}\text{Sum of the lengths of}\\ \text{uncrossed strings}\end{pmatrix} \tag{7.33}$$

where ℓ_1 is the length of surface 1.

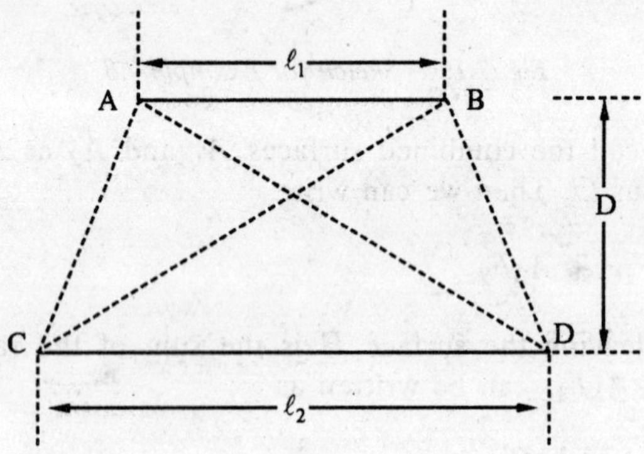

Fig. 7.20 Hottel's cross strings for the determination of shape factor.

The surfaces 1 and 2 are connected by the uncrossed strings AC and BD and the energy which leaves surface 1 and reaches surface 2 does not cross these strings. The energy which leaves surface 1 and reaches surface 2 has to cross the two crossed strings AD and BC (see Fig. 7.20).

EXAMPLE 7.9 Determine the shape factor A_{12} by applying the Hottel's cross-strings method for the surfaces shown in Fig. 7.20.

SOLUTION From the geometry of Fig. 7.20, we get

$$\text{AC} = \text{BD} = \sqrt{\left(\frac{\ell_2 - \ell_1}{2}\right)^2 + D^2}$$

$$\text{AD} = \text{BC} = \sqrt{\left(\frac{\ell_2 + \ell_1}{2}\right)^2 + D^2}$$

We know that

$$2\ell_1 F_{12} = \begin{pmatrix}\text{Sum of the lengths of the}\\ \text{two crossed strings}\end{pmatrix} - \begin{pmatrix}\text{Sum of the lengths of}\\ \text{uncrossed strings}\end{pmatrix}$$

$$\text{or} \quad 2\ell_1 F_{12} = 2\sqrt{\left(\frac{\ell_2 + \ell_1}{2}\right)^2 + D^2} - 2\sqrt{\left(\frac{\ell_2 - \ell_1}{2}\right)^2 + D^2}$$

$$\text{or} \quad F_{12} = \sqrt{\frac{1}{4}\left(1 + \frac{\ell_2}{\ell_1}\right)^2 + \left(\frac{D}{\ell_1}\right)^2} - \sqrt{\frac{1}{4}\left(\frac{\ell_2}{\ell_1} - 1\right)^2 + \left(\frac{D}{\ell_1}\right)^2}$$

In the special case of $\ell_1 = \ell_2 = \ell$, we get

$$F_{12} = \sqrt{1 + \frac{D^2}{\ell^2}} - \frac{D}{\ell}$$

7.8 Electrical Network Analogy–Radiation Heat Transfer between Black Surfaces

The net rate of radiation heat transfer between two black bodies held at temperatures T_1 and T_2 is given by

$$q_{\text{net}} = A_1 F_{12}(E_{b1} - E_{b2}) = A_2 F_{21}(E_{b1} - E_{b2}) \tag{7.28}$$

or

$$q_{\text{net}} = \frac{E_{b1} - E_{b2}}{\frac{1}{A_1 F_{12}}} \tag{7.34}$$

If the black body emissive power $E_b = \sigma T^4$ is viewed as a potential, then the form of Eq. (7.34) suggests an electrical analogy. E_b is analogous to electrical potential, q_{net} is the analogue of electric current and $1/A_1 F_{12}$ is the analogue of electrical resistance. For the surfaces i and j, the *space radiation resistance* R_{ij} between nodes i and j is given by

$$R_{ij} = \frac{1}{A_i F_{ij}} \tag{7.35}$$

As an example of a simple network, consider two large, black parallel walls 1 and 2 at temperatures T_1 and T_2. Then the equivalent electrical network for the radiation energy exchange between the plates is as shown in Fig. 7.21.

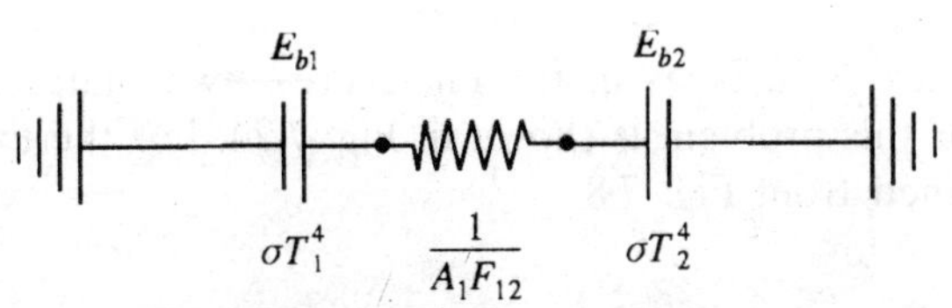

Fig. 7.21 *Equivalent electrical network for a 2-surface enclosure.*

If we consider an enclosure of three black surfaces, then the equivalent electrical circuit will be as shown in Fig. 7.22. Figure 7.23 shows the equivalent electrical circuit for radiation energy exchange in an enclosure of 4 black surfaces.

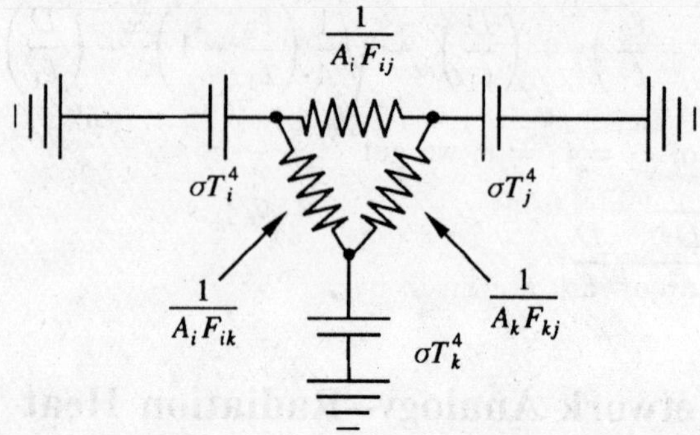

Fig. 7.22 Equivalent electrical circuit for a 3-surface enclosure.

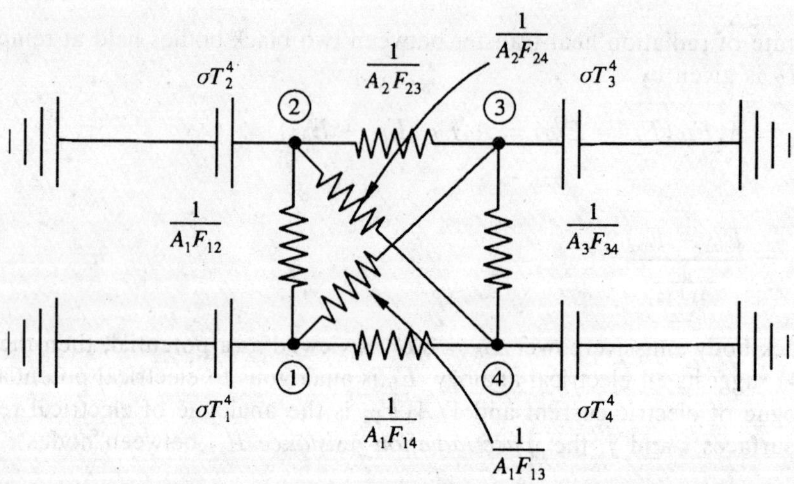

Fig. 7.23 Equivalent electrical circuit for radiation heat exchange in a 4-surface enclosure.

EXAMPLE 7.10 Two parallel black plates 0.5 m × 1 m are separated by a distance of 0.5 m. One plate is maintained at 1200 K while the other plate is held at 600 K. These plates are placed in a large room, the walls of which are maintained at 300 K. The plates exchange heat with each other and with the room. Determine the net heat transfer to each plate and to the room.

SOLUTION There are 3 surfaces in the enclosure—two plates and the room. The equivalent circuit for this problem is shown in Fig. 7.24. For the two plates, the shape factors can be obtained from Fig. 7.8.

$$\frac{x}{D} = \frac{0.5}{0.5} = 1 \quad \text{and} \quad \frac{y}{D} = \frac{1}{0.5} = 2$$

Hence

$$F_{12} = F_{21} = 0.29$$

$$F_{12} + F_{13} = 1 \quad \text{or} \quad F_{13} = 1 - F_{12} = 1 - 0.29 = 0.71$$

$$F_{21} + F_{23} = 1 \quad \text{or} \quad F_{23} = 1 - F_{21} = 1 - 0.29 = 0.71$$

$$A_1 = A_2 = 0.5 \times 1 = 0.5 \text{ m}^2$$

The resistances in the network are given by

$$\frac{1}{A_1 F_{12}} = \frac{1}{0.5 \times 0.29} = 6.897$$

$$\frac{1}{A_1 F_{13}} = \frac{1}{0.5 \times 0.71} = 2.817$$

$$\frac{1}{A_2 F_{23}} = \frac{1}{0.5 \times 0.71} = 2.817$$

$$E_{b1} = \sigma T_1^4 = 5.669 \times 10^{-8}(1200)^4 = 117.552 \text{ kW/m}^2$$

$$E_{b2} = \sigma T_2^4 = 5.669 \times 10^{-8}(600)^4 = 7.347 \text{ kW/m}^2$$

$$E_{b3} = \sigma T_3^4 = 5.669 \times 10^{-8}(300)^4 = 0.459 \text{ kW/m}^2$$

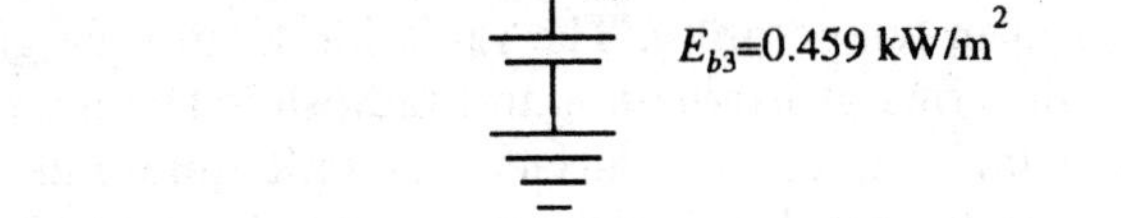

Fig. 7.24 Radiation network for Example 7.10.

The net heat transfer from plate 1 is given by

$$E = \frac{E_{b1} - E_{b2}}{6.897} + \frac{E_{b1} - E_{b3}}{2.817}$$

or

$$q_1 = \frac{(117.552 - 7.347)10^3}{6.897} + \frac{(117.552 - 0.459)10^3}{3.817} = 57.545 \text{ kW}$$

The net heat transfer from plate 2 is given by

$$q_2 = \frac{E_{b2} - E_{b1}}{6.897} + \frac{E_{b2} - E_{b3}}{2.817}$$

or

$$q_2 = \frac{(7.347 - 117.552) \times 10^3}{6.897} + \frac{(7.347 - 0.459)10^3}{2.817} = -13.534 \text{ kW}$$

The total heat received by the room is given by

$$q_3 = \frac{E_{b1} - E_{b3}}{2.817} + \frac{E_{b2} - E_{b3}}{2.817}$$

or

$$q_3 = \frac{(117.552 - 0.459 + 7.347 - 0.459) \times 10^3}{2.817} = 44.011 \text{ kW}$$

7.9 Radiation Heat Exchange between Gray Bodies

Radiation heat exchange between black bodies is easy to analyze since all the radiation intercepted by a black body is absorbed. With real surfaces, the radiation falling on a body can be reflected back and forth several times and their absorptivity and emissivity are less than unity. The real surfaces are not perfectly diffuse and hence the intensity of emitted radiation is not constant in all directions. For a diffuse surface the intensity is constant and does not depend on the direction. Such a diffuse surface obeys Lambert's cosine law. In this section we shall assume that all the real surfaces are diffuse and uniform in temperature. The radiation from such gray surfaces can be treated conveniently in terms of irradiation and radiosity. The *irradiation* G is defined as the total radiation incident upon a surface per unit time and per unit area. The *radiosity* J is defined as the total radiation which leaves a surface per unit time and per unit area. For gray bodies we shall assume that the irradiation and radiosity are uniform over the surface. The radiosity is the sum of the radiation emitted, reflected and transmitted. For opaque bodies which do not transmit any radiation, the radiosity is given by

$$J = \rho G + \varepsilon E_b \tag{7.36}$$

In Eq. (7.36) the first term on the right-hand side denotes the radiation reflected and the second term denotes the radiation emitted. For an opaque body, the reflectivity is given by

$$\rho = 1 - \alpha = 1 - \varepsilon \tag{7.37}$$

Hence

$$J = (1 - \varepsilon)G + \varepsilon E_b \tag{7.38}$$

or

$$G = \frac{J - \varepsilon E_b}{1 - \varepsilon} \tag{7.39}$$

The net rate at which radiation is leaving a gray surface per unit time, is equal to the difference between the radiosity and irradiation. That is,

$$q = A(J - G) = A\left(J - \frac{J - \varepsilon E_b}{1 - \varepsilon}\right) = \frac{A\varepsilon}{1 - \varepsilon}(E_b - J) \tag{7.40}$$

or

$$q = \frac{E_b - J}{(1 - \varepsilon)/A\varepsilon} \tag{7.41}$$

Equation (7.41) gives the net radiant energy leaving a gray surface in terms of its black body emissive power E_b and its radiosity. If the denominator of the right-hand side of Eq. (7.41) is considered as the *surface resistance to radiation* heat transfer and the numerator as the potential difference, then one can draw an electrical network analogy for radiation heat transfer from gray surfaces.

Now consider the radiation energy exchange between two gray surfaces A_1 and A_2 as shown in Fig. 7.25. The surface A_1 is at uniform temperature T_1 while the surface A_2 is at uniform temperature T_2. The fraction of energy leaving surface A_1 that is intercepted by surface A_2 is $J_1A_1F_{12}$. Similarly, the fraction of the radiant energy leaving surface A_2 that is intercepted by surface A_1 is $J_2A_2F_{21}$. Then the net radiant energy exchange between the two surfaces is given by

$$q = J_1A_1F_{12} - J_2A_2F_{21} \tag{7.42}$$

The reciprocity relation gives $A_1F_{12} = A_2F_{21}$. Therefore,

$$q = (J_1 - J_2)A_1F_{12} = (J_1 - J_2)A_2F_{21}$$

or

$$q = \frac{J_1 - J_2}{1/A_1F_{12}} \tag{7.43}$$

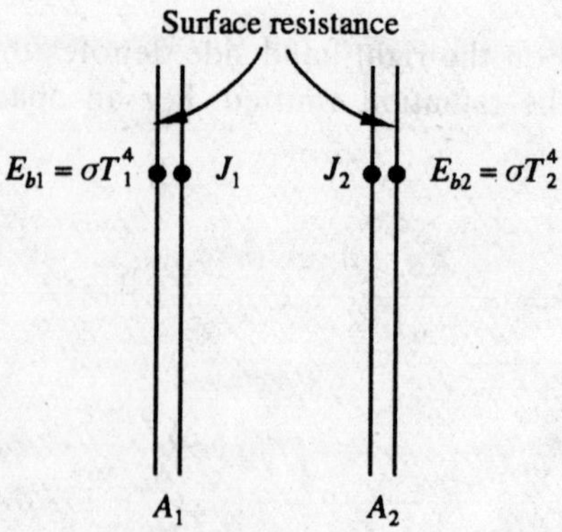

Fig. 7.25 An enclosure formed by two gray surfaces A_1 and A_2.

The radiation leaving the gray surface A_1 is given by

$$q_1 = \frac{A_1 \varepsilon_1}{1 - \varepsilon_1}(E_{b1} - J_1) \tag{7.44}$$

The radiation leaving the gray surface A_2 is given by

$$q_2 = \frac{A_2 \varepsilon_2}{1 - \varepsilon_2}(E_{b2} - J_2) \tag{7.45}$$

The conservation of energy requires that

$$q_1 = q_2 = q \tag{7.46}$$

Eliminating the radiosities J_1 and J_2 from Eqs. (7.43) – (7.45) we get

$$q = \frac{E_{b1} - E_{b2}}{\dfrac{1 - \epsilon_1}{A_1 \epsilon_1} + \dfrac{1}{A_1 F_{12}} + \dfrac{1 - \epsilon_2}{A_2 \epsilon_2}} \tag{7.47}$$

Knowing the temperatures and emissivities of the surfaces A_1 and A_2, we can calculate q from Eq. (7.47).

The equivalent electrical circuit for the enclosure consisting of two gray surfaces can be drawn as shown in Fig. 7.26.

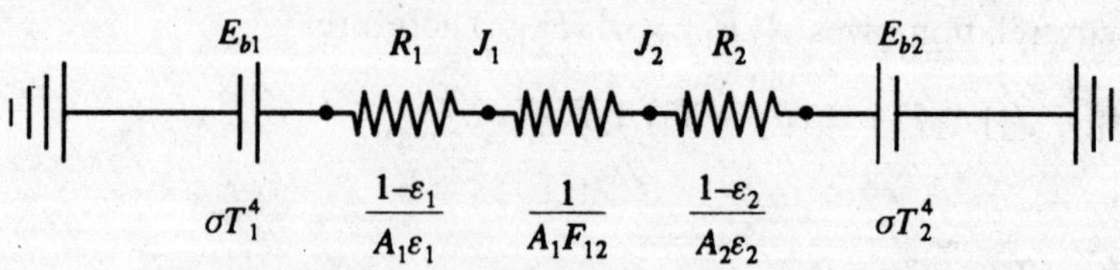

Fig. 7.26 Equivalent electrical circuit for a two-gray surface enclosure.

The surface resistance to radiation for the surface i is given by

$$R_i = \frac{1 - \varepsilon_i}{A_i \varepsilon_i} \tag{7.48}$$

The thermal radiation network for a three-gray surface enclosure is shown in Fig. 7.27.

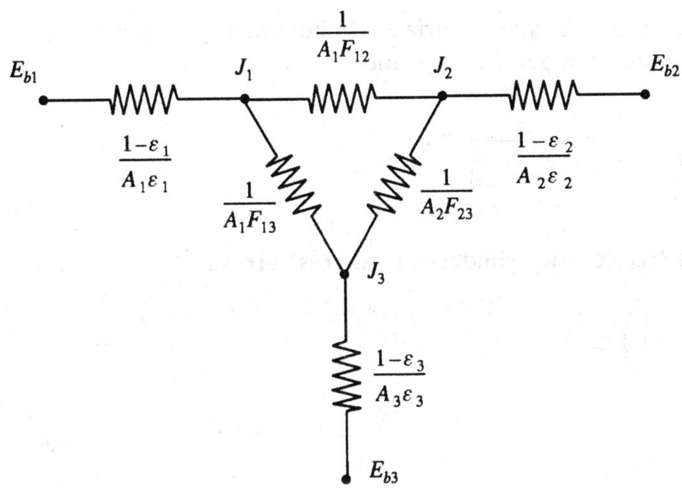

Fig. 7.27 Radiation network for three-gray surface enclosure.

One can define the *radiation transfer factor* or *gray body shape factor* $\mathcal{F}_{12}$ for the radiation heat transfer between two gray bodies as

$$q = A_1 \mathcal{F}_{12} (\sigma T_1^4 - \sigma T_2^4) \tag{7.49}$$

We know that the net radiation heat transfer between two gray bodies is given by Eq. (7.47). Comparing Eq. (7.47) with Eq. (7.49) we get

$$q = \frac{\sigma(T_1^4 - T_2^4)}{\dfrac{1 - \varepsilon_1}{A_1 \varepsilon_1} + \dfrac{1}{A_1 F_{12}} + \dfrac{1 - \varepsilon_2}{A_2 \varepsilon_2}} = A_1 \mathcal{F}_{12} \sigma (T_1^4 - T_2^4) = \frac{\sigma(T_1^4 - T_2^4)}{1/A_1 \mathcal{F}_{12}}$$

or

$$\frac{1}{A_1 \mathcal{F}_{12}} = \frac{1 - \varepsilon_1}{A_1 \varepsilon_1} + \frac{1}{A_1 F_{12}} + \frac{1 - \varepsilon_2}{A_2 \varepsilon_2} \tag{7.50}$$

or

$$\mathcal{F}_{12} = \frac{1}{\left(\dfrac{1}{\varepsilon_1} - 1\right) + \dfrac{1}{F_{12}} + \dfrac{A_1}{A_2}\left(\dfrac{1}{\varepsilon_2} - 1\right)} \tag{7.51}$$

For the special case of an enclosure consisting of two large parallel plates facing each other, $F_{12} = 1$ and $A_1 = A_2$. Then

$$\mathcal{F}_{12} = \frac{1}{\dfrac{1}{\varepsilon_1} + \dfrac{1}{\varepsilon_2} - 1} \tag{7.52}$$

For the special case of a convex surface A_1 bounded by a surface A_2, like concentric cylinders or spheres, we get $F_{12} = 1$ and

$$\mathcal{F}_{12} = \frac{1}{\dfrac{1}{\varepsilon_1} + \dfrac{A_1}{A_2}\left(\dfrac{1}{\varepsilon_2} - 1\right)} \tag{7.53}$$

If the surfaces (concentric cylinders or spheres) are such that $A_2 >> A_1$, then

$$\frac{A_1}{A_2}\left(\frac{1}{\varepsilon_2} - 1\right) \simeq 0$$

Then,

$$\mathcal{F}_{12} = \varepsilon_1 \tag{7.54}$$

Therefore, when a small object is placed within a room, the emissivity of the room (ε_2) does not have any influence on the heat flow from the object to the room. Any energy reflected by the room is mostly incident on itself and the room effectively acts as a black body.

EXAMPLE 7.11 Suppose the parallel plates of Example 7.10 are non-black and their emissivities are 0.5. Determine the net heat transfer from each plate and to the room.

SOLUTION The radiation network for the three-gray surface enclosure of this example is shown in Fig. 7.28. We have

$$A_1 = A_2 = 0.5 \text{ m}^2$$

$$\frac{1}{A_1 F_{12}} = 6.897$$

$$\frac{1}{A_1 F_{13}} = 2.817$$

$$\frac{1}{A_2 F_{23}} = 2.817$$

$$E_{b1} = 117.552 \text{ kW/m}^2$$

$$E_{b2} = 7.347 \text{ kW/m}^2$$

$$E_{b3} = 0.459 \text{ kW/m}^2 \text{ [see Example 7.10]}$$

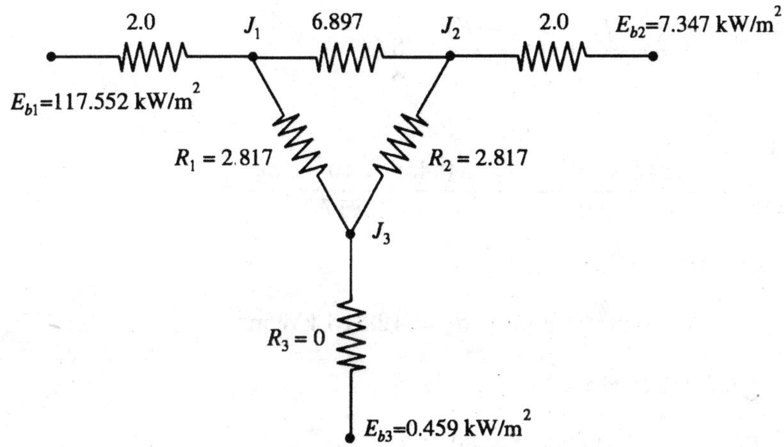

Fig. 7.28 Thermal radiation network for Example 7.11.

The surface resistances to radiation are given by

$$R_1 = \frac{1-\epsilon_1}{A_1\epsilon_1} = \frac{1-0.5}{0.5 \times 0.5} = 2.0$$

$$R_2 = \frac{1-\varepsilon_2}{A_2\epsilon_2} = \frac{1-0.5}{0.5 \times 0.5} = 2.0$$

Since the area of the room A_3 is very large,

$$R_3 = \frac{1-\epsilon_3}{A_3\epsilon_3} = 0$$

Since $R_3 = 0$, we have $J_3 = E_{b3}$.

To calculate the heat flows at each surface we should determine the radiosities J_1 and J_2. The sum of the heat currents entering each node should be zero. Therefore,

For node J_1:

$$\frac{E_{b1} - J_1}{2.0} + \frac{J_2 - J_1}{6.897} + \frac{E_{b3} - J_1}{2.817} = 0$$

or

$$\frac{117.552 \times 10^3 - J_1}{2.0} + \frac{J_2 - J_1}{6.897} + \frac{0.459 \times 10^3 - J_1}{2.817} = 0 \qquad \text{(A)}$$

For node J_2:

$$\frac{J_1 - J_2}{6.897} + \frac{E_{b2} - J_2}{2.0} + \frac{E_{b3} - J_2}{2.817} = 0$$

or

$$\frac{J_1 - J_2}{6.897} + \frac{7.347 \times 10^3 - J_2}{2.0} + \frac{0.459 \times 10^3 - J_2}{2.817} = 0 \qquad \text{(B)}$$

Solving Eqs. (A) and (B), we get

$$J_1 = 60.723 \text{ kW/m}^2 \quad \text{and} \quad J_2 = 12.315 \text{ kW/m}^2$$

Heat lost by plate 1 is given by

$$q_1 = \frac{E_{b1} - J_1}{2.0} = \frac{(117.552 - 60.723) \times 10^3}{2.0} = 28.415 \text{ kW}$$

Heat received by the room is given by

$$q_3 = \frac{J_1 - E_{b3}}{2.817} + \frac{J_2 - E_{b3}}{2.817}$$

$$q_3 = \frac{(60.723 - 0.459 + 12.315 - 0.459) \times 10^3}{2.817} = 25.602 \text{ kW}$$

EXAMPLE 7.12 Liquid Helium at 4.2 K is stored in a dewar flask of ID = 0.48 m and OD = 0.5 m. The dewar flask can be treated as a spherical vessel. The outer surface of the inner vessel and the inner surface of the outer vessel are well polished and the emissivity of these surfaces is 0.05. The space between the two vessels is thoroughly evacuated. The inner surface of the dewar flask is at 4.2 K while the outer surface is at 300 K. Estimate the rate of vaporization of Helium if the enthalpy of vaporization of Helium is 83.3 J/mol.

SOLUTION There is no conduction heat transfer, since the space between the two spherical shells is completely evacuated. The surfaces are real and hence they can be treated as gray diffuse surfaces. The thermal radiation network is shown in Fig. 7.29.

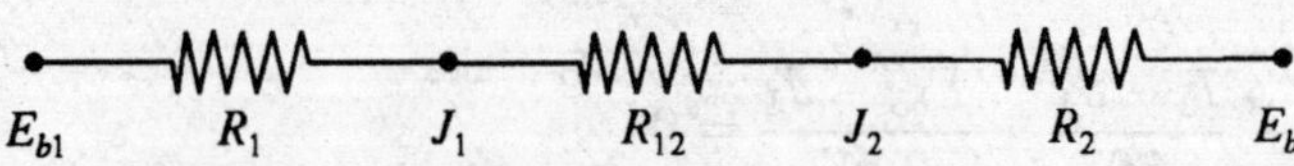

Fig. 7.29 Radiation network for Example 7.12.

For the inner sphere

$$A_1 = 4\pi r_i^2 = \pi D_i^2 = \pi(0.48)^2 = 0.7238 \text{ m}^2$$

$$E_{b1} = \sigma T_1^4 = 5.669 \times 10^{-8}(4.2)^4 = 0.1746 \times 10^{-4} \text{ W/m}^2$$

$$R_1 = \frac{1-\varepsilon_1}{\varepsilon_1 A_1} = \frac{1-0.05}{0.05 \times 0.7238} = 26.25$$

For the outer sphere

$$A_2 = \pi D_0^2 = \pi(0.5)^2 = 0.7854 \text{ m}^2$$

$$E_{b2} = \sigma T_2^4 = 5.669 \times 10^{-8}(300)^4 = 459.189 \text{ W/m}^2$$

$$R_2 = \frac{1-\epsilon_2}{\epsilon_2 A_2} = \frac{1-0.05}{0.05 \times 0.7854} = 24.19$$

$$R_{12} = \frac{1}{A_1 F_{12}} = \frac{1}{0.7238 \times 1} = 1.3816$$

$$\sum R = 26.25 + 24.19 + 1.3816 = 51.8216$$

The total heat transfer between the surfaces is given by

$$q = \frac{E_{b1} - E_{b2}}{\sum R} = \frac{0.1764 \times 10^{-4} - 459.189}{51.8216} = -8.86 \text{ W}$$

Rate of vaporization of Helium

$$= \frac{8.86}{83.3} = 0.1064 \text{ mol/s} = 0.1064 \times 4 = 0.426 \text{ g/s}$$

(Molar mass of Helium = 4 g/mol)

7.10 Radiation Shields

One method of reducing radiant heat transfer between two surfaces is to polish or mirror the surfaces such that they are highly reflective. This method is invariably used in thermos flasks. An alternative method is to use radiation shields in between the heat exchange surfaces. These shields provide additional resistance in the heat flow path so that the overall rate of heat transfer is reduced. Consider two infinite parallel plates shown in Fig. 7.30 (a). The radiation heat transfer between the two parallel plates is given by

$$q = \frac{E_{b1} - E_{b2}}{\dfrac{1-\varepsilon_1}{A_1\varepsilon_1} + \dfrac{1}{A_1F_{12}} + \dfrac{1-\varepsilon_2}{A_2\varepsilon_2}} \tag{7.47}$$

For the two parallel plates, $A_1 = A_2 = A$ and $F_{12} = 1$. Then Eq. (7.47) reduces to

$$\frac{q}{A} = \frac{E_{b1} - E_{b2}}{\dfrac{1}{\varepsilon_1} + \dfrac{1}{\varepsilon_2} - 1} = \frac{\sigma(T_1^4 - T_2^4)}{\dfrac{1}{\epsilon_1} + \dfrac{1}{\epsilon_2} - 1} \tag{7.55}$$

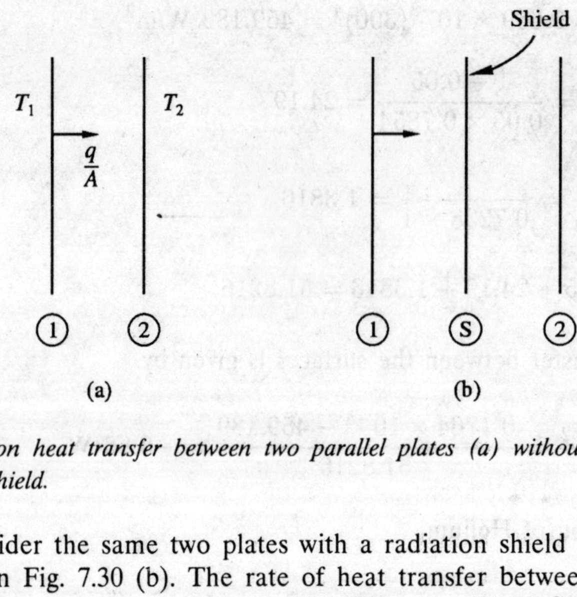

Fig. 7.30 Radiation heat transfer between two parallel plates (a) without a radiation shield. (b) with radiation shield.

Now, let us consider the same two plates with a radiation shield placed in between them as shown in Fig. 7.30 (b). The rate of heat transfer between the plate 1 and shield S is equal to the rate of heat transfer from the shield S to plate 2. Let T_s represent the temperature of the shield and ε_s represent the emissivity of the shield. Thus the overall radiation heat transfer with the shield is

$$\left(\frac{q}{A}\right)_{1-s} = \left(\frac{q}{A}\right)_{s-2} = \left(\frac{q}{A}\right)^1$$

or

$$\left(\frac{q}{A}\right)^1 = \frac{\sigma(T_1^4 - T_s^4)}{\dfrac{1}{\varepsilon_1} + \dfrac{1}{\varepsilon_s} - 1} = \frac{\sigma(T_s^4 - T_2^4)}{\dfrac{1}{\varepsilon_s} + \dfrac{1}{\varepsilon_2} - 1} \tag{7.56}$$

In Eq. (7.56), the temperature of the shield T_s is unknown. Let us consider the simple case in which the emissivities of all the surfaces are the same. That is,

$$\varepsilon_1 = \varepsilon_2 = \varepsilon_s \tag{7.57}$$

Then we get

$$T_s^4 = \frac{T_1^4 + T_2^4}{2} \tag{7.58}$$

$$\left(\frac{q}{A}\right)^1 = \frac{\sigma(T_1^4 - T_2^4)}{2\left(\frac{2}{\varepsilon} - 1\right)} \tag{7.59}$$

When $\varepsilon_1 = \varepsilon_2 = \varepsilon$, Eq. (7.55) reduces to

$$\frac{q}{A} = \frac{\sigma(T_1^4 - T_2^4)}{(2/\varepsilon) - 1} \tag{7.60}$$

Equations (7.59) and (7.60) show that $(q/A)^1$, the rate of radiation heat transfer with one radiation shield is one half of (q/A), the rate of radiation heat transfer without a shield.

The radiation network for two parallel plates without a radiation shield is shown in Fig. 7.31, and the radiation network for the same two parallel plates with one radiation shield placed in between the plates is presented in Fig. 7.32.

A comparison of Figs. 7.31 and 7.32 shows that the introduction of one radiation shield adds three additional resistances in the heat flow path—two surface resistances (one for each surface of the shield) and one space resistance. Therefore insertion of a radiation shield reduces the radiation heat transfer rate.

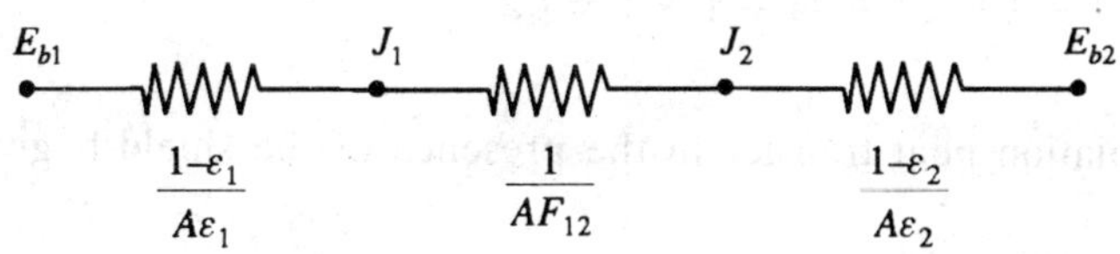

Fig. 7.31 *Radiation network for two plates without a radiation shield.*

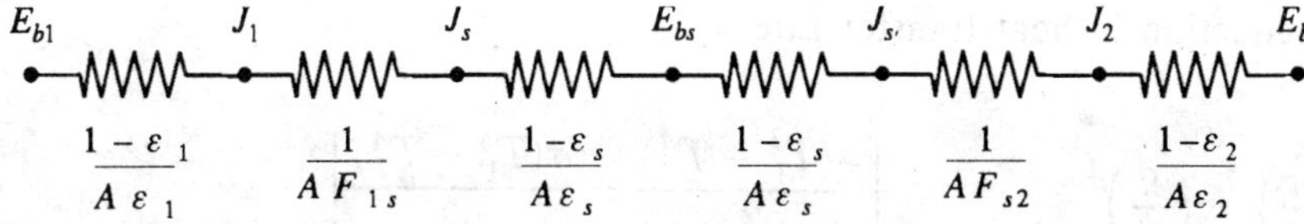

Fig. 7.32 *Radiation network for two parallel plates with one radiation shield placed in between them.*

EXAMPLE 7.13 Two very large parallel plates with emissivities 0.5 exchange heat. Determine the percentage reduction in the heat transfer rate if a polished aluminium radiation shield (ε = 0.04) is placed in between the plates.

SOLUTION Let T_1 and T_2 represent the temperatures of the plates. In the absence of the radiation shield, the heat transfer rate per unit area of the plate is given by

$$\frac{q}{A} = \frac{\sigma(T_1^4 - T_2^4)}{\dfrac{1}{\varepsilon_1} + \dfrac{1}{\varepsilon_2} - 1} = \frac{\sigma(T_1^4 - T_2^4)}{\dfrac{1}{0.5} + \dfrac{1}{0.5} - 1} = \frac{\sigma\left(T_1^4 - T_2^4\right)}{3} \tag{A}$$

When the radiation shield is placed in between the plates, we have

$$F_{1s} = F_{s2} = 1$$

The resistances are (see Fig. 7.32)

$$\frac{1-\varepsilon_1}{\varepsilon_1} = \frac{1-\varepsilon_2}{\varepsilon_2} = \frac{1-0.5}{0.5} = 1$$

$$\frac{1-\varepsilon_s}{\varepsilon_s} = \frac{1-0.04}{0.04} = 24$$

The total resistance is given by

$$\frac{1-\varepsilon_1}{\varepsilon_1} + \frac{1}{F_{1s}} + \frac{1-\varepsilon_s}{\varepsilon_s} + \frac{1-\varepsilon_s}{\varepsilon_s} + \frac{1}{F_{s2}} + \frac{1-\varepsilon_2}{\varepsilon_2}$$

$$= 1 + 1 + 24 + 24 + 1 + 1 = 52$$

The rate of radiation heat transfer in the presence of the shield is given by

$$\left(\frac{q}{A}\right)^1 = \frac{\sigma(T_1^4 - T_2^4)}{52} \tag{B}$$

Percent reduction in heat transfer rate =

$$\frac{\left(\dfrac{q}{A}\right) - \left(\dfrac{q}{A}\right)^1}{\left(\dfrac{q}{A}\right)} \times 100 = \frac{\left[\dfrac{\sigma(T_1^4 - T_2^4)}{3} - \dfrac{\sigma(T_1^4 - T_2^4)}{52}\right]}{\dfrac{\sigma(T_1^4 - T_2^4)}{3}} \times 100 = 92.86$$

7.11 Summary

Thermal radiation is that electromagnetic radiation emitted by a body as a result of its temperature. The electromagnetic radiation propagates with the speed of light (C). The wavelength of radiation (λ) is defined as the ratio of the propagation velocity to the frequency (ν). Thermal radiation wavelengths range from 0.1 to 100 μ m.

The total amount of radiation emitted by a body per unit time and per unit area is called emissive power (E). The emissive power depends on the temperature and characteristics of the body. At a given temperature, the quantity of radiation emitted per unit wavelength is different at various wavelengths. The emissive power has a maximum value at a particular wavelength for a given temperature.

When radiation energy is incident on a body, it is partially reflected, partially absorbed and partially transmitted. The reflectivity (ρ) is defined as the fraction of the incident radiation reflected from the surface of the body. The absorptivity (α) is defined as the fraction of the incident radiation absorbed by the body. The transmissivity (τ) is defined as the fraction of the incident radiation transmitted through the body. The bodies which do not transmit radiation are called opaque.

$$\rho + \alpha + \tau = 1 \qquad \text{for all bodies}$$

$$\rho + \alpha = 1 \qquad \text{for opaque bodies}$$

The reflection of thermal radiation may be regular or diffuse. In regular reflection, the angle of incidence is equal to the angle of reflection. Regular reflection can be observed from smooth or highly polished surfaces. The reflection of radiation from a rough surface occurs in all directions, and such a reflection is called diffuse.

The black body or an ideal radiator is defined as a body which absorbs all the radiation incident upon it and does not reflect or transmit. The black body emits the maximum possible thermal radiation at all wavelengths. The black body is a theoretical concept and is used as a standard of perfection against which the radiation characteristics of real bodies are compared.

The emissivity (ε) of a body is defined as the ratio of the emissive power of a body to the emissive power of a black body. Kirchoff's identity gives $\varepsilon = \alpha$. The emissive power of a black body (E_b) at temperature T is given by the Stefan–Boltzmann law

$$E_b = \sigma T^4$$

where $\sigma = 5.669 \times 10^{-8}$ W/m^2 K^4 = Stefan–Boltzmann constant.

The Stefan–Boltzmann law gives the total emissive power of a black body. It does not provide the distribution of energy among the different wavelengths. The emissive power of a black body as a function of wavelength is given by Planck's law

$$E_{b\lambda} = \frac{c_1 \lambda^{-5}}{\exp\left(\dfrac{c_2}{\lambda T}\right) - 1}$$

where $E_{b\lambda}$ is the monochromatic emissive power of a black body. The wavelength at which monochromatic emissive power is a maximum shifts to shorter wavelengths as the temperature of the black body increases. Wein's displacement law gives the relationship between the wavelength λ_{max} at which $E_{b\lambda}$ is maximum and the temperature as

$$\lambda_{max}T = 2897.6 \ \mu\text{m K}$$

The intensity of radiation I from a surface is defined as the radiant energy propagated in a particular direction per unit solid angle and per unit area of the surface as projected on a plane perpendicular to the direction of propagation. The total emissive power of a diffuse surface is given by

$$E = \pi I$$

If the monochromatic emissivity ε of a body is constant over the entire wavelength spectrum, then the body is said to be gray. The emissive power (E_g) of a gray body is given by

$$E_g = \varepsilon\sigma T^4$$

The fraction of the radiant energy leaving the surface A_i which reaches the surface A_j is called the shape factor F_{ij} or view factor or angle factor or configuration factor. The shape factor is given by

$$A_1F_{12} = A_2F_{21} = \int_{A1}\int_{A2} \frac{\cos\phi_1 \cos\phi_2}{\pi r^2} dA_1 dA_2$$

where ϕ_1 and ϕ_2 are the angles between the normal to the surfaces and the line joining the surfaces A_1 and A_2. The distance between the surfaces is denoted by r. The relation

$$A_1F_{12} = A_2F_{21}$$

is called the reciprocity relation.

The net rate of radiation heat transfer between two black surfaces (q_{net}) is given by

$$q_{net} = (E_{b1} - E_{b2})A_1F_{12} = (E_{b1} - E_{b2})A_2F_{21}$$

The shape factor can be determined by evaluating the double integral, which is a tedious task. The shape factors for a number of geometries have been evaluated and the results are available in the form of graphs. The shape factors for some two- and three-dimensional configurations are also available in the form of equations. The use of shape factor formulae and graphs can be extended by the use of the shape factor algebra. The shape factor algebra involves the judicious use of reciprocity theorem and the principle of conservation of energy. In an enclosure of n surfaces the shape factors must satisfy the condition

$$\sum_{j=1}^{n} F_{ij} = 1$$

If the surface j is subdivided into m subsurfaces, then

$$F_{ij} = \sum_{k=1}^{m} F_{ik}$$

F_{ii} is called the self-viewing factor and it is non-zero only for a concave surface. The shape factors of two-dimensional configurations can also be determined by the Hottel's string rule.

The net rate of radiation heat transfer between two black surfaces is given by

$$q = \frac{E_{b1} - E_{b2}}{1/A_1F_{12}}$$

The form of the above equation suggests that it is analogous to electrical current flow through an electrical circuit. q is analogous to electric current, $1/A_1F_{12}$ is the analogue of electrical resistance and $(E_{b1} - E_{b2})$ is the analogue of electric potential difference. For the surfaces i and j, the space radiation resistance R_{ij} is given by

$$R_{ij} = \frac{1}{A_iF_{ij}}$$

Using the analogy, one can draw an equivalent electrical circuit for a given radiation network.

The radiation from gray surfaces can be treated in terms of irradiation and radiosity. The irradiation G is defined as the total radiation incident upon a surface per unit time and per unit area. The radiosity J is defined as the total radiation which leaves a surface per unit time and per unit area. The radiosity is the sum of the radiation emitted, reflected and transmitted. For opaque bodies the radiosity is given by

$$J = \rho G + \epsilon E_b$$

For an opaque body, the reflectivity ρ is given by

$$\rho = 1 - \alpha = 1 - \varepsilon$$

and hence $J = (1 - \varepsilon)G + \varepsilon E_b$

Then, the net rate at which radiation is leaving a gray surface per unit time is equal to the difference between the radiosity and irradiation. That is,

$$q = A(J - G) = A\left(J - \frac{J - \varepsilon E_b}{1 - \varepsilon}\right) = \frac{A\varepsilon}{1 - \varepsilon}(E_b - J)$$

or

$$q = \frac{E_b - J}{(1 - \varepsilon)/A\varepsilon}$$

The surface resistance to radiation is given by $(1 - \varepsilon)/A\varepsilon$. The net radiation energy exchange between two gray surfaces is given by

$$q = \frac{E_{b1} - E_{b2}}{\dfrac{1 - \varepsilon_1}{A_1\varepsilon_1} + \dfrac{1}{A_1 F_{12}} + \dfrac{1 - \varepsilon_2}{A_2\varepsilon_2}}$$

The above equation can be rewritten as

$$q = A_1\mathcal{F}_{12}(E_{b1} - E_{b2})$$

where the radiation transfer factor or gray body shape factor $\mathcal{F}_{12}$ is given by

$$\mathcal{F}_{12} = \frac{1}{\left(\dfrac{1}{\varepsilon_1} - 1\right) + \dfrac{1}{F_{12}} + \dfrac{A_1}{A_2}\left(\dfrac{1}{\varepsilon_2} - 1\right)}$$

For the special case of an enclosure consisting of two large parallel plates facing each other,

$$\mathcal{F}_{12} = \frac{1}{\dfrac{1}{\varepsilon_1} + \dfrac{1}{\varepsilon_2} - 1}$$

and for the special case of a convex surface A_1 bounded by a surface A_2, like concentric cylinders or spheres where $A_2 \gg A_1$, we get $\mathcal{F}_{12} = \varepsilon_1$.

One method of reducing radiant heat transfer between two surfaces is to use radiation shields in between the heat exchange surfaces. For the special case of two parallel plates of equal area and equal emissivities facing each other, it can be shown that the rate of radiation heat transfer with one radiation shield in between the plates will be just half of that which would be experienced if there was no radiation shield.

Review Questions

1. How does thermal radiation differ from other types of electromagnetic radiation?
2. What is the relation between wavelength and frequency ?
3. Define emissive power of a body.
4. Define the terms—reflectivity, transmissivity and absorptivity.
5. What is meant by an opaque body ?
6. How does regular or specular reflection differ from diffuse reflection?
7. What is a black body ?
8. Define emissivity of a body.
9. What is Kirchoff's identity ?
10. What is Stefan–Boltzmann law ?
11. What is Planck's law ?
12. What is Wien's displacement law ?
13. How does an increase in temperature of the black body affect the wavelength at which monochromatic emissive power is maximum ?
14. What is the use of radiation functions ?
15. Define intensity of radiation.
16. What is a gray body ?
17. What is the physical significance of shape factor ?
18. What are the other names for shape factor ?
19. What is reciprocity theorem ?
20. Explain Hottel's string rule.
21. Define space radiation resistance.
22. Define the terms—irradiation and radiosity.
23. What is surface resistance to radiation ?
24. What is gray body shape factor ?
25. What purpose does a radiation shield serve?

Problems

7.1 Calculate the black body emissive power E_b and the peak wavelength λ_{max} for surfaces at 300 K, 1000 K and 2000 K.

7.2 Derive Wein's displacement law from Planck's law.

7.3 Assuming that the Sun is a black body at 5800 K, estimate the percent radiation in the wavelength range 0.1 to 2.5μm.

7.4 A furnace is provided with a glass window which transmits 90 percent of the incident radiation in the wavelength range 0.35 to 2.7 μm. Estimate the percent of the furnace's radiation which the glass will transmit, assuming that the furnace is a black body at 1800 K.

7.5 Two parallel concentric disks have d_1 = 10 cm and d_2 = 4 cm and are spaced 6 cm apart. Determine the shape factors F_{12} and F_{21}.

7.6 Two parallel black plates 0.5 m × 1 m are kept 2 m apart and they face each other. One plate is held at 600 K while the other plate is maintained at 400 K. Calculate the radiant heat exchange between the plates.

7.7 A domestic oven has the shape of a cubical box of side length 0.5 m. Determine the shape factors (a) between the bottom surface and one side wall and (b) between the top and bottom surfaces.

7.8 A bathroom of 2 m × 3 m has a ceiling height of 3 m. A small drain is located in the floor in the middle of the bathroom. Determine the fraction of energy leaving the ceiling which strikes the drain.

7.9 Determine the shape factor F_{12} for the surfaces shown in Fig.7.33.

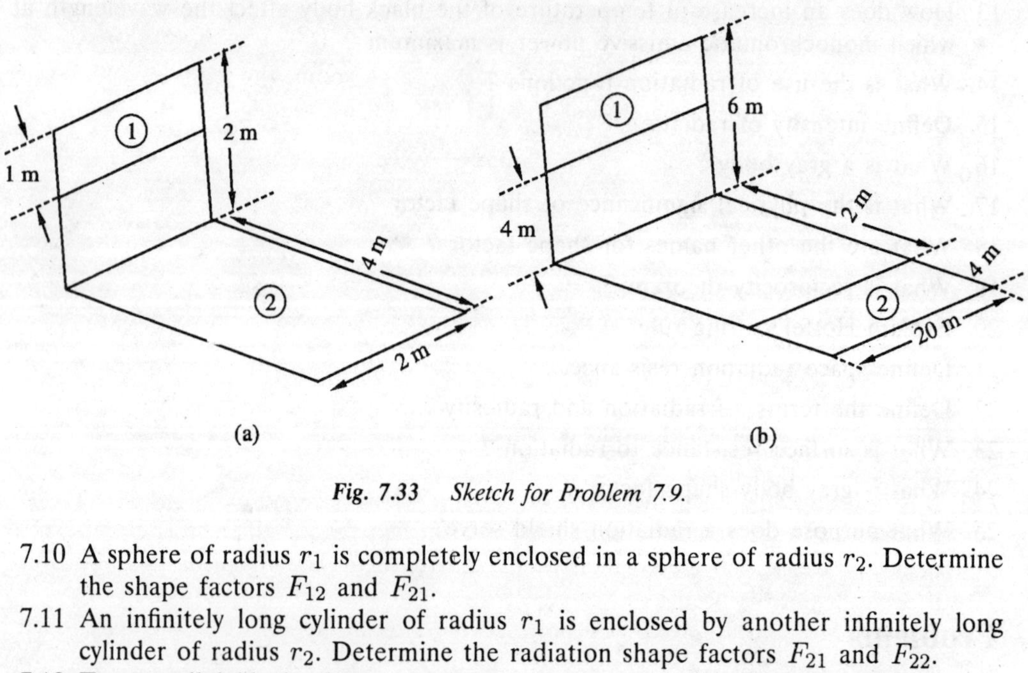

Fig. 7.33 Sketch for Problem 7.9.

7.10 A sphere of radius r_1 is completely enclosed in a sphere of radius r_2. Determine the shape factors F_{12} and F_{21}.

7.11 An infinitely long cylinder of radius r_1 is enclosed by another infinitely long cylinder of radius r_2. Determine the radiation shape factors F_{21} and F_{22}.

7.12 Two parallel black plates of 0.3 m × 0.6 m are separated by a distance of 2 m. One plate is maintained at 1000 K while the other plate is held at 800 K. These plates are placed inside a large hall, the walls of which are at 300 K. The plates exchange heat with each other and with the hall only. Draw the radiation network and indicate the thermal resistances on the network.

7.13 Determine the net heat transfer from each plate and to the room for Problem 7.12.

7.14 Determine the shape factor for the geometrical configuration shown in Fig.7.34.

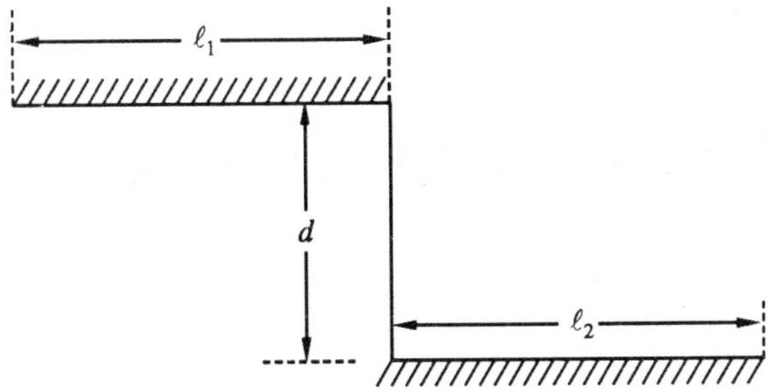

Fig. 7.34 Sketch for Problem 7.14.

7.15 In Problem 7.12, assume that the plates are non-black and their emissivities are 0.35 and determine the net heat transfer rate from each plate and to the room.

7.16 Draw the equivalent electrical circuit for Problem 7.15.

7.17 Two parallel real surfaces of size 0.3 m × 0.5 m are placed 1 m apart. One plate with emissivity 0.3 is held at 300°C while the other plate with emissivity 0.45 is held at 200°C. These plates are placed in a large room, the walls of which are at 10°C. Determine the net heat transfer rate to the room.

7.18 Draw the equivalent electrical circuit for Problem 7.17.

7.19 A steel pipe of 2 cm OD is maintained at 300°C by passing superheated steam through it. The emissivity of the pipe's surface is 0.6. The pipe is passing through a room which is at 17°C. Determine the rate of radiative heat transfer from the pipe to the room per unit area of the pipe wall.

7.20 A spherical vessel of ID = 28 cm is filled with ice at 0°C. This vessel is placed in another spherical vessel of ID = 30 cm and the space between the two vessels is completely evacuated. If the emissivity of the inner vessel is 0.1 and that of the outer vessel is 0.05, and the outer vessel is at a temperature of 40°C, determine the amount of ice which melts in 10 hours. The latent heat of fusion of water at 0°C is 333.3 kJ/kg.

7.21 It is desired to estimate the percentage reduction in the radiative heat transfer by the addition of a radiation shield in between two parallel plates. The emissivity of each plate is 0.4. Determine the percent reduction in radiation heat transfer if a black plate is used as a radiation shield in between the plates.

7.22 Rework Problem 7.21 if a well-polished aluminium plate with $\varepsilon = 0.05$ is used as a radiation shield.

7.23 Two large parallel plates held at temperatures T_1 and T_2 are facing each other and are set apart. If n radiation shields are inserted between the plates while the plates are still maintained at T_1 and T_2, determine the ratio of the radiation energy transfer rate with shields to the radiation energy transfer in the absence of the shields, assuming that the emissivities of the plates and shields are equal.

CHAPTER EIGHT

PRINCIPLES OF CONVECTION

Learning Objectives

This chapter will enable the readers to

- define the convection heat transfer coefficient and understand its importance.
- understand the relation between fluid flow and heat transfer.
- understand natural convection and forced convection.
- define the Nusselt number and understand its physical significance.
- know several methods of estimating convection heat transfer coefficient.
- perform dimensional analysis.
- know the Buckingham π-theorem and its limitations.
- correlate the experimental data in terms of dimensionless groups and obtain empirical relations.
- define several dimensionless groups which deal with heat transfer.
- know about the principle of similarity.
- derive the continuity equation, momentum equation and energy equation for a laminar boundary layer.
- define hydrodynamic and thermal boundary layer thickness.
- know the Reynolds–Colburn analogy between fluid friction and heat transfer.
- obtain exact mathematical solutions to laminar boundary layer equations for flow over a flat plate.
- understand the approximate analysis of boundary layer equations by the integral method.
- know the analogy between heat and momentum transfer.
- know the analysis of turbulent boundary layer and to estimate turbulent boundary layer thickness.
- state the universal velocity profile.
- understand the laminar and turbulent flows in a tube.
- to define bulk temperature or cup mixing temperature.
- understand the correlations for estimating convection heat transfer coefficient for flow through a tube.

8.1 Convection Heat Transfer Coefficient

We have discussed conduction and radiation heat transfer in the earlier chapters. In order to simplify the analysis, we had eliminated, as much as possible, the problems related to convection. In practice, there are several situations in which heat is transferred from the surface of a conducting body to the surrounding fluid by convection. The rate of heat transfer q by convection from the surface of a body to the surrounding fluid is given by

$$q = hA(T_s - T_\infty) \tag{8.1}$$

where

T_s = temperature of the surface

T_∞ = temperature of the fluid

A = surface area of the body

h = convection heat transfer coefficient.

Equation (8.1) is a definition of the convection heat transfer coefficient and it is not a law of heat transfer. The convection heat transfer coefficient depends on the geometry of the system, the thermal properties of the fluid and the characteristics of the fluid flow. The value of the convection heat transfer coefficient depends on the location where the temperature of the fluid is measured, and h does not remain constant over the entire surface of the body. Though Eq. (8.1) is applicable in evaluating the rate of convection heat transfer, it does not provide any information regarding the mechanism of heat flow. The convection heat transfer analysis requires a knowledge of fluid dynamics in addition to energy balance. The transfer of the heat between a solid surface and the surrounding fluid takes place by a combination of conduction and mass transport. If the surface of the body is at a higher temperature than the fluid, the heat flows first by conduction from the surface of the body to the fluid particles adjacent to the surface of the body. These heated fluid particles are subsequently carried away by the flowing fluid to regions of low temperature, where energy is again transferred by conduction from the hot fluid particles to cold fluid particles. Thus the convection heat transfer is closely coupled with fluid motion. Hence, it is necessary to understand the flow of fluids, before attempting to understand convection heat transfer.

8.2 Fluid Flow and Heat Transfer

The flow of the fluid can be laminar or turbulent. In *laminar* or *streamline flow*, the fluid moves in layers and each fluid particle follows along a smooth and continuous path. The fluid particles in each layer follow one another and the fluid particles in one layer do not mix with the fluid particles in the other layers. In contrast to this laminar flow, the motion of the fluid particles in turbulent flow is quite irregular and mixing occurs. Some fluid particles move with higher velocity, while some other fluid particles move with lower velocity, although the aggregate or statistical average velocity of the fluid remains constant. The paths and velocities of individual fluid particles will be different. When a fluid flows in laminar motion along a surface at a temperature which is greater than that of the fluid, heat is transferred by conduction from the surface to the fluid particles which are adjacent to the surface, and then heat is transferred between the fluid layers by molecular conduction. On the other hand, in turbulent flow the conduction heat transfer is aided by eddies which carry lumps of fluid particles from one region to another region causing mixing. As the rate of mixing or turbulence is increased, the rate of heat transfer also increases.

The fluid motion can be induced by two different processes. If there is a temperature gradient in the fluid, it causes a density gradient which in turn sets the fluid into motion. The heat transfer in a fluid whose motion is induced by density difference is called *natural* or *free convection* heat transfer. The fluid motion can also be induced by an external agent such as a pump or blower. When the motion of the fluid is caused by an external agent, the heat transfer is called *forced convection* heat transfer.

Consider the flow over a flat plate as shown in Fig. 8.1. When the fluid flows along a surface, the fluid particles in the vicinity of the surface are slowed down by virtue of viscous forces. The fluid particles adjacent to the surface stick to it and their velocity will be zero. Other fluid particles sliding over them will be retarded because of interaction between the fast moving and slow moving fluid layers. Beginning at the leading edge of the plate, the effects of the viscous forces extend into the body of the fluid.

The viscosity of the fluid is defined by

$$\tau = \mu \frac{du}{dy} \tag{8.2}$$

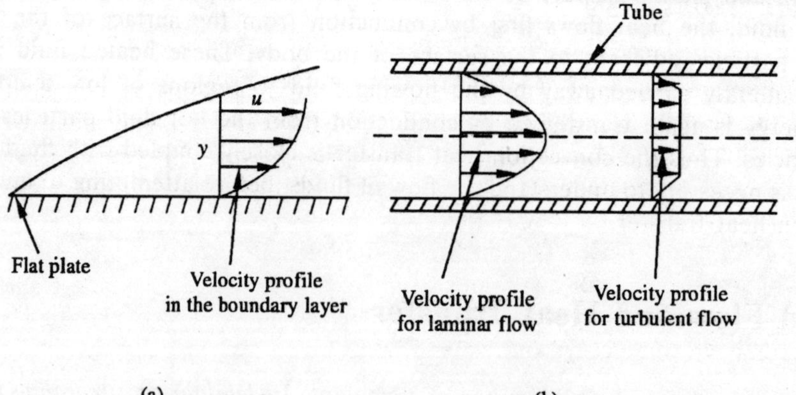

Fig. 8.1 Velocity profile for flow (a) over a flat plate. (b) through a tube.

where

τ = shear stress between the fluid layers

μ = dynamic viscosity

u = velocity.

The transition from laminar to turbulent flow occurs in a tube if the Reynolds number is greater than 2100, where the Reynolds number (Re) is given by

$$Re = \frac{DV\rho}{\mu} \tag{8.3}$$

where

D = diameter of the tube

ρ = density of the fluid.

For flow over a flat plate, the transition from laminar flow to turbulent flow occurs when

$$Re_x = \frac{u_\infty x \rho}{\mu} > 5 \times 10^5$$

where

x = the distance from the leading edge

u_∞ = free-stream velocity of the fluid.

8.3 Nusselt Modulus

The conduction and the mass transport play an important role in convection heat transfer. If the velocity of the fluid is small, the heat transfer is not aided by eddies and mixing, and is largely governed by conduction from the surface to the fluid. Since the thermal conductivity of liquids is very small, the heat transfer coefficient in natural convection will be low. If the fluid velocity is large, turbulence occurs and mixing of matter takes place. The mixing of fluid from regions of high temperature with the fluid from regions of low temperature increases the rate of heat transfer and conduction becomes less important. At the interface of the solid boundary and the fluid, heat flows by conduction and we can write

$$q = -kA\left(\frac{\partial T}{\partial y}\right)_{y=0} \tag{8.4}$$

The convection heat transfer coefficient is defined by Eq. (8.1). From Eqs. (8.1) and (8.4), we can write

$$\frac{h\ell}{k} = \frac{-\left(\dfrac{\partial T}{\partial y}\right)_{y=0}}{(T_w - T_\infty)/\ell} = \left[\frac{\partial\left(\dfrac{T_w - T}{T_w - T_\infty}\right)}{\partial(y/\ell)}\right]_{y=0} = Nu \tag{8.5}$$

where

Nu = Nusselt modulus

ℓ = a significant length which specifies the geometry of the solid body.

Equation (8.5) shows that the Nusselt modulus physically signifies the ratio of the temperature gradient at the surface to a reference temperature gradient $(T_w - T_\infty)/\ell$. The Nusselt modulus or Nusselt number is a convenient measure of convection heat transfer coefficient and can be evaluated from the relation

$$Nu = h\ell/k \tag{8.6}$$

8.4 Estimation of Convection Heat Transfer Coefficient

The general methods for the estimation of convection heat transfer coefficients are

1. Dimensional analysis coupled with experimental data.
2. Exact mathematical analysis of the boundary layer equations.
3. Approximate analysis of the boundary layer equations.
4. Analogy between heat, mass and momentum transfer.

All the above four methods contribute some knowledge of the convection heat transfer and yet no single method can provide complete solutions to all the problems. Each of these methods have some limitations.

The dimensional analysis is mathematically simple and finds a wide range of applications. The main drawback or limitation of this method is that it does not contribute to the understanding of the heat transfer mechanism, and the results obtained by this method are incomplete and require experimental data. However, it facilitates the extension of the range of application of experimental data by correlating them in terms of dimensionless groups.

The exact mathematical analysis requires the simultaneous solution of boundary layer equations dealing with the motion of fluid and the transfer of energy in the moving fluid. This necessitates a thorough knowledge of the mechanism so that the phenomena can be adequately described by mathematical equations. This is a serious limitation of the method because equations describing the fluid flow and heat transfer are known only for laminar flow. Even in the case of laminar flow, the equations are quite complicated and the solutions can be obtained for some simple systems only. The exact mathematical analysis is important because the assumptions made in the course of analysis can be verified for their validity through experimentation. The results obtained through this analysis form a basis against which the simpler solutions obtained through approximate methods can be compared.

The approximate analysis of boundary layer equations uses the integral Methods and avoids the detailed mathematical description of the flow. In this method, simple equations are used to describe the velocity and temperature distributions in the boundary layer. The problem is analyzed on a macroscopic level by applying the equations of motion and energy to the fluid contained in the boundary layer. This method is relatively simple and yields solutions to the problems which cannot be analyzed by an exact mathematical analysis.

For analyzing turbulent flow, the analogy between heat, mass and momentum transfer method is useful. The advantage of this method is that the transfer mechanism can be described in terms of simple equations, even though it is not possible to write mathematical equations describing the flow and temperature distribution in turbulent flow. The results predicted by this method are found to be substantially in agreement with the experimental results.

8.5 Dimensional Analysis

The dimensional analysis differs from other methods in the sense that it does not give equations which can be solved. This method combines the variables into dimensionless groups which allow us to extend the range of applicability of experimental data. In practice, the convection heat transfer coefficients are calculated from empirical equations which are obtained by correlating the experimental data in terms of dimensionless groups. The limitations of the dimensional analysis are (a) it is necessary to know apriori the variables which influence the phenomena and (b) it does not provide any knowledge of the mechanism.

8.5.1 Buckingham π-Theorem

The Buckingham π-theorem is used to determine the number of independent dimensionless groups which are necessary to describe the phenomenon in a mathematical expression. The Buckingham π-theorem states that *the number of independent dimensionless groups which can be formed by combining the physical variables of a problem is equal to the total number of physical quantities (n) minus the number of primary dimensions (m) needed to express the dimensional formulae of the n physical quantities.*

The equation which expresses the relationship along the variables is of the form

$$F(\pi_1, \pi_2, \pi_3 ...) = 0 \tag{8.7}$$

where $\pi_1, \pi_2, \pi_3, \ldots$ are the dimensionless groups.

To express the dimensional formulae of the physical quantities, one should select the primary dimensions. We shall use the primary dimensions of mass (M), length (ℓ), time (t) and temperature (T). The dimensional formula of a physical quantity is derived from the definition of the physical quantity or a law which governs it. For example, the velocity is equal to the distance travelled per unit time. That is, the dimensional formula of velocity (u) is given by ℓ/t. The dimensional formulae and the symbols of physical quantities which occur frequently in the analysis of heat transfer are presented in Table 8.1.

Table 8.1 *Dimensional formulae of some physical quantities*

Physical Quantity	Symbol	Dimensional Formula	Units
Mass	M	M	kg
Length	ℓ	ℓ	m
Time	t	t	s
Temperature	T	T	K
Density	ρ	M/ℓ^3	kg/m^3
Velocity	u	ℓ/t	m/s
Acceleration	a	ℓ/t^2	m/s^2
Force	F	$M\ell/t^2$	N
Pressure	P	$M/\ell t^2$	Pa
Work	W	$M\ell^2/t^2$	J
Power	$\dot{W}$	$M\ell^2/t^3$	W
Heat	Q	$M\ell^2/t^2$	J
Internal energy	u	ℓ^2/t^2	J/kg
Enthalpy	h	ℓ^2/t^2	J/kg
Specific heat	c	ℓ^2/t^2T	J/kg K
Dynamic viscosity	μ	$M/\ell t$	kg/m s
Kinematic viscosity	ν	ℓ^2/t	m^2/s
Thermal conductivity	k	$M\ell/t^3T$	W/m K
Thermal resistance	R	$Tt^3/M\ell^2$	K/W
Thermal diffusivity	α	ℓ^2/t	m^2/s
Convection heat transfer coefficient	h	M/t^3T	W/m^2 K

The method of determining the dimensionless groups for a given phenomenon is illustrated below through an example.

EXAMPLE 8.1 The convection heat transfer coefficients for flow of a fluid through a tube have been experimentally determined. Using the Buckingham π-theorem develop an expression for correlating the experimental data.

SOLUTION Let us list the physical quantities along with the dimensional formulae which are likely to influence the convection heat transfer coefficient as given below.

S.No	Physical Quantity	Symbol	Dimensional Formula
1.	Tube diameter	D	ℓ
2.	Thermal conductivity of fluid	k	$M\ell/t^3T$
3.	Velocity of fluid	u	ℓ/t
4.	Density of fluid	ρ	M/ℓ^3
5.	Viscosity of fluid	μ	$M/\ell t$
6.	Specific heat of fluid	c_p	ℓ^2/t^2T
7.	Convection heat transfer coefficient	h	M/t^3T

The number of physical quantities $n = 7$

The number of primary dimensions needed to express the physical quantities $m = 4$

Therefore, the number of dimensionless groups $= n - m = 7 - 4 = 3$.

Let us denote the dimensionless groups as π_1, π_2 and π_3. To determine the dimensionless groups, let us write π as a product of the physical quantities, each raised to an unknown power as

$$\pi = D^a k^b u^c \rho^d \mu^e c_p^f h^g \tag{A}$$

Then substitute the dimensional formulae in Eq. (A) to obtain

$$\pi = \ell^a \left(\frac{M\ell}{t^3T}\right)^b \left(\frac{\ell}{t}\right)^c \left(\frac{M}{\ell^3}\right)^d \left(\frac{M}{\ell t}\right)^e \left(\frac{\ell^2}{t^2T}\right)^f \left(\frac{M}{t^3T}\right)^g \tag{B}$$

Since π is a dimensionless group, the exponents of each primary dimension are equal to zero. Hence, we compare the exponents of primary dimensions from the left-hand side and right-hand side of Eq. (B). Then we obtain

For M: $$b + d + e + g = 0 \tag{C}$$

For ℓ: $$a + b + c - 3d - e + 2f = 0 \tag{D}$$

For t: $$-3b - c - e - 2f - 3g = 0 \tag{E}$$

For T: $$-b - f - g = 0 \tag{F}$$

Thus we obtain four Eqs. (C)–(F) in seven unknowns $-a$, b, c, d, e, f and g. Hence any four unknowns can be expressed in terms of the other three. Therefore, we can arbitrarily choose values for three of the unknowns. Since the convection heat transfer coefficient h is to be correlated, let us choose its exponent g as unity. To simplify the algebraic expressions let us also choose $c = 0$ and $d = 0$. Then Eqs. (C)–(F) reduce to

$$b + e + 1 = 0 \tag{G}$$

$$a + b - e + 2f = 0 \tag{H}$$

$$-3b - e - 2f - 3 = 0 \tag{I}$$

$$-b - f - 1 = 0 \tag{J}$$

Thus we have a set of four Eqs. (G)–(J) in four unknowns a, b, e and f. Solving these four equations simultaneously, we get

$$a = 1; \quad b = -1; \quad e = 0; \quad \text{and} \quad f = 0$$

Hence, the first dimensionless group is given by

$$\pi_1 = \frac{hD}{k} = Nu$$

To determine the second dimensionless group let us choose $g = 0$ so that h will not appear again. Let us also choose $a = 1$ and $f = 0$. By repeating the above procedure we obtain $b = 0$; $c = 1$; $d = 1$ and $e = 1$.

Then the second dimensionless group is given by

$$\pi_2 = \frac{DV\rho}{\mu} = Re = \text{ Reynolds number}$$

To determine the third dimensionless group let us choose $g = 0$, $c = 0$ and $e = 0$ and repeat the above procedure. This yields the third dimensionless group as

$$\pi_3 = \frac{c_p\mu}{k} = Pr = \text{ Prandtl number}$$

Then, the functional relationship between the dimensionless groups can be written as

$$F(Nu, Re, Pr) = 0$$

or

$$Nu = \phi(Re)\psi(Pr) = c(Re)^m(Pr)^n \tag{K}$$

where c, m; and n are constants. Now Eq. (K) can be used to correlate the experimental data. The dimensionless groups of importance in convection heat transfer are presented in Table 8.2.

Table 8.2 *Dimensionless groups characterizing momentum transfer and convection heat transfer*

Dimensionless Group	Definition
Friction coefficient	$C_f = \dfrac{\tau_w}{\rho u^2/2}$
Euler number	$Eu = \dfrac{\Delta P}{\rho u^2}$
Reynolds number	$Re = \dfrac{Du\rho}{\mu}$
Nusselt number	$Nu = \dfrac{h\ell}{k}$
Prandtl number	$Pr = \dfrac{c_p\mu}{k}$
Stanton number	$St = \dfrac{h}{\rho c_p u}$
Grashof number	$Gr = \dfrac{\beta \Delta T g \ell^3}{\nu^2}$
Peclet number	$Pe = \dfrac{u\ell}{\alpha}$
Rayleigh number	$Ra = \dfrac{\beta \Delta T g \ell^3}{\nu\alpha}$
Boussinesq number	$Bo = \dfrac{\beta \Delta T g \ell^3}{\alpha^2}$
Brinkman number	$Br = \dfrac{u^2\mu}{k\Delta T}$

8.5.2 Correlation of Experimental Data

Suppose the convection heat transfer coefficients for the flow of fluids through a tube are measured covering a wide range of velocities. The experimental data can be presented as shown in Fig. 8.2, where h is plotted against velocity v of the fluid. This figure permits one to determine the value of h for a given velocity of the fluid. However, the correlated data as presented in Fig. 8.2 cannot be used if the diameter of the tube is different or if the density or viscosity of the fluid is different from that used in the experimental condition. Unless the data is correlated in a better way it would be necessary to perform additional experiments with different diameters of the tube and by varying the density and viscosity of the fluid. By coupling the variables into dimensionless groups, the same data can be correlated in a better way as shown in Fig. 8.3, where Nu is plotted against Re. Such a correlation allows the prediction of h for different diameters of the tube.

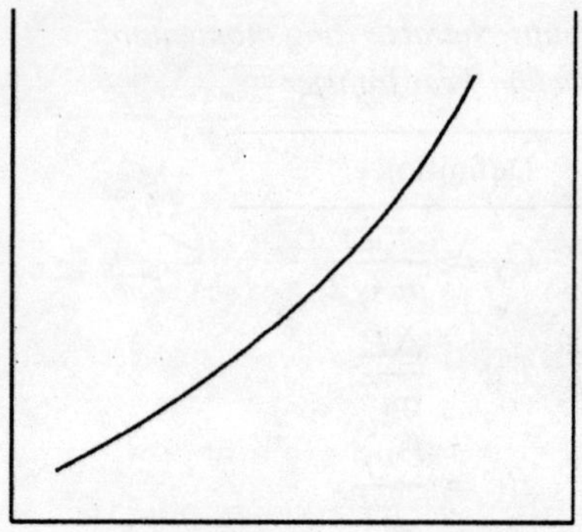

Fig. 8.2 Variation of convection heat transfer coefficient with the velocity of the fluid.

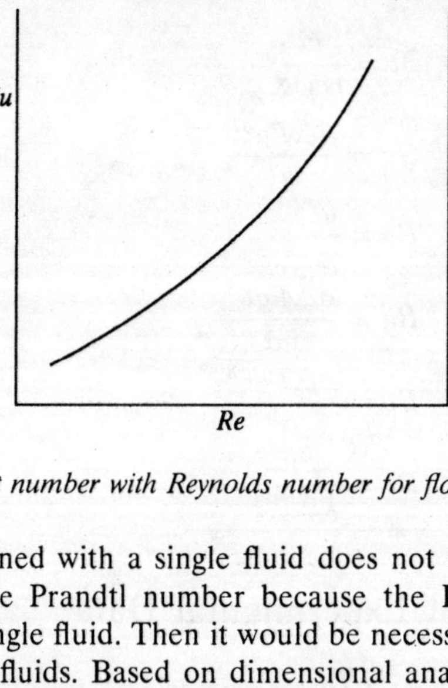

Fig. 8.3 Variation of Nusselt number with Reynolds number for flow of a fluid through a tube.

Experimental data obtained with a single fluid does not reveal the dependence of the Nusselt number on the Prandtl number because the Prandtl number does not change appreciably for a single fluid. Then it would be necessary to perform additional experiments with different fluids. Based on dimensional analysis, we have seen that

$$Nu = c(Re)^m(Pr)^n \tag{8.8}$$

or

$$\ln Nu = \ln c + m\ln(Re) + n\ln(Pr)$$

Therefore, one can plot $\ln(Nu)$ versus $\ln(Pr)$ for a particular value of Re and determine the value of n. Having determined the value of n one can plot $\ln(Nu/Pr^n)$ versus Re to determine the values of c and m. Thus one can correlate the experimental data so that it can be used over a wide range of velocities and for different types of fluids. For laminar flow of a fluid over a flat plate, the convection heat transfer coefficient is given by

$$Nu = 0.332 Re^{1/2} Pr^{1/3} \tag{8.9}$$

8.5.3 Principle of Similarity

The correlation obtained above can be explained by the *principle of similarity* or *model law*. The principle of similarity states that the behaviour of two systems will be similar if the ratios of their linear dimensions, forces, velocities, etc. are the same. If the ratio of inertia forces to viscous forces is the same in the two fluids, the velocity fields will be similar in geometrically similar systems. The Reynolds number is the ratio of inertia forces to viscous forces and hence similar flow conditions will prevail in two systems for a given value of Reynolds number. The Prandtl number is the ratio of the kinematic viscosity to thermal diffusivity. That is,

$$Pr = \frac{\text{Kinematic viscosity}}{\text{Thermal diffusivity}} = \frac{\nu}{\alpha} = \frac{\mu/\rho}{k/\rho c_p} = \frac{c_p \mu}{k} \tag{8.10}$$

The kinematic viscosity affects the velocity distribution and the thermal diffusivity affects the temperature profile. Therefore, the Prandtl number relates the temperature distribution to the velocity distribution. Hence, for the given values of Re and Pr, in geometrically similar systems, the temperature distribution will be similar. The Nusselt number is the ratio of the temperature gradient at the surface and fluid interface to a reference temperature gradient. Therefore, in geometrically similar systems, the values of the Nusselt number will be the same for given values of Reynolds number and Prandtl number.

8.5.4 Limitation of the Buckingham π-Theorem

The Buckingham π-theorem holds good as long as the set of simultaneous equations, formed by equating the exponents of each primary dimension to zero, is linearly independent. However, if the set of equations are linearly dependent, that is if any equation is a linear combination of one or more equations, then the Buckingham π-theorem is to be modified as

$$\begin{Bmatrix} \text{Number of dimen-} \\ \text{sionless groups} \end{Bmatrix} = \begin{Bmatrix} \text{Total number of} \\ \text{physical quantities} \end{Bmatrix} - \begin{Bmatrix} \text{Number of linearly} \\ \text{independent equations} \end{Bmatrix} \tag{8.11}$$

EXAMPLE 8.2 It is desired to correlate the data obtained on the temperature–time history of a lump of steel, originally at a uniform temperature T_0, which is suddenly exposed to a convection environment at T_∞. Apply the Buckingham π-theorem and develop a relation among the pertinent dimensionless groups to correlate the data.

SOLUTION Let us list the physical quantities, with their symbols and dimensional formulae, which are likely to affect the temperature–time history of the lump of steel as shown below.

S.No.	Physical quantity	Symbol	Dimensional formula
1.	Temperature of the lump of steel above the environment temperature at any time t.	$T - T_\infty$	T
2.	Initial temperature of the lump of steel above the environment temperature	$T_0 - T_\infty$	T
3.	Time	t	t
4.	Surface conductance between the lump of steel and the environment	hA	$(M\ell^2/t^3T)$
5.	Thermal capacitance of the lump of steel	ρcV	$(M\ell^2/t^2T)$

$$\text{Number of physical quantities } n = 5$$

$$\text{Number of primary dimensions } m = 4$$

Then according to the Buckingham π-theorem, we get

$$\text{Number of dimensionless groups } = n - m = 5 - 4 = 1$$

Now, the question that naturally arises is how does one correlate only one dimensionless group? Let us proceed further to check whether we are likely to get any additional information or not. We write π as a product of the variables, each raised to an unknown power, as

$$\pi = (T - T_\infty)^a (T_0 - T_\infty)^b t^c (hA)^d (\rho cV)^e \tag{A}$$

Substituting the dimensional formulae in Eq. (A), we get

$$\pi = T^a T^b t^c \left(\frac{M\ell^2}{t^3T}\right)^d \left(\frac{M\ell^2}{t^2T}\right)^e \tag{B}$$

Since π is dimensionless, we get the following relations by comparing the exponents of the primary dimensions on the left-hand side and right-hand side of Eq. (B).

$$\text{For } M: \quad d + e = 0 \tag{C}$$

$$\text{For } \ell: \quad 2d + 2e = 0 \tag{D}$$

For t : $\quad c - 3d - 2e = 0 \qquad$ (E)

For T : $\quad a + b - d - e = 0 \qquad$ (F)

Now we find that Eqs. (C) and (D) are not linearly independent. Therefore, we have only three linearly independent equations. Hence we use the modified Buckingham π-theorem to obtain the number of dimensionless groups.

$$\left\{\begin{array}{l}\text{No. of dimension-}\\ \text{less groups}\end{array}\right\} = \left\{\begin{array}{l}\text{Total no.}\\ \text{of variables}\end{array}\right\} - \left\{\begin{array}{l}\text{No. of linearly}\\ \text{independent Eqs.}\end{array}\right\} = 5 - 3 = 2$$

There are three Eqs. (C), (E) and (F) and five unknowns a, b, c, d and e. Therefore, we have to arbitrarily assign some values to two of the unknowns and determine the values of the remaining three unknowns. Since the temperature– time history of the lump of steel is desired, let us choose $a = 1$ and $c = 0$. Then the equations reduce to

$$d + e = 0$$

$$-3d - 2e = 0$$

$$1 + b - d - e = 0$$

Then the above three equations can be simultaneously solved to obtain

$$b = -1; \quad d = 0 \quad \text{and} \quad e = 0$$

Therefore, the first dimensionless group is given by

$$\pi_1 = \frac{T - T_\infty}{T_0 - T_\infty}$$

To obtain the second dimensionless group we set $a = 0$ so that the temperature difference $(T - T_\infty)$ will not appear again, and choose $c = 1$. By repeating the above procedure we obtain the second dimensionless group as

$$\pi_2 = \frac{hAt}{\rho c V}$$

Then the data can be correlated as

$$F(\pi_1, \pi_2) = 0$$

$$\text{or} \quad \pi_1 = f(\pi_2) \quad \text{or} \quad \frac{T - T_\infty}{T_0 - T_\infty} = f\left(\frac{hAt}{\rho c V}\right)$$

8.6 Laminar Boundary Layer on a Flat Plate

Based on dimensional analysis we found that

$$Nu = \phi(Re)\psi(Pr)$$

In order to determine the functional relationship it is necessary either to correlate the experimental data or to obtain an analytical solution of the problem. In this section we shall discuss the analytical approach. To illustrate the method we shall consider the simplest problem of heat transfer between a flat plate and a fluid flowing parallel to its surface. To simplify the analysis, we shall assume that

1. The fluid is incompressible.
2. The flow is in steady-state.
3. Viscous-shear forces in the y-direction are negligible.
4. There is no pressure variation in the direction perpendicular to the plate.
5. The flow is two-dimensional, that is the velocity distribution is the same in any plane perpendicular to the z-axis.
6. The fluid is not affected by heat flow.

These assumptions are good approximations to forced convection in flow over surfaces of streamlined bodies or in the inlet regions of tubes.

8.6.1 The Continuity Equation

To derive the equations which govern the flow of fluid in a boundary layer, consider an elementary control volume as shown in Fig. 8.4. The control volume has the shape of a parallelopiped with dimensions $dx \times dy \times 1$.

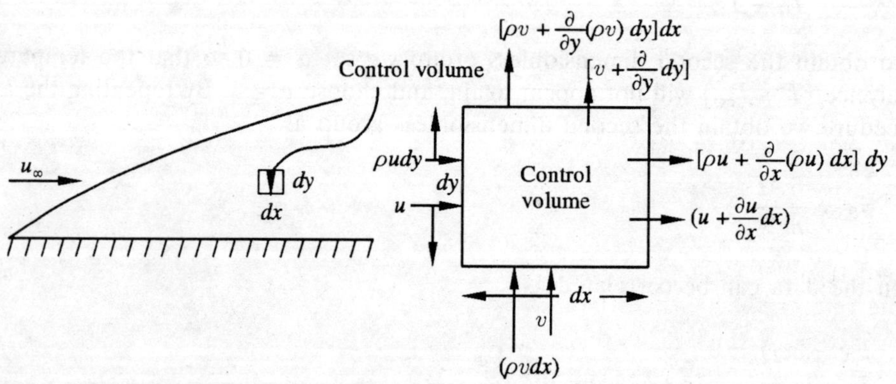

Fig. 8.4 Elementary control volume. u and v denote the velocity components in the x and y directions, respectively.

In steady-state, the principle of conservation of mass gives that during a time interval dt, the mass of fluid entering the control volume is equal to the mass of fluid leaving the control volume.

Mass of fluid entering the control volume through the left face, per unit time

$$= (\rho u dy)$$

Mass of fluid entering the control volume through the lower face, per unit time

$$= (\rho v dx)$$

Mass of fluid leaving the control volume through the right face, per unit time

$$= \left[\rho u + \frac{\partial}{\partial x}(\rho u)dx\right] dy$$

Mass of fluid leaving the control volume through the upper face, per unit time

$$= \left[\rho v + \frac{\partial(\rho v)}{\partial y}dy\right] dx$$

Therefore, the principle of conservation of mass gives

$$\rho u dy + \rho v dx = \rho u dy + \frac{\partial(\rho u)}{\partial x}dxdy + \rho v dx + \frac{\partial(\rho v)}{\partial y}dxdy$$

or

$$\frac{\partial u}{\partial x} + \frac{\partial v}{\partial y} = 0 \tag{8.12}$$

Equation (8.12) is generally known as the *continuity equation* for incompressible, steady, two-dimensional flow.

8.6.2 The Momentum Equation

In addition to the conservation of mass, the flow of fluid is governed by Newton's second law of motion, which states that the sum of forces acting on a body is equal to the rate of change of its momentum. That is,

$$\sum F = \frac{d(mv)}{dt} \tag{8.13}$$

The x-momentum, per unit time, entering from the left face of the control volume

$$= (\rho u dy)u = \rho u^2 dy$$

The x-momentum, per unit time, entering from the lower face of the control volume

$$= (\rho v dx)u = \rho u v dx$$

The x-momentum, per unit time, leaving through the right face of the control volume

$$= \left(\rho u + \frac{\partial}{\partial x}(\rho u)dx\right)\left(u + \frac{\partial u}{\partial x}dx\right)dy = \left[\rho u^2 + \rho u\frac{\partial u}{\partial x}dx + u\frac{\partial(\rho u)}{\partial x}dx\right]dy$$

Similarly, the x-momentum, per unit time, leaving through the top face of the control volume

$$= \left[\rho u v + \rho v\frac{\partial u}{\partial y}dy + u\frac{\partial(\rho v)}{\partial y}dy\right]dx$$

Therefore, the change in momentum, per unit time, of the fluid in the control volume

$$= \left[\rho u\frac{\partial u}{\partial x} + \rho v\frac{\partial u}{\partial y} + \rho u\left(\frac{\partial u}{\partial x} + \frac{\partial v}{\partial y}\right)\right]dxdy \tag{8.14}$$

Substituting the continuity Eq. (8.12) in Eq. (8.14), we get the change in momentum, per unit time, of the fluid in the control volume

$$= \left[\rho u\frac{\partial u}{\partial x} + \rho v\frac{\partial u}{\partial y}\right]dxdy \tag{8.15}$$

The change in momentum is the result of the forces acting on the surface of the control volume. In general, three types of forces, namely (1) body forces such as gravity (2) dynamic forces such as pressure and (3) frictional forces such as viscous-shear, act on the body. In forced convection, the body forces due to gravity are negligible and for flow over a flat plate, the pressure in the flow field is uniform. Therefore, we should consider the viscous-shear forces only.

The shear stress in a plane parallel to the plate and perpendicular tc the y-axis is given by

$$\tau_{yx} = \mu\frac{du}{dy} \tag{8.16}$$

where τ_{yx} is the shear stress per unit area.

At the lower face the shear acting on the fluid in the control volume

$$= \tau_{yx} dx = \mu \frac{du}{dy} dx$$

At the upper face the shear acting on the fluid in the control volume

$$= \left[\tau_{yx} + \frac{\partial}{\partial y}(\tau_{yx}) dy\right] dx = \left[\mu \frac{du}{dy} + \frac{\partial}{\partial y}\left(\mu \frac{\partial u}{\partial y}\right) dy\right] dx$$

Therefore, the net viscous-shear force in the direction of motion

$$= \left[\tau_{yx} + \frac{\partial(\tau_{yx})}{\partial y} dy - \tau_{yx}\right] dx = \frac{\partial}{\partial y}\left(\mu \frac{\partial u}{\partial y}\right) dx dy = \mu \frac{\partial^2 u}{\partial y^2} dx dy$$

Now, equating the net viscous-shear force to the increase in the momentum per unit time, we get

$$u \frac{\partial u}{\partial x} + v \frac{\partial u}{\partial y} = \nu \frac{\partial^2 u}{\partial y^2} \tag{8.17}$$

which is the momentum equation of the laminar boundary layer.

8.6.3 The Energy Equation

We have discussed the fluid dynamics of a laminar boundary layer in Sections 8.6.1 and 8.6.2. We shall now derive the energy equation which can be used to determine the rate of heat transfer by convection. The energy equation or the equation governing the temperature distribution in the boundary layer will be obtained by applying the first law of thermodynamics to the control volume shown in Fig. 8.5.

In steady-state, the first law of thermodynamics for the control volume (CV) shown in Fig. 8.5 can be written as

$$\begin{bmatrix}\text{Energy convected}\\ \text{through the}\\ \text{left face of CV}\end{bmatrix} + \begin{bmatrix}\text{Energy convected}\\ \text{through bottom}\\ \text{face of CV}\end{bmatrix} + \begin{bmatrix}\text{Energy conducted}\\ \text{through bottom}\\ \text{face of CV}\end{bmatrix} + \begin{bmatrix}\text{Net viscous}\\ \text{work done}\\ \text{on CV}\end{bmatrix}$$

$$= \begin{bmatrix}\text{Energy convected}\\ \text{out of CV through}\\ \text{right face}\end{bmatrix} + \begin{bmatrix}\text{Energy convected}\\ \text{out of CV through}\\ \text{top face}\end{bmatrix} + \begin{bmatrix}\text{Energy conducted}\\ \text{out through top}\\ \text{face of CV}\end{bmatrix} \tag{8.18}$$

Energy convected into the control volume through the left face

$$= (\rho u dy) c_p T \tag{8.19}$$

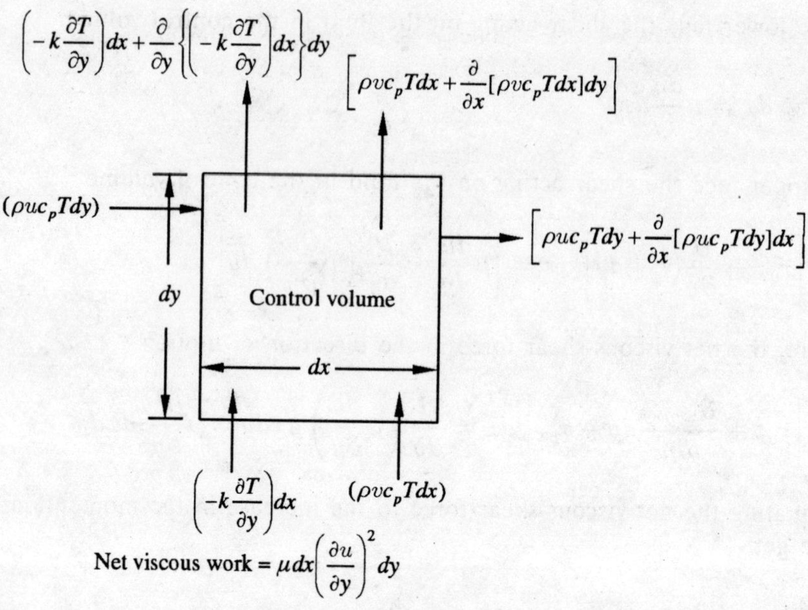

Fig. 8.5 *Control volume in the boundary layer.*

Energy convected into the control volume through the bottom face

$$= (\rho v dx)c_pT \tag{8.20}$$

Energy conducted into the control volume through the bottom face

$$= -\left(k\frac{\partial T}{\partial y}\right)dx \tag{8.21}$$

Energy convected out of the control volume through the right face

$$= (\rho u dy)c_pT + \frac{\partial}{\partial x}(\rho u dy c_pT)dx \tag{8.22}$$

Energy convected out of the control volume through the top face

$$= (\rho v dx)c_pT + \frac{\partial}{\partial y}(\rho v dx c_pT)dy \tag{8.23}$$

Energy conducted out of the control volume through the top face

$$= -\left(k\frac{\partial T}{\partial y}\right)dx + \frac{\partial}{\partial y}\left(-k\frac{\partial T}{\partial y}dx\right)dy \tag{8.24}$$

The viscous work done can be evaluated as a product of the net viscous-shear force and the distance through which this force moves in unit time.

$$\text{Viscous-shear force} = \text{(Shear stress)(Area } dx) = \mu \frac{\partial u}{\partial y} dx \tag{8.25}$$

$$\text{Distance moved in unit time} = \frac{\partial u}{\partial y} dy \tag{8.26}$$

Hence, the net viscous work done on control volume

$$= \mu \left(\frac{\partial u}{\partial y} \right)^2 dxdy \tag{8.27}$$

Substituting Eqs. (8.19) to (8.24) and (8.27) in Eq. (8.18), we get

$$\rho c_p \left[u \frac{\partial T}{\partial x} + v \frac{\partial T}{\partial y} + T \left(\frac{\partial u}{\partial x} + \frac{\partial v}{\partial y} \right) \right] dxdy$$

$$= k \frac{\partial^2 T}{\partial y^2} dxdy + \mu \left(\frac{\partial u}{\partial y} \right)^2 dxdy \tag{8.28}$$

Substituting the continuity equation

$$\frac{\partial u}{\partial x} + \frac{\partial v}{\partial y} = 0 \tag{8.12}$$

in Eq. (8.28) and dividing by ρc_p, we get

$$u \frac{\partial T}{\partial x} + v \frac{\partial T}{\partial y} = \alpha \frac{\partial^2 T}{\partial y^2} + \frac{\mu}{\rho c_p} \left(\frac{\partial u}{\partial y} \right)^2 \tag{8.29}$$

The second term on the right-hand side of Eq. (8.29) representing the viscous work term is of importance only at high velocities, but for low subsonic flow the frictional term is small compared to the other terms and can be neglected. With this simplification Eq. (8.29) reduces to

$$u \frac{\partial T}{\partial x} + v \frac{\partial T}{\partial y} = \alpha \frac{\partial^2 T}{\partial y^2} \tag{8.30}$$

Equation (8.30) is the required energy equation for the laminar boundary layer.

8.7 Exact Mathematical Solution of Boundary Layer Equations

8.7.1 Hydrodynamic Boundary Layer Thickness

The continuity equation (8.12) and the momentum equation (8.17) should be simultaneously solved to obtain the velocity distribution, boundary layer thickness and friction force at the wall. To solve these equations, let us define the stream function $\psi(x, y)$ such that

$$u = \frac{\partial \psi}{\partial y} \quad \text{and} \quad v = -\frac{\partial \psi}{\partial x} \tag{8.31}$$

Then introduce the new variable

$$\eta = \frac{y}{\sqrt{\nu x / u_\infty}} = y\sqrt{\frac{u_\infty}{\nu x}} \tag{8.32}$$

Let

$$\psi = \sqrt{\nu x u_\infty} f(\eta) \tag{8.33}$$

where $f(\eta)$ is a dimensionless stream function. Then

$$u = \frac{\partial \psi}{\partial y} = \frac{\partial \psi}{\partial \eta}\frac{\partial \eta}{\partial y} = u_\infty \frac{df}{d\eta} \tag{8.34}$$

$$v = -\frac{\partial \psi}{\partial x} = \frac{1}{2}\sqrt{\frac{\nu u_\infty}{x}}\left(\frac{df}{d\eta}\eta - f\right) \tag{8.35}$$

Expressing $(\partial u/\partial x)$; $(\partial u/\partial y)$ and $(\partial^2 u/\partial y^2)$ in terms of η and substituting in the momentum equation, we get

$$f\frac{d^2 f}{d\eta^2} + 2\frac{d^3 f}{d\eta^3} = 0 \tag{8.36}$$

Equation (8.36) is an ordinary, non-linear third-order differential equation. This equation can be solved subject to the following boundary conditions:

$$\text{At } y = 0;\ u = 0; \quad \text{or} \quad \text{at } \eta = 0;\ \frac{df}{d\eta} = 0 \tag{8.37}$$

$$\text{At } y = 0;\ v = 0; \quad \text{or} \quad \text{at } \eta = 0;\ f = 0 \tag{8.38}$$

$$\text{At } y \to \infty; \ \frac{\partial u}{\partial y} = 0; \quad \text{or} \quad \text{at } \eta \to \infty; \ \frac{\partial f}{\partial \eta} = 1 \tag{8.39}$$

The solution to Eq. (8.36) was obtained by Blasius and the results are presented in Fig. 8.6.

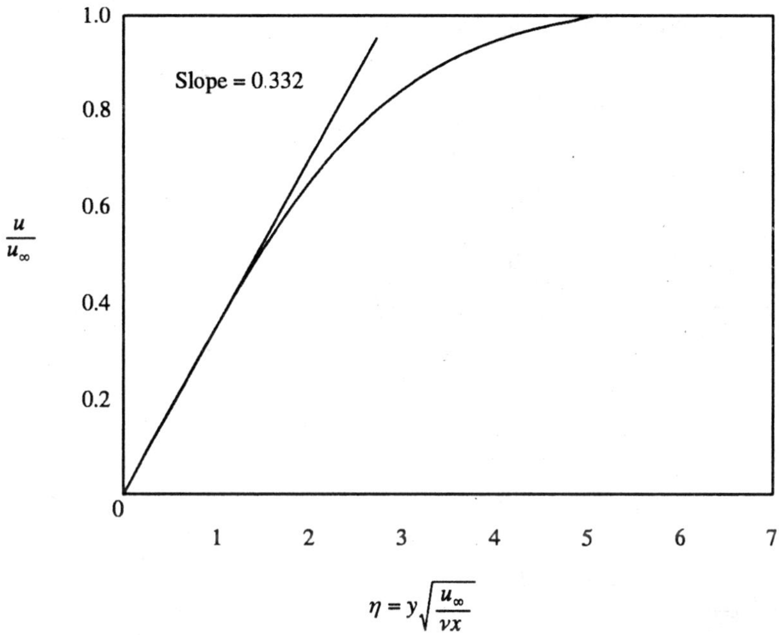

Fig. 8.6 *Velocity profile in laminar boundary layer at* $\eta = 0; \frac{du}{d\eta} = 0.332$.

The slope of the curve in Fig. 8.6 has a value of 0.332 at $\eta = 0$. The velocity u reaches 99 percent of the free-stream velocity u_∞ at $\eta = 5$. We define the hydrodynamic boundary layer thickness δ as that distance from the surface at which the local velocity u reaches 99 percent of the free-stream velocity u_∞.

$$\eta = y\sqrt{\frac{u_\infty}{\nu x}} = y\sqrt{\frac{\rho u_\infty}{\mu x}} = \frac{y}{x}\sqrt{\frac{\rho u_\infty x}{\mu}} = 5 \tag{8.40}$$

Therefore

$$\delta = \frac{5x}{\sqrt{\rho u_\infty x/\mu}} = \frac{5x}{\sqrt{Re_x}} \tag{8.41}$$

where

$$Re_x = \frac{\rho u_\infty x}{\mu} = \text{local Reynolds number.}$$

Equation (8.41) shows that at the leading edge ($x = 0$), the boundary layer thickness δ is zero and increases with increasing x, which is in qualitative agreement with the description of boundary layer growth. At any given value of x, the thickness of the boundary layer is inversely proportional to the square root of the local Reynolds number. Therefore, the boundary layer thickness decreases with increasing velocity.

The shear force at the wall can be determined by substituting the velocity gradient at $y = 0$ in Eq. (8.16). From Fig. 8.6, we find that

$$\frac{du}{d\eta} = 0.332 \quad \text{at} \quad \eta = 0 \tag{8.42}$$

or

$$\left[\frac{\partial(u/u_\infty)}{\partial\left(y\sqrt{\frac{u_\infty}{\nu x}}\right)}\right]_{\eta=0} = \left[\frac{\partial(u/u_\infty)}{\partial\left(\frac{y}{x}\sqrt{\frac{\rho u_\infty x}{\mu}}\right)}\right]_{\eta=0}$$

$$= \left[\frac{\partial(u/u_\infty)}{\partial\left(\frac{y}{x}\sqrt{Re_x}\right)}\right]_{\eta=0} = 0.332 \tag{8.43}$$

or

$$\left(\frac{\partial u}{\partial y}\right)_{y=0} = 0.332\frac{u_\infty}{x}\sqrt{Re_x} \tag{8.44}$$

Therefore,

$$\tau_{yx} = \mu\left(\frac{\partial u}{\partial y}\right)_{y=0} = 0.332\frac{\mu u_\infty}{x}\sqrt{Re_x} \tag{8.45}$$

Equation (8.45) shows that τ is very large at the leading edge and decreases with increasing distance from the leading edge. The shear stress at the wall may be expressed in terms of the *friction coefficient* C_f, which is defined as

$$\tau_w = C_f\frac{\rho u_\infty^2}{2} \tag{8.46}$$

From (8.45) and (8.46) we obtain the *local drag* or friction coefficient as

$$\frac{C_{fx}}{2} = \frac{0.332}{\sqrt{Re_x}} \tag{8.47}$$

The variation of the local friction coefficient with local Reynolds number is shown in Fig. 8.7.

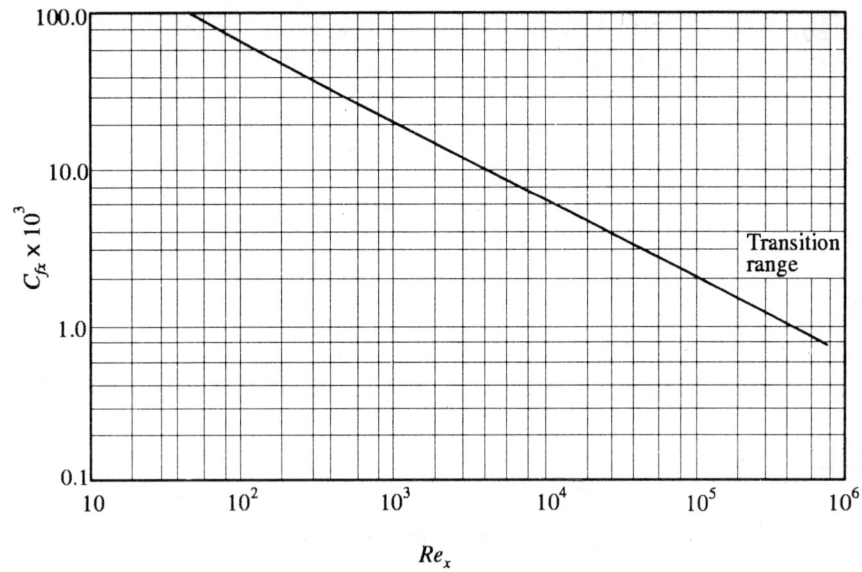

Fig. 8.7 Variation of local friction coefficient with local Reynolds number for laminar flow over a flat plate.

For many practical applications, the average friction coefficient for a plate of length ℓ is more important than the local friction coefficient. The average friction coefficient is given by

$$\frac{\bar{C_f}}{2} = \frac{1}{\ell}\int_0^{\ell} \frac{C_{fx}}{2} dx = \frac{0.664}{\sqrt{\dfrac{u_\infty \rho \ell}{\mu}}} \tag{8.48}$$

Thus, the average friction coefficient $\bar{C_f}$ is equal to twice the value of the local friction coefficient $C_{f\ell}$ at $x = \ell$.

8.7.2 Thermal Boundary Layer Thickness

The momentum equation for the laminar boundary layer is given by

$$u\frac{\partial u}{\partial x} + v\frac{\partial u}{\partial y} = \nu\frac{\partial^2 u}{\partial y^2} \tag{8.17}$$

and the energy equation for the laminar boundary layer is given by

$$u\frac{\partial T}{\partial x} + v\frac{\partial T}{\partial y} = \alpha\frac{\partial^2 T}{\partial y^2} \tag{8.30}$$

The two equations (8.17) and (8.30) are similar. Therefore, the solution for the velocity distribution $u(x, y)$ is also a solution for the temperature distribution $T(x, y)$ if $\nu = \alpha$ and if the temperature of the plate T_w is constant. The boundary conditions which the temperature distribution must satisfy are

$$\text{At} \quad y = 0;\ T = T_w \quad \text{or} \quad \frac{T - T_w}{T_\infty - T_w} = 0 \tag{8.49}$$

$$\text{At} \quad y \to \infty;\ T = T_\infty \quad \text{or} \quad \frac{T - T_w}{T_\infty - T_w} = 1 \tag{8.50}$$

where T_∞ is the free-stream temperature. The condition that $\nu = \alpha$ implies that

$$\frac{\nu}{\alpha} = \frac{c_p \mu}{k} = Pr = 1 \tag{8.51}$$

That is, for $Pr = 1$, the temperature distribution will be identical to the velocity distribution. In other words, the transfer of heat is analogous to the transfer of momentum if $Pr = 1$.

The *thermal boundary layer thickness* δ_t is defined as the distance from the surface at which the temperature difference between the wall and the fluid reaches 99 percent of the free-stream value. The theoretically estimated temperature profiles in the boundary layer are presented in Fig. 8.8 for different values of Pr.

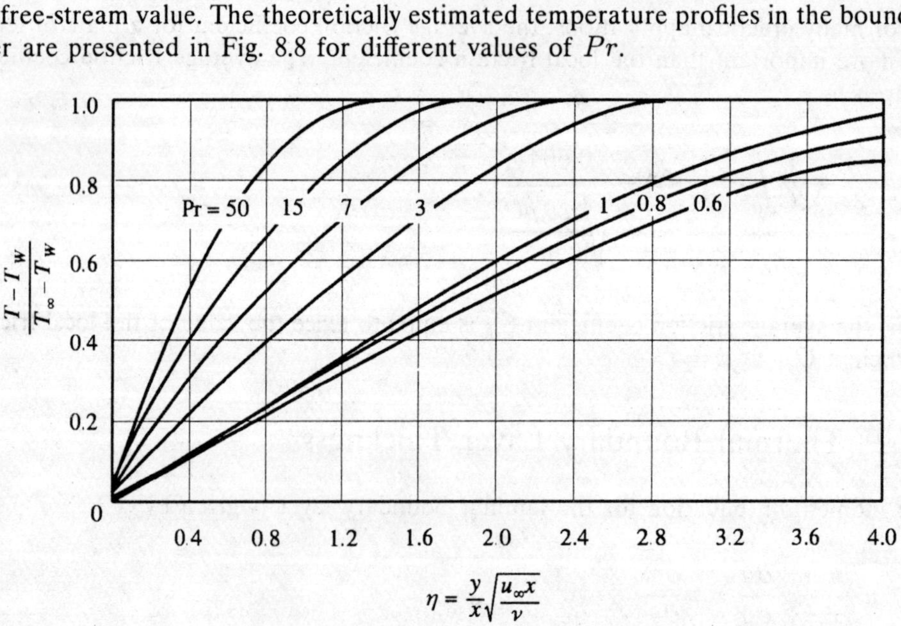

Fig. 8.8 *Temperature distribution in a fluid flowing over a flat plate.*

It can be observed from Fig. 8.8 that the thermal boundary layer thickness is larger than the hydrodynamic boundary layer thickness for fluids having Pr less than unity, but smaller for fluids having Pr greater than unity. The relationship between the thermal and hydrodynamic boundary layer thicknesses is given by

$$\delta_t = \delta(Pr)^{-1/3} \tag{8.52}$$

The data presented in Fig. 8.8 is replotted in Fig. 8.9 after incorporating the correction factor as given by Eq. (8.52). It can be observed from Fig. 8.9 that

$$\left[\frac{\partial\left(\dfrac{T - T_w}{T_\infty - T_w}\right)}{\partial\left(\dfrac{y}{x}\sqrt{Re_x}Pr^{1/3}\right)}\right]_{y=0} = 0.332 \tag{8.53}$$

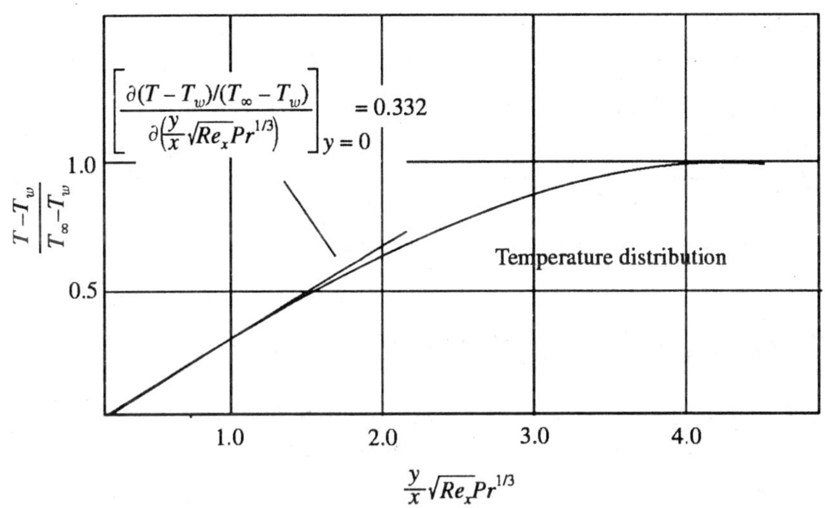

Fig. 8.9 Temperature distribution in a fluid flowing over a flat plate.

8.7.3 Evaluation of Convection Heat Transfer Coefficient

We can now determine the rate of heat transfer by convection and the convection heat transfer coefficient. The temperature gradient at the surface of the plate is given by

$$\left[\frac{\partial\left(\dfrac{T - T_w}{T_\infty - T_w}\right)}{\partial\left(\dfrac{y}{x}\sqrt{Re_x}Pr^{1/3}\right)}\right]_{y=0} = 0.332$$

or

$$\left(\frac{\partial T}{\partial y}\right)_{y=0} = \frac{0.332}{x} Re_x^{1/2} Pr^{1/3} (T_\infty - T_w) \tag{8.54}$$

Then the local rate of heat transfer by convection per unit area, is given by

$$\frac{q}{A} = -k\left(\frac{\partial T}{\partial y}\right)_{y=0} = -\frac{0.332k}{x} Re_x^{1/2} Pr^{1/3} (T_\infty - T_w) = h_x(T_w - T_\infty) \tag{8.55}$$

or the local convection heat transfer coefficient h_x is given by

$$h_x = 0.332 Re_x^{1/2} Pr^{1/3} \frac{k}{x} \tag{8.56}$$

or

$$Nu_x = \frac{h_x x}{k} = 0.332 Re_x^{1/2} Pr^{1/3} \tag{8.57}$$

where Nu_x = local Nusselt number.

The average value of the Nusselt number $h\ell/k$ is given by

$$Nu = \frac{h\ell}{k} = \frac{1}{k}\int_0^\ell h_x dx = 0.664 Re_\ell^{1/2} Pr^{1/3} \tag{8.58}$$

A useful relation between the local Nusselt number Nu_x and the friction coefficient C_{fx} can be obtained by dividing Eq. (8.57) with $Re_x Pr^{1/3}$.

$$\left(\frac{Nu_x}{Re_x Pr}\right) Pr^{2/3} = \frac{0.332 Re_x^{1/2} Pr^{1/3} Pr^{2/3}}{Re_x Pr} = \frac{0.332}{\sqrt{Re_x}} = \frac{C_{fx}}{2}$$

or

$$St_x Pr^{2/3} = \frac{C_{fx}}{2} \tag{8.59}$$

where

$$St_x = \frac{Nu_x}{Re_x Pr} = \text{Stanton number} \tag{8.60}$$

Equation (8.59) is known as the Reynolds–Colburn analogy between fluid friction and heat transfer in laminar flow.

8.8 Approximate Analysis of Boundary Layer by Integral Method

8.8.1 Hydrodynamic Boundary Layer Thickness

We have studied the exact mathematical analysis of boundary layer equations in Section 8.7 and observed that it is difficult to obtain a solution even for a simple geometry such as a flat plate. The mathematical difficulties encountered in an exact solution can be circumvented by an approximate analysis. The solutions obtained by approximate analysis agree reasonably well with the solutions obtained by an exact analysis. This approximate analysis is due to von Karman. von Karman suggested that the equations of motion and heat transfer can be written for the aggregate mass in the control volume instead of writing the equations for a differential control volume.

Consider the control volume bounded by the planes AB, CD and the solid wall as shown in Fig. 8.10. The control volume is sufficiently high such that $H > \delta$.

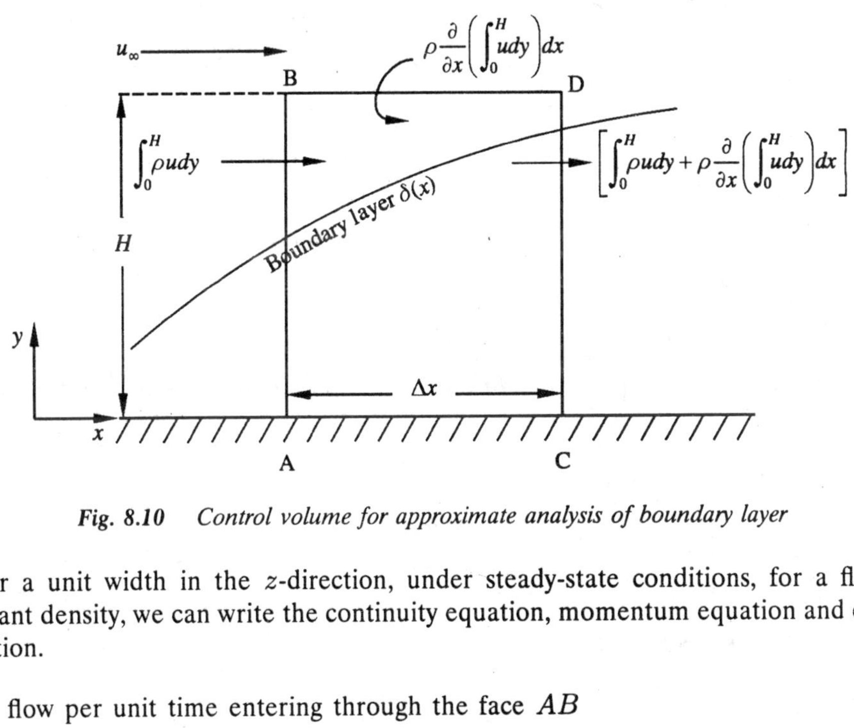

Fig. 8.10 Control volume for approximate analysis of boundary layer

For a unit width in the z-direction, under steady-state conditions, for a fluid of constant density, we can write the continuity equation, momentum equation and energy equation.

Mass flow per unit time entering through the face AB

$$= \int_0^H \rho u dy \tag{8.61}$$

Mass flow per unit time leaving through the face CD

$$= \int_0^H \rho u dy + \frac{\partial}{\partial x}\left(\int_0^H \rho u dy\right) dx \tag{8.62}$$

No mass can enter the control volume through the wall and the conservation of mass must satisfy. Therefore, the additional mass which is appearing in (8.62) must enter the control volume through face BD.

Rate of momentum inflow through face AB

$$= \int_0^H \rho u^2 dy \tag{8.63}$$

Rate of momentum outflow through face CD

$$= \int_0^H \rho u^2 dy + \frac{d}{dx}\left(\int_0^H \rho u^2 dy\right) dx \tag{8.64}$$

The velocity of the fluid at $y = H$ is equal to the free-stream velocity u_∞. Therefore,

Rate of x-momentum inflow through face BD

$$= u_\infty \frac{d}{dx}\left(\int_0^H \rho u dy\right) dx \tag{8.65}$$

The increase of x momentum per unit time of the fluid

$$= \text{Rate of } x \text{ momentum outflow} - \text{Rate of } x \text{ momentum inflow}$$

$$= \int_0^H \rho u^2 dy + \frac{d}{dx}\left(\int_0^H \rho u^2 dy\right) dx - \int_0^H \rho u^2 dy - u_\infty \frac{d}{dx}\left(\int_0^H \rho u dy\right) dx$$

$$= \frac{d}{dx}\left(\int_0^H \rho u^2 dy\right) dx - u_\infty \frac{d}{dx}\left(\int_0^H \rho u dy\right) dx \tag{8.66}$$

We know that

$$d(ab) = adb + bda$$

or

$$adb = d(ab) - bda$$

Let

$$u_\infty = a \quad \text{and} \quad \int_0^H \rho u dy = b$$

Then

$$u_\infty \frac{d}{dx}\left(\int_0^H \rho u dy\right) dx = \frac{d}{dx}\left(u_\infty \int_0^H \rho u dy\right) dx - \frac{du_\infty}{dx}\left(\int_0^H \rho u dy\right) dx$$

$$= \frac{d}{dx}\left(\int_0^H \rho u u_\infty dy\right) dx - \frac{du_\infty}{dx}\left(\int_0^H \rho u dy\right) dx \quad (8.67)$$

(Since u_∞ is a constant)

Substituting Eq. (8.67) in Eq. (8.66), we get

Net increase in x-momentum

$$= \frac{d}{dx}\left(\int_0^H \rho u^2 dy\right) dx - \frac{d}{dx}\left(\int_0^H \rho u u_\infty dy\right) dx + \frac{du_\infty}{dx}\left(\int_0^H \rho u dy\right) dx$$

$$= -\rho \frac{d}{dx}\left(\int_0^H (u_\infty - u) u dy\right) dx + \frac{du_\infty}{dx}\left(\int_0^H \rho u dy\right) dx \quad (8.68)$$

The increase in the x momentum flux is equal to the sum of the forces, in the x direction, acting on the surface of the control volume.

The force on the face AB is the pressure force $= PH$ (8.69)

The force on the face $CD = \left[P + \left(\frac{dP}{dx}\right) dx\right] H$ (8.70)

Shear force at the wall $AC = -\tau_w dx = -\mu dx \left(\frac{\partial u}{\partial y}\right)_{y=0}$ (8.71)

There is no shear force at the face BD since the velocity gradient is zero outside the boundary layer.

Therefore, the net force acting on control volume

$$= \left[-\tau_w - \frac{dP}{dx} H\right] dx \quad (8.72)$$

Since the net force acting on the element is equal to the net increase in the momentum, we get from Eqs. (8.68) and (8.72)

$$-\tau_w - \frac{dP}{dx} H = -\rho \frac{d}{dx} \int_0^H (u_\infty - u) u dy + \frac{du_\infty}{dx} \int_0^H \rho u dy \quad (8.73)$$

Equation (8.73) is known as the von Karman momentum integral equation of the boundary layer for incompressible flow.

The Bernoullis equation relates the pressure and free-stream velocity as

$$P + \frac{\rho u_\infty^2}{2} = \text{Constant}$$

or

$$\frac{dP}{dx} = -\rho u_\infty \frac{du_\infty}{dx} \tag{8.74}$$

Since the boundary layer is very thin, the pressure may be assumed to be constant at any given location x throughout the boundary layer. Hence

$$\frac{dP}{dx} = 0 = -\rho u_\infty \frac{du_\infty}{dx} \tag{8.75}$$

Substituting Eq. (8.75) in Eq. (8.73), the integral boundary layer equation reduces to

$$\rho \frac{d}{dx} \int_0^\delta (u_\infty - u) u dy = \tau_w = \mu \left(\frac{du}{dy} \right)_{y=0} \tag{8.76}$$

The upper limit on the integral in Eq. (8.76) has been changed to δ because the integral is zero for $y > \delta$, since $u = u_\infty$ for $y > \delta$.

If the velocity profile is known, it can be substituted in Eq. (8.76) to obtain an expression for the boundary layer thickness. The velocity profile must satisfy the following conditions:

$$\text{At} \quad y = 0; \quad u = 0 \tag{8.77}$$

$$\text{At} \quad y = \delta; \quad u = u_\infty \tag{8.78}$$

$$\text{At} \quad y = \delta; \quad \frac{du}{dy} = 0 \tag{8.79}$$

$$\text{At} \quad y = 0; \quad \frac{d^2u}{dy^2} = 0 \text{ for constant pressure} \tag{8.80}$$

The simplest function one can choose to satisfy the above conditions is a polynomial with four arbitrary constants. That is,

$$u = a + by + cy^2 + dy^3 \tag{8.81}$$

Applying the four conditions [Eqs. (8.77) to (8.80)] to Eq. (8.81), one can obtain

$$\frac{u}{u_\infty} = \frac{3}{2}\frac{y}{\delta} - \frac{1}{2}\left(\frac{y}{\delta}\right)^3 \tag{8.82}$$

Substituting the velocity distribution given by Eq. (8.82) in the integral boundary layer equation (8.76), we get

$$\frac{d}{dx}\left[\rho u_\infty^2 \int_0^\delta \left(\frac{3}{2}\frac{y}{\delta} - \frac{1}{2}\left(\frac{y}{\delta}\right)^3\right)\left(1 - \frac{3}{2}\frac{y}{\delta} + \frac{1}{2}\left(\frac{y}{\delta}\right)^3\right) dy\right] = \mu\left(\frac{du}{dy}\right)_{y=0}$$

$$= \frac{3}{2}\frac{\mu u_\infty}{\delta} \tag{8.83}$$

Integrating the above expression on the left-hand side of Eq. (8.83), we get

$$\frac{d}{dx}\left(\frac{39}{280}\rho u_\infty^2 \delta\right) = \frac{3}{2}\frac{\mu u_\infty}{\delta}$$

or

$$\delta d\delta = \frac{140}{13}\frac{\mu}{\rho u_\infty}dx = \frac{140}{13}\frac{\nu}{u_\infty}dx$$

or

$$\frac{\delta^2}{2} = \frac{140}{13}\frac{\nu x}{u_\infty} + \text{const} \tag{8.84}$$

Since $\delta = 0$ at $x = 0$, Eq. (8.84) reduces to

$$\delta = 4.64\sqrt{\frac{\nu x}{u_\infty}} \tag{8.85}$$

or

$$\delta = \frac{4.64x}{\sqrt{\dfrac{u_\infty x \rho}{\mu}}} = \frac{4.64x}{\sqrt{Re_x}} \tag{8.86}$$

Based on exact mathematical analysis of the boundary layer equations we have obtained

$$\delta = \frac{5x}{\sqrt{Re_x}} \tag{8.41}$$

This shows that the result obtained by approximate analysis is quite satisfactory.

8.8.2 Thermal Boundary Layer Thickness

Consider the control volume shown in Fig. 8.11 and make an energy balance for the aggregate of fluid particles within the control volume. To simplify the problem, let us ignore the shear work due to frictional forces along the wall which are negligible. Let us also assume that the properties of the fluid are constant.

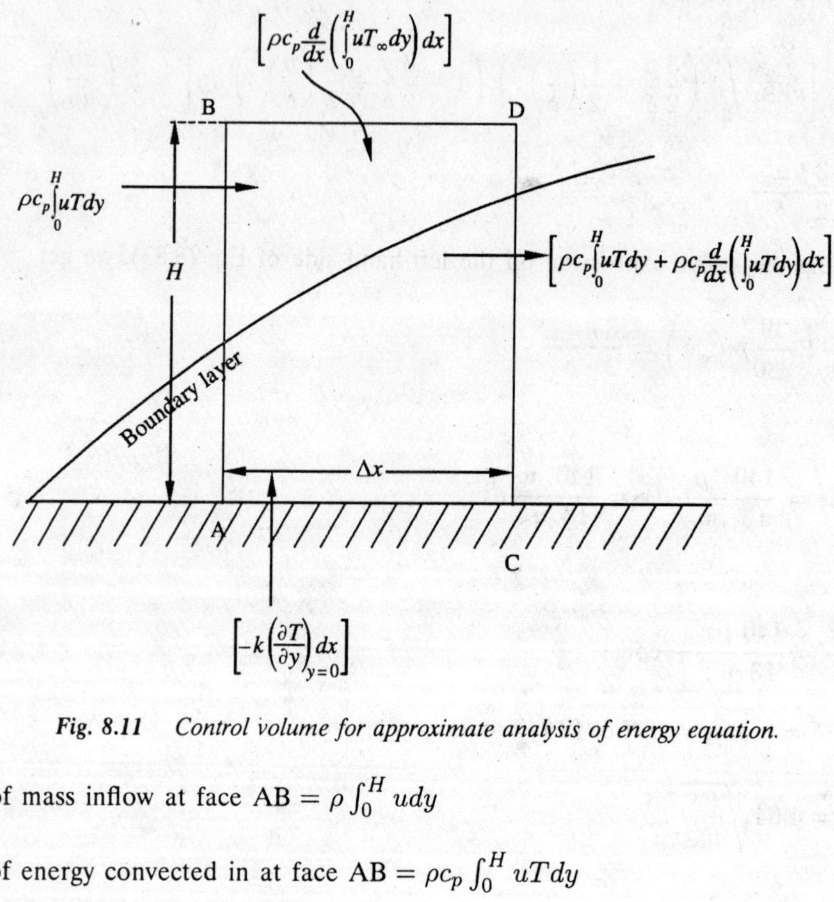

Fig. 8.11 Control volume for approximate analysis of energy equation.

Rate of mass inflow at face AB $= \rho \int_0^H u dy$ (8.87)

Rate of energy convected in at face AB $= \rho c_p \int_0^H uT dy$ (8.88)

Rate of mass outflow at face CD

$$= \rho \left[\int_0^H u dy + \frac{d}{dx} \left(\int_0^H u dy \right) dx \right] \tag{8.89}$$

Rate of energy convected out at face CD

$$= \rho c_p \int_0^H uT dy + \rho c_p \frac{\partial}{\partial x} \left(\int_0^H uT dy \right) dx \tag{8.90}$$

$$\text{Rate of mass inflow at face BD} = \rho \frac{\partial}{\partial x}\left(\int_0^H u dy\right) dx \tag{8.91}$$

$$\text{Rate of energy convected in at face BD} = \rho c_p \frac{\partial}{\partial x}\left(\int_0^H u T_\infty dy\right) dx \tag{8.92}$$

$$\text{Rate of energy convected in at face AC} = -k \left(\frac{\partial T}{\partial y}\right)_{y=0} dx \tag{8.93}$$

In steady-state, the rate of energy inflow must be equal to the rate of energy outflow from the control volume. Therefore, equating the net rate of energy inflow and net rate of energy outflow, we get

$$\frac{\partial}{\partial x}\int_0^H (T_\infty - T) u dy = \frac{k}{\rho c_p}\left(\frac{\partial T}{\partial y}\right)_{y=0} \tag{8.94}$$

The value of the integral on the left-hand side of Eq. (8.94) is zero for $y > \delta_t$. Therefore, we can replace the upper limit of the integral to δ_t. Then Eq. (8.94) can be rewritten as

$$\frac{\partial}{\partial x}\int_0^{\delta_t} (T_\infty - T) u dy = \frac{k}{\rho c_p}\left(\frac{\partial T}{\partial y}\right)_{y=0} \tag{8.95}$$

Equation (8.95) is the integral energy equation for the boundary layer.

If the temperature distribution and the velocity distribution are known, then they can be substituted in Eq. (8.95). The velocity distribution is given by Eq. (8.82). The temperature distribution must satisfy the following conditions:

$$\text{At} \quad y = 0; \quad T = T_w \tag{8.96}$$

$$\text{At} \quad y = \delta_t; \quad \frac{\partial T}{\partial y} = 0 \tag{8.97}$$

$$\text{At} \quad y = \delta_t; \quad T = T_\infty \tag{8.98}$$

$$\text{At} \quad y = 0; \quad \frac{\partial^2 T}{\partial y^2} = 0 \tag{8.99}$$

since the temperature gradient must be linear where heat flows by conduction.

As in the case of velocity profile, one can assume a polynomial with arbitrary constants and determine the values of the constants so that Eqs. (8.96) to (8.99) are satisfied and obtain the temperature profile as

$$\frac{T - T_w}{T_\infty - T_w} = \frac{3}{2}\frac{y}{\delta_t} - \frac{1}{2}\left(\frac{y}{\delta_t}\right)^3 \tag{8.100}$$

Substituting the velocity profile given by Eq. (8.82) and the temperature profile given by Eq. (8.100) in the integral energy equation (8.95), we get for the left-hand side of Eq. (8.95) as

$$\int_0^{\delta_t} (T_\infty - T)u dy = \int_0^{\delta_t} [(T_\infty - T_w) - (T - T_w)]\, u dy$$

$$= (T_\infty - T_w)u_\infty \int_0^{\delta_t} \left[1 - \frac{3}{2}\frac{y}{\delta_t} + \frac{1}{2}\left(\frac{y}{\delta_t}\right)^3\right]\left[\frac{3}{2}\frac{y}{\delta} - \frac{1}{2}\left(\frac{y}{\delta}\right)^3\right] dy \tag{8.101}$$

After performing the required integration, we get

$$\int_0^{\delta_t} (T_\infty - T)u dy = (T_\infty - T_w)u_\infty \delta \left(\frac{3}{20}\xi^2 - \frac{3}{280}\xi^4\right) \tag{8.102}$$

where

$$\xi = \delta t/\delta \tag{8.103}$$

For fluids having $Pr > 1, \xi < 1$ and hence the second term on the right-hand side of Eq. (8.102) becomes negligible. Using this approximation, Eq. (8.102) reduces to

$$\int_0^{\delta_t} (T_\infty - T)u dy = (T_\infty - T_w)u_\infty \delta \frac{3}{20}\xi^2 \tag{8.104}$$

Substituting Eq. (8.104) in Eq. (8.95), we get

$$\frac{3}{20}u_\infty (T_w - T_\infty)\xi^2 \frac{\partial \delta}{\partial x} = \alpha \left(\frac{\partial T}{\partial y}\right)_{y=0} = \frac{3}{2}\alpha \frac{T_w - T_\infty}{\delta \xi}$$

or

$$\alpha = \frac{1}{10}u_\infty \xi^3 \delta \frac{\partial \delta}{\partial x} \tag{8.105}$$

From Eq. (8.86) we get

$$\delta \frac{\partial \delta}{\partial x} = 10.75 \frac{\nu}{u_\infty}$$

Therefore

$$\xi^3 = \frac{10}{10.75}\frac{\alpha}{\nu}$$

or $\delta_t = 0.9\delta Pr^{-1/3}$ (8.106)

Based on the exact mathematical analysis of the boundary layer we have seen that

$$\delta_t = \delta(Pr)^{-1/3} \tag{8.52}$$

It can be observed that the approximate analysis of the boundary layer by the integral method provided a result which is in close agreement with that obtained by an exact analysis.

The rate of heat transfer by convection from the plate, per unit area, is given by

$$\frac{q}{A} = -k\left(\frac{\partial T}{\partial y}\right)_{y=0} = -\frac{3}{2}\frac{k}{\delta_t}(T_\infty - T_w) \tag{8.107}$$

Substituting for δ_t from Eq. (8.106) and for δ from Eq. (8.86) in Eq. (8.107), we get

$$\frac{q}{A} = -\frac{3}{2}\frac{k}{x}\frac{Pr^{1/3}Re_x^{1/2}}{0.9 \times 4.64}(T_\infty - T_w)$$

$$= 0.36\frac{k}{x}Re_x^{1/2}Pr^{1/3}(T_w - T_\infty) = h_x(T_w - T_\infty) \tag{8.108}$$

or

$$h_x = 0.36Re_x^{1/2}Pr^{1/3}\frac{k}{x} \tag{8.109}$$

or

$$Nu_x = 0.36Re_x^{1/2}Pr^{1/3} \tag{8.110}$$

A comparison of Eqs. (8.109) and (8.110) with Eqs. (8.56) and (8.57) obtained from exact analysis shows that the results obtained from approximate analysis are in good agreement with those obtained by the exact analysis. This illustrates the usefulness of the approximate boundary layer analysis.

8.9 Analogy between Heat and Momentum Transfer

We have noticed that the temperature and flow fields are related for flow over a flat plate. Now we shall develop an expression by which the frictional resistance can be related to heat transfer. The shear stress at the wall is given by

$$\tau_w = C_f\frac{\rho u_\infty^2}{2} \tag{8.111}$$

The shear stress can also be calculated from the relation

$$\tau_w = \mu \left(\frac{\partial u}{\partial y}\right)_{y=0} \tag{8.112}$$

For a laminar boundary layer, the velocity distribution is given by

$$\frac{u}{u_\infty} = \frac{3}{2}\frac{y}{\delta} - \frac{1}{2}\left(\frac{y}{\delta}\right)^3 \tag{8.82}$$

Substituting for $\left(\frac{\partial u}{\partial y}\right)_{y=0}$ from Eq. (8.82) in Eq. (8.112), we get

$$\tau_w = \frac{3}{2}\frac{\mu u_\infty}{\delta} \tag{8.113}$$

The boundary layer thickness is given by

$$\delta = 4.64\sqrt{\frac{\nu x}{u_\infty}} \tag{8.85}$$

Substituting Eq. (8.85) in Eq. (8.113), we have

$$\tau_w = \frac{3}{2}\left(\frac{\mu u_\infty}{4.64}\right)\sqrt{\frac{u_\infty}{\nu x}} \tag{8.114}$$

From Eq. (8.111) and (8.114), we get

$$\frac{C_{fx}}{2} = \frac{3}{2}\left(\frac{\mu u_\infty}{4.64}\right)\sqrt{\frac{u_\infty}{\nu x}}\frac{1}{\rho u_\infty^2} = \frac{0.323}{\sqrt{Re_x}} \tag{8.115}$$

The exact solution of the boundary layer equation gives

$$\frac{C_{fx}}{2} = 0.332 Re_x^{-1/2} \tag{8.116}$$

We know that the local Nusselt number is given by

$$Nu_x = \frac{h_x x}{k} = 0.332 Re_x^{1/2} Pr^{1/3} \tag{8.57}$$

This equation can be rearranged as

$$\frac{Nu_x}{Re_x Pr} = 0.332 Re_x^{-1/2} Pr^{-2/3} \tag{8.117}$$

or

$$St_x Pr^{2/3} = 0.332 Re_x^{-1/2} \tag{8.118}$$

where

$$St = \frac{Nu}{RePr} = \text{Stanton number} \tag{8.119}$$

Comparing Eqs. (8.116) and (8.118), we observe that the right-hand side of both the equations are the same and hence

$$St_x Pr^{2/3} = \frac{C_{fx}}{2} \tag{8.120}$$

Equation (8.120) is known as the Reynolds–Colburn analogy which relates the friction coefficient to the heat transfer coefficient. This analogy provides a means of estimating the convection heat transfer coefficient by measuring the frictional coefficient under conditions in which no heat transfer is involved.

8.10 Turbulent Boundary Layer

Consider the turbulent boundary layer shown in Fig. 8.12. A very thin region near the plate surface has a laminar character and this layer is called the *laminar sublayer*. Above this laminar sublayer some turbulence occurs, but the viscous action is still important in this region called the *buffer layer*. Above the buffer layer, the flow is fully

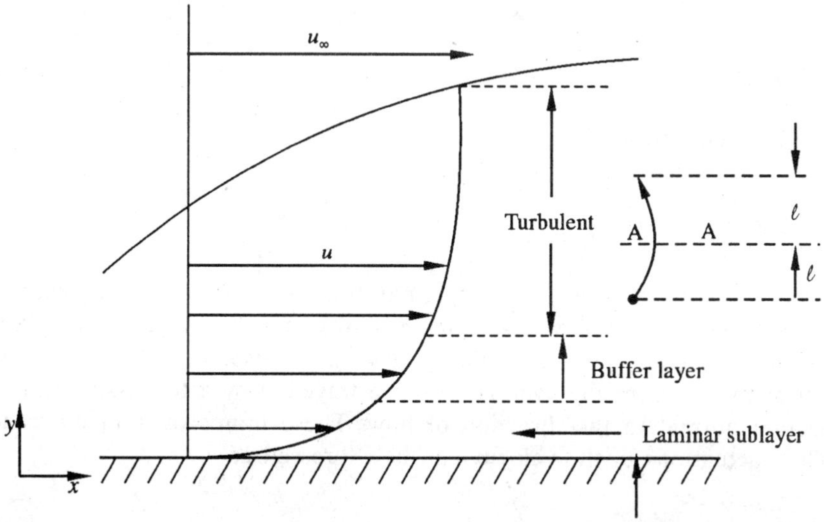

Fig. 8.12 *Velocity profile in turbulent boundary layer on a flat plate.*

turbulent and the momentum and energy transfer mainly occurs via the movement of macroscopic lumps of matter from one region to another region. In the fully developed turbulent zone, we use *eddy diffusivity for momentum* ε_m and *eddy diffusivity for heat* ε_H, instead of ν and α. The eddy properties may be large compared to the molecular properties ν and α. The physical mechanism of heat transfer in turbulent flow is assumed to be similar to that in laminar flow when we replace the molecular properties with eddy properties.

The instantaneous velocity in a turbulent flow can be expressed as

$$u = \bar{u} + u' \tag{8.121}$$

where

u = instantaneous velocity

$\bar{u}$ = mean velocity

u' = fluctuation above the mean velocity.

For steady-flow conditions, the mean value of u' will be zero. Similarly, we can write for the y-component velocity as

$$v = \bar{v} + v'$$

If a lump of macroscopic mass crosses a given plane shown as A–A in Fig. 8.12 due to turbulence, the rate of mass transport per unit area of the plane is $\rho v'$. This mass undergoes a change in the x-component velocity also. Then the net momentum flux per unit area in the x-direction is given by $\rho v'u'$, and this is equal to the turbulent shear stress at the plane. Then the average turbulent shear stress is given by

$$\tau_t = -\overline{\rho v'u'} \tag{8.122}$$

We define the eddy diffusivity ε_m as

$$\tau_t = -\overline{\rho v'u'} = \rho\varepsilon_m \frac{du}{dy} \tag{8.123}$$

In the kinetic theory of gases, the molecular transport properties are estimated by introducing the concept of *mean free path* to represent the average distance a particle travels before it encounters a collision. In a similar sense, we introduce the *Prandtl mixing length* to represent the average distance travelled by a turbulent lump of fluid in a direction normal to the direction of flow. These lumps of fluid move across a plane. The fluctuations in the velocity can be expressed as

$$u' = \ell \frac{\partial u}{\partial y} \tag{8.124}$$

where ℓ = Prandtl mixing length. Assuming that u' and v' would be similar, we can write

$$\tau_t = -\rho\overline{v'u'} = \rho\ell^2\left(\frac{\partial u}{\partial y}\right)^2 = \rho\varepsilon_m\frac{\partial u}{\partial y} \tag{8.125}$$

Therefore, the eddy diffusivity is given by

$$\varepsilon_m = \ell^2\frac{\partial u}{\partial y} \tag{8.126}$$

Prandtl proposed a hypothesis that the mixing length is proportional to the distance from the wall. That is,

$$\ell = Ky \tag{8.127}$$

where K is a constant. Substituting Eq. (8.127) in Eq. (8.125), we get

$$\tau = \rho K^2y^2\left(\frac{\partial u}{\partial y}\right)^2 \tag{8.128}$$

or

$$u = \frac{1}{K}\sqrt{\frac{\tau}{\rho}}\ln y + c \tag{8.129}$$

where c is a constant of integration. The shear stress at the wall can be expressed as a sum of molecular and turbulent parts. That is,

$$\tau = \rho(\nu + \varepsilon_m)\frac{\partial u}{\partial y} \tag{8.130}$$

The velocity profile in a turbulent boundary layer is usually given by the so called *universal velocity profile* which is expressed in terms of two dimensionless coordinates:

$$u^+ = \frac{u}{\sqrt{\tau/\rho}} \tag{8.131}$$

$$y^+ = \frac{\sqrt{\tau/\rho y}}{\nu} \tag{8.132}$$

The universal velocity profile is given by

Laminar sublayer: $0 < y^+ < 5$; $u^+ = y^+$ (8.133)

Buffer layer: $5 < y^+ < 30$; $u^+ = 5\ln y^+ - 3.05$ (8.134)

Turbulent layer: $30 < y^+ < 400$; $u^+ = 2.5\ln y^+ + 5.5$ (8.135)

8.11 Turbulent Heat Transfer

Turbulent heat transfer is analogous to turbulent momentum transfer and is given by

$$\left(\frac{q}{A}\right)_t = -\rho c_p \varepsilon_H \frac{\partial T}{\partial y} \tag{8.136}$$

and in the region where both molecular and turbulent energy transfer are present,

$$\frac{q}{A} = -\rho c_p (\alpha + \varepsilon_H) \frac{\partial T}{\partial y} \tag{8.137}$$

Analyses similar to universal velocity profile have been carried out to predict the turbulent boundary layer heat transfer. For our purposes, the Colburn analogy between fluid friction and heat transfer is easier to apply and yields results which are in good agreement with experimental results. In the turbulent flow region $\varepsilon_m >> \nu$ and $\varepsilon_H >> \alpha$. We define the turbulent Prandtl number Pr_t as

$$Pr_t = \frac{\varepsilon_m}{\varepsilon_H} \tag{8.138}$$

In the turbulent region we need experimental values of C_f to make use of the Colburn analogy. The literature provides the following information for the friction coefficient:

$$5 \times 10^5 < Re_x < 10^7 \ : C_{fx} = 0.0592 Re_x^{-1/5} \tag{8.139}$$

$$10^7 < Re_x < 10^9 \ : C_{fx} = 0.370 (\log Re_x)^{-2.584} \tag{8.140}$$

Now, applying the Colburn analogy that $StPr^{2/3} = C_f/2$, we obtain

$$5 \times 10^5 < Re_x < 10^7: \quad St_x Pr^{2/3} = 0.0296 Re_x^{-1/5} \tag{8.141}$$

$$10^7 < Re_x < 10^9: \quad St_x Pr^{2/3} = 0.185 (\log Re_x)^{-2.584} \tag{8.142}$$

8.12 Turbulent Boundary Layer Thickness

Several experimental results indicate that the velocity profile in a turbulent boundary layer is given by

$$\frac{u}{u_\infty} = \left(\frac{y}{\delta}\right)^{1/7} \tag{8.143}$$

To determine the boundary layer thickness, the momentum integral boundary layer equation (8.76) can be solved with the velocity profile given by Eq. (8.143). For $Re_x < 10^7, C_{fx}$ is given by Eq. (8.139). Therefore from Eqs.(8.111) and (8.139), we get

$$\tau_w = 0.0296 \left(\frac{\nu}{u_\infty x}\right)^{1/5} \rho u_\infty^2 \tag{8.144}$$

Carrying out the analysis further, we get

$$\frac{\delta}{x} = 0.381 Re_x^{-1/5} \tag{8.145}$$

if the boundary layer is fully turbulent from the leading edge of the plate.

$$\frac{\delta}{x} = 0.381 Re_x^{-1/5} - 10,256 Re_x^{-1} \tag{8.146}$$

if the boundary layer follows a laminar pattern up to $Re(critical) = 5 \times 10^5$ and a turbulent pattern thereafter.

8.13 Laminar Flow in a Tube

Consider the laminar flow of a fluid through a tube. Let

T_w = wall temperature

r_0 = radius of the tube

u_0 = velocity of the fluid at the centre of the tube

T_c = temperature of the fluid at the centre of the tube

To derive the velocity distribution in the tube, consider the control volume shown in Fig. 8.13.

A force balance on the contol volume gives

$$\pi r^2 dP = \tau 2\pi r dx = 2\pi r \mu dx \frac{du}{dr}$$

or $$du = \frac{1}{2\mu}\frac{dP}{dx} r dr$$

or

$$u = \frac{1}{4\mu}\frac{dP}{dx}r^2 + c_1 \tag{8.147}$$

The constant c_1 can be evaluated from the boundary condition

$$u = 0 \quad \text{at} \quad r = r_0 \tag{8.148}$$

Therefore,

$$u = \frac{1}{4\mu}\frac{dP}{dx}(r^2 - r_0^2) \tag{8.149}$$

The velocity at the centre of the tube is given by

$$u_0 = -\frac{r_0^2}{4\mu}\frac{dP}{dx} \tag{8.150}$$

Hence, the velocity profile in the tube is given by

$$\frac{u}{u_0} = 1 - \left(\frac{r}{r_0}\right)^2 \tag{8.151}$$

Equation (8.151) is the well known parabolic velocity Distribution for laminar flow in a tube.

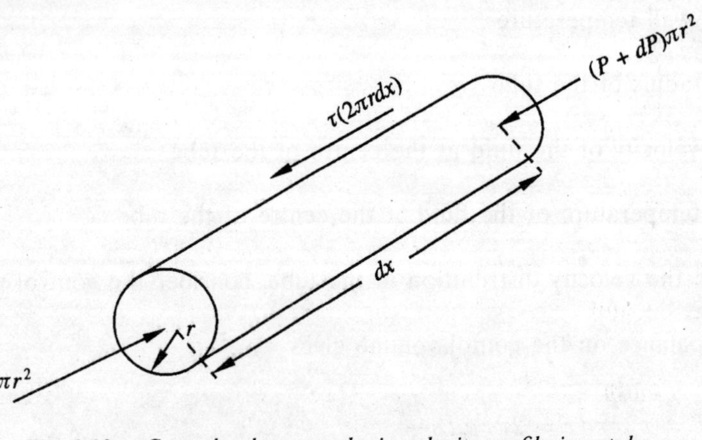

Fig. 8.13 Control volume to obtain velocity profile in a tube.

Now, let us consider the control volume shown in Fig. 8.13 to determine the temperature distribution for the laminar flow of a fluid through a tube.

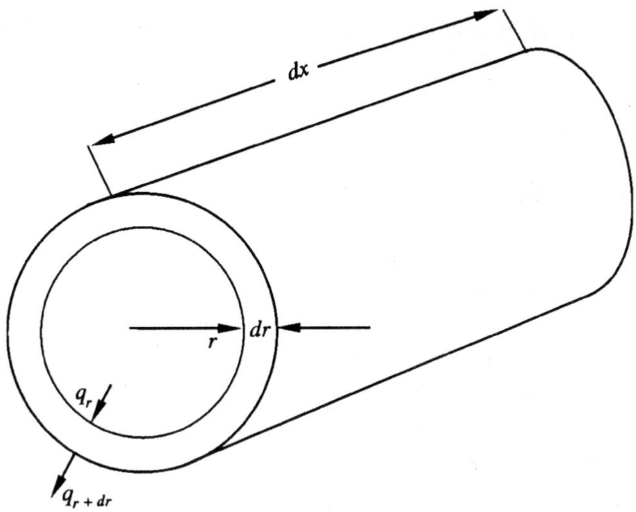

Fig. 8.14 Control volume to determine temperature distribution for laminar flow of a fluid through a tube.

Consider the annular element of radius r and thickness dr shown in Fig. 8.14.

The rate of heat flow by conduction into the element $= dq_r = -k 2\pi r dx \dfrac{\partial T}{\partial r}$

The rate of heat flow by conduction out of the element

$$= dq_{r+dr} = -k2\pi(r + dr)dx \left(\frac{\partial T}{\partial r} + \frac{\partial^2 T}{\partial r^2} dr \right)$$

The rate of heat flow by convection out of the element $= 2\pi r dr \rho c_p u \dfrac{\partial T}{\partial x} dx$

At steady-state,

Net energy flow into the element = Net energy flow out of the element

or

$$r\rho c_p u \frac{\partial T}{\partial x} dx dr = k \left(\frac{\partial T}{\partial r} + r \frac{\partial^2 T}{\partial r^2} \right) dx dr$$

or

$$\frac{1}{ur} \frac{\partial}{\partial r} \left(r \frac{\partial T}{\partial r} \right) = \frac{1}{\alpha} \frac{\partial T}{\partial x} \tag{8.152}$$

Let us assume that there is a constant heat flux at the wall. That is,

$$\frac{dq_w}{dx} = 0$$

Therefore, the average fluid temperature linearly changes with x. That is,

$$\frac{\partial T}{\partial x} = \text{constant}$$

This means that the temperature profiles will be similar at various x values along the tube. The boundary conditions which must be satisfied by Eq. (8.152) are

$$\frac{\partial T}{\partial r} = 0 \quad \text{at} \quad r = 0 \tag{8.153}$$

$$k\left(\frac{\partial T}{\partial r}\right)_{r=r_0} = q_w = \text{constant} \tag{8.154}$$

To obtain the solution of Eq. (8.152), we substitute the velocity distribution given by Eq. (8.151) in Eq. (8.152). Then

$$\frac{\partial}{\partial r}\left(r\frac{\partial T}{\partial r}\right) = \frac{1}{\alpha}\frac{\partial T}{\partial x}u_0\left(1 - \frac{r^2}{r_0^2}\right)r \tag{8.155}$$

Integrating Eq. (8.155), we get

$$r\frac{\partial T}{\partial r} = \frac{1}{\alpha}\frac{\partial T}{\partial x}u_0\left(\frac{r^2}{2} - \frac{r^4}{4r_0^2}\right) + c_1 \tag{8.156}$$

Integrating Eq. (8.156) again, we get

$$T = \frac{1}{\alpha}\frac{\partial T}{\partial x}u_0\left(\frac{r^2}{4} - \frac{r^4}{16r_0^2}\right) + c_1 \ln r + c_2 \tag{8.157}$$

Applying the boundary condition (8.153) to Eq. (8.157), we get

$$c_1 = 0 \tag{8.158}$$

We know that at $r = 0$, $T = T_c$. Therefore, from Eq. (8.157), we get

$$c_2 = T_c$$

Hence, the solution of Eq. (8.152) is given by

$$T - T_c = \frac{1}{\alpha}\frac{\partial T}{\partial x}\frac{u_0 r_0^2}{4}\left[\left(\frac{r}{r_0}\right)^2 - \frac{1}{4}\left(\frac{r}{r_0}\right)^4\right] \tag{8.159}$$

Bulk Temperature

In tube flow there is nothing like a free-stream flow as in the case of flow over a flat plate. The convection heat transfer coefficient is defined by the equation

$$q = hA(T_w - T_\infty)$$

where T_∞ is a reference temperature. Thus, the numerical value of the heat transfer coefficient depends on the choice of the reference temperature. In tube flow, the temperature of the fluid varies in the direction of flow and in the direction of heat flow. One possibility is to use the centre temperature as the reference temperature. However, it is difficult to measure the centre temperature and moreover it does not indicate the change in the energy of the fluid flowing through the tube. Therefore, it is a common practice to use the *bulk temperature or cup mixing temperature* as the reference temperature. The bulk temperature T_b is the temperature which the fluid, passing a cross-sectional area of the tube during a given time, would have if the fluid were collected and mixed in a cup. That is, the bulk temperature T_b is given by

$$T_b = \frac{\int_0^{r_0} \rho 2\pi r dr u c_p T}{\int_0^{r_0} \rho 2\pi r dr u c_p} \tag{8.160}$$

Hence, for tube flow the convection heat transfer coefficient is defined by the relation

$$q = hA(T_w - T_b) \tag{8.161}$$

Substituting the temperature distribution given by Eq. (8.159) in Eq. (8.160), we get

$$T_b = T_c + \frac{7}{96}\frac{u_0 r_0^2}{\alpha}\frac{\partial T}{\partial x} \tag{8.162}$$

and the wall temperature is given by

$$T_w = T_c + \frac{3}{16}\frac{u_0 r_0^2}{\alpha}\frac{\partial T}{\partial x} \tag{8.163}$$

Now, the heat transfer coefficient can be calculated from

$$q = hA(T_w - T_b) = kA\left(\frac{\partial T}{\partial r}\right)_{r=r_0}$$

or

$$h = \frac{k\left(\dfrac{\partial T}{\partial r}\right)_{r=r_0}}{T_w - T_b} \tag{8.164}$$

The temperature gradient at the wall can be obtained from Eq. (8.156) as

$$\left(\frac{\partial T}{\partial r}\right)_{r=r_0} = \frac{u_0}{\alpha}\frac{\partial T}{\partial x}\left(\frac{r}{2} - \frac{r^3}{4r_0^2}\right)_{r=r_0} = \frac{u_0 r_0}{4\alpha}\frac{\partial T}{\partial x} \tag{8.165}$$

Substituting for T_w, T_b and $(\partial T/\partial r)_{r=r_0}$ from Eqs. (8.162), (8.163) and (8.165) in Eq. (8.164), we get

$$h = \frac{24}{11}\frac{k}{r_0} = \frac{48}{11}\frac{k}{d_0}$$

or

$$Nu_d = \frac{hd_0}{k} = 4.364 \tag{8.166}$$

where d_0 is the diameter of the tube. The relation (8.166) is found to be in agreement with the exact calculations.

8.14 Turbulent Flow in a Tube

To determine a relation for the heat transfer coefficient for turbulent flow in a tube, one should have a knowledge of the velocity profile and temperature distribution. Instead of an exact analysis we shall use an approximate analysis relating the momentum transfer and heat transfer.

In laminar flow, the heat transfer rate across a fluid element is given by

$$\frac{q}{A} = -k\frac{\partial T}{\partial y}$$

Dividing the above equation by ρc_p, we get

$$\frac{q}{\rho c_p A} = -\alpha\frac{\partial T}{\partial y} \tag{8.167}$$

In turbulent flow the heat transfer rate can be written as [see Eq. (8.137)]

$$\frac{q}{\rho c_p A} = -(\alpha + \varepsilon_H)\frac{\partial T}{\partial y} \tag{8.168}$$

The shear stress in turbulent flow is given by [see Eq. (8.130)]

$$\frac{\tau}{\rho} = (\nu + \varepsilon_m)\frac{\partial u}{\partial y} \tag{8.169}$$

We assume that if $\varepsilon_m = \varepsilon_H$ and $\nu = \alpha$ or $Pr = 1$, the heat and momentum are transported at the same rate. Dividing Eq. (8.168) by Eq. (8.169), we get

$$\frac{q}{c_p A \tau} du = -dT \tag{8.170}$$

Integrating Eq. (8.170) between the wall conditions and the bulk conditions, we get

$$\frac{q_w}{A_w \tau_w c_p} \int_0^{u_\infty} du = -\int_{T_w}^{T_b} dT$$

or

$$\frac{q_w u_m}{A_w \tau_w c_p} = T_w - T_b \tag{8.171}$$

The shear stress τ_w can be calculated from

$$\tau_w = \frac{\Delta P(\pi d_0^2)}{4\pi d_0 \ell} = \frac{\Delta P}{4}\frac{d_0}{\ell} \tag{8.172}$$

The pressure drop ΔP can be expressed in terms of friction factor f as

$$\Delta P = f\frac{\ell}{d}\rho\frac{u_m^2}{2} \tag{8.173}$$

From Eqs. (8.172) and (8.173), we can write

$$\tau_w = \frac{f}{8}\rho u_m^2 \tag{8.174}$$

The heat transfer at the wall is given by

$$q_w = hA_w(T_w - T_b) \tag{8.175}$$

Substituting for τ_w and q_w from Eqs. (8.174) and (8.175) in Eq. (8.171), we obtain

$$St = \frac{h}{\rho c_p u_m} = \frac{Nu_d}{Re_d Pr} = \frac{f}{8} \tag{8.176}$$

Equation (8.176) is known as the *Reynolds analogy for tube flow*. For flow in smooth tubes, an empirical formula for the friction factor is given by

$$f = \frac{0.316}{Re_d^{1/4}} \tag{8.177}$$

Using Eq. (8.177) in Eq. (8.174), one obtains

$$\frac{Nu_d}{Re_d Pr} = 0.0395\, Re_d^{-1/4}$$

or

$$Nu_d = 0.0395\, Re_d^{3/4} \quad (\text{Since } Pr = 1\,) \tag{8.178}$$

In Section 8.9 we discussed the analogy between heat and momentum transfer and observed the Prandtl number dependence as $Pr^{2/3}$. This dependence is also found to be approximately true for turbulent tube flow. Hence, we can modify Eqs.(8.176) and (8.178), so that the relations will be true for fluids with $Pr \neq 1$, as

$$St Pr^{2/3} = f/8 \tag{8.179}$$

$$Nu_d = 0.0395\, Re_d^{3/4} Pr^{1/3} \tag{8.180}$$

For practical purposes, a more correct relation for predicting convection heat transfer coefficient is given by the Dittus–Boelter equation

$$Nu_d = 0.023\, Re^{0.8}\, Pr^{0.4} \tag{8.181}$$

8.15 Summary

The rate of convection heat transfer q from a surface at T_w to a fluid at T_∞ is given by

$$q = hA(T_w - T_\infty)$$

The convection heat transfer coefficient h depends on the geometry of the system, the thermal properties of the fluid and the characteristics of the fluid flow. The transfer of heat between a solid surface and the surrounding fluid takes place by a combination of conduction and mass transport. The convection heat transfer is closely coupled with fluid motion. Hence, it is necessary to understand the flow characteristics of the fluid.

The flow of a fluid can be laminar or turbulent. In laminar or streamline flow, the fluid moves in layers and each fluid particle follows a smooth and continuous path. The motion of the fluid particles in turbulent flow is quite irregular and mixing occurs. In laminar flow, heat is transferred by molecular conduction from one layer to another. In turbulent flow the conduction is aided by eddies which cause mixing. If the turbulence is increased, the rate of mixing increases and hence the rate of heat transfer also increases. The fluid motion can be induced due to density gradients and the heat transfer in such cases is called natural or free convection heat transfer. The fluid motion can also be induced by an external agent like a pump or blower. The heat transfer in such cases is called forced convection heat transfer.

For flow over a flat plate, the transition from laminar to turbulent flow occurs when

$$Re_x = \frac{u_\infty x \rho}{\mu} > 5 \times 10^5$$

where

u_∞ = the free-stream velocity

x = the distance from the leading edge

Re_x = Reynolds number.

The Nusselt modulus or Nusselt number signifies the ratio of the temperature gradient at the surface to a reference temperature gradient, and is given by

$$Nu = \frac{h\ell}{\kappa} = \left[\frac{\partial\left(\dfrac{T_w - T}{T_w - T_\infty}\right)}{\partial(y/\ell)}\right]_{y=0}$$

where

Nu = Nusselt number

ℓ = a significant length which specifies the geometry of the solid body

T_w = wall temperature

T_∞ = free-stream temperature.

The Nusselt number is a convenient measure of convection heat transfer coefficient. The convection heat transfer coefficients can be estimated by

1. Dimensional analysis coupled with experimental data.
2. Exact mathematical analysis of the boundary layer equations.
3. Approximate analysis of the boundary layer equations.
4. Analogy between heat, mass and momentum transfer.

All these methods contribute some knowledge of the convection heat transfer, but no single method can provide complete solutions to all the problems.

The dimensional analysis is mathematically simple and finds a wide range of applications. However, it does not contribute to the understanding of the heat transfer mechanism, and the results obtained by this method are incomplete and requires experimental data. The exact mathematical analysis requires a thorough knowledge of the mechanism so that the phenomena can be described by mathematical equations. This method of analysis is important because the results obtained by this method form a basis against which the simpler solutions obtained through approximate methods can be compared. The approximate analysis uses integral methods and avoids the detailed mathematical description of the flow. This method is relatively simple and provides solutions to problems which cannot be easily solved by exact mathematical analysis. The analogy between heat and momentum transfer provides solutions which are in close agreement with the experimental values.

The dimensional analysis does not give equations which can be solved, but it combines the variables into dimensionless groups which allow the extension of the range of applicability of experimental data. This method provides a means of obtaining empirical equations for correlating the data. The dimensional analysis is based on the Buckingham π-theorem which states that the number of independent dimensionless groups which can be formed by combining the variables is equal to the total number of physical quantities minus the number of primary dimensions needed to express the dimensional formulae of the physical quantities.

The principle of similarity or model law states that the behaviour of two systems will be similar if the ratios of their linear dimensions, forces, velocities, etc. are the same. The Reynolds number is the ratio of inertia forces to viscous forces and hence similar flow conditions will prevail in two systems for a given value of Reynolds number. The Prandtl number is given by

$$Pr = \frac{\text{Kinematic viscosity}}{\text{Thermal diffusivity}} = \frac{\nu}{\alpha} = \frac{\mu/\rho}{k/\rho c_p} = \frac{c_p \mu}{k}$$

The Prandtl number relates the temperature distribution to the velocity distribution. For given values of Re and Pr in geometrically similar systems, the temperature distribution will be similar.

The simplest problem that can be analyzed by the exact mathematical analysis is the transfer of heat from a flat plate and a fluid flowing parallel to its surface. The continuity equation for an incompressible fluid is given by

$$\frac{\partial u}{\partial x} + \frac{\partial v}{\partial y} = 0$$

The momentum equation is given by

$$u\frac{\partial u}{\partial x} + v\frac{\partial u}{\partial y} = \nu\frac{\partial^2 u}{\partial y^2}$$

The energy equation for a laminar boundary layer is given by

$$u\frac{\partial T}{\partial x} + v\frac{\partial T}{\partial y} = \alpha\frac{\partial^2 T}{\partial y^2}$$

The hydrodynamic boundary layer thickness δ is defined as that distance from the surface at which the local velocity u reaches 99 percent of the free-stream velocity u_∞ . The laminar boundary layer thickness as given by an exact analysis is

$$\delta = \frac{5x}{\sqrt{Re_x}}$$

The thickness of the boundary layer decreases with increasing velocity at any given value of x.

The friction coefficient C_{fx} is given by

$$\frac{C_{fx}}{2} = \frac{0.332}{\sqrt{Re_x}}$$

The average friction coefficient is given by

$$\frac{\overline{C}_f}{2} = \frac{0.664}{\sqrt{(u_\infty \rho \ell)/\mu}}$$

The thermal boundary layer thickness δ_t is defined as the distance from the surface at which the temperature difference between the wall and the fluid reaches 99 percent of the free-stream value. The relationship between the thermal and hydrodynamic boundary layer thicknesses is given by

$$\delta_t = \delta Pr^{-1/3}$$

The local convection heat transfer coefficient h_x can be obtained from

$$Nu_x = \frac{h_x x}{k} = 0.332\, Re_x^{1/2} Pr^{1/3}$$

and the average value of the convection heat transfer coefficient can be obtained from

$$Nu = \frac{h\ell}{k} = 0.664\, Re_\ell^{1/2} Pr^{1/3}$$

The relation between the local Nusselt number Nu_x and the friction coefficient C_{fx} is given by

$$St_x Pr^{2/3} = \frac{C_{fx}}{2}$$

which is known as the Reynolds–Colburn analogy between fluid friction and heat transfer in laminar flow. The Stanton number is given by

$$St_x = \frac{Nu_x}{Re_x Pr}$$

The approximate analysis of laminar boundary layer equations give the velocity distribution as

$$\frac{u}{u_\infty} = \frac{3}{2}\frac{y}{\delta} - \frac{1}{2}\left(\frac{y}{\delta}\right)^3$$

$$\delta = \frac{4.64x}{\sqrt{Re_x}}$$

$$\delta_t = 0.9\delta Pr^{-1/3}$$

$$Nu_x = 0.36\, Re_x^{1/2} Pr^{1/3}$$

$$\frac{C_{fx}}{2} = \frac{0.323}{\sqrt{Re_x}}$$

For analyzing turbulent flow, we use eddy diffusivity for momentum ε_m and eddy diffusivity for heat ε_H in place of ν and α. The Prandtl mixing length ℓ is the average distance travelled by a turbulent lump of fluid in a direction normal to the direction of flow. The eddy diffusivity ε_m is given by

$$\varepsilon_m = \ell^2 \frac{\partial u}{\partial y}$$

The universal velocity profile is given by

Laminar sublayer: $0 < y^+ < 5; \quad u^+ = y^+$

Buffer layer: $5 < y^+ < 30; \quad u^+ = 5 \ln y^+ - 3.05$

Turbulent layer: $30 < y^+ < 400; \quad u^+ = 2.5 \ln y^+ + 5.5$

where

$$u^+ = \frac{u}{\sqrt{\tau/\rho}}$$

$$y^+ = \frac{\sqrt{\tau/\rho y}}{\nu}$$

The velocity profile in a turbulent boundary layer is given by

$$\frac{u}{u_\infty} = \left(\frac{y}{\delta}\right)^{1/7}$$

The turbulent boundary layer thickness is given by

$$\frac{\delta}{x} = 0.381\, Re_x^{-1/5}$$

if the boundary layer is fully turbulent from the leading edge, and

$$\frac{\delta}{x} = 0.381\, Re_x^{-1/5} - 10,256\, Re_x^{-1}$$

if the boundary layer follows a laminar pattern upto the $Re = 5 \times 10^5$ and a turbulent pattern thereafter.

The velocity profile for a laminar flow in a tube is given by

$$\frac{u}{u_0} = 1 - \left(\frac{r}{r_0}\right)^2$$

where u_0 is the velocity of the fluid at the centre of the tube. The convection heat transfer coefficient for a tube flow is defined as

$$q = hA(T_w - T_b)$$

where the bulk temperature T_b is defined as

$$T_b = \frac{\int_0^{r_0} \rho 2\pi r dr u c_p T}{\int_0^{r_0} \rho 2\pi r dr u c_p}$$

The bulk temperature is also known as the cup mixing temperature.

The convection heat transfer coefficient for laminar flow in a tube can be estimated from the relation

$$Nu_d = \frac{h d_0}{k} = 4.364$$

where d_0 is the tube diameter.

The convection heat transfer coefficient for turbulent flow in a tube can be estimated from the relation

$$Nu_d = 0.0395 Re_d^{3/4} Pr^{1/3}$$

For practical purposes, a more correct relation proposed by Dittus–Boelter is given by

$$Nu_d = 0.023\, Re^{0.8} Pr^{0.4}$$

Review Questions

1. How is the convection heat transfer coefficient defined?
2. Which factors affect the value of the convection heat transfer coefficient between a surface and a fluid?
3. What is meant by laminar flow?
4. Distinguish between natural convection and forced convection.
5. Define Reynolds number and what its physical significance is?
6. What is the critical Reynolds number for flow through a tube and for flow of a fluid over a flat plate?
7. What is the physical significance of Nusselt number?
8. What are the advantages and drawbacks of dimensional analysis?
9. What are the limitations of exact mathematical analysis of boundary layer problems?
10. Why is it important to understand the exact mathematical analysis of boundary layer problems?
11. How does the approximate analysis differ from the exact analysis of boundary layer problems?
12. What is the Buckingham π-theorem?
13. Give the dimensional formulae for kinematic viscosity, thermal diffusivity and convection heat transfer coefficient.
14. Define Prandtl number and Stanton number.
15. How does the dimensional analysis aid in correlating the experimental data?
16. How does one modify the Buckingham π-theorem if some of the equations in the set of equations formed by equating the exponents of each primary dimension to zero are linearly dependent?
17. What is meant by hydrodynamic boundary layer?
18. What is the continuity equation for incompressible, steady, two-dimensional flow?
19. What is the momentum equation for the laminar boundary layer on a flat plate?
20. What assumptions are made in deriving the momentum equation for the laminar boundary layer on a flat plate?
21. What is the energy equation for the laminar boundary layer on a flat plate?
22. Define the hydrodynamic boundary layer thickness.
23. What is the relation between hydrodynamic boundary layer thickness and local Reynolds number?
24. How does the hydrodynamic boundary layer change with the distance from the leading edge for laminar flow?
25. How does the hydrodynamic boundary layer thickness change with increasing velocity of the fluid?
26. Define the local friction coefficient C_{fx}.
27. What is the relation between the local friction coefficient and local Reynolds number in a laminar boundary layer?
28. Define the average friction coefficient $\overline{C}_f$.

29. What is meant by thermal boundary layer?
30. Define the thermal boundary layer thickness.
31. What is the relation between thermal and hydrodynamic boundary layer thicknesses?
32. What is the Reynolds–Colburn analogy between fluid friction and heat transfer in laminar flow?
33. What is the velocity profile used in the approximate analysis of laminar boundary layer equation?
34. Compare the hydrodynamic boundary layer thickness predicted by the approximate analysis with that given by an exact mathematical analysis for a laminar boundary layer.
35. Compare the thermal boundary layer thickness for laminar flow as predicted by the approximate analysis with that given by an exact mathematical analysis.
36. What is Prandtl mixing length?
37. What is the universal velocity profile?
38. What is the velocity distribution in a turbulent boundary layer?
39. What is the relation between hydrodynamic turbulent boundary layer thickness and local Reynolds number?
40. What is the velocity profile for laminar flow through a tube?
41. Define the bulk temperature. How is it used?
42. What is the value of the Nusselt number for laminar flow through a tube?
43. Define the friction factor f.
44. What is the Reynolds analogy for turbulent flow through a tube?
45. State the empirical relation to predict the friction factor for flow through smooth tubes.

Problems

8.1 An infinite plate of thickness 2ℓ is initially at a uniform temperature T_i. The plate is suddenly exposed to a convection environment at T_∞ with convection heat transfer coefficient h. It is required to correlate the experimental data of the centre temperature–time history. Develop the necessary expression through dimensional analysis.

8.2 Rework Problem 8.1 and develop a relation to correlate the temperature–time history at several distances from the centre of the plate.

8.3 A semi-infinite solid is initially at a uniform temperature T_i. At time $t = 0$, one surface of the solid is suddenly changed to T_∞ and we would like to determine an expression for the temperature distribution in the solid as a function of time. Develop the necessary expression through dimensional analysis.

8.4 Air at 100 kPa and 300 K flows over a flat plat at a speed of 1 m/s. Calculate the hydrodynamic boundary layer thickness, using the solution obtained by an exact mathematical analysis, at distances of 20 cm and 50 cm from the leading edge. The viscosity of air at 300 K is 1.85×10^{-5} kg/m s.

8.5 Air at 100 kPa and 300 K is flowing over a flat plate at a velocity of 1 m/s. If the plate is 1 m long and at 500 K, calculate the hydrodynamic boundary layer thickness at a distance of 50 cm from the leading edge. Also calculate the local friction coefficient and the average friction coefficient for x = 50 cm. For air at 100 kPa and 300 K, ρ = 1.092 kg/m^3; C_p = 1.014 kJ/kg K; $\nu = 17.6 \times 10^{-6}$ m^2/s, k = 0.0265 W/m K and Pr = 0.71.

8.6 For Problem 8.5, calculate the thickness of the thermal boundary layer and the local convection heat transfer coefficient at x =50 cm.

8.7 For Problem 8.5, calculate the average convection heat transfer coefficient and rate of heat transfer by convection for x = 50 cm.

8.8 Determine the hydrodynamic boundary layer thickness for laminar flow over a flat plate by means of approximate analysis of the boundary layer. Assume that the velocity distribution in the boundary layer is given by

$$u/u_\infty = y/\delta$$

8.9 The velocity profile in the laminar boundary layer for flow over a flat plate can be assumed to follow the relation

$$u = a + by + cy^2 + dy^3$$

The velocity profile has to satisfy the boundary conditions given by Eqs. (8.77) to (8.80). Using these boundary conditions, determine the velocity profile.

8.10 If one assumes the relation $u = a + by + cy^2$ to express the velocity profile in the laminar boundary layer for flow over a flat plate, determine the velocity profile which satisfies the boundary conditions given by Eqs. (8.77) to (8.79).

8.11 Assume that the velocity profile in the laminar boundary layer for flow over a flat plate is given by

$$\frac{u}{u_\infty} = 2\left(\frac{y}{\delta}\right) - \left(\frac{y}{\delta}\right)^2$$

Using the approximate analysis of the boundary layer equation, derive a relation to estimate the hydrodynamic boundary layer thickness.

8.12 Air at 300 K and 100 kPa flows over a flat plate at a velocity of 2 m/s. If the plate is 1 m wide and 3 m long and is maintained at 400 K, calculate the rate of heat transfer from the plate using (a) the exact solution and (b) the approximate solution of the boundary layer equations. See Problem 8.5 for data.

8.13 Mercury (ρ = 13,600 kg/m^3; $\nu = 0.114 \times 10^{-6}$ m^2/s; C_p = 0.1394 kJ/kg K; k = 8.7 W/m K) at 25°C is flowing through a tube of 2 cm diameter with $Re = 10^4$. Calculate the convection heat transfer coefficient.

8.14 Freon-12 (ρ = 1518 kg/m^3; c_p = 0.88 kJ/kg K; $\nu = 0.28 \times 10^{-6}$ m^2/s and k = 0.07 W/m K) at −40°C is flowing through a tube of 2 cm diameter with $Re = 5 \times 10^4$. Calculate the convection heat transfer coefficient using the Dittus–Boelter correlation.

CHAPTER NINE

NATURAL CONVECTION AND FORCED CONVECTION

Learning Objectives

This chapter will enable the readers to

- derive the equation of motion for the free convection boundary layer.
- understand the integral method of solution for the free convection boundary layer problem.
- gain familiarity with several empirical relations for predicting the free convection heat transfer from flat plates, horizontal cylinders and spheres.
- understand the order-of-magnitude analysis.
- gain familiarity with empirical relations for predicting forced convection heat transfer coefficients for laminar flow and turbulent flow through tubes.
- define hydraulic diameter and Peclet number.
- learn the empirical correlations to predict the heat transfer coefficients for fluid flow across cylinders and spheres.
- know the empirical correlations for predicting forced convection heat transfer coefficients for laminar and turbulent flow across tube banks.

9.1 Introduction

Natural or free convection occurs from a surface when a body is placed in a fluid at a different temperature. As a result of the temperature difference between the body and the fluid, heat exchanges between them and causes a change in the density of the fluid layers adjacent to the surface. This difference in density leads to the motion of the fluid. The movement of the fluid results from the buoyancy forces. The buoyancy forces which give rise to the free convection currents are called *body forces* and the associated heat transfer is called *natural* or *free convection*. In free convection the intensity of mixing of the fluid is generally small and hence the free convection heat transfer coefficients are lower than the forced convection heat transfer coefficients. Many devices like electrical transformers, heating elements of electric furnaces, radiators used for room heating, etc., transfer energy by natural convection.

9.2 Natural Convection Heat Transfer on a Vertical Flat Plate

Consider the vertical flat plate shown in Fig. 9.1. When the plate is heated, a boundary layer forms near the surface and the fluid velocity at the interface will be zero. At a given distance from the bottom of the plate, the local upward velocity of the fluid increases with increasing distance from the surface to a maximum value, and then decreases with increasing distance. Finally the fluid velocity approaches zero at the edge of the boundary layer, since the free-stream velocity is zero in free convection. In free convection the initial boundary layer is laminar, but at some distance from the leading edge turbulent eddies will be formed, depending on the fluid properties and the temperature difference between the surface and the fluid.

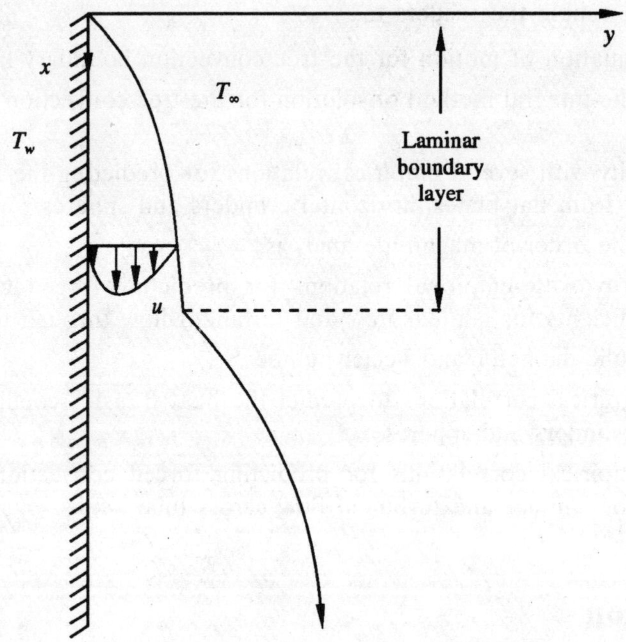

Fig. 9.1 Boundary layer on a vertical flat plate.

To analyze the free convection heat transfer we have to obtain the differential equation describing the motion of the fluid in the boundary layer. Let us choose the x-coordinate along the plate and the y-coordinate perpendicular to the plate. Consider the differential control volume shown in Fig. 9.2.

The fluid in the vicinity of the surface will be heated and its density changes. Hence the body force (the force per unit mass) on a unit volume in the heated portion of the fluid is less than that in the unheated fluid. This imbalance causes the heated fluid to rise. In addition to this buoyant force, pressure forces and frictional forces also act on the fluid in motion. Thus, the total force acting on the volume element $dxdydz$ consists of

1. The force due to pressure gradient

$$= Pdydz - \left(P + \frac{\partial P}{\partial x}dx\right) dydz = -\frac{\partial P}{\partial x}dxdydz$$

2. The body force $= -\rho g dxdydz$

3. The frictional shearing force

$$= -\tau_{yx}dxdz + \left(\tau_{yx} + \frac{\partial \tau_{yx}}{\partial y}dy\right) dxdz = \mu\frac{\partial^2 u}{\partial y^2}dxdydz$$

since

$$\tau_{yx} = \mu\frac{\partial u}{\partial y}$$

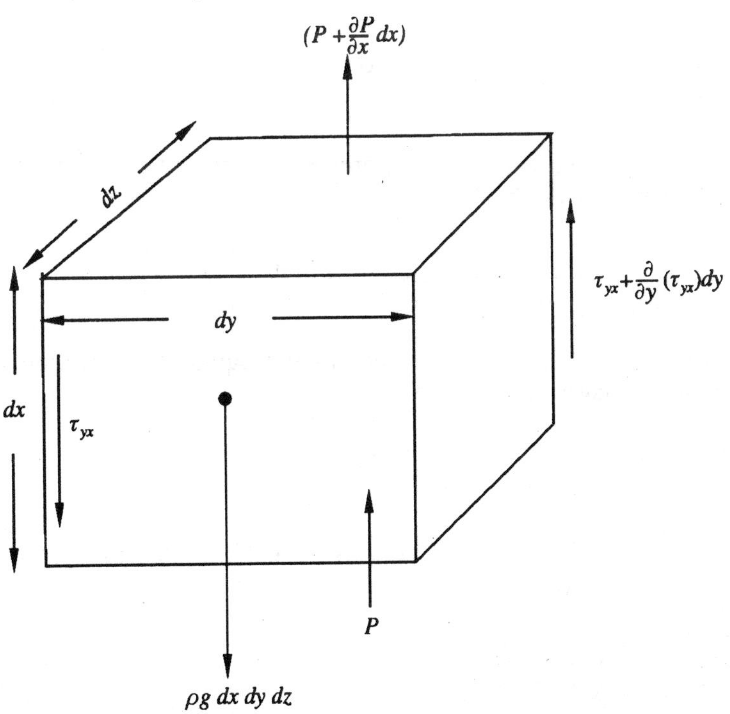

Fig. 9.2 *Control volume showing the forces acting on the fluid.*

Therefore, the net force acting on the fluid element

$$= \left(-\frac{\partial P}{\partial x} - \rho g + \mu \frac{\partial^2 u}{\partial y^2} \right) dx dy dz \tag{9.1}$$

The rate of change of momentum of the fluid element

$$= \rho \left(u \frac{\partial u}{\partial x} + v \frac{\partial u}{\partial y} \right) dx dy dz \tag{9.2}$$

Applying Newton's second law, we get

$$\rho \left(u \frac{\partial u}{\partial x} + v \frac{\partial u}{\partial y} \right) = -\frac{\partial P}{\partial x} - \rho g + \mu \frac{\partial^2 u}{\partial y^2} \tag{9.3}$$

The pressure gradient in the x direction results from the change in the elevation. The unheated fluid far removed from the plate is in hydrostatic equilibrium. That is,

$$\frac{\partial P}{\partial x} = -\rho_\infty g \tag{9.4}$$

Substituting Eq. (9.4) in Eq. (9.3), we get

$$\rho \left(u \frac{\partial u}{\partial x} + v \frac{\partial u}{\partial y} \right) = g(\rho_\infty - \rho) + \mu \frac{\partial^2 u}{\partial y^2} \tag{9.5}$$

The density difference $(\rho_\infty - \rho)$ can be expressed in terms of the coefficient of volume expansion β, which is given by

$$\beta = \frac{1}{v} \left(\frac{\partial v}{\partial T} \right)_p = \frac{1}{v_\infty} \left(\frac{v - v_\infty}{T - T_\infty} \right) = \frac{\rho_\infty - \rho}{\rho (T - T_\infty)} \tag{9.6}$$

Substituting Eq. (9.6) in Eq. (9.5), we obtain the equation of motion for the free convection boundary layer as

$$\rho \left(u \frac{\partial u}{\partial x} + v \frac{\partial u}{\partial y} \right) = g \rho \beta (T - T_\infty) + \mu \frac{\partial^2 u}{\partial y^2} \tag{9.7}$$

The energy equation for the free convection is the same as that for a forced convection and is given by [see Eq. (8.30)]

$$\rho c_p \left(u \frac{\partial T}{\partial x} + v \frac{\partial T}{\partial y} \right) = k \frac{\partial^2 T}{\partial y^2} \tag{9.8}$$

To obtain the solution of the equation of motion, one can adopt the integral method instead of an exact mathematical analysis.

The integral momentum equation for free convection is given by

$$\frac{d}{dx}\left(\int_0^\delta \rho u^2 dy\right) = -\tau_w + \int_0^\delta \rho g\beta(T - T_\infty)dy$$

$$= -\mu\left(\frac{\partial u}{\partial y}\right)_{y=0} + \int_0^\delta \rho g\beta(T - T_\infty)dy \tag{9.9}$$

The integral form of the energy equation for free convection is given by

$$\frac{d}{dx}\left[\int_0^\delta u(T - T_\infty)dy\right] = -\alpha\left(\frac{dT}{dy}\right)_{y=0} \tag{9.10}$$

The velocity profile should satisfy the following boundary conditions:

$$u = 0 \quad \text{at} \quad y = 0 \tag{9.11}$$

$$u = 0 \quad \text{at} \quad y = \delta \tag{9.12}$$

$$\frac{\partial u}{\partial y} = 0 \quad \text{at} \quad y = \delta \tag{9.13}$$

$$\frac{\partial^2 u}{\partial y^2} = -g\beta\frac{T_w - T_\infty}{\nu} \quad \text{at} \quad y = 0 \tag{9.14}$$

The temperature distribution must satisfy the following conditions:

$$T = T_w \quad \text{at} \quad y = 0 \tag{9.15}$$

$$T = T_\infty \quad \text{at} \quad y = \delta \tag{9.16}$$

$$\frac{\partial T}{\partial y} = 0 \quad \text{at} \quad y = \delta \tag{9.17}$$

Then, following the analysis one can obtain the temperature distribution as

$$\frac{T - T_\infty}{T_w - T_\infty} = \left(1 - \frac{y}{\delta}\right)^2 \tag{9.18}$$

Assuming that the velocity can be represented by a polynomial in y as

$$\frac{u}{u_\infty} = a + by + cy^2 + dy^3 \tag{9.19}$$

and by determining the constants a, b, c and d in Eq. (9.19) from the boundary conditions (9.11) to (9.14), the velocity profile can be expressed as

$$\frac{u}{u_\infty} = \frac{y}{\delta}\left(1 - \frac{y}{\delta}\right)^2 \tag{9.20}$$

Using the velocity and temperature distributions it will be possible to solve the integral momentum equation and integral energy equation to obtain

$$\frac{\delta}{x} = 3.93 Pr^{-1/2}(0.952 + Pr)^{1/4} Gr_x^{-1/4} \tag{9.21}$$

where

$$Gr_x = \frac{g\beta(T_w - T_\infty)x^3}{\nu^2} = \text{Grashof number} \tag{9.22}$$

The Grashof number represents the ratio of the buoyancy forces to the viscous forces in free convection. It plays a role similar to that played by the Reynolds number in forced convection. The Grashof number is used as a criterion for transition from laminar to turbulent boundary layer. For air in free convection on a vertical flat plate, the critical Grashof number is approximately 4×10^8.

$$Nu_x = \frac{2x}{\delta}$$

or

$$Nu_x = 0.508 Pr^{1/2}(0.952 + Pr)^{-1/4} Gr_x^{1/4} \tag{9.23}$$

Equation (9.23) gives the local heat transfer coefficient. The average heat transfer coefficient can then be obtained as

$$\overline{h} = \frac{1}{\ell}\int_0^\ell h_x dx \tag{9.24}$$

Substituting Eq. (9.23) in Eq. (9.24), we obtain

$$\overline{h} = \frac{4}{3} h_{x=\ell} \tag{9.25}$$

9.3 Empirical Relations for Natural Convection Heat Transfer Coefficients

The analysis of free convection heat transfer on a vertical flat plate presented in Section 9.2 shows that the Nusselt number for free convection is a function of Gr and Pr. That is,

$$Nu = \phi(Gr)\psi(Pr)$$

It has been found that the average free convection heat transfer coefficients can be represented as

$$\overline{Nu_f} = C(Gr_f Pr_f)^m \tag{9.26}$$

where the subscript f indicates that the dimensionless groups are evaluated at the film temperature T_f, given by

$$T_f = \frac{T_w + T_\infty}{2} \tag{9.27}$$

The product of the Grashof and Prandtl numbers is called the Rayleigh number.

$$Ra = (Gr)(Pr) \tag{9.28}$$

The characteristic dimension x to be used in the Nusselt number and Grashof number depend on the geometry of the body. The characteristic dimension for a vertical flat plate is the length ℓ and for a horizontal cylinder it is the diameter d. The functional relationships of Eq. (9.26) to be used for different geometries are presented in Table 9.1, where the constants c and m are given. These constants can be used if the surface is at isothermal conditions. For vertical planes and cylinders, if $0.1 < Gr_f Pr_f < 10^4$, the Nu_f is presented as a function of $(Gr_f Pr_f)$ in Fig. 9.3. For horizontal cylinders, the dependence of Nu_f on $Gr_f Pr_f$ for $10^{-5} < Gr_f Pr_f < 10^4$ is presented in Fig. 9.4.

Table 9.1 *Constants for use with Eq. (9.26) for isothermal surfaces.*

Geometry	c	m	Range of applicability for $(Gr_f Pr_f)$
Vertical planes and cylinders	0.59	1/4	10^4–10^9
	0.021	2/5	10^9–10^{13}
Horizontal cylinders	0.4	0	0–10^{-5}
	0.53	1/4	10^4–10^9
	0.13	1/3	10^9–10^{12}
Upper surface of heated plates	0.54	1/4	2×10^4–8×10^6
or lower surface of cooled plates	0.15	1/3	8×10^6–10^{11}
Lower surface of heated plates or upper surface of cooled plates	0.27	1/4	10^5–10^{11}

The characteristic dimension for a horizontal plate can be taken as the length of a side for a square plate, the average of the two dimensions for a rectangular plate and $0.9d$ for a circular disk.

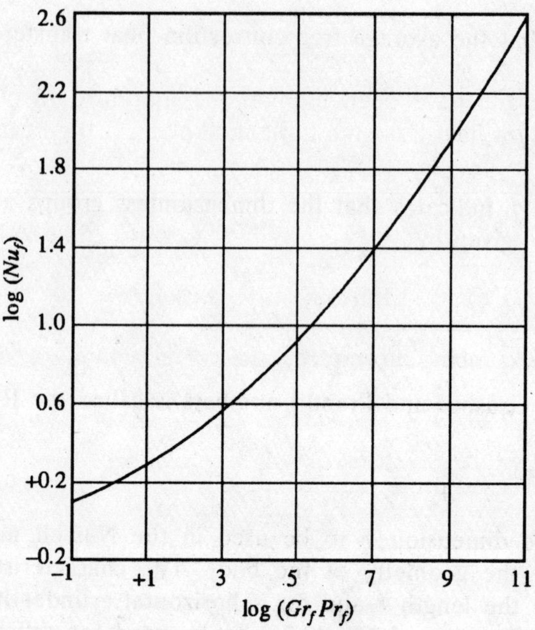

Fig. 9.3 *The dependence of Nu_f on $Gr_f Pr_f$ for vertical planes and cylinders if* $0.1 < Gr_f Pr_f < 10^4$.

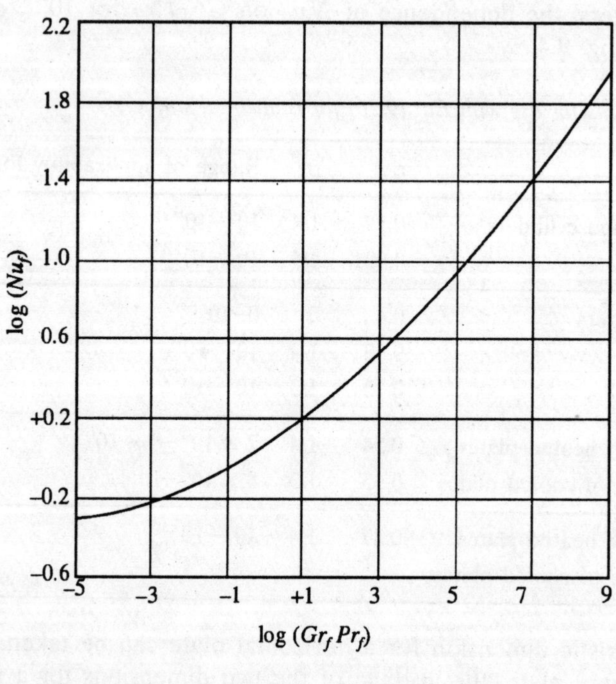

Fig. 9.4 *Dependence of Nu_f on $Gr_f Pr_f$ for horizontal cylinders if* $10^{-5} < Gr_f Pr_f < 10^4$.

Constant heat flux

The experimental results have been used in the literature to obtain the following correlations for constant heat flux on a horizontal plate. If the heated surface is facing upwards,

$$\overline{Nu_\ell} = 0.13(Gr_\ell Pr)^{1/3} \quad \text{for} \quad Gr_\ell Pr < 2 \times 10^8 \tag{9.29}$$

$$\overline{Nu_\ell} = 0.16(Gr_\ell Pr)^{1/3} \quad \text{for} \quad 2 \times 10^8 < Gr_\ell Pr < 10^{11} \tag{9.30}$$

If the heated surface is facing downwards, the correlation is given by

$$\overline{Nu_\ell} = 0.58(Gr_\ell Pr)^{1/5} \quad \text{for} \quad 10^6 < Gr_\ell Pr < 10^{11} \tag{9.31}$$

In Eqs. (9.29) to (9.31), all properties except β should be determined at temperature T_ℓ, which is given by

$$T_\ell = T_w - 0.25(T_w - T_\infty) \tag{9.32}$$

and the average wall temperature T_w is related to the heat flux as

$$\overline{h} = \frac{q_w}{T_w - T_\infty} \tag{9.33}$$

The experimental results for free convection from vertical and inclined surfaces under constant heat flux conditions have been correlated in terms of the modified Grashof number Gr^* defined as

$$Gr^* = Gr_x Nu_x = \frac{g\beta q_w x^4}{k\nu^2} \tag{9.34}$$

where q_w is the wall heat flux (W/m^2). In terms of the modified Grashof number, the boundary layer transition was observed to begin at $3\times 10^{12} < Gr_x^* Pr < 4\times 10^{13}$ and end at $2 \times 10^{13} < Gr_x^* Pr < 10^{14}$.

The flow becomes fully turbulent at $Gr_x^* Pr = 10^{14}$. The local heat transfer coefficients can be calculated from the relations

$$Nu_{xf} = \frac{hx}{k} = 0.60(Gr_x^* Pr_f)^{1/5} \tag{9.35}$$

if $10^5 < Gr_x^* < 10^{11}$ (laminar flow).

$$Nu_x = 0.17(Gr_x^* Pr)^{1/4} \tag{9.36}$$

if $2 \times 10^{13} < Gr_x^* Pr < 10^{16}$ (turbulent flow).

For the laminar region, the average heat transfer coefficient is given by

$$\bar{h} = \frac{1}{\ell}\int_0^{\ell} h_x dx = \frac{5}{4}h_{x=\ell} \tag{9.37}$$

It may be noted that the variation of h_x with x will be

$$h_x \sim x^{-1/4} \text{ for the laminar region} \tag{9.38}$$

$$h_x = \text{constant, with } x \text{ for the turbulent region.} \tag{9.39}$$

9.4 Free Convection from Horizontal Cylinders and Spheres

The free convection heat transfer coefficients from horizontal cylinders can be calculated from Eq. (9.26), with the constants c and m presented in Table 9.1. In addition to Eq. (9.26), one can use the following empirical equations also for calculating the heat transfer coefficients:

$$Nu_d = 0.36 + \frac{0.518(Gr_d Pr)^{1/4}}{[1+(0.559/Pr)^{9/16}]^{4/9}} \tag{9.40}$$

if $10^{-6} < GrPr < 10^9$

$$\overline{Nu}^{1/2} = 0.60 + 0.387\left(\frac{GrPr}{[1+(0.559/Pr)^{9/16}]^{16/9}}\right)^{1/6} \tag{9.41}$$

if $10^{-5} < GrPr < 10^{12}$

In Eqs. (9.40) and (9.41), all the properties are to be evaluated at the film temperature.

The empirical equation for the estimation of the free convection heat transfer coefficient from the sphere to air is given by

$$Nu_f = \frac{\bar{h}d}{k_f} = 2 + 0.392 Gr_f^{1/4} \quad \text{for} \quad 1 < Gr_f < 10^5 \tag{9.42}$$

A more general expression for predicting the free convection heat transfer coefficient from spheres to liquids is

$$Nu_f = 2 + 0.50(Gr_f Pr_f)^{1/4} \quad \text{for} \quad 3 \times 10^5 < GrPr < 8 \times 10^8 \tag{9.43}$$

9.5 Forced Convection Heat Transfer

The analytical solution of boundary layer problems for the prediction of the forced convection heat transfer coefficient is discussed in Chapter 8. It is unfortunate that analytical solutions cannot be obtained for several convection problems, and it is necessary to obtain the values of the convection heat transfer coefficients through experiments. The experimental results are usually presented either in the form of charts or empirical relations. If an analytical solution is available for a similar problem, the correlation of experimental data can be carried out easily by assuming a functional form with the relevant dimensionless groups. If an analytical solution for a similar problem is not available, it will be quite difficult to correlate the experimental data. In such cases one can apply dimensional analysis and obtain appropriate dimensionless groups, and assume a functional relationship with unknown constants and exponents. Then the experimental data can be correlated to determine the values of the constants and exponents in the assumed functional relationship. In general, it is better to attempt an order-of-magnitude analysis as described below, based on a knowledge of the governing equations, to determine the significant variables for correlating the experimental data.

Let us consider the problem of determining the boundary layer thickness for flow over a flat plate, to illustrate the method of the order-of-magnitude analysis. The continuity equation for the laminar boundary layer is given by [see Eq. (8.12)]

$$\frac{\partial u}{\partial x} + \frac{\partial v}{\partial y} = 0 \tag{9.44}$$

and the momentum equation is given by [see Eq. (8.17)]

$$u\frac{\partial u}{\partial x} + v\frac{\partial u}{\partial y} = \nu\frac{\partial^2 u}{\partial y^2} \tag{9.45}$$

In the boundary layer, the velocity u of the fluid is of the order of magnitude of the free-stream velocity u_∞, and the y dimension is of the order of magnitude of the boundary layer thickness δ. That is,

$$u \sim u_\infty \quad \text{and} \quad y \sim \delta \tag{9.46}$$

Then the continuity equation (9.44) can be rewritten in an approximate form as

$$\frac{u_\infty}{x} + \frac{v}{\delta} \approx 0$$

or

$$v \approx \frac{u_\infty \delta}{x} \tag{9.47}$$

The momentum Eq. (9.45) can be rewritten in an approximate form as

$$u_\infty \frac{u_\infty}{x} + v\frac{u_\infty}{\delta} \approx \nu \frac{u_\infty}{\delta^2} \tag{9.48}$$

Substituting Eq. (9.47) in Eq. (9.48), we get

$$\frac{u_\infty^2}{x} + u_\infty \frac{\delta}{x}\frac{u_\infty}{\delta} \approx \nu \frac{u_\infty}{\delta^2}$$

or

$$\delta^2 \sim \frac{\nu x}{u_\infty} \quad \text{or} \quad \delta \sim \sqrt{\frac{\nu x}{u_\infty}}$$

or

$$\frac{\delta}{x} \sim \sqrt{\frac{\nu}{u_\infty x}} = \frac{1}{\sqrt{Re_x}} \tag{9.49}$$

Equation (9.49) shows that the hydrodynamic boundary layer thickness can be correlated with $\sqrt{Re_x}$ and x.

9.6 Empirical Correlations for Tube Flow

The analytical solution of fully developed laminar flow through tubes was discussed in Section 8.13. If the fluid properties change considerably with temperature or the flow is undeveloped, it will be quite difficult to obtain an analytical solution. In such cases, the empirical correlations are of great advantage in estimating the convection heat transfer coefficients. In this section, some important and useful correlations for the estimation of forced convection heat transfer coefficients for flow through tubes are presented.

The Dittus–Boelter equation given below is commonly used for the estimation of heat transfer coefficients for fully developed turbulent flow through tubes.

$$Nu_d = 0.023 Re_d^{0.8} Pr^n \quad \text{for} \quad 0.6 < Pr < 100 \tag{9.50}$$

with the exponent n of the Prandtt number having the values

$n = 0.4$ for heating of the fluid

$n = 0.3$ for cooling of the fluid

The properties of the fluid for use in Eq. (9.50) are to be evaluated at the average fluid bulk temperature. To take into account the property variations of the fluid, Sieder and Tate recommended the equation

$$Nu_d = 0.027 Re_d^{0.8} Pr^{1/3} \left(\frac{\mu}{\mu_w}\right)^{0.14} \tag{9.51}$$

where μ_w is the viscosity of the fluid at the wall temperature.

In the entrance region, the turbulent flow is not fully developed and in such cases it is advisable to use the relation

$$Nu_d = 0.036 Re_d^{0.8} Pr^{1/3} \left(\frac{d}{\ell}\right)^{0.055} \quad \text{for} \quad 10 < \frac{\ell}{d} < 400 \tag{9.52}$$

For laminar flow through tubes, Sieder and Tate recommended the expression

$$\begin{aligned} Nu_d &= 1.86 (Re_d Pr)^{1/3} \left(\frac{d}{\ell}\right)^{1/3} \left(\frac{\mu}{\mu_w}\right)^{0.14} \quad \text{for} \quad Re_d Pr \frac{d}{\ell} > 10 \\ &= 1.86 (Pe)^{1/3} (d/\ell)^{1/3} (\mu/\mu_w)^{0.14} = 1.86 (Gz)^{1/3} (\mu/\mu_w)^{0.14} \end{aligned} \tag{9.53}$$

The product of Reynolds number and Prandtl number is called Peclet number Pe, that is,

$$Pe = Re_d Pr = \frac{d u \rho c_p}{k} \tag{9.54}$$

$$Gz = (Re)(Pr)(d/\ell) = \text{Graetz number}$$

Equation (9.53) cannot be used for extremely long tubes since it would give a value of zero for the heat transfer coefficient. This equation is valid for

$$(Re_d)(Pr)\left(\frac{d}{\ell}\right) > 10 \tag{9.55}$$

If the fluid flows through a tube which is not circular in cross-section, the heat transfer coefficient can be calculated based on the *hydraulic diameter* d_H defined as

$$d_H = \frac{4A}{p} \tag{9.56}$$

where

A = cross-sectional area of the flow

P = wetted perimeter.

9.7 Correlations for Flow Across Cylinders and Spheres

Engineers deal with situations in which heat transfers by forced convection between tubes, tube bundles and spheres, and fluid flowing perpendicularly to the axes of these bodies. The boundary layer development on these bodies determines the heat transfer coefficients. As long as the boundary layer remains laminar, it is possible to estimate the heat transfer coefficients by a method similar to the boundary layer analysis discussed in Chapter 8. If the free-stream velocity is large, boundary layer separation takes place on the reverse side of the cylinder. Because of the complex nature of flow process around cylinders it is not possible to estimate, through analysis, the heat transfer coefficients in cross flow. The correlations of experimental data for gases and liquids indicate that the average heat transfer coefficients for flow across cylinders can be evaluated from the relation

$$\frac{hd}{k_f} = c\left(\frac{u_\infty d}{\nu_f}\right)^n Pr_f^{1/3} \tag{9.57}$$

The properties in Eq. (9.57) are to be evaluated at the film temperature, and the values of the constants c and n are presented in Table 9.2.

Table 9.2 *Constants for use with Eq. (9.57) for heat transfer from cylinders with circular cross-section*

c	n	Re_d
0.989	0.330	0.4–4
0.911	0.385	4–40
0.683	0.466	40–4000
0.193	0.618	4000–40,000
0.0266	0.805	40,000–400,000

The following relation suggested by Fand can be used to estimate the heat transfer coefficients over a wide range of Reynolds number:

$$Nu_f = (0.35 + 0.56 Re_f^{0.52}) Pr_f^{0.3} \quad \text{for} \quad 10^{-1} < Re_f < 10^5 \tag{9.58}$$

In Eq. (9.58), the properties are to be estimated at the film temperature.

To estimate the heat transfer coefficients from non-circular cylinders, Eq. (9.57) can be employed with the constants presented in Table 9.3.

In the Reynolds number range from 25 to 100,000, the heat transfer coefficients from spheres to a flowing gas can be estimated by the following equation suggested by McAdams:

$$\frac{hd}{k_f} = 0.37\left(\frac{u_\infty d}{\nu_f}\right)^{0.6} \quad \text{for} \quad 25 < Re_d < 100,000 \tag{9.59}$$

Table 9.3 Constants for use with Eq. (9.57) for heat transfer from non-circular cylinders

Geometry	c	n	Re_{df}
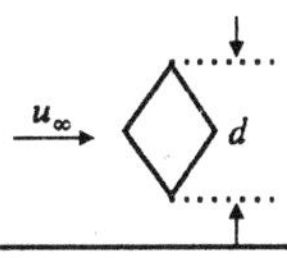	0.246	0.588	$5 \times 10^3 - 10^5$
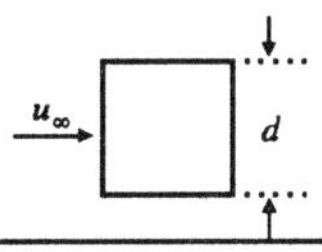	0.102	0.675	$5 \times 10^3 - 10^5$
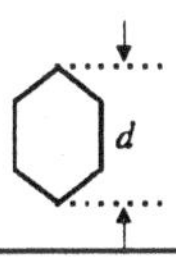	0.160 0.0385	0.638 0.782	$5 \times 10^3 - 1.95 \times 10^4$ $1.95 \times 10^4 - 10^5$
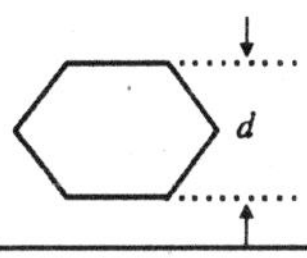	0.153	0.638	$5 \times 10^3 - 10^5$

The Achenbach relation given below can be used to estimate the heat transfer coefficients from spheres to a gas over a wider range of Reynolds number:

$$Nu = 2 + (0.25 + 3 \times 10^{-4} Re^{1.6})^{1/2} \quad \text{for} \quad 100 < Re < 3 \times 10^5 \tag{9.60}$$

For flow of liquids past spheres, Kramers relation given below can be used:

$$\frac{hd}{k_f} Pr_f^{-0.3} = 0.97 + 0.68 \left(\frac{u_\infty d}{\nu_f} \right)^{0.5} \quad \text{for} \quad 1 < Re_d < 2000 \tag{9.61}$$

Whitaker proposed the following correlation which is applicable for the flow of gases and liquids past spheres:

$$Nu = 2 + (0.4 Re_d^{1/2} + 0.06 Re_d^{2/3}) Pr^{0.4} \left(\frac{\mu_\infty}{\mu_w} \right)^{1/4} \tag{9.62}$$

Equation (9.62) is applicable in the range $3.5 < Re_d < 8 \times 10^4$ and $0.7 < Pr < 380$. All the properties in Eq. (9.62) are to be evaluated at the free-stream temperature.

9.8 Flow Across Tube Banks

The estimation of convection heat transfer coefficients between a bank of tubes and a fluid flowing at right angles to the tubes is very important in the design and analysis of

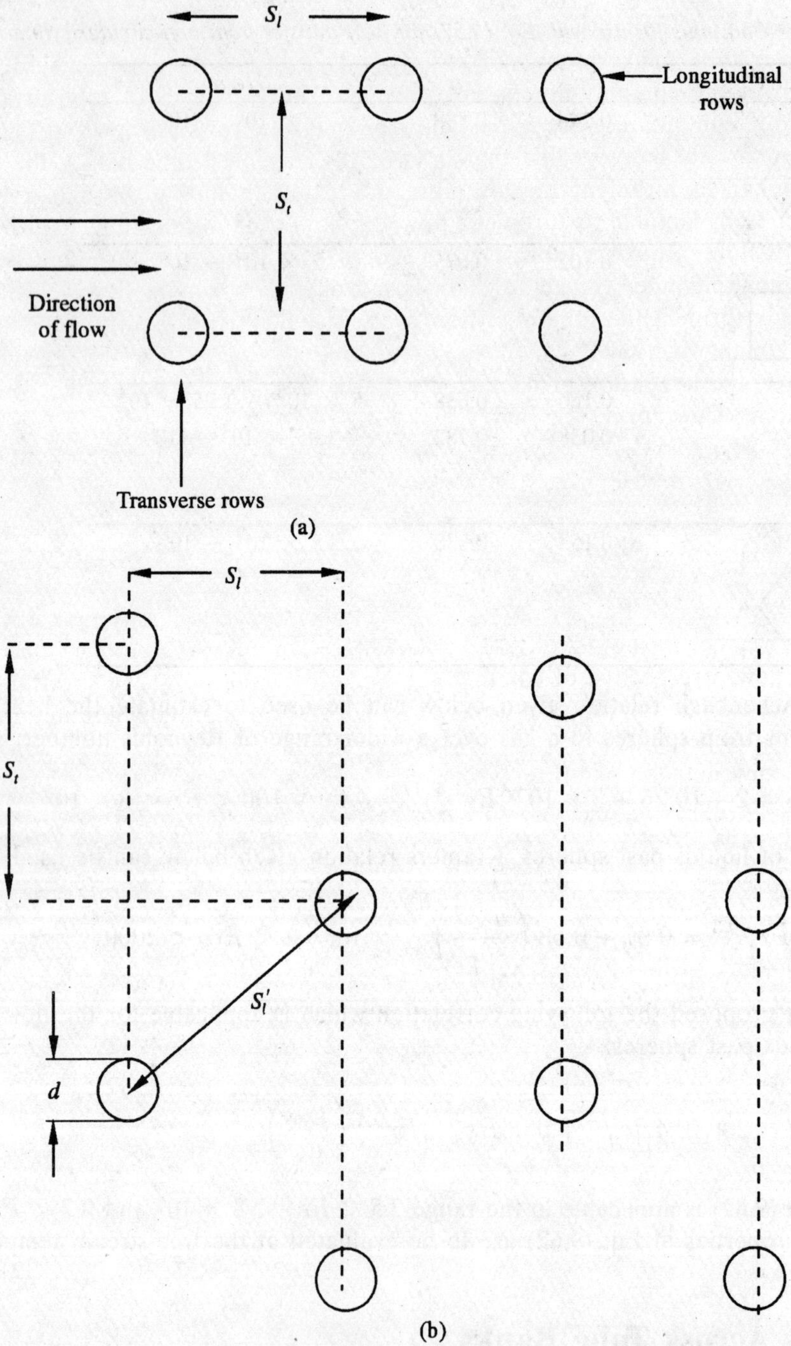

Fig. 9.5 (a) In-line tube arrangement. (b) Staggered tube arrangement.

heat exchangers. The heat transfer inflow over tube banks depends largely on the flow characteristics, which in turn depends on the velocity of the fluid and the arrangement of the tubes. The equations available for the estimation of heat transfer coefficients inflow over tube banks are based entirely on experimental data because analytical treatment of the complex flow is not possible. The two types of tube arrangements which are in common use are *in-line* and *staggered* tube rows as shown in Fig. 9.5. The experimental data has revealed that inflow over staggered tube banks, the transition from laminar to turbulent flow is more gradual than inflow through a tube. The transition from laminar to turbulent flow for flow over in-line tube banks is similar to that for flow through a tube. The transition from laminar to turbulent flow begins at a Reynolds number (of about 200), based on the maximum flow velocity, that is the velocity through the minimum flow area. The flow is fully turbulent at Reynolds number greater than 6000.

The average heat transfer coefficients for the entire tube bank are correlated by an equation of the form [see Eq. (9.57)]

$$\overline{Nu} = c(Re)^m (Pr)^n$$

which is used to correlate the data for flow over a single tube. To use this equation for flow over a tube bank it is necessary to select a reference velocity, since the fluid velocity changes along the path. The reference velocity used in the calculation of Reynolds number for flow over tube banks is based on the *minimum free area* available for fluid flow, irrespective of whether the minimum area occurs in the transverse or diagonal openings. For flows normal to in-line tube banks, the minimum free-flow area per unit length of tube is $(S_t - d)$, where S_t is the *transverse pitch* or the distance between tube centres in adjacent longitudinal rows. The reference velocity or maximum flow velocity for in-line tube banks is given by

$$u_{\max} = u_\infty \left(\frac{S_t}{S_t - d} \right) \tag{9.63}$$

For the staggered tube rows, the same maximum flow velocity is given by

$$u_{\max} = \frac{u_\infty (S_t/2)}{[(S_t/2)^2 + S_\ell^2]^{1/2} - d} \tag{9.64}$$

where S_ℓ is the *longitudinal pitch* or the centre-to-centre distance between adjacent transverse rows of tubes.

To take into account the effect of tube arrangement on the heat transfer coefficient, the following forms of the dimensionless correlation equations are used:

$$\frac{\bar{h}d}{k_f} = 0.33 C_H \left(\frac{G_{\max} d}{\mu_f} \right)^m Pr_f^{1/3} \tag{9.65}$$

$$j = \frac{\bar{h}}{c_p G_{\max}} Pr_b^{2/3} \left(\frac{\mu_w}{\mu_b} \right)^{0.14} = \phi \left(\frac{G_{\max} d}{\mu_b} \right) \tag{9.66}$$

where

$\bar{h}$ = average heat transfer coefficient for a tube bank of 10 or more transverse rows

$G_{\max}$ = mass flow rate per unit of minimum flow area

j = Colburn j factor.

The subscripts f, b and w in Eqs. (9.65) and (9.66) denote the conditions at film, bulk and wall, respectively.

For turbulent flow ($Re_{\max} \geq 6000$) over tube banks, irrespective of whether they are in-line or staggered, the experimental data was found to correlate quite satisfactorily with the form of Eq. (9.65) and resulted in

$$\frac{\bar{h}d}{k_f} = 0.33C_H\left(\frac{G_{\max}d}{\mu_f}\right)^{0.6} Pr_f^{0.3} \tag{9.67}$$

if the tube bank has 10 or more transverse rows. For pitch-to-diameter ratios between 1.25 and 1.5, the range of practical interest for heat exchangers, the value of C_H is approximately equal to unity. Hence, for engineering applications Eq. (9.67) with $C_H = 1$ can be used to predict the heat transfer coefficients for turbulent flow across tube banks. To estimate the average heat transfer coefficient for tube banks with less than 10 transverse rows, the value of $\bar{h}$ obtained from Eq. (9.67) should be multiplied by the appropriate ratio (h_N/h) presented in Table 9.4.

Table 9.4 *Ratio (h_N/h) to be used with Eq. (9.67) if the number of transverse rows are less than 10*

	Number of transverse rows (N)									
h_N/h	1	2	3	4	5	6	7	8	9	10
Staggered tubes	0.68	0.75	0.83	0.89	0.92	0.95	0.97	0.98	0.99	1.0
In-line tubes	0.64	0.80	0.87	0.90	0.92	0.94	0.96	0.98	0.99	1.0

In the laminar flow range ($(G_{\max}d/\mu_b) < 200$) and in the transition flow range ($200 < (G_{\max}d/\mu_b) < 6000$), a relation of the type given by Eq. (9.66) was found to correlate the experimental data quite satisfactorily. The results are presented in Fig. 9.6 in which the Colburn j factor is plotted against the bulk Reynolds number ($(G_{\max}d/\mu_b)$).

The variation of the heat transfer coefficient for laminar flow with N, the number of tube rows in the direction of flow, can be expressed by the relation

$$\frac{j_N}{j} = N^{0.18} \tag{9.68}$$

where j_N = Colburn j factor with N tube rows

j = Colburn j factor with 10 tube rows.

The value of the Colburn j factor with 10 tube rows can be read from Fig. 9.6, and if the number of tube rows is less than 10, then j_N can be estimated from Eq. (9.68).

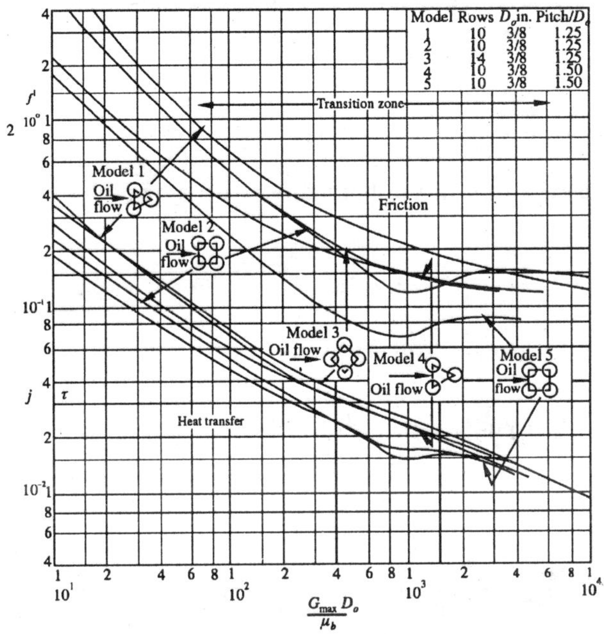

Fig. 9.6 *Variation of Colburn j factor with Reynolds number.*

9.9 Summary

Natural or free convection occurs from a surface when a body is placed in a fluid at a different temperature. As a result of the temperature difference between the body and the fluid, heat exchanges between them and causes a change in the density of the fluid layers adjacent to the surface. This difference in density induces fluid motion.

If a heated vertical flat plate is placed in contact with a fluid, a boundary layer forms near the surface. The fluid velocity at the surface will be zero. The local upward velocity of the fluid increases with increasing distance from the surface to a maximum value and then decreases to zero. The momentum equation for free convection boundary layer is

$$\rho\left(u\frac{\partial u}{\partial x}+v\frac{\partial u}{\partial y}\right)=g\rho\beta(T-T_\infty)+\mu\frac{\partial^2 u}{\partial y^2}$$

and the energy equation is given by

$$\rho c_p \left(u\frac{\partial T}{\partial x} + v\frac{\partial T}{\partial y} \right) = k\frac{\partial^2 T}{\partial y^2}$$

The integral method of analysis can be used to obtain the solution to the free convection boundary layer equations. The temperature distribution in the boundary layer is

$$\frac{T - T_\infty}{T_w - T_\infty} = \left(1 - \frac{y}{\delta}\right)^2$$

and the velocity distribution in the boundary layer is given by

$$\frac{u}{u_\infty} = \frac{y}{\delta}\left(1 - \frac{y}{\delta}\right)^2$$

The boundary layer thickness is given by

$$\frac{\delta}{x} = 3.93 Pr^{-1/2}(0.952 + Pr)^{1/4} Gr_x^{-1/4}$$

where

$$Gr = \frac{g\beta(T_w - T_\infty)x^3}{\nu^2} = \text{Grashof number.}$$

The local heat transfer coefficient is given by

$$Nu_x = 0.508 Pr^{1/2}(0.952 + Pr)^{-1/4} Gr_x^{1/4}$$

and the average heat transfer coefficient is

$$\overline{h} = \frac{4}{3} h_{x=\ell}$$

The free convection heat transfer coefficients from vertical planes, cylinders and horizontal cylinders can be estimated by the relation

$$\overline{Nu_f} = c(Gr_f Pr_f)^m = c(Ra)^m$$

where

$Ra = GrPr$ = Rayleigh number

c, m = constants

and the subscript f denotes that the dimensionless groups are evaluated at the film temperature T_f given by

$$T_f = \frac{T_w + T_\infty}{2}$$

Similarly, empirical correlations are available to predict the free convection heat transfer coefficients from horizontal cylinders and spheres.

The Dittus–Boelter equation

$$Nu_d = 0.023 Re_d^{0.8} Pr^n$$

where

$$n = 0.4 \quad \text{for heating of the fluid}$$

$$n = 0.3 \quad \text{for cooling of the fluid}$$

can be used to predict the forced convection heat transfer coefficients for fully developed turbulent flow through tubes. To take into account the variation in the fluid properties with temperature, Seider and Tate proposed the relation

$$Nu_d = 0.027 Re_d^{0.8} Pr^{1/3} \left(\frac{\mu}{\mu_w}\right)^{0.14}$$

For laminar flow through tubes, the Seider–Tate correlation is

$$Nu_d = 1.86(Pe)^{1/3}\left(\frac{d}{\ell}\right)^{1/3}\left(\frac{\mu}{\mu_w}\right)^{0.14} = 1.86(Gz)^{1/3}\left(\frac{\mu}{\mu_w}\right)^{0.14}$$

where

$Pe = (Re)(Pr)$ = Peclet number

$Gz = (Re)(Pr)(d/\ell)$ = Graetz number.

If the fluid flows through a non-circular tube, the heat transfer coefficient can be calculated based on the hydraulic diameter d_H defined as

$$d_H = \frac{4A}{p}$$

where

A = cross-sectional area of the flow

P = wetted perimeter.

For flow across cylinders and spheres, the forced convection heat transfer coefficients can be evaluated from the relation

$$\frac{hd}{k_f} = c\left(\frac{u_\infty d}{\nu_f}\right)^n Pr^{1/3}$$

with appropriate values of c and n depending on Re_d.

For turbulent flow across tube banks ($Re_{\max} \geq 6000$), the forced convection heat transfer coefficient can be evaluated by the relation

$$\frac{hd}{k_f} = 0.33\left(\frac{G_{\max} d}{\mu_f}\right)^{0.6} Pr_f^{0.3}$$

where $G_{\max}$ = mass flow rate per unit of minimum flow area. If the number of tube rows is less than 10, the value of h calculated from the above relation is to be corrected by multiplying with (h_N/h), which depends on N, the number of tube rows.

For laminar flow ($G_{\max} d/\mu_b < 200$) and in transition flow range ($200 < G_{\max} d/\mu_b < 6000$) across tube banks, the forced convection heat transfer coefficient can be estimated from the Colburn j factor which depends on $Re_{\max}$. The Colburn j factor is given by

$$j = \frac{h}{c_p G_{\max}} Pr_b^{2/3}\left(\frac{\mu_w}{\mu_b}\right)^{0.14}$$

The value of h estimated from the above relation has to be corrected by multiplying with (j_N/j) if the number of tube rows is less than 10. The correction factor is given by

$$\frac{j_N}{j} = N^{0.18}$$

Review Questions

1. Distinguish between natural convection and forced convection.
2. What are the forces responsible for fluid motion in natural convection?
3. Specify the boundary conditions which must be satisfied by the velocity distribution for a free convection boundary layer on a vertical flat plate.
4. State the boundary conditions which must be satisfied by the temperature distribution in the boundary layer for free convection heat transfer between a vertical flat plate and a fluid.

5. State the velocity distribution and temperature distribution for free convection heat transfer on a vertical flat plate.
6. Define Grashof number and what its physical significance is?
7. What is the relation between local heat transfer coefficient and average heat transfer coefficient for free convection from a vertical flat plate ?
8. Define Rayleigh number.
9. What is the film temperature?
10. What is the functional form between the dimensionless groups that can be used to correlate the natural convection heat transfer coefficients?
11. What is the Dittus–Boelter equation? When does it apply?
12. What is the Sieder–Tate relation? How is it different from the Dittus– Boelter equation?
13. What is the Peclet number?
14. What is the Graetz number?
15. What is the hydraulic diameter? When is it used?
16. What form of equation is used to calculate the heat transfer coefficients for flow across cylinders?
17. What is Colburn j factor?
18. What is the criterion for transition from laminar to turbulent boundary layer in free convection on a vertical flat plate?
19. Why is an analytical solution of a free convection problem more involved than the forced convection problem?
20. How is a modified Grashof number defined for a constant heat flux condition on a vertical plate?

Problems

9.1 A large vertical plate 5 m high is maintained at 100°C and exposed to air at 30°C. Calculate the convection heat transfer coefficient.

9.2 A vertical isothermal flat plate at 100°C is exposed to air at 20°C and 100 kPa. Plot the free convection velocity profiles as a function of distance from the plate surface at x positions of 20 cm and 40 cm.

9.3 Plot the free convection boundary layer thickness as a function of x for the vertical plate of Problem 9.2. Consider the laminar portion only.

9.4 A vertical flat plate is maintained at 100°C and exposed to air at 25°C. At a distance of 30 cm from the leading edge of the plate the boundary layer thickness is 3 cm. Estimate the thickness of the boundary layer at a distance of 50 cm from the leading edge.

9.5 A vertical plate 2 m × 2 m is maintained at 80°C and exposed to air at 10°C and atmospheric pressure. Calculate the heat lost by convection from the plate.

9.6 If the plate of Problem 9.5 is exposed to helium at the same conditions, determine the heat lost by the plate.

9.7 A large vertical plate 2 m high is maintained at 100°C. This plate is exposed to air at 10°C. Determine the heat transfer coefficient at heights of 0.5 m, 1.0 m, and 2 m . Also calculate the average heat transfer coefficient for thc plate.

9.8 A vertical metal plate 1 m × 1 m is receiving a constant heat flux of 600 W/m^2. One side of the plate is insulated and the other side is exposed to air at 10°C and atmospheric pressure. What will be the average temperature of the plate?

9.9 A 25 cm × 25 cm vertical plate fitted with an electrical heater which provides a constant heat flux of 1 kW/m^2 is immersed in water at 10°C . Calculate the heat transfer coefficient and the average temperature of the plate.

9.10 A horizontal rod of 5 cm diameter is maintained at a constant surface temperature of 70°C by submerging it in water at 20°C. Calculate the heat lost by free convection per unit length of the rod.

9.11 A copper wire of 1 mm diameter is maintained at a constant temperature of 75°C by passing electric current. The wire is exposed to atmospheric air at 25°C. Calculate the rate of heat transfer by convection, per metre length of the wire.

9.12 Steam is flowing through a horizontal pipe of 3 cm OD which is exposed to atmospheric air at 20°C. The surface temperature of the pipe is 150°C. Calculate the heat lost by free convection per metre length of pipe.

9.13 Condensing steam at 120°C is to be used inside a 5 cm diameter horizontal pipe to provide heating in a room where the ambient air is at 5°C. The total heating requirement is 3 kW. What length of pipe would be required to provide the necessary heating?

9.14 A horizontal rod of 2 cm diameter is maintained at a temperature of 70°C by passing electric current while it is immersed in glycerine at 10°C. Calculate the rate of heat transfer per unit length of the rod.

9.15 Calculate the rate of free convection heat transfer from a 5 cm diameter sphere, the surface of which is maintained at 90°C while it is exposed to atmospheric air at 20°C.

9.16 A sphere of 5 cm diameter is maintained at 50°C and submerged in water at 10°C. Calculate the heat transfer coefficient.

9.17 Air at 100 kPa and 100°C is heated while it is flowing through a 2 cm ID tube at a velocity of 10 m/s. Calculate the convection heat transfer coefficient.

9.18 Water at an average bulk temperature of 10°C flows inside a horizontal smooth tube of 5 mm diameter with a velocity of 10 cm/s. The surface of the tube is maintained at 70°C and the tube is 1 metre long. Calculate the rate of heat transfer to the water.

9.19 Water flowing through a tube of 1 cm diameter is heated from 20°C to 30°C while the tube surface is maintained at 70°C. If the mass flow rate of water through the tube is 0.5 kg/s, calculate the length of the tube.

9.20 Liquid Freon-12 enters a 5 mm diameter tube at 0°C and flows with a Reynolds number of 600. The outer surface of the tube is maintained at a constant temperature of 40°C. If the liquid leaves the tube at 15°C, calculate the length of the tube.

9.21 Calculate the convection heat transfer coefficient for water flowing at a velocity of 3 m/s in an annulus formed by a 2 cm OD tube and 3 cm ID tube. The water enters the tube at 30°C.

9.22 Air at atmospheric pressure and 30°C flows across a tube bank consisting of 100 tubes of 1.25 cm OD arranged in a staggered manner in 20 rows. The pitch-to-diameter ratio is 1.5 and the tube wall is maintained at 60°C. Calculate the convection heat transfer coefficient if the free-stream velocity of air is 5 m/s.

CHAPTER TEN

CONDENSATION AND BOILING HEAT TRANSFER

Learning Objectives

A careful study of this chapter will enable the readers to

- distinguish between film condensation and drop-wise condensation.
- understand the Nusselt analysis of laminar film condensation on a vertical plate.
- calculate the local and average heat transfer coefficients for film condensation on a vertical plate.
- estimate the heat transfer coefficients for film condensation on the outside of tubes and tube banks.
- predict the average heat transfer coefficient for turbulent film condensation on vertical plates.
- estimate the heat transfer coefficients for film condensation inside horizontal tubes for condensing refrigerant vapors.
- extend the film condensation analysis to superheated vapors.
- understand the various regimes of boiling heat transfer.
- predict the heat flux for nucleate pool boiling.
- predict the critical heat flux for nucleate boiling.

10.1 Introduction

In the earlier chapters we have discussed the convection heat transfer in homogeneous single phase systems. The convection heat transfer processes associated with phase change, that is during condensation and boiling, are also important. In thermal power plants one is interested in converting liquid water into steam in a boiler and in the conversion of low pressure steam (steam leaving the turbine) into water in a condenser. An engineer should have a good knowledge of the heat transfer processes associated with condensation and boiling, in order to design the appropriate heat transfer equipment. Similarly, the condensation and boiling heat transfer are important in the design of condensers and evaporators in vapor compression refrigeration cycles. The phenomena associated with condensation and boiling heat transfer are much more complex than the convection heat transfer in a homogeneous single phase. In this chapter we shall study the phenomena of condensation and boiling heat transfer and a few important correlations to estimate the convection heat transfer coefficients.

10.2 Condensation Heat Transfer

Suppose a vertical flat plate is exposed to a condensable vapor. If the temperature of the plate is less than the saturation temperature of the vapor, the vapor condenses on the plate surface and flows down under the action of gravity. If the condensate does not wet the plate's surface, the liquid droplets fall from the surface. This process is termed *drop-wise condensation*. On the other hand, if the condensate wets the plate's surface, a smooth film will be formed on the surface. The thickness of the film grows in size as the liquid flows down the surface. This process is called *film condensation.* The liquid film presents additional resistance for heat transfer from the plate to the vapor. In drop-wise condensation, most of the plate's surface is uncovered and hence the thermal resistance for heat transfer is low. Therefore, the convection heat transfer coefficient in drop-wise condensation will be much higher than in the case of film condensation. In view of the higher values of convection heat transfer coefficients for drop-wise condensation, it is desirable to maintain the condition of the plate's surface to have drop-wise condensation. Since most of the surfaces get wet with the condensate, it is quite difficult to have drop-wise condensation. In order to maintain the drop-wise condensation, some surface coatings can be used.

10.2.1 Film Condensation

Nusselt analyzed the film condensation and developed relations for the estimation of convection heat transfer coefficients. To illustrate the classical approach of Nusselt, let us consider a vertical plate at constant temperature T_w on which a vapor is condensing as shown in Fig. 10.1. Let T_s denote the saturation temperature of the vapor. A continuous film of liquid flows downward under the action of gravity. The thickness of the film increases as more and more vapor condenses at the liquid-vapor interface.

Let us choose the positive x-direction measured downward. Let the film thickness be represented by δ. The vapor is at temperature T_s at the liquid-vapor interface. Assume that a linear temperature profile exists between the condensing vapor and the surface. That is, the temperature profile is given by

$$\frac{dT}{dy} = \frac{T_s - T_w}{\delta} \tag{10.1}$$

The weight of the fluid element of thickness dx between y and δ per unit depth

$$= \rho g(\delta - y)dx$$

The viscous shear force at $y = \mu(du/dy)dx$

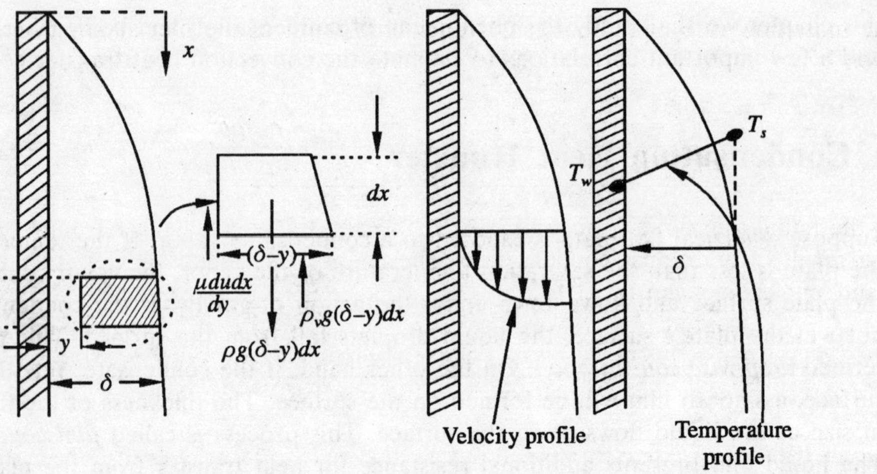

Fig. 10.1 Film condensation on a vertical flat plate.

Buoyancy force $= \rho_v g(\delta - y)dx$

where

ρ = density of liquid film

ρ_v = density of the vapor

u = velocity of the liquid film

μ = viscosity of the fluid

g = acceleration due to gravity.

Under steady-state conditions, a force balance on the element of the fluid gives

$$\rho g(\delta - y)dx = \mu \frac{du}{dy}dx + \rho_v g(\delta - y)dx$$

or $$(\delta - y)(\rho - \rho_v)g = \mu \frac{du}{dy} \tag{10.2}$$

The boundary condition is

$$u = 0 \quad \text{at} \quad y = 0 \tag{10.3}$$

Integrating Eq. (10.2) and using the boundary condition (10.3), we obtain

$$u_y = \frac{(\rho - \rho_v)g}{\mu}\left(\delta y - \frac{1}{2}y^2\right) \tag{10.4}$$

The mass flow rate of the condensate per unit depth of the film at any position x is given by

$$\dot{m} = \int_{o}^{\delta} \rho \left[\frac{(\rho - \rho_v)g}{\mu} (\delta y - \frac{1}{2} y^2) \right] dy = \frac{\rho(\rho - \rho_v)g\delta^3}{3\mu} \tag{10.5}$$

As the liquid film flows from x to $x + dx$, the film thickness changes from δ to $\delta + d\delta$ as a result of additional condensation. The change in condensate flow rate between x and $x + dx$ is given by

$$\frac{d}{dx} \left[\frac{\rho(\rho - \rho_v)g\delta^3}{3\mu} \right] dx = \frac{d}{d\delta} \left[\frac{\rho(\rho - \rho_v)g\delta^3}{3\mu} \right] \frac{d\delta}{dx} dx = \frac{\rho(\rho - \rho_v)g\delta^2}{\mu} d\delta \tag{10.6}$$

The change in condensate flow rate is due to the energy transferred from the condensing vapor to the wall. That is, the heat transferred from the condensing vapor to the wall is equal to the increased mass flow rate times the latent heat of vaporization. Therefore,

$$\frac{\rho(\rho - \rho_v)g\delta^2 d\delta}{\mu} h_{fg} = k dx \frac{T_s - T_w}{\delta} \tag{10.7}$$

where

h_{fg} = latent heat of vaporization

k = thermal conductivity of the condensate.

Equation (10.7) can be integrated, and by using the boundary condition that

$$\delta = 0 \quad \text{at} \quad x = 0$$

we get

$$\delta_x = \left[\frac{4\mu k x (T_s - T_w)}{g h_{fg} \rho(\rho - \rho_v)} \right]^{1/4} \tag{10.8}$$

The heat transfer across the condensate layer is by conduction. We can express the local heat transfer coefficient h_x as

$$h_x dx (T_w - T_s) = -k dx \frac{T_s - T_w}{\delta}$$

or

$$h_x = \frac{k}{\delta_x} \tag{10.9}$$

From Eqs. (10.8) and (10.9), we get

$$h_x = \left[\frac{\rho(\rho - \rho_v)gh_{fg}k^3}{4\mu x(T_s - T_w)}\right]^{1/4} \tag{10.10}$$

The local Nusselt number Nu_x is then given by

$$Nu_x = \frac{hx}{k} = \left[\frac{\rho(\rho - \rho_v)gh_{fg}x^3}{4\mu x(T_s - T_w)}\right]^{1/4} \tag{10.11}$$

An inspection of Eq. (10.10) shows that the heat transfer coefficient for film condensation decreases with increasing distance from the top as the film becomes thicker and thicker. The thickening of the condensate film is similar to the growth of a boundary layer on a flat plate. It can also be observed from Eq. (10.10) that an increase in the temperature difference $(T_s - T_w)$ causes a decrease in the heat transfer coefficient.

The average value of the heat transfer coefficient $\bar{h}$ for a vapor condensing on a plate of length ℓ is obtained by integrating the local value h_x over the plate and by dividing by the length.

$$\bar{h} = \frac{1}{\ell}\int_o^\ell h_x dx = \frac{4}{3}h_{x=\ell} \tag{10.12}$$

or

$$\bar{h} = 0.943\left[\frac{\rho(\rho - \rho_v)gh_{fg}k_f^3}{\ell\mu_f(T_s - T_w)}\right]^{1/4} \tag{10.13}$$

where the subscript f denotes that the properties should be evaluated at the film temperature T_f given by

$$T_f = \frac{T_s + T_w}{2} \tag{10.14}$$

Rohsenow performed a refined analysis of this problem and obtained results which are in better agreement with the experimental data if $Pr > 0.5$ and $c(T_s - T_w)/h_{fg} \leq 1.0$. Similar results can also be obtained by replacing h_{fg} by h'_{fg} in Eqs. (10.10)–(10.13), where h'_{fg} is defined as

$$h'_{fg} = h_{fg} + 0.68c(T_s - T_w) \tag{10.15}$$

where c is the specific heat of the condensate.

10.2.2 Film Condensation on Tubes and Tube Banks

The average heat transfer coefficient for laminar film condensation on the outside of a sphere or a horizontal tube can be evaluated from the relation

$$\bar{h} = A\left[\frac{\rho(\rho-\rho_v)gh_{fg}k_f^3}{\mu_f d(T_s-T_w)}\right]^{1/4} \tag{10.16}$$

where

A = 0.815 for a sphere

A = 0.725 for a tube

d = diameter.

If condensation occurs on a horizontal tube bank with N tubes so arranged that the condensate from one tube flows directly onto the tube below, the average heat transfer coefficient for the system can be obtained by replacing d with Nd in Eq. (10.16).

If a plate or a cylinder is inclined at an angle θ with the horizontal, Eqs. (10.10)–(10.16) can be used by replacing g with its component parallel to the heat transfer surface. That is, by replacing g with $g\sin\theta$.

A fluid flowing over a surface undergoes a transition from laminar to turbulent flow. Therefore, the motion of the condensate becomes turbulent when its Reynolds number exceeds a critical value of about 1800. If the plate on which condensation occurs is sufficiently large, or there is a sufficient amount of condensate flow, turbulence may appear in the condensate film. The Reynolds number of the condensate film is defined as

$$Re_f = \frac{d_H\rho V}{\mu_f} = \frac{4A\rho V}{P\mu_f} = \frac{4\dot{m}}{P\mu_f} \tag{10.17}$$

where

d_H = hydraulic diameter

A = flow cross-sectional area

P = wetted perimeter

V = average velocity of flow

$\dot{m}$ = ρAV = mass flow of the condensate.

The wetted perimeter P for a vertical tube is equal to πd, and for a vertical plate of unit depth $P = 1$. The Reynolds number is sometimes expressed in terms of the mass flow per unit depth of plate Γ as

$$Re_f = \frac{4\Gamma}{\mu_f} \tag{10.18}$$

The mass flow rate $\dot{m}$ can be related to the heat transfer coefficient by

$$q = \bar{h}A(T_s - T_w) = \dot{m}h_{fg} \tag{10.19}$$

or

$$\dot{m} = \frac{q}{h_{fg}} = \frac{\bar{h}A(T_s - T_w)}{h_{fg}}$$

and

$$Re_f = \frac{4\dot{m}}{P\mu_f} = \frac{4\bar{h}A(T_s - T_w)}{h_{fg}P\mu_f} \tag{10.20}$$

For a plate of length ℓ and width w, the flow cross-sectional area $A = \ell w$ and the wetted perimeter $P = w$. For a plate, Eq. (10.20) reduces to

$$Re_f = \frac{4\bar{h}\ell(T_s - T_w)}{h_{fg}\mu_f} \tag{10.21}$$

For laminar flow of condensate film, the experimental results are found to be in good agreement with the results predicted from the theory. In practice, for turbulent flow it has been found that the theoretical results are approximately 20% lower than the experimental values if ripples develop in the film. In the light of this observation, McAdams suggests the following expression for predicting the average heat transfer coefficient for film condensation on plates, if the flow is turbulent:

$$\bar{h} = 1.13\left[\frac{\rho(\rho - \rho_v)gh_{fg}k^3}{\ell\mu(T_s - T_w)}\right]^{1/4} \tag{10.22}$$

EXAMPLE 10.1 A vertical plate of length 1 m and width 0.50 m at 92°C is exposed to saturated steam at 100°C. Calculate the average heat transfer coefficient and the mass of steam condensed in one second.

SOLUTION The value of the average heat transfer coefficient depends on whether the flow is laminar or turbulent. To check whether the flow is laminar or turbulent, we have to calculate the Reynolds number which depends on the mass flow. The mass flow in turn depends on the heat transfer coefficient. Hence, to solve the problem let us assume that the flow is laminar and estimate the heat transfer coefficient. Then we shall use the calculated heat transfer coefficient to calculate the Reynolds number to check whether our assumption is correct or not.

For laminar flow, the average heat transfer coefficient is given by

$$\bar{h} = 0.943\left[\frac{\rho(\rho-\rho_v)gh_{fg}k_f^3}{\ell\mu_f(T_s-T_w)}\right]^{1/4}$$

The film temperature

$$T_f = \frac{T_s+T_w}{2} = \frac{100+92}{2} = 96°\text{C}$$

At 96°C, k_f =0.68 W/m K; $\mu_f = 293.4\times10^{-6}$ N s/m^2 and ρ = 96 kg/m^3 (obtained by interpolation).

At 100°C

$$h_{fg} = 2257 \text{ kJ/kg}$$

For water, $\rho_v \ll \rho$. Hence, we can write

$$\rho(\rho-\rho_v) = \rho^2$$

Then

$$\bar{h} = 0.943\left[\frac{\rho^2 gh_{fg}k_f^3}{\ell\mu_f(T_s-T_w)}\right]^{1/4}$$

$$= 0.943\left[\frac{(961)^2\times9.8\times(2257\times10^3)(0.68)^3}{1\times293.4\times10^{-6}(100-92)}\right]^{1/4} = 6820 \text{ W/m}^2\text{ K}$$

Now, let us calculate the Reynolds number to check whether the flow is laminar or not.

$$Re_f = \frac{4\bar{h}\ell(T_s-T_w)}{h_{fg}\mu_f} = \frac{4\times6820\times1(100-92)}{2257\times10^3\times293.4\times10^{-6}} = 329.6$$

Since the Reynolds number is less than 1800, the flow is laminar. Hence our assumption is correct.

$$q = \bar{h}A(T_s-T_w) = 6820\times1\times0.5(100-92) = 27280 \text{ W}$$

Mass flow rate of condensate $\dot{m}$ is given by

$$\dot{m} = \frac{q}{h_{fg}} = \frac{27280}{2257 \times 10^3} = 0.012 \text{ kg/s}$$

EXAMPLE 10.2 A tube of 2 m length and 25 mm OD is to be used to condense saturated steam at 100°C while the tube surface is maintained at 92°C. Estimate the average heat transfer coefficient and the rate of condensation of steam if the tube is kept horizontal. The steam condenses on the outside of the tube.

SOLUTION

$$T_f = \frac{T_s + T_w}{2} = \frac{100 + 92}{2} = 96°\text{C}$$

At 96°C

$$k_f = 0.68 \text{ W/m K}; \ \mu_f = 293.4 \times 10^{-6} \text{ N s/m}^2$$

and $\rho = 961$ kg/m^3; $h_{fg} = 2257$ kJ/kg at 100°C (see Example 10.1)

The average heat transfer coefficient for laminar film condensation on the outside of a horizontal tube is given by

$$\bar{h} = 0.725 \left[\frac{\rho(\rho - \rho_v) g h_{fg} k_f^3}{\mu_f d(T_s - T_w)}\right]^{1/4} \quad \text{(see Eq. (10.16))}$$

Since $\rho_v \ll \rho$, we can approximate $\rho(\rho - \rho_v)$ as ρ^2. Then

$$\bar{h} = 0.725 \left[\frac{\rho^2 g h_{fg} k_f^3}{\mu_f d(T_s - T_w)}\right]^{1/4}$$

$$= 0.725 \left[\frac{(961)^2 \times 9.8 \times (2257 \times 10^3)(0.68)^3}{293.4 \times 10^{-6} \times 25 \times 10^{-3}(100 - 92)}\right]^{1/4} = 13.187 \text{ kW/m}^2 \text{ K}$$

$$q = \bar{h}A(T_s - T_w)$$

$$= 13.187 \times 10^3 \times \pi \times 25 \times 10^{-3} \times 2(100 - 92) = 16.571 \text{ kW}$$

Rate of condensation m is given by

$$\dot{m} = \frac{q}{h_{fg}} = \frac{16.57 \times 10^3}{2257 \times 10^3} = 7.34 \times 10^{-3} \text{ kg/s}$$

EXAMPLE 10.3 Rework Example 10.2 if the tube is kept in the vertical position.

SOLUTION If the tube is in the vertical position, it can be treated as a vertical plate of area $A = \pi d\ell$, and Eq. (10.13) can be used to determine the average value of the heat transfer coefficient.

$$T_f = \frac{T_s + T_w}{2} = \frac{100 + 92}{2} = 96°\text{C}$$

$$\rho(\rho - \rho_v) \simeq \rho^2$$

At 96°C

$$k_f = 0.68 \text{ W/m K}; \ \mu_f = 293.4 \times 10^{-6} \text{ N s/m}^2$$

and $\rho = 961$ kg/m^3; $h_{fg} = 2257$ kJ/kg at 100°C [see Example 10.1]

$$\bar{h} = 0.943 \left[\frac{\rho(\rho - \rho_v) g h_{fg} k_f^3}{\ell \mu_f (T_s - T_w)}\right]^{1/4} = 0.943 \left[\frac{\rho^2 g h_{fg} k_f^3}{\ell \mu_f (T_s - T_w)}\right]^{1/4}$$

$$= 0.943 \left[\frac{(961)^2 \times 9.8 \times 2257 \times 10^3 (0.68)^3}{2 \times 293.4 \times 10^{-6}(100 - 92)}\right]^{1/4} = 5735 \text{ W/m}^2 \text{ K}$$

$$q = \bar{h}A(T_s - t_w) = 5735 \times \pi \times 25 \times 10^{-3} \times 2(100 - 92) = 7207 \text{ W}$$

Rate of condensation $\dot{m}$ is given by

$$\dot{m} = \frac{q}{h_{fg}} = \frac{7207}{2257 \times 10^3} = 3.19 \times 10^{-3} \text{ kg/s}$$

10.2.3 Film Condensation inside Horizontal Tubes

Condensation inside tubes is of considerable importance in refrigeration and air-conditioning systems where the refrigerant flows through the tubes. The flow rate of condensable vapor through the tubes strongly influences the heat transfer coefficient, which in turn influences the rate of accumulation of liquid in the tubes. The analysis of condensation inside tubes is quite complex. Chato obtained the following expression for condensation of refrigerants at low vapor velocities inside horizontal tubes:

$$\bar{h} = 0.555 \left[\frac{\rho(\rho - \rho_v) g h_{fg} k^3}{\ell d (T_s - T_w)}\right]^{1/4} \tag{10.23}$$

This equation is valid for low vapor Reynolds number Re_v which should be evaluated at the inlet conditions to the tube.

$$Re_v = \frac{d\rho_v V}{\mu_v} < 35,000 \tag{10.24}$$

10.2.4 Condensation of Superheated Vapors

The equations developed earlier are strictly applicable to the condensation of saturated vapors. However, the same equations can also be used with reasonable accuracy for condensation of superheated vapors. The heat transfer rate from a superheated vapor to a wall at T_w is given by

$$q = \bar{h}A(T_s' - T_w) \tag{10.25}$$

where T_s' is the saturation temperature corresponding to the pressure of the superheated vapor.

10.3 Drop-wise Condensation

If the surface on which condensation takes place is coated with an agent such that the condensate does not wet the surface, the vapor condenses in drops rather than as a continuous film. In drop-wise condensation, a large part of the surface is not covered by a liquid film which acts as an additional resistance for heat transfer. Therefore, in drop-wise condensation the heat transfer coefficients can be as high as 4 to 8 times that of film condensation.

10.4 Boiling Heat Transfer

Heat transfer to boiling liquids is a convection process involving a phase change from liquid to vapor. The phenomenon of boiling heat transfer is considered to be more complex than the phenomenon of convection heat transfer without phase change. An examination of the boiling phenomenon with the help of high speed photography has revealed distinct boiling regimes in which the heat transfer mechanisms are quite different.

To understand the characteristics of various boiling regimes, let us consider a simple system consisting of an electrically heated wire which is completely submerged in a pool of liquid at saturation temperature. Boiling in such situations is called *pool boiling*. An example of pool boiling is the boiling of water in a kettle placed on a stove. If the temperature of the heating surface is not much above the boiling point of the liquid, heat transfer to the liquid occurs by natural convection. If the temperature of the heating surface is increased, one can observe the formation of vapor bubbles at certain places, known as nucleation sites, on the surface. Initially, the vapor bubbles will be small and they condense before reaching the surface. If the temperature of the heating surface is raised further, the number of bubbles and their size increase and these bubbles rise to the free surface of the liquid.

The different regimes of boiling are shown in Fig. 10.2, where the heat flux is plotted against excess temperature, $\Delta T_x = (T_w - T_s)$.

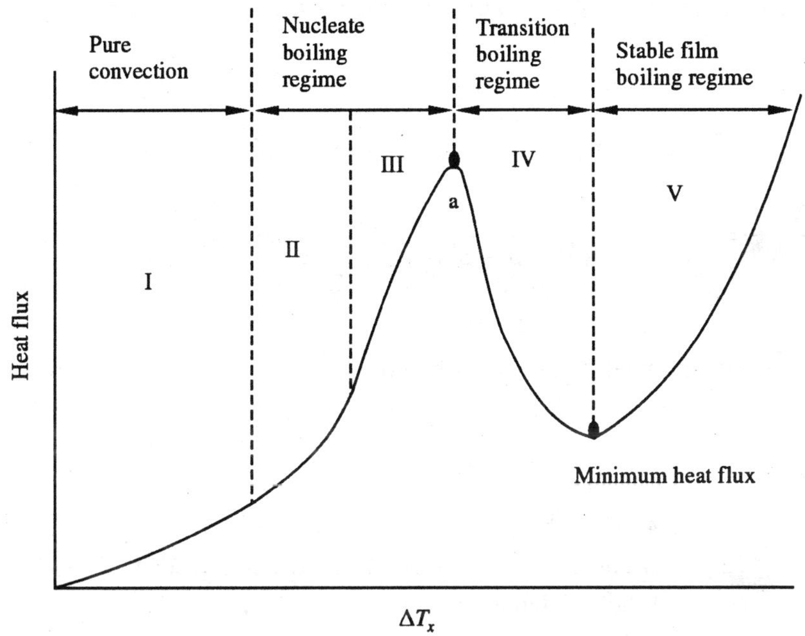

Fig. 10.2 Different regimes of boiling heat transfer.

In region I, pure convection heat transfer takes place. The liquid near the heated surface is slightly superheated and the superheated liquid rises to the liquid-vapor interface due to natural convection currents. The heat transfer rates in this region can be estimated by treating the mechanism as natural convection. In Region II, bubbles form at the nucleation sites on the surface and the bubbles rise through the liquid. As the temperature of the surface is increased further, bubbles form more rapidly and rise to the surface of the liquid, and the heat flux rises more rapidly with increasing excess temperature as indicated in region III. Regions II and III represent the nucleate boiling regime. With further increase in excess temperature, the rate of formation of bubbles will be so high that the bubbles blanket the heating surface and the heat flux reduces. The heat flux reaches a maximum value at a particular excess temperature as shown in Fig. 10.2. This maximum heat flux is called *critical heat flux* and occurs at *critical excess temperature*. After the critical excess temperature, the bubbles merge into slugs and form a vapor film over the heating surface. Then heat transfer is mainly by conduction through the vapor film before the liquid can boil. Therefore, in region IV the heat flux reduces with increasing excess temperature, and this region represents the beginning of the film boiling regime. This regime represents a transition from nucleate boiling to film boiling. In the stable film boiling regime, region V, the surface temperature required will be high for a given heat flux.

If the heating surface is an electrically heated wire at point a (that is at critical heat flux) the wire becomes unstable since a small increase in excess temperature results in a decrease in the heat flux. If the wire is unable to dissipate the heat flux corresponding to the critical heat flux, its temperature rises and it melts down.

10.4.1 Nucleate Pool Boiling

Using experimental data, Rohsehow developed the correlation

$$\frac{c\Delta T_x}{h_{fg}Pr^n} = C_{sf}\left[\frac{q'}{\mu h_{fg}}\sqrt{\frac{\sigma}{g(\rho-\rho_v)}}\right]^{0.33} \tag{10.26}$$

where

c = specific heat of saturated liquid

q' = heat flux for nucleate boiling

μ = viscosity of liquid

σ = surface tension of the liquid to vapor interface

C_{sf} = empirical constant which depends on the nature of the heating surface-fluid combination

n = 1.0 for water and 1.7 for other fluids

Pr = Prandtl number of the saturated liquid.

The values of the vapor–liquid surface tension for water at some selected temperatures are presented in Table 10.1, and the values of C_{sf} for various liquid surface combinations are presented in Table 10.2.

Table 10.1 *Vapor–liquid surface tension for water*

Saturation temperature (°C)	20	40	60	80	100	150	200
Surface tension ($\sigma \times 10^3$) N/m	72.9	69.5	66.1	62.7	58.9	48.7	37.8

10.4.2 Critical Heat Flux

The Rohsenow correlation [Eq. (10.26)] correlates the data for all types of nucleate boiling processes, including pool boiling of saturated liquids and sub-cooled liquids and also for boiling of sub-cooled and saturated liquids flowing by natural or forced convection in tubes. However, this correlation does not indicate the excess temperature at which critical heat flux is reached. Zuber has developed the following expression for the peak heat flux q'' in nucleate boiling:

$$q'' = \frac{\pi}{24}\rho_v^{1/2} h_{fg}[\sigma g(\rho - \rho_v)]^{1/4} \tag{10.27}$$

where σ is the vapor–liquid surface tension. This relation is in good agreement with experimental data. It can be observed from the above equation that the peak heat flux is larger for fluids with larger latent heats of vaporization.

Table 10.2 *Values of C_{sf} for some liquid surface combinations*

Fluid/Heating surface	C_{sf}
Water/Copper	0.013
Water/Brass	0.006
Water/Teflon coated stainless steel	0.0058
Water/Polished stainless steel	0.0132
Water/Platinum	0.013
Ethanol/Chromium	0.0027
Carbon tetrachloride/Copper	0.013
Benzene/Chromium	0.01

EXAMPLE 10.4 It is desired to boil water at atmospheric pressure on a copper surface which is electrically heated. Estimate the heat flux from the surface to the water, if the surface is maintained at 110°C.

SOLUTION For water at 100°C, h_{fg} = 2257 kJ/kg; ρ = 958.4 kg/m^3; c = 4.211 kJ/kg K, μ = 277.5×10^{-6} N s/m^2 and Pr = 1.75. From Table 10.1, σ = 58.9×10^{-3} N/m and from Table 10.2, C_{sf} = 0.013

The Rohsenow correlation is given by

$$\frac{c\Delta T_x}{h_{fg} Pr^n} = C_{sf}\left[\frac{q'}{\mu h_{fg}}\sqrt{\frac{\sigma}{g(\rho - \rho_v)}}\right]^{0.33}$$

Neglect ρ_v in comparison to ρ and $n = 1$ for water. Substituting the values in the above equation we get

$$\frac{4.211 \times 10^3 (110 - 100)}{2257 \times 10^3 \times 1.75} = 0.013\left[\frac{q'}{277.5 \times 10^{-6} \times 2257 \times 10^3}\sqrt{\frac{58.9 \times 10^{-3}}{9.8 \times 958.4}}\right]^{0.33}$$

or $0.82 = [3.998 \times 10^{-6} q']^{0.33}$

or $q' = 137.91 \text{ kW/m}^2$

Therefore, heat flux from surface = 137.91 kW/m²

EXAMPLE 10.5 Estimate the peak heat flux for Example 10.4.

SOLUTION The Zuber correlation for the peak heat flux is given by

$$q'' = \frac{\pi}{24}\rho_v^{1/2} h_{fg}[\sigma g(\rho - \rho_v)]^{1/4}.$$

At 100°C, $\rho_v = 0.5977$ kg/m³ (from steam table). See Example 10.4 for other data.

$$q'' = \frac{\pi}{24}(0.5977)^{1/2} 2257 \times 10^3 \left[58.9 \times 10^{-3} \times 9.8 \times (958.4 - 0.5977)\right]^{1/4}$$

or $q'' = 1.108 \text{ MW/m}^2$.

10.5 Summary

If a vertical plate at a temperature less than the saturation temperature is exposed to a saturated vapor, the vapor condenses on the plate and the condensate flows down under the action of gravity. If the condensate does not wet the surface, the liquid droplets fall from the surface. This process is called drop-wise condensation. If the condensate wets the surface, a smooth film of condensate will form on the surface. The film thickness increases as the condensate flows down, due to additional condensation. This process is called film condensation. The condensate film presents additional thermal resistance for heat transfer and hence the heat transfer coefficient for film condensation will be lower than that for drop-wise condensation.

Nusselt has analyzed the laminar film condensation on a vertical plate and arrived at the following expressions:

$$\delta_x = \left[\frac{4\mu k x (T_s - T_w)}{g h_{fg} \rho(\rho - \rho_v)}\right]^{1/4}$$

$$h_x = \left[\frac{p(\rho - \rho_v) g h_{fg} k^3}{4\mu x (T_s - T_w)}\right]^{1/4}$$

$$\bar{h} = 0.943\left[\frac{\rho(\rho - \rho_v) g h_{fg} k^3}{\ell \mu_f (T_s - T_w)}\right]^{1/4}$$

where

δ_x = condensate film thickness at a distance x from the top edge of a vertical plate of length ℓ

μ = viscosity of saturated liquid

k = thermal conductivity of saturated liquid

T_s = saturation temperature of condensable vapor

T_w = surface temperature of plate

h_{fg} = latent heat of vaporization of liquid

g = acceleration due to gravity

ρ = density of saturated liquid

ρ_v = density of vapor

h_x = local heat transfer coefficient

$\bar{h}$ = average heat transfer coefficient.

The average heat transfer coefficient for laminar film condensation on the outside of a sphere or a horizontal tube can be calculated from the relation

$$\bar{h} = A\left[\frac{\rho(\rho - \rho_v)gh_{fg}k_f^3}{\mu_f d(T_s - T_w)}\right]^{1/4}$$

where A = 0.815 for a sphere and 0.725 for a tube, and d is the diameter. The same equation by replacing d with Nd can be used to estimate the average heat transfer coefficient for film condensation on a horizontal tube bank with N tubes so arranged that the condensate from one tube flows directly onto the tube below.

The Reynolds number for condensate film is defined as

$$Re_f = \frac{d_H \rho V}{\mu_f} = \frac{4\dot{m}}{P\mu_f}$$

where

d_H = hydraulic diameter

P = wetted perimeter

V = average velocity of flow

$\dot{m}$ = mass flow rate of the condensate.

The subscript f indicates that the properties should be evaluated at the film temperature T_f given by $T_f = (T_s + T_w)/2$. The critical Reynolds number for transition from laminar to turbulent film condensation is 1800.

The Chato's correlation for condensation of refrigerants at low vapor velocities inside horizontal tubes is given by

$$\bar{h} = 0.555 \left[\frac{\rho(\rho - \rho_v) g h_{fg} k^3}{\mu d (T_s - T_w)} \right]^{1/4}$$

If a heated surface is completely submerged in a pool of saturated liquid, then the boiling process is called pool boiling. If the temperature of the heating surface is not much above the saturation temperature of the liquid, the heat transfer occurs primarily by natural convection. The temperature difference between the heating surface and the saturation temperature of the liquid is called excess temperature. If the excess temperature is increased, vapor bubbles will be formed at the nucleation sites and these bubbles rise to the liquid-vapor interface. This type of boiling is called nucleate boiling. In the nucleate boiling regime, the heat flux rapidly increases with increasing excess temperature and reaches a maximum value, called the critical heat flux. A further increase in excess temperature leads to rapid formation of bubbles, and these bubbles merge into slugs and form a vapor film over the heating surface. Then heat transfer is mainly by conduction through the vapor film and the heat flux reduces. This stage is called the transition boiling regime. If the excess temperature is further increased, a continuous vapor film will be formed on the heating surface and this regime is called the stable film boiling regime.

The heat flux in nucleate boiling can be estimated from the Rohsenow correlation given by

$$\frac{c \Delta T_x}{h_{fg} Pr^n} = C_{sf} \left[\frac{q'}{\mu h_{fg}} \sqrt{\frac{\sigma}{g(\rho - \rho_v)}} \right]^{0.33}$$

where

c = specific heat of saturated liquid

C_{sf} = constant, which depends on the nature of heating surface-fluid combination

Pr = Prandtl number of saturated liquid

q' = heat flux for nucleate boiling

σ = surface tension of the liquid-to-vapor interface

n = 1.0 for water and 1.7 for other fluids.

The critical heat flux q'' can be estimated from the Zuber correlation given by

$$q'' = \frac{\pi}{24}\rho_v^{1/2}h_{fg}[\sigma g(\rho - \rho_v]^{1/4}$$

Review Questions

1. What is meant by film condensation and drop-wise condensation?
2. Why are the convection heat transfer coefficients for drop-wise condensation greater than the heat transfer coefficients for film condensation?
3. In film condensation, how does the local heat transfer coefficient vary with the distance from the top of the plate?
4. Does the local heat transfer coefficient for film condensation increase with increasing temperature difference between the condensing vapors and the plate?
5. How is the Reynolds number of the condense film defined?
6. What is the critical Reynolds number for transition from laminar flow to turbulent flow in condensation of vapors on a vertical plate?
7. How is the mass flow rate of the condensate related to the average heat transfer coefficient?
8. What is the relation between Reynolds number and average heat transfer coefficient for film condensation on a vertical plate?
9. Is it possible to use the relations which were developed for condensation of saturated vapors to predict the average heat transfer coefficients for condensation of superheated vapors?
10. What is pool boiling?
11. Distinguish between nucleate boiling and film boiling.
12. What is critical heat flux?
13. Why is the heat flux in nucleate boiling much higher than in film boiling?
14. Discuss the various regimes of boiling heat transfer.

Problems

10.1 A vertical plate at 96°C is exposed to saturated steam at 100°C. If the steam condenses on the plate, calculate the condensate film thickness at a distance of 0.1 m from the top of the plate.

10.2 Saturated steam at 145°C is condensing on a vertical plate which is maintained at 135°C. Estimate the condensate film thickness at a distance of 0.40 m from the top edge of the plate.

10.3 Estimate the local heat transfer coefficient at a distance of 0.1 m from the top of the plate for Problem 10.1.

10.4 Calculate the local heat transfer coefficient and local Nusselt number at a distance of 0.30 m from the top edge of a 1 m long plate which is exposed to condensing steam at 110°C. The plate is held at a temperature of 104°C.

10.5 Estimate the average heat transfer coefficient for Problem 10.4.

10.6 A 1 m long vertical plate of width 0.30 m at 96°C is exposed to saturated steam at 100°C. Estimate the average heat transfer coefficient and the mass flow rate of the condensate for film condensation.

10.7 Calculate the Reynolds number for the condensate film of Problem 10.6 and verify whether the flow is laminar or not.

10.8 A 0.5 m × 0.5 m plate is inclined at an angle of 30° with the vertical and exposed to saturated steam at 100°C. The plate is maintained at 96°C. Calculate the average heat transfer coefficient and mass flow rate of the condensate.

10.9 Rework Problem 10.8 if the plate is inclined at an angle of 45° with the vertical.

10.10 A vertical plate 0.5 m×0.5 m is exposed to saturated ammonia at 32°C and the plate surface is maintained at 28°C. Calculate the average heat transfer coefficient and mass flow rate of the condensate. For ammonia at 30°C, ρ_v = 7.157 kg/m^3 and h_{fg}(at 32°C) = 1166.2 kJ/kg.

10.11 Saturated steam at 100°C condenses on the outside of a vertical tube of 0.2 m diameter. The tube is 5 m long and is maintained at 96°C. Calculate the amount of steam condensed in one hour.

10.12 Saturated steam at 100°C condenses on the outside of a horizontal tube of 0.20 m diameter and 2 m length. The tube is maintained at 92°C. Estimate the mass of steam condensed in one hour.

10.13 Saturated steam at 30°C is to be condensed on the outside of a bank of 10 × 10 horizontal tubes 25 mm in diameter. The tube surface temperature is maintained at 20°C. Estimate the mass of steam condensed in one hour if the length of each tube 1 m.

10.14 Saturated Freon-12 at 35°C is condensed inside a horizontal 10 mm diameter tube at a low vapor velocity. The tube wall is maintained at 25°C. Estimate the mass of Freon-12 condensed in one hour per meter of tube length. For Freon-12 at 35°C, h_{fg} = 131.328 kJ/kg and ρ_v = 42.68 kg/m^3.

10.15 It is desired to boil water at atmospheric pressure in a circular pan of 0.30 m diameter having a copper bottom. The temperature of the bottom copper plate is maintained at 110°C. Estimate the heat transferred to water in 10 minutes.

10.16 An electrically heated platinum wire is submerged in saturated liquid water at 100°C. Determine the heat flux if the excess temperature is 6°C.

10.17 A kettle with a flat copper bottom of 0.30 m diameter is to be used to boil 2 kg water at atmospheric pressure in 20 minutes. Determine the temperature at which the copper bottom of the kettle should be maintained.

10.18 It is desired to boil 1 kg water at atmospheric pressure in 10 minutes in a brass vessel of 0.30 m diameter. Determine the temperature at which the vessel bottom should be maintained.

10.19 A stainless steel vessel of 0.25 m diameter is used for boiling water at atmospheric pressure. If the bottom surface of the vessel is maintained at 105°C, estimate the heat transfer rate from the bottom surface to water.

10.20 A teflon coated stainless steel vessel of 0.25 m diameter is used for boiling water at atmospheric pressure. Estimate the heat transfer rate from the bottom surface of the vessel to water, if the bottom surface is maintained at 105°C.

10.21 Estimate the peak heat flux for Problem 10.19.

10.22 Determine the peak heat flux for Problem 10.20.

CHAPTER ELEVEN

HEAT EXCHANGERS

Learning Objectives

A careful study of this chapter will enable the readers to

- know the important factors in the design of heat exchangers.
- know the types of heat exchangers in common use.
- define the overall heat transfer coefficient for a heat exchanger.
- calculate the Log Mean Temperature Difference (LMTD).
- derive a relation for estimating LMTD for parallel-flow heat exchangers.
- know about the correction factor to LMTD for various types of heat exchangers.
- define the heat exchanger effectiveness.
- learn about the relation between effectiveness and Number of Transfer Units (NTU).
- know about fouling factors.

11.1 Introduction

A device which transfers heat from one fluid to another fluid is called a *heat exchanger*. The simplest type of heat exchanger is an open feed water heater used in thermal power plants. In an open feed water heater, a stream of steam is directly mixed with cold water and the mixture leaves the unit at a uniform temperature. However, in most of the situations one is not interested in mixing both the fluids but would like to transfer heat from a hot fluid to another cold fluid. There are several types of devices or heat exchangers in which one fluid is separated from the other by a wall or partition through which heat flows. The simplest type of heat exchanger is a double-pipe heat exchanger in which one fluid flows through the inner pipe while the other fluid flows through the annular space between the inner and outer pipes. In the design of a heat exchanger one should take into account the following important factors:

- *The thermal analysis.* In this phase of design one is primarily concerned with the estimation of heat transfer area for the transfer of heat at a specified rate for the given flow rates and temperatures of the fluids.

- *The mechanical design.* In this phase of design one takes into account the operating temperatures, pressures, corrosive nature of the fluids, pressure drop, thermal stresses, etc.
- *The design for manufacture.* In this phase of design one concentrates on the selection of material, seals, enclosures, cost of manufacturing and the manufacturing procedures.

A discussion of all the above aspects is beyond the scope of this book and we limit our discussion to the thermal analysis only. Moreover, we limit our discussion to heat exchangers in which heat is transferred by conduction and convection only. In order to achieve maximum economy, the heat exchangers are usually selected from the available standard items in the market. The selection of the heat exchanger has to be based on the thermal analysis, cost, physical size and corrosive nature of the fluids.

11.2 Types of Heat Exchangers

The simplest type of heat exchanger is a double-pipe heat exchanger shown in Fig. 11.1. It consists of a tube located concentrically inside another tube which acts as a shell. In this arrangement both the fluids traverse the heat exchanger only once. If both the fluids flow in the same direction, the heat exchanger is called a *parallel-flow type* or *co-current type*. If the fluids flow in opposite directions, the heat exchanger is called a *counter-current type*.

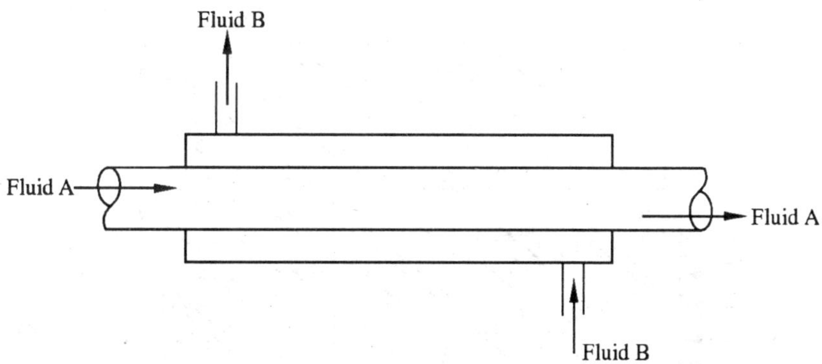

Fig. 11.1 Schematic of a double-pipe heat exchanger.

The shell-and-tube heat exchangers shown in Fig. 11.2 are widely used in chemical process industries. One fluid flows through the inside tubes while the other fluid flows through the shell. This type of heat exchanger has fixed tube plates at each end and the tubes are welded into the plates. The shell-and-tube heat exchangers can be used for small temperature differences between the hot and cold fluids because they cannot withstand high thermal stresses. To ensure that the shell-side fluid flows across the tubes to enhance the heat transfer rates, baffles are placed in the shell. Depending on the arrangement at the ends of the heat exchanger, one or more tube passes can be utilized. A shell-and-tube heat exchanger with two tube passes and one shell pass is shown in Fig. 11.2.

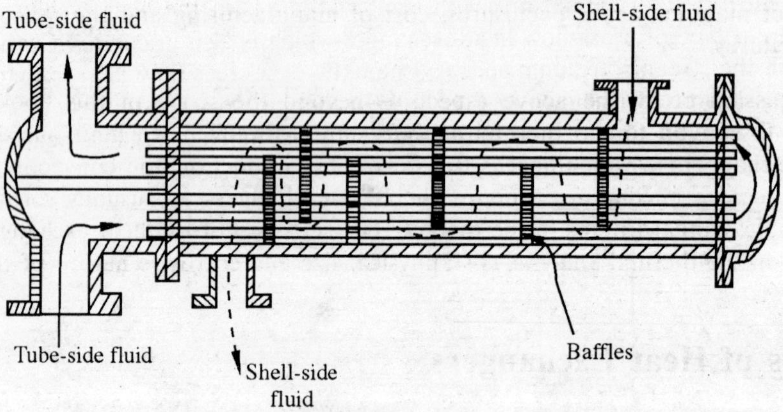

Fig. 11.2 *Schematic of a shell-and-tube heat exchanger with two tube passes and one shell pass.*

If the two fluids flowing along the heat transfer surface move at right angles to each other, then the heat exchanger is of the *cross-flow type*. The cross-flow heat exchangers are usually used in air or gas heating and cooling applications. A schematic of the cross-flow heat exchanger is shown in Fig. 11.3 in which a gas flows across a

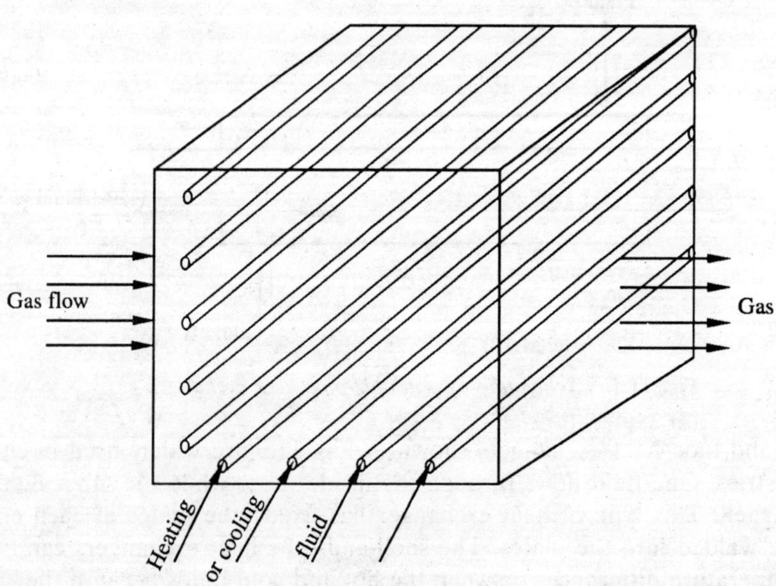

Fig. 11.3 *Schematic of a cross-flow heat exchanger in which one stream is mixed while the other is unmixed.*

tube-bank, while another fluid flows through the tubes. In this case, the gas stream is mixed and its temperature will be uniform across any section and varies only in the direction of flow. This stream is said to be a *mixed stream*. The fluid in the tubes is said to be an *unmixed stream* because the fluid is contained in separate tubular channels and the fluid cannot mix with itself while it is in the heat exchanger.

A different type of cross-flow heat exchanger which is also in common use is one in which both the streams are unmixed. A schematic of a cross-flow heat exchanger with both streams unmixed is shown in Fig. 11.4. In this type of cross-flow heat exchanger, a gas flows across a finned tube bank and remains unmixed. Such a heat exchanger is used in air-conditioning equipment.

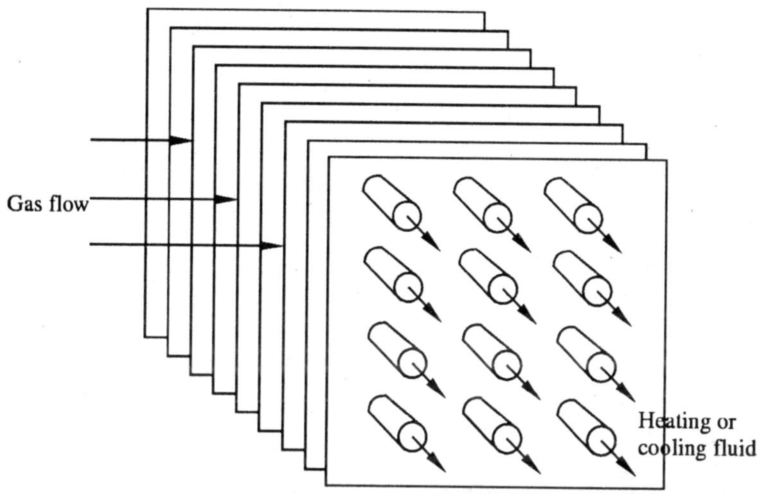

Fig. 11.4 *Schematic of a cross-flow heat exchanger with both streams unmixed.*

11.3 Overall Heat Transfer Coefficient

Suppose a fluid at temperature T_1 flowing through a tube of inner radius r_i and outer radius r_o exchanges heat with a fluid at temperature T_2 which is flowing outside the tube. Then the thermal resistances for heat transfer are:

$$\text{Thermal resistance for convection at the inner surface} = \frac{1}{h_i A_i}.$$

$$\text{Thermal resistance for conduction (tube wall)} = \frac{\ln(r_o/r_i)}{2\pi k \ell}.$$

$$\text{Thermal resistance for convection at the outer surface} = \frac{1}{h_o A_o}.$$

Therefore, the total thermal resistance is

$$\sum R = \frac{1}{h_i A_i} + \frac{\ln(r_o/r_i)}{2\pi k \ell} + \frac{1}{h_o A_o} \tag{11.1}$$

The rate of heat transfer is given by

$$q = \frac{\Delta T}{\sum R} = \frac{T_1 - T_2}{\dfrac{1}{h_i A_i} + \dfrac{\ln(r_o/r_i)}{2\pi k \ell} + \dfrac{1}{h_o A_o}} \tag{11.2}$$

The overall heat transfer by combined conduction and convection is frequently expressed in terms of the overall heat transfer coefficient U defined as

$$q = UA\Delta T \tag{11.3}$$

From Eqs. (11.2) and (11.3) we get

$$UA = \frac{1}{\dfrac{1}{h_i A_i} + \dfrac{\ln(r_o/r_i)}{2\pi k \ell} + \dfrac{1}{h_o A_o}} \tag{11.4}$$

The overall heat transfer coefficient may be based on either the inside area A_i or the outside area A_o of the tube.

$$U_i = \frac{1}{\dfrac{1}{h_i} + \dfrac{A_i \ln(r_o/r_i)}{2\pi k \ell} + \dfrac{A_i}{A_o}\dfrac{1}{h_o}} \tag{11.5}$$

$$U_o = \frac{1}{\dfrac{A_o}{A_i}\dfrac{1}{h_i} + \dfrac{A_o \ln(r_o/r_i)}{2\pi k \ell} + \dfrac{1}{h_o}} \tag{11.6}$$

EXAMPLE 11.1 In a double-pipe heat exchanger, oil flows through a 1.5 m long copper tube of 18 mm ID and 24 mm OD while water flows through the annulus. The convective heat transfer coefficient between the oil and the tube is 500 W/m^2 K and the convective heat transfer coefficient between the tube and water is 800 W/m^2 K. Calculate the overall heat transfer coefficient based on the outside area of the pipe.

SOLUTION Given $r_i = 9 \times 10^{-3}$ m; $r_o = 12 \times 10^{-3}$ m; $\ell = 1.5$ m; $h_i = 500$ W/m^2 K and $h_0 = 800$ W/m^2 K

For copper, $k = 386$ W/m K

The overall heat transfer coefficient is given by

$$U_o = \cfrac{1}{\cfrac{A_o}{A_i}\cfrac{1}{h_i} + \cfrac{A_o \ln(r_o/r_i)}{2\pi k\ell} + \cfrac{1}{h_o}}$$

$$= \cfrac{1}{\cfrac{2\pi \times 12 \times 10^{-3}}{2\pi \times 9 \times 10^{-3} \times 500} + \cfrac{2\pi \times 12 \times 10^{-3} \times 1.5 \ln\left(\cfrac{12 \times 10^{-3}}{9 \times 10^{-3}}\right)}{2\pi \times 386 \times 1.5} + \cfrac{1}{800}}$$

$$= 254.7 \text{ W/m}^2 \text{ K}$$

11.4 Log Mean Temperature Difference (LMTD)

In a heat exchanger, the temperatures of the heating and cooling fluids do not in general, remain constant, but vary from point to point along the length of the heat exchanger. Since the temperature difference between the two fluids keeps changing, the rate of heat transfer also changes along the length of the heat exchanger. The temperature profiles in single pass heat exchangers are shown in Fig. 11.5. An inspection of Fig. 11.5(a) shows that the temperature difference in a parallel-flow heat exchanger is large at the inlet point and small at the exit. Fig. 11.5 (b) shows that the temperature difference between the two fluids, in a counter-current heat exchanger, can be fairly constant.

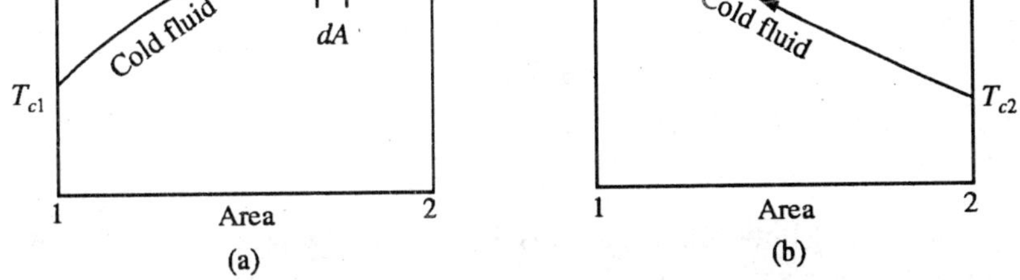

Fig. 11.5 *Temperature profile in (a) a parallel-flow heat exchanger. (b) counter-current heat exchanger.*

The rate of heat transfer can be calculated from the relation

$$q = UA\Delta T \tag{11.3}$$

Since ΔT changes from point to point in a heat exchanger, we propose to use ΔT_m, a suitable mean temperature difference between the two ends of a heat exchanger, and rewrite Eq. (11.3) as

$$q = UA\Delta T_m \tag{11.7}$$

The rate of heat transfer through an element of area dA is given by

$$dq = -\dot{m}_h c_h dT_h = \dot{m}_c c_c dT_c \tag{11.8}$$

where the subscripts h and c denote the hot and cold fluids, respectively. The rate of heat transfer can also be expressed as

$$dq = U(T_h - T_c)dA \tag{11.9}$$

From Eq. (11.8) we obtain

$$dT_h = \frac{-dq}{\dot{m}_h c_h} \tag{11.10}$$

$$dT_c = \frac{dq}{\dot{m}_c c_c} \tag{11.11}$$

$$dT_h - dT_c = d(T_h - T_c) = -dq\left(\frac{1}{\dot{m}_h c_h} + \frac{1}{\dot{m}_c c_c}\right) \tag{11.12}$$

Substituting Eq. (11.9) in Eq. (11.12), we get

$$d(T_h - T_c) = -U(T_h - T_c)dA\left(\frac{1}{\dot{m}_h c_h} + \frac{1}{\dot{m}_c c_c}\right)$$

or

$$\frac{d(T_h - T_c)}{T_h - T_c} = -U\left(\frac{1}{\dot{m}_h c_h} + \frac{1}{\dot{m}_c c_c}\right) dA \tag{11.13}$$

Integrating Eq. (11.13) between conditions 1 and 2 corresponding to the two ends of the heat exchanger, we get

$$\ln\left(\frac{T_{h2} - T_{c2}}{T_{h1} - T_{c1}}\right) = -UA\left(\frac{1}{\dot{m}_h c_h} + \frac{1}{\dot{m}_c c_c}\right) \tag{11.14}$$

The energy balance gives

$$q = \dot{m}_h c_h (T_{h1} - T_{h2}) = \dot{m}_c c_c (T_{c2} - T_{c1})$$

or

$$\dot{m}_h c_h = \frac{q}{T_{h1} - T_{h2}} \tag{11.15}$$

$$\dot{m}_c c_c = \frac{q}{T_{c2} - T_{c1}} \tag{11.16}$$

Substituting Eqs. (11.15) and (11.16) in Eq. (11.14), we get

$$\ln\left(\frac{T_{h2} - T_{c2}}{T_{h1} - T_{c1}}\right) = \frac{-UA}{q}(T_{h1} - T_{h2} + T_{c2} - T_{c1})$$

or

$$q = UA\frac{(T_{h2} - T_{c2}) - (T_{h1} - T_{c1})}{\ln\left(\dfrac{T_{h2} - T_{c2}}{T_{h1} - T_{c1}}\right)} = UA\Delta T_m \tag{11.17}$$

where the ΔT_m, the Log Mean Temperature Difference (LMTD), is given by

$$\Delta T_m = \frac{(T_{h2} - T_{c2}) - (T_{h1} - T_{c1})}{\ln\left(\dfrac{T_{h2} - T_{c2}}{T_{h1} - T_{c1}}\right)} = \text{LMTD} \tag{11.18}$$

We have obtained the above relation for LMTD for a double-pipe heat exchanger. This derivation for LMTD involves two important assumptions: (i) The fluids have constant specific heats, and (ii) The convection heat transfer coefficients are constant throughout the heat exchanger. Equation (11.18) for LMTD is applicable for both parallel-flow and counter-current heat exchangers.

If the temperature difference $(T_{h1} - T_{c1})$ is not more than 50 percent greater than $(T_{h2} - T_{c2})$, the arithmetic mean temperature difference will be approximately equal to the LMTD, and it can be used to simplify the calculations.

EXAMPLE 11.2 It is desired to cool oil from 120°C to 50°C using a double-pipe heat exchanger. The cooling water enters the heat exchanger at 20°C and leaves it at 40°C. Calculate the LMTD for (a) a parallel-flow heat exchanger, and (b) for a counter-current heat exchanger.

SOLUTION The temperature profile, for the oil and water are shown in Fig. 11.6.

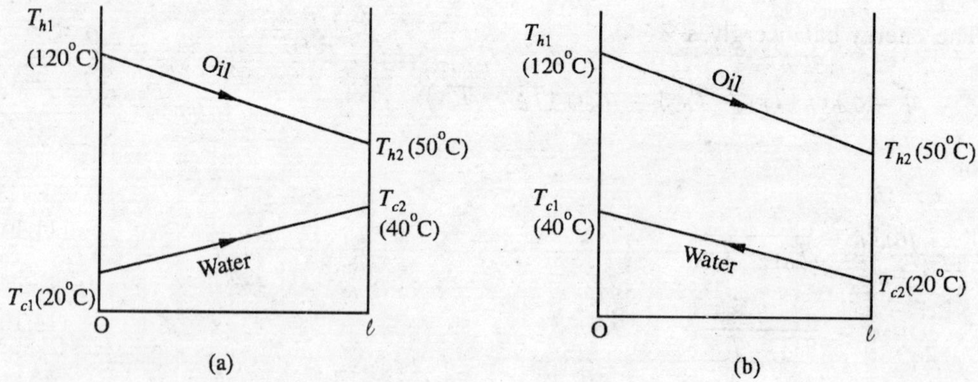

Fig. 11.6 Temperature profiles for the oil and water in (a) parallel-flow. (b) counter-current double-pipe heat exchanger.

$$\text{LMTD} = \frac{(T_{h2} - T_{c2}) - (T_{h1} - T_{c1})}{\ln\left(\dfrac{T_{h2} - T_{c2}}{T_{h1} - T_{c1}}\right)}$$

(a) parallel-flow heat exchanger

$$\text{LMTD} = \frac{(50 - 40) - (120 - 20)}{\ln \dfrac{50 - 40}{120 - 20}} = 39.09°\text{C}$$

(b) counter-current heat exchanger

$$\text{LMTD} = \frac{(50 - 20) - (120 - 40)}{\ln \dfrac{50 - 20}{120 - 40}} = 50.98°\text{C}$$

For more complex heat exchangers like shell-and-tube heat exchangers with several shell passes and several tube passes, and for cross-flow heat exchangers, the mathematical derivation of an expression for the mean temperature difference is quite complex. The usual practice is to modify the LMTD by including a correction factor. Then the rate of heat transfer can be written as

$$q = UAF\Delta T_m \tag{11.19}$$

where ΔT_m = LMTD for a counter-current heat exchanger.

The values of the correction factor F are presented in Figs. 11.7 to 11.10 for different types of heat exchangers. In Figs.11.7 to 11.10, the values of F are plotted as a function of the parameters P and R. The parameters P and R are defined as

$$P = \frac{t_2 - t_1}{T_1 - t_1} \tag{11.20}$$

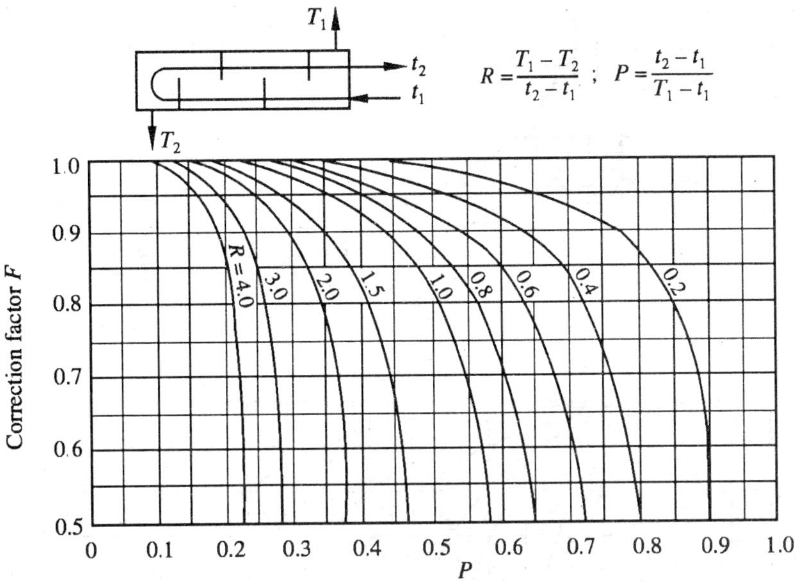

Fig. 11.7 *Correction factor to counter-flow LMTD for heat exchanger with one shell pass and two or a multiple of two tube passes.*

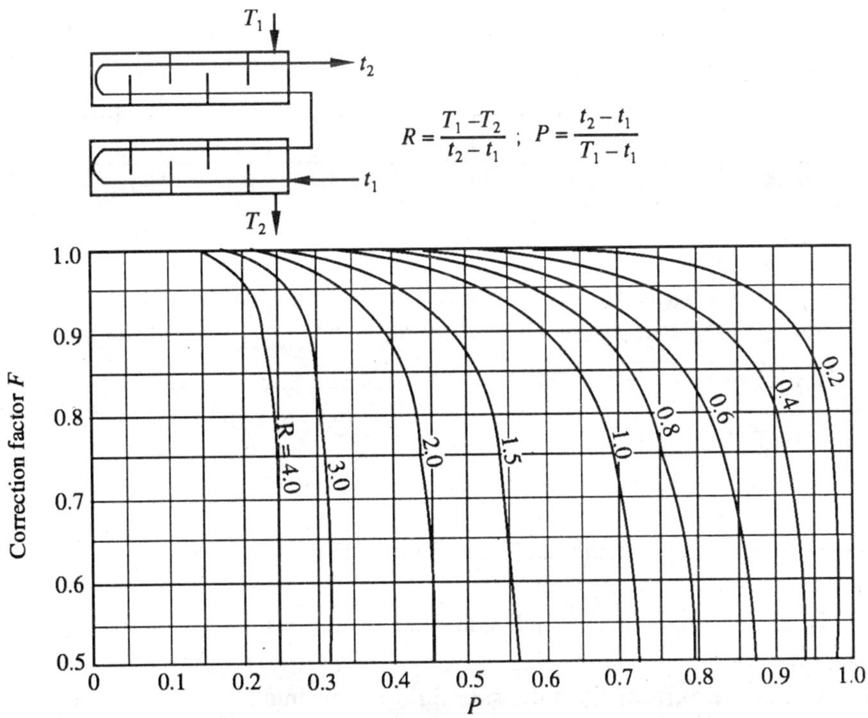

Fig. 11.8 *Correction factor to counter-flow LMTD for heat exchangers with two shell passes and a multiple of two tube passes.*

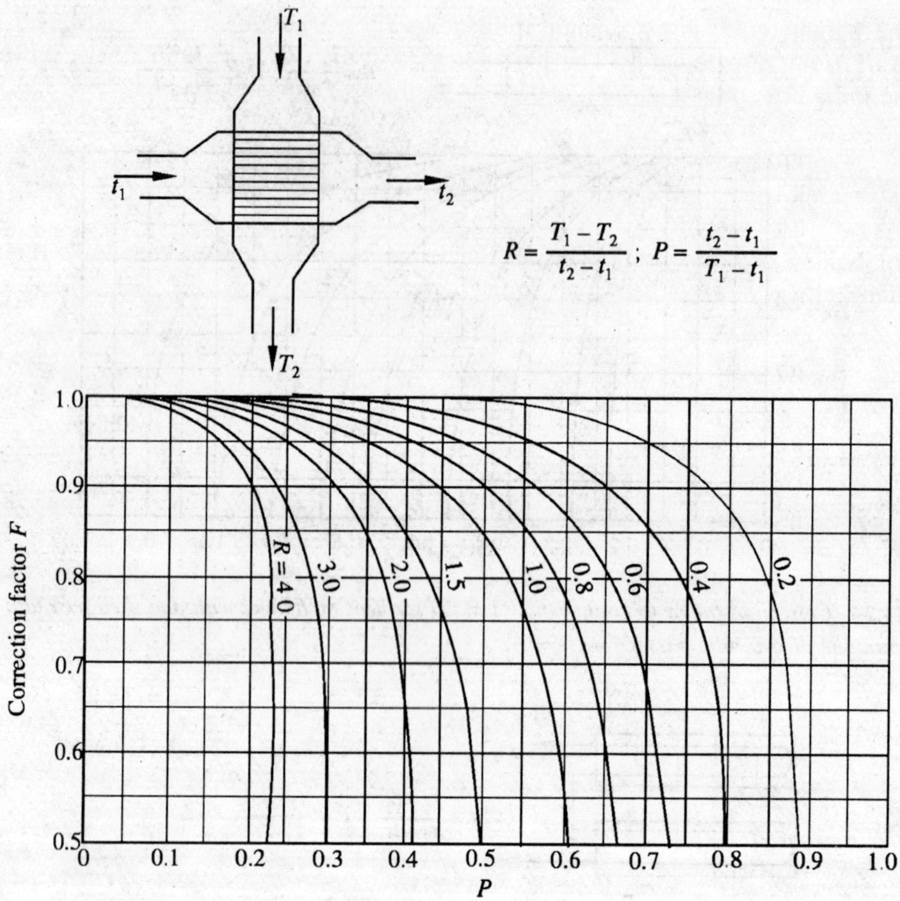

Fig. 11.9 Correction factor to counter-flow LMTD for cross-flow heat exchangers—fluid on shell-side mixed and the other fluid unmixed—one tube pass.

$$R = \frac{T_1 - T_2}{t_2 - t_1} \tag{11.21}$$

where

T_1 = temperature of the shell-side fluid at the inlet

T_2 = temperature of the shell-side fluid at the outlet

t_1 = temperature of the tube-side fluid at the inlet

t_2 = temperature of the tube-side fluid at the outlet.

The parameter P gives an indication of the heating or cooling effectiveness and can vary from zero to unity. The parameter R is equal to the ratio of the products of the mass flow rates and heat capacities of the two fluids. That is,

$$R = \frac{(\dot{m}c)_{\text{tubefluid}}}{(\dot{m}c)_{\text{shellfluid}}} = \frac{T_1 - T_2}{t_2 - t_1} \tag{11.22}$$

For boiling or condensation, when the temperature of one of the fluids remains constant, either P or R becomes zero. Then $F = 1.0$.

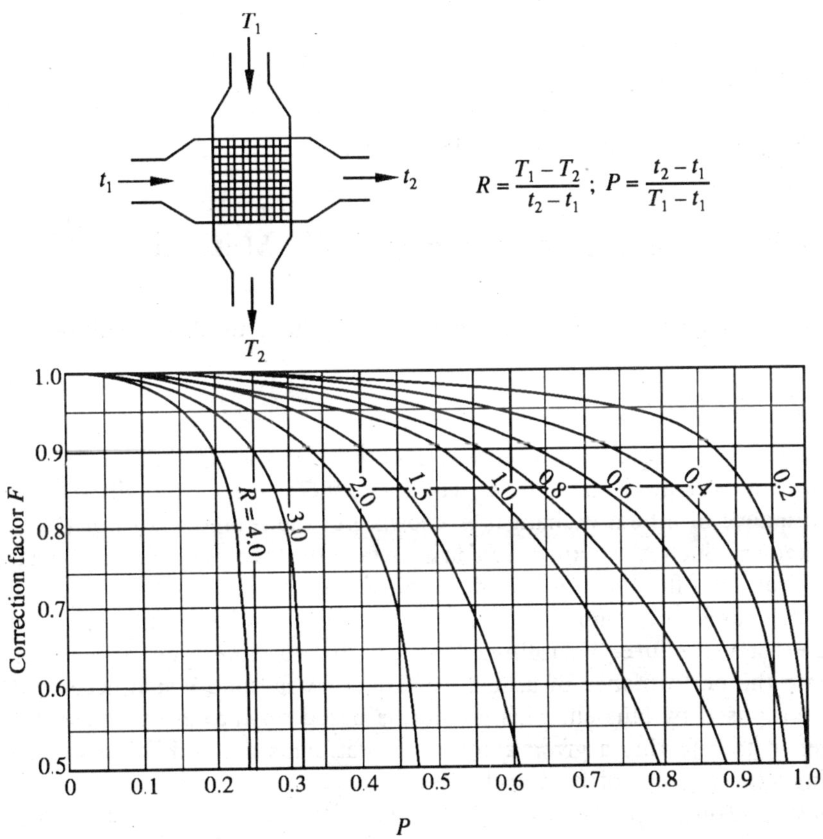

Fig. 11.10 Correction factor to counter-flow LMTD for cross-flow heat exhangers—both fluids unmixed—with a one tube pass.

EXAMPLE 11.3 It is desired to use a shell-and-tube heat exchanger with one shell pass and two tube passes to cool oil from 120°C to 50°C. The oil flows through the tubes while the cooling water flows through the shell. The cooling water enters the heat exchanger at 20°C and leaves it at 40°C. Determine the correction factor to LMTD.

SOLUTION Given

$$t_1 = 120°\text{C};\ t_2 = 50°\text{C};\ T_1 = 20°\text{C and } T_2 = 40°\text{C}$$

The parameters P and R are given by

$$P = \frac{t_2 - t_1}{T_1 - t_1} = \frac{50 - 120}{20 - 120} = 0.7$$

$$R = \frac{T_1 - T_2}{t_2 - t_1} = \frac{20 - 40}{50 - 120} = 0.286$$

Then read $F = 0.9$ at $P = 0.7$ and $R = 0.286$ from Fig. 11.7.

Hence, the correction factor for LMTD = 0.9.

11.5 Heat Exchanger Effectiveness–NTU Method

In the thermal analysis of the heat exchangers, we have used the relation

$$q = UA\Delta T_m \tag{11.17}$$

This form of the equation can be conveniently used if all the terminal temperatures are known for evaluating the LMTD. However, there are several situations in which the performance of a heat exchanger is known, but the temperatures of the exit fluids are not known. Such a situation can arise when the heat exchanger is tested for a given flow rate of a fluid, and it is desired to use a heat exchanger with a different flow rate. In such cases, the temperatures of the exit fluid can be determined by a tedious trial-and-error procedure. Therefore, it is desirable to circumvent the use of LMTD to evaluate the performance of a heat exchanger. A method which accomplishes this task was proposed by Nusselt. This method is based on the effectiveness of the heat exchanger in transferring a given amount of heat. This method can also be used for comparing various types of heat exchangers.

The *heat exchanger effectiveness* ε is defined as the ratio of the actual rate of heat transfer in a given heat exchanger to the maximum possible rate of heat transfer. That is,

$$\varepsilon = \frac{\text{actual rate of heat transfer}}{\text{maximum possible rate of heat transfer}} \tag{11.23}$$

The maximum possible rate of heat transfer will be obtained in a counter-current heat exchanger of infinite heat transfer area. In such a heat exchanger, the temperature of the exit cold fluid is equal to the temperature of the inlet hot fluid if $\dot{m}_c c_c < \dot{m}_h c_h$.

If $\dot{m}_h c_h < \dot{m}_c c_c$, the temperature of the exit hot fluid is equal to the temperature of the inlet cold fluid. Depending on the relative values of $\dot{m}_c c_c$ and $\dot{m}_h c_h$, either the cold fluid or the hot fluid would undergo the maximum temperature difference. The fluid having the minimum value of $(\dot{m}c)$ will undergo the maximum temperature difference. Hence, the maximum possible rate of heat transfer is given by

$$q_{\max} = (\dot{m}c)_{\min}(T_{h\,in} - T_{c\,in}) \tag{11.24}$$

The minimum fluid can be either the hot or cold fluid, depending on the mass flow rates and specific heats.

For the parallel-flow heat exchanger,

$$\varepsilon_h = \frac{\dot{m}_h c_h (T_{h1} - T_{h2})}{\dot{m}_h c_h (T_{h1} - T_{c1})} = \frac{T_{h1} - T_{h2}}{T_{h1} - T_{c1}} \tag{11.25}$$

$$\varepsilon_c = \frac{\dot{m}_c c_c (T_{c2} - T_{c1})}{\dot{m}_c c_c (T_{h1} - T_{c1})} = \frac{T_{c2} - T_{c1}}{T_{h1} - T_{c1}} \tag{11.26}$$

where

ε_h = effectiveness of heat exchanger if the hot fluid is the minimum fluid (that is, $\dot{m}c$ is minimum for hot fluid)

ε_c = effectiveness of heat exchanger if the cold fluid is the minimum fluid.

For the counter-current heat exchanger,

$$\varepsilon_h = \frac{\dot{m}_h c_h (T_{h1} - T_{h2})}{\dot{m}_h c_h (T_{h1} - T_{c2})} = \frac{T_{h1} - T_{h2}}{T_{h1} - T_{c2}} \tag{11.27}$$

$$\varepsilon_c = \frac{\dot{m}_c c_c (T_{c1} - T_{c2})}{\dot{m}_c c_c (T_{h1} - T_{c2})} = \frac{T_{c1} - T_{c2}}{T_{h1} - T_{c2}} \tag{11.28}$$

We can now generalize Eqs. (11.25) to (11.28) and express the effectiveness ε as

$$\varepsilon = \frac{\Delta T \text{ (minimum fluid)}}{\text{Maximum temperature difference in heat exchanger}} \tag{11.29}$$

Once the effectiveness of a heat exchanger is known, the rate of heat transfer can be directly determined from the equation

$$q = \varepsilon(\dot{m}c)_{\min}(T_{h1} - T_{c1}) \quad \text{for co-current} \tag{11.30}$$

$$q = \varepsilon(\dot{m}c)_{\min}(T_{h1} - T_{c2}) \quad \text{for counter-current} \tag{11.31}$$

since Eqs. (11.25) and (11.26) give

$$\varepsilon(\dot{m}c)_{\min}(T_{h1} - T_{c1}) = \dot{m}_h c_h (T_{h1} - T_{h2}) = \dot{m}_c c_c (T_{c2} - T_{c1}) \tag{11.32}$$

We can derive an expression for the effectiveness in the case of a parallel-flow heat exchanger as given below.

We can rewrite Eq. (11.14) as

$$\ln\left(\frac{T_{h2} - T_{c2}}{T_{h1} - T_{c1}}\right) = -UA\left(\frac{1}{\dot{m}_h c_h} + \frac{1}{\dot{m}_c c_c}\right) = \frac{-UA}{\dot{m}_c c_c}\left(1 + \frac{\dot{m}_c c_c}{\dot{m}_h c_h}\right) \tag{11.33}$$

or

$$\frac{T_{h2} - T_{c2}}{T_{h1} - T_{c1}} = \exp\left[\frac{-UA}{\dot{m}_c c_c}\left(1 + \frac{\dot{m}_c c_c}{\dot{m}_h c_h}\right)\right] \tag{11.34}$$

For the parallel-flow heat exchanger, energy balance gives

$$q = \dot{m}_h c_h (T_{h1} - T_{h2}) = \dot{m}_c c_c (T_{c_2} - T_{c_1})$$

or

$$T_{h2} = \frac{\dot{m}_c c_c}{\dot{m}_h c_h}(T_{c1} - T_{c2}) + T_{h1} \tag{11.35}$$

Substituting Eq. (11.35) in Eq. (11.34), we get

$$\frac{T_{h2} - T_{c2}}{T_{h1} - T_{c1}} = \frac{\dfrac{\dot{m}_c c_c}{\dot{m}_h c_h}(T_{c1} - T_{c2}) + T_{h1} - T_{c2}}{T_{h1} - T_{c1}}$$

$$= \exp\left[\frac{-UA}{\dot{m}_c c_c}\left(1 + \frac{\dot{m}_c c_c}{\dot{m}_h c_h}\right)\right] \tag{11.36}$$

The LHS of Eq. (11.36) can be rewritten as

$$\frac{(T_{h1} - T_{c1}) + \dfrac{\dot{m}_c c_c}{\dot{m}_h c_h}(T_{c1} - T_{c2}) + (T_{c1} - T_{c2})}{T_{h1} - T_{c1}}$$

$$= 1 + \left(\frac{T_{c1} - T_{c2}}{T_{h1} - T_{c1}}\right)\left(1 + \frac{\dot{m}_c c_c}{\dot{m}_h c_h}\right) \tag{11.37}$$

If the cold fluid is the minimum fluid, we have

$$\varepsilon_c = \frac{T_{c2} - T_{c1}}{T_{h1} - T_{c1}} \tag{11.26}$$

Substituting Eq. (11.26) in Eq. (11.37), we get

$$\frac{(T_{h1} - T_{c1}) + \dfrac{\dot{m}_c c_c}{\dot{m}_h c_h}(T_{c1} - T_{C2}) + (T_{c1} - T_{c2})}{T_{h1} - T_{c1}} = 1 - \varepsilon_c\left(1 + \frac{\dot{m}_c c_c}{\dot{m}_h c_h}\right) \tag{11.38}$$

From Eqs. (11.36) to (11.38), we get

$$\frac{T_{h2} - T_{c2}}{T_{h1} - T_{c1}} = 1 - \varepsilon_c\left(1 + \frac{\dot{m}_c c_c}{\dot{m}_h c_h}\right) = \exp\left[\frac{-UA}{\dot{m}_c c_c}\left(1 + \frac{\dot{m}_c c_c}{\dot{m}_h c_h}\right)\right]$$

or

$$\varepsilon_c = \frac{1 - \exp\left[\dfrac{-UA}{\dot{m}_c c_c}\left(1 + \dfrac{\dot{m}_c c_c}{\dot{m}_h c_h}\right)\right]}{1 + \dfrac{\dot{m}_c c_c}{\dot{m}_h c_h}} \tag{11.39}$$

It can be shown that the effectiveness of a parallel-flow heat exchanger, if the hot fluid is the minimum fluid, is given by

$$\varepsilon_h = \frac{1 - \exp\left[\dfrac{-UA}{\dot{m}_h c_h}\left(1 + \dfrac{\dot{m}_h c_h}{\dot{m}_c c_c}\right)\right]}{1 + \dfrac{\dot{m}_h c_h}{\dot{m}_c c_c}} \tag{11.40}$$

Equations(11.39) and (11.40) are similar except that $\dot{m}_c c_c$ and $\dot{m}_h c_h$ are interchanged. As a consequence, we can write the effectiveness ε as

$$\varepsilon = \frac{1 - \exp\left[\dfrac{-UA}{C_{\min}}\left(1 + \dfrac{C_{\min}}{C_{\max}}\right)\right]}{1 + \dfrac{C_{\min}}{C_{\max}}} \tag{11.41}$$

where $C = \dot{m}c$ = capacity rate. (11.42)

It is possible to carry out a similar analysis for a counter-current heat exchanger and obtain the following expression for effectiveness:

$$\varepsilon = \frac{1 - \exp\left[\frac{-UA}{C_{\min}}\left(1 - \frac{C_{\min}}{C_{\max}}\right)\right]}{1 - \left(\frac{C_{\min}}{C_{\max}}\right)\exp\left[\frac{-UA}{C_{\min}}\left(1 - \frac{C_{\min}}{C_{\max}}\right)\right]} \tag{11.43}$$

Equations(11.41) and (11.43) show that the effectiveness of a heat exchanger is a function of two dimensionless parameters:

(a) the ratio of the capacity rates $\dfrac{C_{\min}}{C_{\max}}$, and

(b) the ratio of the overall conductance to the minimum capacity rate $\dfrac{UA}{C_{\min}}$.

The dimensionless parameter $(UA/C_{\min})$ is called the *number of heat transfer units* or NTU for short, since it is a measure of the heat transfer size of the heat exchanger. A similar analysis can be carried out to determine the effectiveness of a heat exchanger for the flow conditions of interest. Kays and London have presented the results through convenient graphs, and some of their results are presented in Figs. 11.11 to 11.16.

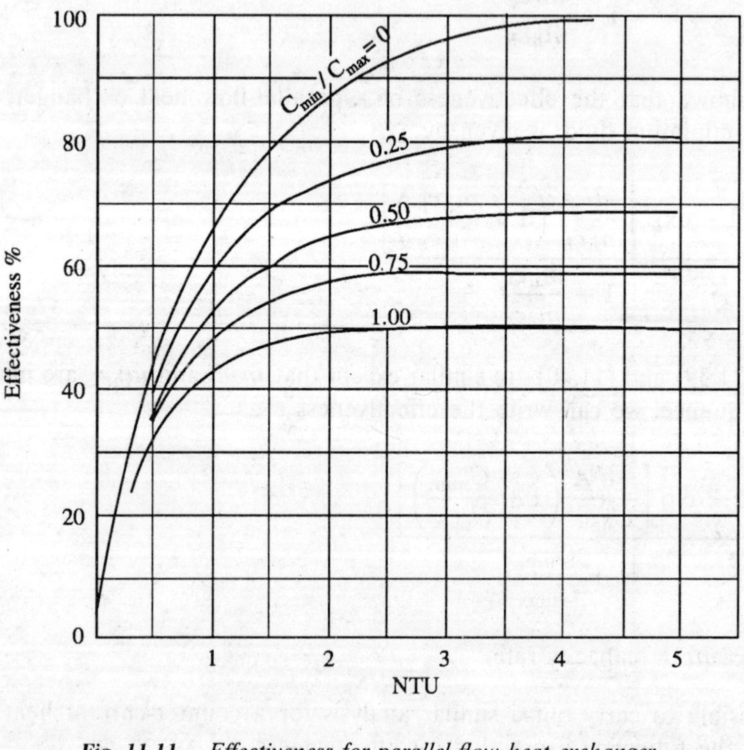

Fig. 11.11 Effectiveness for parallel-flow heat exchanger.

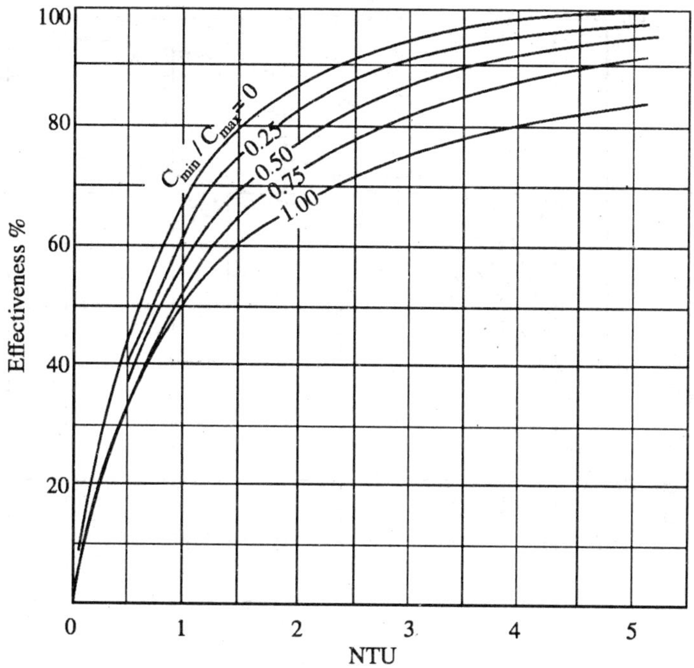

Fig. 11.12 Effectiveness for counter-current heat exchanger.

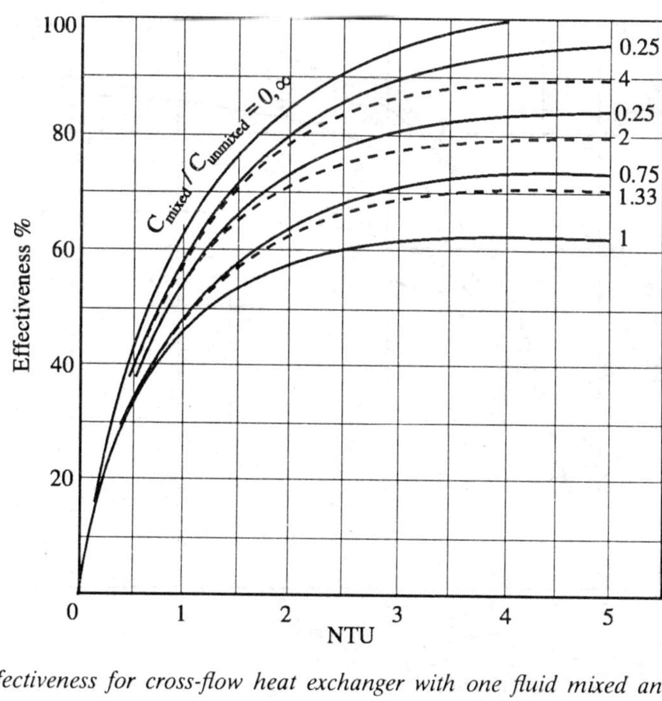

Fig. 11.13 *Effectiveness for cross-flow heat exchanger with one fluid mixed and the other fluid unmixed.*

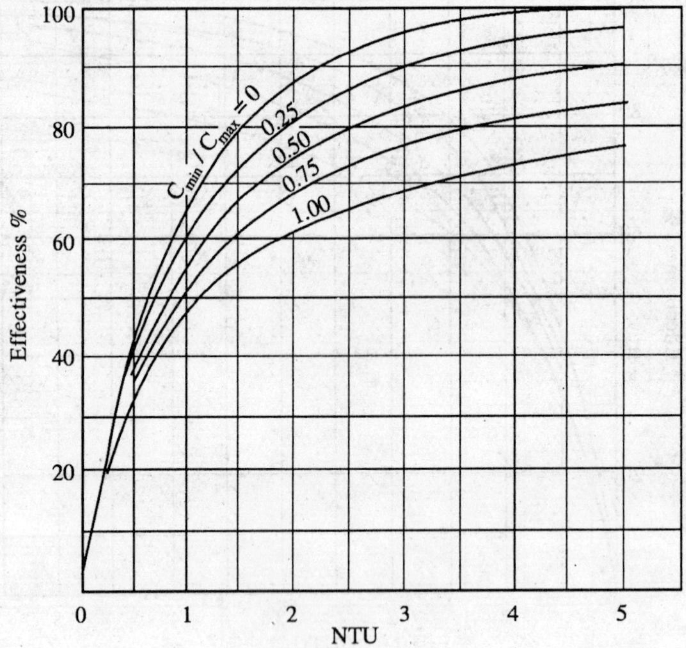

Fig. 11.14 Effectiveness for cross-flow heat exchanger with both fluids unmixed.

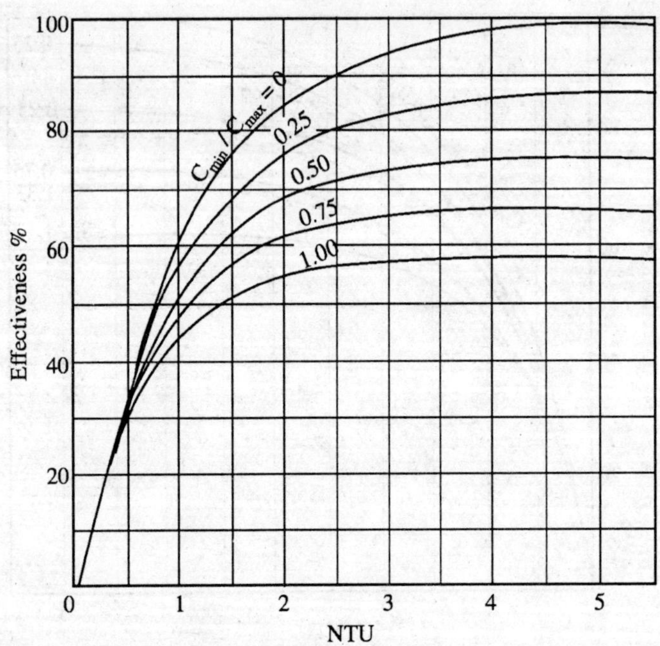

Fig. 11.15 Effectiveness for shell-and-tube heat exchanger with one shell pass and two or multiples of two tube passes.

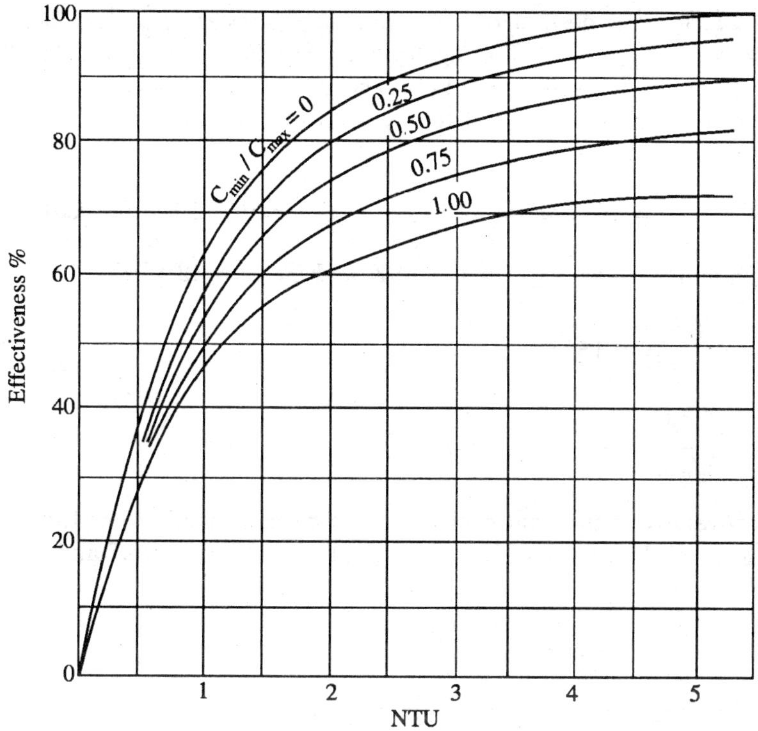

Fig. 11.16 Effectiveness for shell and tube heat exchangers with two shell passes and four or multiples of four tube passes.

EXAMPLE 11.4 In a chemical plant, 1 kg/s of oil (C_p = 2.8 kJ/kg K) is cooled from 120°C to 50°C in a shell-and-tube heat exchanger with one shell pass and two tube passes. The heat transfer area of the heat exchanger is 4 m^2. The overall heat transfer coefficient of the heat exchanger is known to be 800 W/m^2 K. The cooling water at 20°C enters the heat exchanger at a rate of 2.67 kg/s. Calculate the effectiveness of the heat exchanger and the number of heat transfer units.

SOLUTION $\dot{m}_c c_c = 2.67 \times 4.2 = 11.2$ kJ/K; $\dot{m}_h c_h = 1 \times 2.8 = 2.8$ kJ/K

$\dot{m}_h c_h$ is minimum. Hence $C_{\min} = 2.8$ kJ/K and $C_{\max} = 11.2$ kJ/K.

$$\text{NTU} = \frac{UA}{C_{\min}} = \frac{800 \times 4}{2.8 \times 10^3} = 1.143$$

$$\frac{C_{\min}}{C_{\max}} = \frac{2.8}{11.2} = 0.25$$

Then read $\varepsilon = 60\%$ at NTU = 1.143 and $\dfrac{C_{\min}}{C_{\max}} = 0.25$ from Fig. 11.15.

Therefore, the effectiveness of the heat exchanger = 60%.

EXAMPLE 11.5 It is desired to use a double-pipe counter-current heat exchanger to cool 3 kg/s of oil (C_p = 2.1 kJ/kg K) from 120°C. Cooling water at 20°C enters the heat exchanger at a rate of 10 kg/s. The overall heat transfer coefficient of the heat exchanger is 600 W/m^2 K and the heat transfer area is 6 m^2. Calculate the exit temperatures of oil and water.

SOLUTION $\dot{m}_c c_c = 10 \times 4.2 = 42$ kJ/K; $\dot{m}_h c_h = 3 \times 2.1 = 6.3$ kJ/K

$\dot{m}_h c_h$ is minimum. Hence $C_{\min} = 6.3$ kJ/K and $C_{\max} = 42$ kJ/K

$$\frac{C_{\min}}{C_{\max}} = \frac{6.3}{42} = 0.15$$

$$\text{NTU} = \frac{UA}{C_{\min}} = \frac{600 \times 6}{6.3 \times 10^3} = 0.57$$

The effectiveness of the heat exchanger can be calculated either from Eq. (11.43) or from Fig. 11.12. Read $\varepsilon = 42\%$ from Fig. 11.12 at NTU = 0.57 and $C_{\min}/C_{\max} =$ 0.15. Alternatively,

$$\varepsilon = \frac{1 - \exp\left[\frac{-UA}{C_{\min}}\left(1 - \frac{C_{\min}}{C_{\max}}\right)\right]}{1 - \frac{C_{\min}}{C_{\max}} \exp\left[\frac{-UA}{C_{\min}}\left(1 - \frac{C_{\min}}{C_{\max}}\right)\right]}$$

$$= \frac{1 - \exp[-0.57(1 - 0.15)]}{1 - 0.15\exp[-0.57(1 - 0.15)]} = 0.42$$

The hot fluid is the minimum fluid, since $\dot{m}_h C_h$ is minimum. Then

$$\varepsilon_h = \frac{T_{h1} - T_{h2}}{T_{h1} - T_{c2}} \quad \text{or} \quad 0.42 = \frac{120 - T_{h2}}{120 - 20}$$

or $T_{h2} = 78°\text{C}$

Therefore, the exit temperature of oil = 78°C.

The rate of heat transfer q is given by

$$q = \varepsilon C_{\min}(T_{h1} - T_{c2}) = 0.42 \times 6.3 \times 10^3(120 - 20) = 264.6 \text{ kW}$$

$$\dot{m}_c c_c (T_{c1} - T_{c2}) = 10 \times 4.2 \times 10^3 (T_{c1} - 20) = 264.6 \text{ kW}$$

or $T_{c1} = 26.3°\text{C}$

Hence the exit temperature of cooling water = 26.3°C.

11.6 Fouling Factors

After a period of operation, the heat transfer surface in a heat exchanger in a process industry gets coated with various deposits. Hence the performance of heat exchangers which are in use for some time cannot be predicted from a thermal analysis. The deposits on the heat transfer surface may be rust or scale or anything else, and this coating presents an additional resistance to the heat flow. The coating results in a decrease in the performance of the heat exchanger. The effect is usually referred to as *fouling*. The heat exchanger manufacturer can specify the performance of a clean heat exchanger only. The thermal resistance of the deposit or the *fouling resistance* can be obtained only from actual tests. If performance tests are made on a clean heat exchanger and the test is repeated after the unit has been in service for a sufficiently long time, the fouling resistance R_f or *fouling factor* can be determined from the relation

$$R_f = \frac{1}{U_d} - \frac{1}{U} \tag{11.44}$$

where

U = overall heat transfer coefficient for a clean heat exchanger

U_d = overall heat transfer coefficient of the heat exchanger after fouling has occurred.

The fouling factors for various applications have been compiled by the Tubular Exchanger Manufacturers Association and a few representative factors are presented in Table 11.1.

Table 11.1 *Normal fouling factors*

Type of fluid	Fouling factor m^2 K/W
Sea water below 50°C	0.00009
Sea water above 50°C	0.002
Treated boiler feed-water above 50°C	0.0002
Fuel oil	0.0009
Steam	0.00009
Industrial air	0.0004
Refrigerating liquid	0.0002
Alcohol vapors	0.00009

The overall heat transfer coefficient for an unfinned tube can be calculated from the relation

$$U_d = \frac{1}{\dfrac{1}{h_o} + R_o + R_k + \dfrac{R_i A_o}{A_i} + \dfrac{A_o}{h_i A_i}} \tag{11.45}$$

where

U_d = overall heat transfer coefficient based on the outside area of the heat exchanger after fouling has occurred

h_o = convection heat transfer coefficient on the outside of the tube

h_i = convection heat transfer coefficient at the inner surface of the tube

R_o = fouling factor on the outside of the tube

R_i = fouling factor on the inside of the tube

R_k = thermal resistance for conduction offered by the tube

A_o = outside tube surface area

A_i = inside tube surface area.

For a preliminary estimate of heat exchanger size, it is sufficient to know the order of magnitude of the overall heat transfer coefficients under average service conditions. Typical values of the overall heat transfer coefficients are presented in Table 11.2.

Table 11.2 *Typical values of overall heat transfer coefficients.*

Physical Situation	U (W/m^2 K)
Steam to water	990–3400
Steam to oil	55–340
Steam to aqueous solution	570–3400
Steam to gases	28–280
Water to air	55–170
Water to water	850–1560
Water to lubricating oil	110–340
Water to condensing oil vapors	225–570
Water to condensing alcohol	250–680
Water to condensing Freon-12	450–850
Water to condensing ammonia	850–1420
Water to organic solvents	280–850
Water to boiling Freon-12	280–850
Water to gasoline	340–510
Water to brine	560–1130
Crude oil to gas oil	170–310
Light organics to light organics	225–425
Medium organics to medium organics	110–340
Heavy organics to heavy organics	55–225
Heavy organics to light organics	55– 340

11.7 Summary

A heat exchanger is a device used for transferring energy from one fluid to another. The simplest type of heat exchanger is a double-pipe heat exchanger which consists of a tube located concentrically inside another tube. If both the fluids flow in the same direction, the heat exchanger is called a parallel-flow type or co-current type. If the fluids flow in opposite directions, the heat exchanger is called a counter-current type. In chemical process industries, shell-and-tube heat exchangers are in wide use. In these heat exchangers, one fluid flows through the inside tubes while the other fluid flows through the shell. If the two fluids flowing along the heat transfer surface move at right angles to each other, then the heat exchanger is of the cross-flow type. The cross- flow heat exchangers are usually used in gas heating and cooling applications.

The heat transfer rate in a heat exchanger is usually expressed by the relation

$$q = UA\Delta T$$

where U is the overall heat transfer coefficient. The overall heat transfer coefficient may be defined based on either the inside area or the outside area of a tube.

In a heat exchanger, the temperatures of the heating and cooling fluids do not remain constant, but vary from point to point along the length of the heat exchanger. Therefore, the temperature difference between the two fluids changes from point to point in a heat exchanger. Hence, a suitable mean temperature difference, called the Log Mean Temperature Difference (LMTD) is used for evaluating the heat exchanger performance. The LMTD for a double-pipe heat exchanger is defined as

$$\text{LMTD} = \frac{(T_{h2} - T_{c2}) - (T_{h1} - T_{c1})}{\ln\left(\dfrac{T_{h2} - T_{c2}}{T_{h1} - T_{c1}}\right)} = \Delta T_m$$

where the subscripts h and c refer to the hot and cold fluids, respectively, and the subscripts 1 and 2 refer to the two ends of the heat exchanger.

If the temperature difference at the two ends do not differ by more than a factor of two, the arithmetic mean temperature difference will be approximately equal to the LMTD. For complex heat exchangers like shell-and-tube heat exchangers with several shell passes and several tube passes and for cross-flow heat exchangers, the usual practice is to modify the LMTD by including a correction factor F. Then the rate of heat transfer can be expressed as

$$q = UAF\Delta T_m$$

The LMTD can be conveniently used only if all the terminal temperatures are known. However, one comes across situations in which the heat exchanger performance is known for particular flow rates of fluids but the temperatures of the exit fluids are not known. Therefore, it is necessary to circumvent the use of LMTD to evaluate the heat exchanger performance. This is done by introducing the heat exchanger effectiveness ε which is defined as

$$\varepsilon = \frac{\text{actual rate of heat transfer}}{\text{maximum possible rate of heat transfer}}$$

The maximum possible rate of heat transfer will be obtained in a counter-current heat exchanger of infinite heat transfer area.

For the counter-current heat exchanger, the effectiveness is given by

$$\varepsilon = \frac{1 - \exp\left[\dfrac{-UA}{C_{\min}}\left(1 - \dfrac{C_{\min}}{C_{\max}}\right)\right]}{1 - \dfrac{C_{\min}}{C_{\max}} \exp\left[\dfrac{-UA}{C_{\min}}\left(1 - \dfrac{C_{\min}}{C_{\max}}\right)\right]}$$

The effectiveness for a parallel-flow heat exchanger is given by

$$\varepsilon = \frac{1 - \exp\left[\dfrac{-UA}{C_{\min}}\left(1 + \dfrac{C_{\min}}{C_{\max}}\right)\right]}{1 + \dfrac{C_{\min}}{C_{\max}}}$$

where $C = \dot{m}c$.

The dimensionless parameter $(UA/C_{\min})$ is called the number of heat transfer units or NTU.

After a period of operation, the heat transfer surface in a heat exchanger gets coated with deposits or scale and this scale formation results in a decrease in the performance of the heat exchanger. This effect is called fouling. The fouling resistance or fouling factor R_f can be determined from the relation

$$R_f = \frac{1}{U_d} - \frac{1}{U}$$

where

U = overall heat transfer coefficient for a clean heat exchanger

U_d = overall heat transfer coefficient of the heat exchanger after fouling.

Review Questions

1. What is meant by thermal analysis of a heat exchanger?
2. What is a double-pipe heat exchanger?
3. What is meant by a parallel-flow or co-current heat exchanger?

4. What is meant by a counter-current heat exchanger?
5. Distinguish between a shell-and-tube heat exchanger and a cross-flow heat exchanger.
6. Define the overall heat transfer coefficient.
7. Define the Log Mean Temperature Difference (LMTD).
8. When is the LMTD approximately equal to the arithmetic mean temperature difference?
9. Define the effectiveness of a heat exchanger.
10. If $\dot{m}_h c_h < \dot{m}_c c_c$, which fluid would undergo the maximum temperature difference in a heat exchanger?
11. How does one determine the rate of heat transfer in a heat exchanger if its effectiveness is known?
12. What is meant by the minimum fluid?
13. Define the number of heat transfer units(NTU)?
14. When is the LMTD method most applicable to heat exchanger calculations?
15. What advantage does the NTU method have over the LMTD method?
16. How can one determine the fouling factor of a heat exchanger?

Problems

11.1 In a parallel-flow double-pipe heat exchanger, the hot fluid enters the heat exchanger at 90°C and leaves it at 50°C, while the cold fluid enters the heat exchanger at 20°C and leaves it at 40°C. Calculate the LMTD.

11.2 In a counter-current double-pipe heat exchanger, the hot fluid enters at 200°C and leaves at 100°C, while the cold fluid enters the heat exchanger at 20°C and leaves it at 80°C. Calculate the LMTD.

11.3 A counter-current double-pipe heat exchanger with an overall heat transfer coefficient of 800 W/m^2 K is used to cool 10 kg/s of oil from 150°C to 50°C. The cooling water enters the heat exchanger at 25°C and leaves at 40°C. The specific heats of oil and water are 3.4 kJ/kg K and 4.2 kJ/kg K, respectively. Calculate the rate of heat transfer and the heat transfer area of the heat exchanger.

11.4 A double-pipe heat exchanger with an overall heat transfer coefficient of 400 W/m^2 K is to be used to cool 10 kg/s of oil from 110°C to 50°C. Cooling water is available at 20°C and the water should leave the heat exchanger at 40°C. The specific heats of oil and water are 3.1 kJ/kg K and 4.2 kJ/kg K, respectively. Determine the heat transfer area for (a) a counter-current flow heat exchanger and for (b) a parallel-flow heat exchanger.

11.5 In a heat exchanger, cold water enters at 30°C and leaves at 60°C. Dry saturated steam at 100°C enters the heat exchanger and the condensate leaves at 100°C. Calculate the LMTD for (a) a counter-flow double-pipe heat exchanger and (b) a parallel-flow double-pipe heat exchanger. (c) Calculate the modified LMTD for a shell-and tube heat exchanger with one shell pass and two tube passes.

11.6 Calculate the correction factor for LMTD for Problem 11.5 if a shell-and-tube heat exchanger with two shell passes and four tube passes is used.

11.7 It is desired to cool 0.5 kg/s of oil from 105°C to 45°C by using an equal flow rate of cooling water. The cooling water is available at 20°C. The specific heats of oil and water are 2.8 kJ/kg K and 4.2 kJ/kg K. Two double-pipe heat exchangers are available:
Heat exchanger 1: U = 500 W/m^2 K; A = 4.5 m^2
Heat exchanger 2: U = 800 W/m^2 K; A = 2 m^2
Which heat exchanger should be used?

11.8 It is desired to cool 0.5 kg/s of oil (C_p = 2.8 kJ/kg K) from 100°C to 60°C by using cold water at 20°C. The cooling water is to leave the heat exchanger at 40°C. If a cross-flow heat exchanger with both fluids unmixed and having an overall heat transfer coefficient of 600 W/m^2 K is to be employed, determine the heat transfer surface area required.

11.9 It is required to cool 0.5 kg/s of oil from 110°C to 50°C using a counter-current double-pipe heat exchanger. The cooling water is available at 25°C and can leave at 35°C. The oil flows through the inner copper tube of 12 mm diameter, 1.5 m length and water flows through the annulus. The internal diameter of the outer tube is 30 mm. The inner tube of 3 mm thickness is made of copper. Use the Dittus–Boelter equation to estimate the convection heat transfer coefficients and determine the overall heat transfer coefficient of the heat exchanger. For oil, ρ = 864 kg/m^3; C_p = 2.4 kJ/kg K; ν =0.85 × 10^{-4} m^2/s; k = 0.16 W/m K

11.10 A shell-and-tube heat exchanger with one shell pass and two tube passes having a overall heat transfer coefficient of 1000 W/m^{2}K and a heat transfer area of 10 m^2 is to be used to cool 2 kg/s of a liquid (C_p = 3 kJ/kg K) from 200°C using 5 kg/s of cooling water at 25°C. Determine the outlet temperatures of the liquid and cooling water.

11.11 In a counter-current double-pipe heat exchanger, 0.4 kg/s of oil (C_p = 2.2 kJ/kg K) is cooled from 120°C using 1 kg/s of water which enters at 25°C. The heat exchanger has an overall heat transfer coefficient of 600 W/m^2 K and a heat transfer area of 3 m^2. Calculate the exit temperatures of the oil and cooling water. Also determine the total heat transfer rate.

CHAPTER TWELVE

MASS TRANSFER

Learning Objectives

A careful study of this chapter will enable the readers to

- understand the phenomenon of diffusion and mass transfer.
- know the thermodynamic criterion of chemical equilibrium for coexisting phases.
- define the chemical potential.
- know the relation between chemical potential and concentration.
- define mass concentration and molar concentration of species in a multi-component mixture.
- define mass average velocity, molar average velocity and diffusion velocity.
- define mass flux and molar flux.
- understand Fick's law of diffusion.
- derive the general mass diffusion equation.
- analyze steady-state diffusion through a plane membrane.
- analyze diffusion through a stagnant gas medium.
- define convection mass transfer coefficient.
- understand laminar boundary layer for mass transfer for flow of a fluid over a flat plate.
- define Schmidt number, Lewis number and Sherwood number.
- know about the empirical correlations for estimating mass transfer coefficient.

12.1 Introduction

Suppose a crystal of potassium permanganate is dropped into a beaker containing clear water. The potassium permanganate crystal begins to dissolve in water and in the immediate neighbourhood of the crystal one observes a dark purple colour. As time progresses, the colour spreads to the other zones also and a light purple colour can be observed at a distance from the crystal. The concentration of $KMnO_4$ will be large where the colour is dark purple while the concentration of $KMnO_4$ will be low where the colour is light purple. Because of the concentration difference, $KMnO_4$ diffuses from regions of high concentration to regions of low concentration. The movement of a chemical species from a region of high concentration to a region of low concentration is called *diffusion*. The process of transfer of mass as a result of concentration difference is usually called *mass transfer*. We know that all macroscopic material contains a large number of microscopic particles (molecules). These molecules move in all directions with varying velocities. Some molecules move with lower velocity while some other molecules move with higher velocity. The average velocity of the molecules can be determined from a knowledge of the velocity distribution function. In a homogeneous mixture of two chemical species, the average velocity with which species 1 moves will be different from the average velocity with which species 2 moves. Since the velocities of the individual species are different, diffusion of the species occurs.

Consider a tank which is divided into two compartments A and B by a partition. Compartment A contains chlorine gas at temperature T and pressure P, while compartment B contains hydrogen gas at the same temperature and pressure. Suppose the partition between the two compartments is carefully removed without causing any disturbance, and the concentrations of chlorine and hydrogen in both the compartments are monitored, it will be observed that in compartment A the concentration of chlorine decreases and that of hydrogen increases with time, while in compartment B, the concentration of chlorine increases and the concentration of hydrogen decreases with time. That is, chlorine gas diffuses into hydrogen and hydrogen diffuses into chlorine. If the tank is left undisturbed for a sufficiently long time, the concentration of chlorine will be identical in both the compartments. There will not be any further change in the concentration of chlorine and the net diffusion of each species into the other will be zero. From this observation we find that concentration difference acts as a driving force for the diffusion, and the diffusion of a chemical species takes place from a region of high concentration to a region of low concentration.

To verify whether the concentration difference is the sole reason for diffusion or not, let us consider the following situation. Suppose one half of the volume of a closed container is filled with a liquid mixture of benzene and toluene of known composition and the container is maintained at a constant temperature. The chemical species, benzene and toluene, vaporize and the species diffuse into the vapor phase. The diffusion of the chemical species takes place between the liquid and vapor phases. If the concentrations of benzene in the liquid and vapor phases are monitored, it will be observed that the two phases (liquid and vapor) reach a state of equilibrium, after a sufficiently long time, and in that state of equilibrium the concentrations of benzene in the two phases are not identical. There exists a concentration difference between the two phases and yet there is no net diffusion of the chemical species from one phase to another. This observation clearly indicates that concentration difference is not the real driving force for diffusion. Then what causes the diffusion of chemical species or mass transfer between the two phases? This can be answered by analyzing the criterion of chemical equilibrium which is discussed in the following section.

12.2 Criterion of Chemical Equilibrium

Consider a closed system consisting of two phases—liquid phase and vapor phase—each phase containing two chemical species 1 and 2, as shown in Fig. 12.1. The system is held at constant temperature and pressure. From thermodynamics, we know that the Gibbs free energy of the system reaches a minimum value at constant temperature and pressure, in a state of equilibrium. The Gibbs free energy is a function of temperature, pressure and composition. That is, the Gibbs free energy G of the system is given by

$$G = G^{\ell}(T, P, N_1^{\ell}, N_2^{\ell}) + G^{v}(T, P, N_1^{v}, N_2^{v}) \tag{12.1}$$

where N_i^{ℓ} and N_i^{v} denote the mole numbers of species i in the liquid and vapor phases, respectively.

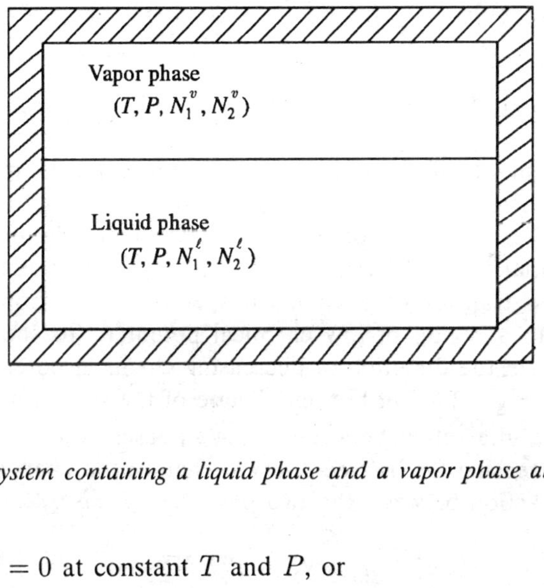

Fig. 12.1 A closed system containing a liquid phase and a vapor phase at constant temperature and pressure.

At equilibrium, $dG = 0$ at constant T and P, or

$$\left(\frac{\partial G^{\ell}}{\partial N_1^{\ell}}\right)_{T,P,N_2^{\ell}} dN_1^{\ell} + \left(\frac{\partial G^{\ell}}{\partial N_2^{\ell}}\right)_{T,P,N_1^{\ell}} dN_2^{\ell} + \left(\frac{\partial G^{v}}{\partial N_1^{v}}\right)_{T,P,N_2^{v}} dN_1^{v}$$

$$+ \left(\frac{\partial G^{v}}{\partial N_2^{v}}\right)_{T,P,N_1^{v}} dN_2^{v} = 0$$

or

$$\mu_1^{\ell} dN_1^{\ell} + \mu_2^{\ell} dN_2^{\ell} + \mu_1^{v} dN_1^{v} + \mu_2^{v} dN_2^{v} = 0 \tag{12.2}$$

where μ_i the chemical potential of species i is given by

$$\mu_i = \left(\frac{\partial G}{\partial N_i}\right)_{T,P,N_j} \tag{12.3}$$

Since the system is closed, the following conditions must be satisfied.

$$N_i = N_i^\ell + N_i^v = \text{constant} \quad \text{for } i = 1, 2$$

or

$$dN_i^\ell = dN_i^v \quad \text{for } i = 1, 2 \tag{12.4}$$

Substituting Eq. (12.4) in Eq. (12.2), we obtain

$$\left(\mu_1^\ell - \mu_1^v\right) dN_1^\ell + \left(\mu_2^\ell - \mu_2^v\right) dN_2^\ell = 0$$

or

$$\mu_i^\ell = \mu_i^v \; for \; i = 1, 2 \tag{12.5}$$

That is, in a state of equilibrium, the chemical potential of species i is identical in both the phases. Hence, the criterion for chemical equilibrium is the equality of chemical potential. The chemical potential is an intensive thermodynamic property. The chemical potential is the driving force for diffusion or mass transfer.

Diffusion or mass transfer of a chemical species occurs from a region of high chemical potential to a region of low chemical potential. To show that diffusion or mass transfer occurs in the direction of decreasing chemical potential, let us consider the system shown in Fig. 12.1. Let the initial state of the system be such that $\mu_1^\ell > \mu_1^v$ and $\mu_2^\ell < \mu_2^v$. As the interaction between the two phases occurs, the system moves in the direction of decreasing Gibbs free energy. That is, the change in the Gibbs free energy, due to interaction between the two phases, is given by

$$\Delta G = \mu_1^\ell \Delta N_1^\ell + \mu_2^\ell \Delta N_2^\ell + \mu_1^v \Delta N_1^v + \mu_2^v \Delta N_2^v < 0 \tag{12.6}$$

Since the system is closed

$$\Delta N_1^\ell = -\Delta N_1^v$$

$$\Delta N_2^\ell = -\Delta N_2^v \tag{12.7}$$

Substituting Eqs. (12.7) in Eq. (12.6), we get

$$\left(\mu_1^\ell - \mu_1^v\right) \Delta N_1^\ell + \left(\mu_2^\ell - \mu_2^v\right) \Delta N_2^\ell < 0 \tag{12.8}$$

By assumption, $\mu_1^\ell > \mu_1^v$ and $\mu_2^\ell < \mu_2^v$. Hence Eq. (12.8) gives $\Delta N_1^\ell < 0$ and $\Delta N_2^\ell > 0$. That is, in the liquid phase the mole numbers of species 1 decrease while the mole numbers of species 2 increase. In other words, the species diffuse in the direction of decreasing chemical potential.

In thermodynamics, the chemical potential is defined as the partial molar Gibbs free energy and is related to fugacity $\hat{f}$ as

$$d\mu_i = RT d \ln \hat{f}_i \tag{12.9}$$

Now let us choose the standard state of a species as the pure species at the system temperature and pressure and denote the standard state fugacity of species i as f_i^o. Integrating Eq. (12.9) from a state in which the species exists in the standard state, to the actual state in a mixture in which its mole fraction is x_i, we get

$$\mu = \mu_i^o + RT \ln \left(\frac{\hat{f}_i}{f_i^o} \right) \tag{12.10}$$

We also know from thermodynamics that in an ideal solution

$$\hat{f}_i = x_i f_i^o \tag{12.11}$$

Hence Eq. (12.10) reduces to

$$\mu_i = \mu_i^o + RT \ln x_i$$

It was stated earlier that in a mixture, the velocities of the individual species are different. There are several ways of calculating the average velocities of the species. It is necessary to calculate the average velocities before attempting to estimate the rate of diffusion. In the following section, the definition of diffusion velocity, concentration and mass flux are presented.

12.3 Concentration, Velocity and Mass Flux

In a multi-component mixture, the concentration of the individual species can be expressed in the following ways:

$$\text{Mass concentration } \rho_i = \frac{\text{mass of species } i}{\text{volume of the solution}} = \frac{m_i}{v}$$

$$\text{Molar concentration } c_i = \frac{\text{moles of species } i}{\text{volume of the solution}} = \frac{N_i}{V} = \left(\frac{m_i}{M_i}\right)\frac{1}{V}$$

or

$$c_i = \frac{\rho_i}{M_i} \tag{12.12}$$

$$\text{Mass fraction } w_i = \frac{\text{mass of species } i}{\text{total mass of solution}} = \frac{m_i}{\sum m_i}$$

$$= \frac{(m_i/V)}{\sum(m_i/V)} = \frac{\rho_i}{\rho} \tag{12.13}$$

$$\text{Mole fraction } x_i = \frac{\text{moles of species } i}{\text{total moles of solution}} = \frac{N_i}{\sum N_i}$$

$$= \frac{N_i/V}{\sum(N_i/V)} = \frac{c_i}{c} \tag{12.14}$$

where

m_i = mass of species i

N_i = moles of species i

M_i = molar mass of species i

V = volume of solution

c = total molar concentration of the solution.

The *local mass average velocity* v of the mixture with respect to the stationary coordinate system is defined as

$$v = \frac{\sum \rho_i v_i}{\sum \rho_i} = \frac{\sum \rho_i v_i}{\rho} \tag{12.15}$$

or

$$\rho v = \sum \rho_i v_i$$

The quantity ρv is the local rate at which mass passes through a unit area through a plane which is perpendicular to the velocity v. The *local molar average velocity* v^* is defined as

$$v^* = \frac{\sum c_i v_i}{\sum c_i} = \frac{\sum c_i v_i}{c} \tag{12.16}$$

and the quantity cv^* represents the local rate at which moles pass through a unit area through a plane perpendicular to v^*. In a system, one is interested in knowing the velocity of a specified species with respect to v or v^* instead of with respect to the stationary coordinate system. This leads to the definition of diffusion velocities as follows:

$$\text{Diffusion velocity of } i \text{ with respect to } v = v_i - v \tag{12.17}$$

$$\text{Diffusion velocity of } i \text{ with respect to } v^* = v_i - v^* \tag{12.18}$$

The diffusion velocities represent the motion of species i relative to the local motion of the fluid.

After discussing the concentrations and velocities, we are now in a position to discuss mass flux and molar flux. The mass of species i that passes through a unit area per unit time is called *mass flux*. The motion of the species can be with reference to the stationary coordinates, to the local mass average velocity v or to the local molar average velocity v^*. That is,

Mass flux relative to the stationary coordinate axis;

$$n_i = \rho_i v_i \tag{12.19}$$

Mass flux relative to the local mass average velocity is given by

$$j_i = \rho_i (v_i - v) \tag{12.20}$$

Similarly, the moles of species i that pass through a unit area, per unit time is called *molar flux*.

Molar flux relative to stationary coordinates

$$= N_i = c_i v_i \tag{12.21}$$

Molar flux relative to local molar average velocity is given by

$$J_i^* = c_i (v_i - v^*) \tag{12.22}$$

EXAMPLE 12.1 How the molar fluxes N_A and J_A^* of species A in a binary mixture related?

SOLUTION We know that

$$J_A^* = c_A (v_A - v^*)$$

$$= c_A v_A - c_A \frac{(c_A v_A + c_B v_B)}{c}$$

$$N_A - x_A (N_A + N_B) = N_A - N x_A$$

or

$$N_A = J_A^* + Nx_A \tag{A}$$

That is, the molar diffusion flux N_A relative to the stationary coordinate axis is equal to the sum of the molar diffusion flux relative to the molar average velocity and the flux associated with the bulk motion of the fluid.

EXAMPLE 12.2 Show that in a binary mixture $J_A^* = -J_B^*$.

SOLUTION We have observed in Example 12.1, that [see Eq. (A)]

$$N_A = J_A^* + Nx_A \tag{A}$$

Similarly we can write

$$N_B = J_B^* + Nx_B \tag{B}$$

Adding Eqs. (A) and (B), we get

$$N_A + N_B = J_A^* + J_B^* + N(x_A + x_B)$$

or

$$N_A + N_B = J_A^* + J_B^* + N_A + N_B$$

or

$$J_A^* + J_B^* = 0$$

or

$$J_A^* = -J_B^*$$

12.4 Fick's Law of Diffusion

Fick's law of diffusion states that the molar diffusion flux of species A is proportional to the concentration gradient, and is given by

$$J_A^* = -D_{AB}\frac{\partial c_A}{\partial y} = -cD_{AB}\frac{\partial x_A}{\partial y} \tag{12.23}$$

where D_{AB} is the diffusion coefficient. Fick's law of diffusion can also be stated in several other forms depending on the choice of the fluxes. Some of the forms in common use are

$$j_A = -\rho D_{AB}\frac{\partial w_A}{\partial y} = -D_{AB}\frac{\partial \rho_A}{\partial y} \tag{12.24}$$

$$j_A = -\frac{c^2}{\rho}M_A M_B D_{AB}\frac{\partial x_A}{\partial y} \tag{12.25}$$

$$N_A = J_A^* + Nx_A = x_A(N_A + N_B) - cD_{AB}\frac{\partial x_A}{\partial y} \tag{12.26}$$

The units of D_{AB} are m^2/s. Note that the units of kinematic viscosity ν and the thermal diffusivity α are also m^2/s. Fick's law of diffusion is analogous to Newton's law of viscosity and Fourier's law of heat conduction given by Eqs. (12.27) and (12.28).

$$\tau_{yx} = -\nu\frac{d}{dy}(\rho v_x) \tag{12.27}$$

$$q = -\alpha\frac{d}{dy}(\rho c_p T) \tag{12.28}$$

$$j_A = -D_{AB}\frac{d}{dy}(\rho_A) \tag{12.29}$$

For the molar diffusion flux of species B, we can write, similar to Eq. (12.23),

$$J_B^* = -cD_{BA}\frac{\partial c_B}{\partial y} \tag{12.30}$$

We have noticed in Example 12.2 that

$$J_A^* = -J_B^* \tag{12.31}$$

or

$$-cD_{AB}\frac{\partial c_A}{\partial y} = cD_{BA}\frac{\partial c_B}{\partial y} = -cD_{BA}\frac{\partial c_A}{\partial y}$$

or

$$D_{AB} = D_{BA} \tag{12.32}$$

From thermodynamic considerations we observed that the chemical potential is the driving force for diffusion or mass transfer. Hence, it is logical to express the molar flux in terms of the chemical potential gradient as

$$J_A^* = -L_{AB}\frac{\partial \mu_A}{\partial y} \tag{12.33}$$

where L_{AB} is a proportionality constant and is known as the *phenomenological coefficient*. From Eq. (12.11) we can write

$$\frac{\partial \mu_A}{\partial y} = RT\frac{\partial ln x_A}{\partial y} = \frac{RT}{x_A}\frac{dx_A}{dy} \tag{12.34}$$

Substituting Eq. (12.34) in (12.33), we obtain

$$J_A^* = -L_{AB}\frac{RT}{x_A}\frac{dx_A}{dy} \tag{12.35}$$

Comparing Eqs. (12.23) and (12.35), we get

$$L_{AB} = \frac{cx_A}{RT}D_{AB} = D_{AB}\frac{c_A}{RT} \tag{12.36}$$

Equation (12.36) gives the relation between the phenomenological coefficient and diffusion coefficient. For ideal gases, the diffusion coefficient D_{AB} varies only slightly with composition. However, the phenomenological coefficient L_{AB} is proportional to the mole fraction [see Eq. (12.36)]. Hence, L_{AB} is not quite convenient to use. As these equations are empirical, we choose the simple forms of Eqs. (12.23) and (12.24) as the most practical, accepting the fact that D_{AB} varies with composition. Of course, D_{AB} is also a function of pressure and temperature.

12.5 General Mass Diffusion Equation in Stationary Media

The general mass diffusion equation for any species i can be derived by adopting the procedure described in Section 4.2 for the derivation of the general heat conduction equation. Consider a homogeneous medium consisting of a binary mixture of species A and B. Let us also assume that the medium is stationary. That is, the mass average velocity v of the mixture as well as the molar average velocity v^* of the mixture are equal to zero. Therefore, mass transfer occurs by diffusion only. Consider a differential control volume $dxdydz$ as shown in Fig. 12.2.

Moles of A entering the left face per unit time

$$= J_{Ax}^* dydz$$

Moles of A leaving the right face per unit time

$$= J_{Ax}^* dydz + \frac{\partial}{\partial x}(J_{Ax}^* dydz)dx$$

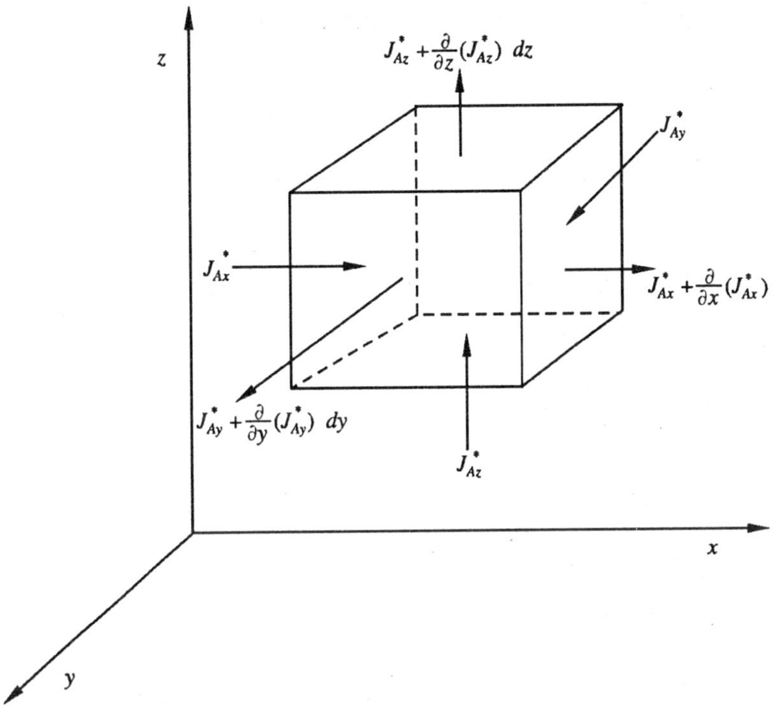

Fig. 12.2 *Control volume for deriving mass diffusion equation.*

Net rate of accumulation of A by diffusion in the x-direction

$$= -\frac{\partial}{\partial x}(J^*_{Ax})dxdydz$$

Similarly, the net rate of accumulation of A by diffusion in the y-direction

$$= -\frac{\partial}{\partial y}(J^*_{Ay})dxdydz$$

Net rate of accumulation of A by diffusion in the z-direction

$$= -\frac{\partial}{\partial z}(J^*_{Az})dxdydz$$

The rate of generation of A due to chemical reaction within the control volume

$$= \dot{G}_A dxdydz$$

where $\dot{G}_A$ = rate of production of A per unit volume.

The rate of change of concentration of A in the control volume

$$= \frac{\partial c_A}{\partial t} dxdydz$$

Substituting the above terms in the mass balance equation, we get

$$\frac{\partial c_A}{\partial t} dxdydz = -\left[\frac{\partial}{\partial x}(J^*_{Ax}) + \frac{\partial}{\partial y}(J^*_{Ay}) + \frac{\partial}{\partial z}(J^*_{Az})\right] dxdydz + \dot{G}_A dxdydz$$

or

$$\frac{\partial c_A}{\partial t} = -\left[\frac{\partial}{\partial x}(J^*_{Ax}) + \frac{\partial}{\partial y}(J^*_{Ay}) + \frac{\partial}{\partial z}(J^*_{Az})\right] + \dot{G}_A \tag{12.37}$$

Substituting Fick's law of diffusion given by Eq. (12.23) in Eq. (12.37), we get

$$\frac{\partial c_A}{\partial t} = D_{AB}\left[\frac{\partial}{\partial x}\left(\frac{\partial c_A}{\partial x}\right) + \frac{\partial}{\partial y}\left(\frac{\partial c_A}{\partial y}\right) + \frac{\partial}{\partial z}\left(\frac{\partial c_A}{\partial z}\right)\right] + \dot{G}_A$$

or

$$\frac{1}{D_{AB}}\frac{\partial c_A}{\partial t} = \frac{\partial^2 c_A}{\partial x^2} + \frac{\partial^2 c_A}{\partial y^2} + \frac{\partial^2 c_A}{\partial z^2} + \frac{\dot{G}_A}{D_{AB}}$$

In the absence of any chemical reaction, the above equation reduces to

$$\frac{1}{D_{AB}}\frac{\partial c_A}{\partial t} = \frac{\partial^2 c_A}{\partial x^2} + \frac{\partial^2 c_A}{\partial y^2} + \frac{\partial^2 c_A}{\partial z^2} \tag{12.38}$$

which is analogous to the heat conduction equation

$$\frac{1}{\alpha}\frac{\partial T}{\partial t} = \frac{\partial^2 T}{\partial x^2} + \frac{\partial^2 T}{\partial y^2} + \frac{\partial^2 T}{\partial z^2} \tag{4.7}$$

12.6 Transient Diffusion in a Semi-infinite Medium

Consider the semi-infinite medium shown in Fig. 12.3, which is initially at a uniform concentration c_{Ai} of species A. If the concentration of A at the surface is suddenly changed to c_{Ao} and held constant at that value, the species diffuses into the medium and the concentration of A changes with time t and location inside the medium. We would like to determine the concentration profile in the medium as a function of time.

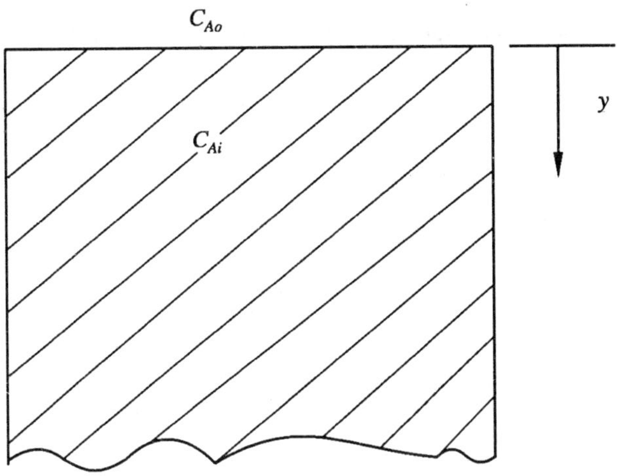

Fig. 12.3 Transient diffusion in a semi-infinite medium.

The differential equation for the concentration distribution of species A is given by

$$\frac{1}{D_{AB}}\frac{\partial c_A}{\partial t} = \frac{\partial^2 c_A}{\partial y^2} \tag{12.39}$$

The initial and boundary conditions are

$$\text{At} \quad t = 0; \quad c_A = c_{Ai} \quad \text{for all } y \tag{12.40}$$

$$\text{At} \quad t > 0 \quad \text{and} \quad y = 0 \quad c_A = c_{Ao} \tag{12.41}$$

Equations (12.39)–(12.41) are similar to Eqs. (5.96)–(5.98) for transient heat flow in a semi-infinite solid. The solution of Eq. (12.39) is given by

$$\frac{c_A - c_{Ao}}{c_{Ai} - c_{Ao}} = \text{erf}\left(\frac{y}{2\sqrt{D_{AB}t}}\right) \tag{12.42}$$

2.7 Steady-State Diffusion through a Plane Membrane

The steady-state diffusion of a species through a plane membrane is analogous to that of steady-state heat conduction through a plane wall. The differential equation which is applicable to this situation is given by

$$\frac{d^2 c_A}{dy^2} = 0 \tag{12.43}$$

This equation is to be solved subject to the boundary conditions

$$c_A = c_{A1} \quad \text{at} \quad y = 0 \tag{12.44}$$

$$c_A = c_{A_2} \quad \text{at} \quad y = \ell$$

The solution of Eq. (12.43) is given by

$$c_A = c_{A1} + (c_{A2} - c_{A1})\frac{y}{\ell} \tag{12.45}$$

The rate of mass transfer through the membrane is given by

$$J_A^* = -D_{AB}\frac{dc_A}{dy} = -D_{AB}\frac{d}{dy}\left[c_{A1} + (c_{A2} - c_{A1})\frac{y}{\ell}\right)$$

$$= D_{AB}\left(\frac{c_{A1} - c_{A2}}{\ell}\right) = \frac{c_{A1} - c_{A2}}{(\ell/D_{AB})} \tag{12.46}$$

The quantity $\left(\dfrac{\ell}{D_{AB}}\right)$ can be called the diffusional resistance.

Composite membranes in series and in parallel can be treated in a manner similar to thermal or electrical resistances in series and in parallel, by making use of the diffusional resistance concept.

12.8 Diffusion through a Stagnant Gas Film

Consider the evaporation of liquid A into a gas B which is assumed to be insoluble in liquid A. Let us assume that the liquid level is held constant at $y = y_1$ as shown in Fig. 12.4. Let x_{A1} denote the mole fraction of component A in the gas phase at the liquid-gas interface. The mole fraction of component A in the gas phase at the top of the tube $(y = y_2)$ is denoted as x_{A2}.

At steady-state, there is no net motion of A across any plane and gas B is stationary. Then the molar diffusion flux N_A, relative to the stationary coordinate axis is given by [see Eq. (A) of Example 12.1]

$$N_A = J_A^* + Nx_A = J_A^* + x_A(N_A + N_B)$$

or

$$N_A = J_A^* + x_A N_A = -cD_{AB}\frac{\partial x_A}{\partial y} + x_A N_A$$

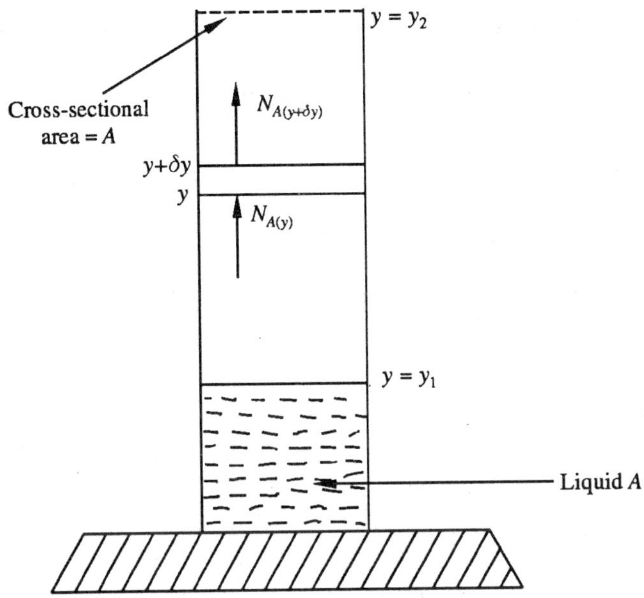

Fig. 12.4 *Diffusion of A through a stagnant gas film.*

or $$N_A = -\frac{cD_{AB}}{1 - x_A}\frac{dx_A}{dy} \tag{12.47}$$

since $N_B = 0$ as the gas B is stationary.

Consider a differential column of height δy at height y as shown in Fig. 12.4. At steady-state, a mass balance on the differential column gives

$$AN_A(y) - AN_A(y + \delta y) = 0$$

or

$$\frac{-dN_A}{dy} = 0 \tag{12.48}$$

From Eqs. (12.47) and (12.48), we get

$$\frac{d}{dy}\left(\frac{cD_{AB}}{1 - x_A}\frac{dx_A}{dy}\right) = 0 \tag{12.49}$$

Let us assume that the system is held at constant temperature and pressure and the gas phase behaves like an ideal gas. At constant temperature and pressure, c is a constant for an ideal gas mixture. Then Eq. (12.49) reduces to

$$\frac{d}{dy}\left(\frac{1}{1 - x_A}\frac{dx_A}{dy}\right) = 0 \tag{12.50}$$

Integrating Eq. (12.50) we get

$$\frac{1}{1-x_A}\frac{dx_A}{dy} = C_1$$

and

$$-\ln(1-x_A) = C_1 y + C_2 \tag{12.51}$$

The constants C_1 and C_2 can be determined from the following boundary conditions:

$$\text{At} \quad y = y_1;\ x_A = x_{A1} \tag{12.52}$$

$$\text{At} \quad y = y_2;\ x_A = x_{A_2} \tag{12.53}$$

From Eqs. (12.51) to (12.53), we get

$$C_1 = \frac{\ln x_{B1} - \ln x_{B2}}{y_2 - y_1} = \frac{\ln(x_{B1}/x_{B2})}{y_2 - y_1} \tag{12.54}$$

$$C_2 = \frac{y_1 \ln x_{B2} - y_2 \ln x_{B1}}{y_2 - y_1} \tag{12.55}$$

Equation (12.55) can be rewritten as

$$C_2 = \frac{y_1 \ln x_{B2} - (y_2 - y_1)\ln x_{B2} - y_1 \ln x_{B1}}{y_2 - y_1}$$

or

$$C_2 = -\frac{y_1}{y_2 - y_1}\ln(x_{B1}/x_{B2}) - \ln x_{B1} \tag{12.56}$$

Substituting the values of C_1 and C_2 from Eqs. (12.54) and (12.56) in Eq. (12.51), we get

$$-\ln(1-x_A) = \frac{y}{y_2 - y_1}\ln(x_{B1}/x_{B2}) - \frac{y_1}{y_2 - y_1}\ln(x_{B1}/x_{B2}) - \ln x_{B1}$$

or

$$\ln x_{B1} - \ln x_B = \left(\frac{y - y_1}{y_2 - y_1}\right)\ln(x_{B1}/x_{B2})$$

or $$\ln\left(\frac{x_{B1}}{x_B}\right) = \ln\left[\left(\frac{x_{B1}}{x_{B2}}\right)^{\frac{y-y_1}{y_2-y_1}}\right]$$

or

$$\frac{x_B}{x_{B1}} = \left(\frac{x_{B2}}{x_{B1}}\right)^{\frac{y-y_1}{y_2-y_1}} \tag{12.57}$$

or

$$\left(\frac{1-x_A}{1-x_{A1}}\right) = \left(\frac{1-x_{A2}}{1-x_{A1}}\right)^{\frac{y-y_1}{y_2-y_1}} \tag{12.58}$$

Then the rate of evaporation of A or the rate of mass transfer can be obtained from

$$[N_A]_{y=y_1} = \left[\frac{-cD_{AB}}{1-x_{A1}}\frac{dx_A}{dy}\right]_{y=y_1} = \left[\frac{cD_{AB}}{x_{B1}}\frac{dx_B}{dy}\right]_{y=y_1}$$

or

$$[N_A]_{y=y_1} = \frac{cD_{AB}}{(y_2-y_1)}\ln\left(\frac{x_{B2}}{x_{B1}}\right) \tag{12.59}$$

It can be shown that the average concentration of B in the region between $y = y_1$ and $y = y_2$ is given by

$$x_{B,\text{avg}} = \frac{x_{B2}-x_{B1}}{\ln(x_{B2}/x_{B1})} = (x_B)_{\ln} \tag{12.60}$$

That is, the average value of x_B is equal to the log mean concentration of B, $(x_B)_{\ln}$. The rate of mass transfer $[N_A]_{y=y_1}$ can also be expressed in terms of the log mean concentration of B as

$$[N_A]_{y=y_1} = \frac{CD_{AB}}{(y_2-y_1)}\frac{(x_{A1}-x_{A2})}{(x_B)_{\ln}} \tag{12.61}$$

EXAMPLE 12.3 Carbon tetrachloride contained in a glass tube of 30 cm diameter is exposed to dry air at 30°C. The liquid level is 30 cm below the open end of the tube. At 30°C the diffusion coefficient between CCl_4 and air is 6×10^{-6} m^2/s and the saturation pressure of CCl_4 is 138 Torr. Calculate the partial pressure of CCl_4 at an elevation 10 cm above the liquid surface and the mass of CCl_4 evaporated in one hour.

SOLUTION

At $y_1 = 0;\ \ p_{A1} = 138$ Torr

or $x_{A1} = p_{A1}/P = \dfrac{138}{760} = 0.1816$

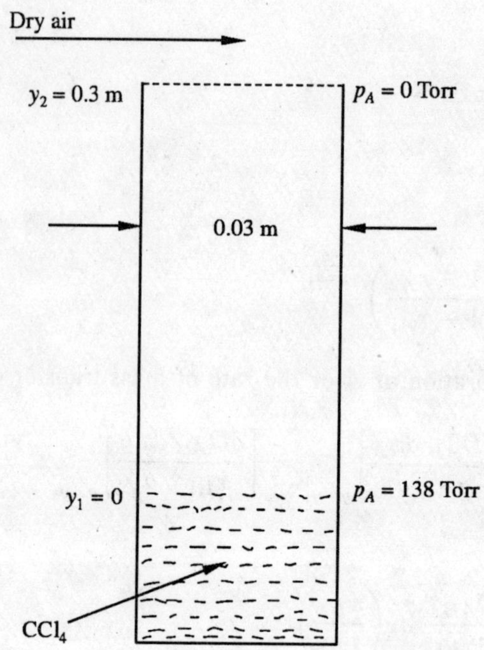

or

$$x_{B1} = 0.8184$$

At $y_2 = 0.3$ m; $p_2 = 0$ Torr

or

$$x_{A2} = 0 \quad \text{or} \quad x_{B2} = 1$$

The concentration profile in the tube is given by

$$\frac{x_B}{x_{B1}} = \left(\frac{x_{B2}}{x_{B1}}\right)^{\frac{y-y_1}{y_2-y_1}}$$

or

$$x_B \text{ (at } y = 0.1 \text{ m)} = 0.8184\left(\frac{1}{0.8184}\right)^{\frac{0.1}{0.3}} = 0.8749$$

or

$$x_A(y = 0.1\,m) = 1 - 0.8749 = 0.1251$$

$$p_A = Px_A = 760 \times 0.1251 = 95.08 \text{ Torr}$$

The rate of evaporation is given by

$$N_A = \frac{cD_{AB}}{(y_2 - y_1})\ln\left(\frac{x_{B2}}{x_{B1}}\right) = \frac{PD_{AB}}{RT(y_2 - y_1)}\ln\left(\frac{x_{B2}}{x_{B1}}\right)$$

$$= \frac{101.325 \times 10^3 \times 6 \times 10^{-6}}{8.314 \times 373 \times 0.3}\ln\left(\frac{1}{0.8184}\right) = 1.3096 \times 10^{-4} \text{ mol/s}$$

$$= 1.3096 \times 10^{-4} \times 154 \times 10^{-3} \times 3600 = 0.0726 \text{ kg/m}^2 \text{ hr}$$

(Molar mass of CCl_4 = 154 kg/kmol)

Cross-sectional area of tube = $A = \frac{\pi}{4}(0.03)^2 = 7.068 \times 10^{-4}\text{m}^2$

Mass of CCl_4 evaporated in one hour = $7.068 \times 10^{-4} \times 0.0726 = 0.513 \times 10^{-4}$ kg $= 0.051$ g

12.9 Mass Transfer Coefficient

In the study of heat transfer, the rate of heat flow is expressed in terms of the convection heat transfer coefficient by the equation

$$q = hA(T_w - T_\infty)$$

In an analogous way we can define mass transfer coefficient by the equation

$$J_A^* = k(c_{A1} - c_{A2}) \tag{12.62}$$

where

J_A^* = molar flux of species A

k = mass transfer coefficient

c_A = concentration of species A.

We have observed that, for steady-state diffusion through a plane membrane

$$J_A^* = \frac{c_{A1} - c_{A2}}{(\ell/D)} \tag{12.46}$$

Comparing Eqs. (12.46) and (12.62), we find

$$k = D/\ell \tag{12.63}$$

and the units of mass transfer coefficient k are m/s.

12.10 Mass Transfer by Convection–Laminar Flow Parallel to a Flat Plate

Consider steady, two-dimensional flow of an incompressible fluid over a flat plate of soluble substance A as shown in Fig. 12.5. The fluid flowing over the flat plate contains a momentum boundary layer in which the velocity of the fluid changes from zero at the surface of the plate to the free stream velocity at the edge of the boundary layer. If the substance A in the plate is soluble in the fluid, a concentration boundary layer also exists in which the concentration of the soluble substance changes from a constant value at the surface to the free stream concentration in the fluid. We have learnt that the momentum boundary layer thickness is different from the thermal boundary layer thickness. Similarly, it can be concluded that the concentration boundary layer thickness will be different from the momentum boundary layer thickness. In the same flow system, it is possible to have simultaneous heat and mass transfer. In such a case there will be three boundary layers—momentum boundary layer, thermal boundary layer and concentration boundary layer. The thicknesses of these boundary layers may differ from each other.

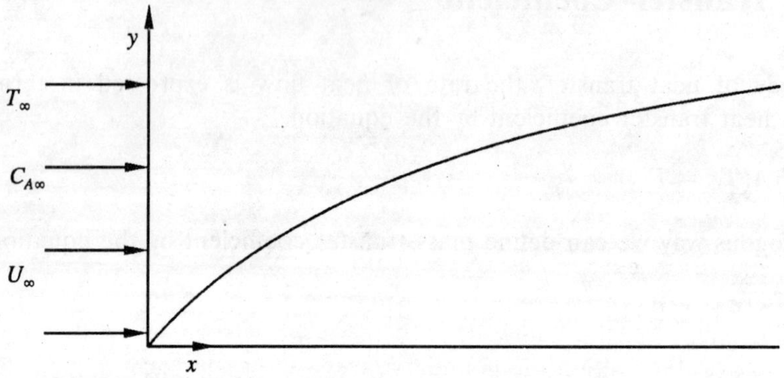

Fig. 12.5 Laminar flow over a flat plate.

The boundary layer equations are given by

Continuity equation:

$$\frac{\partial u}{\partial x} + \frac{\partial v}{\partial y} = 0 \tag{8.12}$$

Momentum equation:

$$u\frac{\partial u}{\partial x} + v\frac{\partial u}{\partial y} = \nu\frac{\partial^2 u}{\partial y^2} \tag{8.17}$$

Energy equation

$$u\frac{\partial T}{\partial x} + v\frac{\partial T}{\partial y} = \alpha\frac{\partial^2 T}{\partial y^2} \tag{8.30}$$

Similarly, we can obtain the continuity equation for species A as

$$u\frac{\partial c_A}{\partial x} + v\frac{\partial c_A}{\partial y} = D\frac{\partial^2 c_A}{\partial y^2} \tag{12.64}$$

The boundary conditions are:

At $y = 0 : u = 0$ and $v = 0$

At $y \to \infty : u = u_\infty$

$$\text{At } y = 0 : T = T_w \quad \text{or} \quad \frac{T - T_w}{T_\infty - T_w} = 0 \tag{8.49}$$

$$\text{At } y \to \infty : T = T_\infty \quad \text{or} \quad \frac{T - T_w}{T_\infty - T_w} = 1 \tag{8.50}$$

$$\text{At } y = 0 : c_A = c_{A0} \quad \text{or} \quad \frac{c_A - c_{A0}}{c_{A\infty} - c_{A0}} = 0 \tag{12.65}$$

$$\text{At } y \to \infty : c_A = c_{A\infty} \quad \text{or} \quad \frac{c_A - c_{A0}}{c_{A\infty} - c_{A0}} = 1 \tag{12.66}$$

Note the similarity between Eqs. (8.17), (8.30) and (12.64) and the boundary conditions. Therefore if $Pr = (\nu/\alpha) = 1$, the solutions to the differential equations (8.17) and (8.30) must be the same. In other words, when $Pr = 1$ the velocity and temperature profiles within the boundary layer will be identical. Similarly if $\nu/D = 1$, the velocity and concentration profiles within the boundary layer will be identical. The dimensionless quantity ν/D is called the Schmidt number.

$$Sc = \frac{\nu}{D} = \frac{\mu}{\rho D_{AB}} \tag{12.67}$$

The Schmidt number plays a role similar to that of the Prandtl number in convection heat transfer, and the Schmidt number is important in problems dealing with convection and mass transfer. When $Pr = Sc = 1$, the solutions to the differential equations (8.17), (8.30) and (12.64) must be the same. That is, the velocity, temperature and concentration profiles in a laminar boundary layer will be same.

While dealing with convection heat transfer, the heat transfer coefficient for flow over a flat plate is expressed as

$$Nu_x = \frac{hx}{k} = f(Re_x, Pr)$$

Similarly, in convection mass transfer the functional relationship is given by

$$\frac{kx}{D} = f(Re_x, Sc) \tag{12.68}$$

The dimensionless quality $\frac{kx}{D}$ is called the Sherwood number.

$$Sh = \frac{kx}{D} \tag{12.69}$$

The local mass transfer coefficient for laminar boundary layer flow over a flat plate is given by the correlation

$$Sh_x = \frac{k_x x}{D} = 0.332(Re_x)^{0.8}(Sc)^{0.33} \tag{12.70}$$

For a turbulent boundary layer flow over a flat plate, the local mass transfer coefficient can be estimated by the correlation

$$Sh_x = \frac{k_x x}{D} = 0.0288(Re_x)^{0.8}(Sc)^{0.33} \tag{12.71}$$

The average value of the mass transfer coefficient is given by

$$\overline{k} = 2k_{x=\ell} \tag{12.72}$$

The temperature and concentration profiles will be similar if $\alpha = D$ or $\alpha/D = 1$. The ratio α/D is called the Lewis number.

$$Le = \frac{\alpha}{D} = \frac{Sc}{Pr} \tag{12.73}$$

The similarities between the governing equations for momentum, heat and mass transfer suggest that the empirical correlations for the mass transfer coefficient will be similar to those for the heat transfer coefficient. Gilliland and Sherwood proposed the following empirical correlation for mass transfer coefficient for turbulent flow through tubes:

$$Sh = 0.023(Re)^{0.83}(Sc)^{0.44} \tag{12.74}$$

Equation (12.71) is valid for $2000 < Re < 35,000$ and $0.6 < Sc < 2.5$. This equation is similar to the Dittus–Boelter equation (9.50).

The Reynold's analogy for flow through a tube is given by

$$StPr^{2/3} = f/8 \tag{8.179}$$

This Reynolds analogy may be extended to mass transfer problems to predict the mass transfer coefficient in terms of the friction factor. The analogy is given by

$$\frac{k}{u}Sc^{2/3} = f/8 \tag{12.75}$$

EXAMPLE 12.4 Dry air at 20°C (ρ = 1.2 kg/m^3; ν = 15 × 10^{-6} m^2/s; D = 4.2× 10^{-5} m^2/s) flows over a flat plate of length 50 cm, which is covered with a thin layer of water, at a velocity of 1 m/s. Estimate the local mass transfer coefficient at a distance of 10 cm from the leading edge and the average mass transfer coefficient.

SOLUTION

$$Re_{x=0.1} = \frac{\rho u x}{\mu} = \frac{ux}{\nu} = \frac{1 \times 0.1}{15 \times 10^{-6}} = 6667$$

$$Sc = \frac{\nu}{D} = \frac{15 \times 10^{-6}}{4.2 \times 10^{-5}} = 0.357$$

Since $Re_x < 5 \times 10^5$, the flow is in the laminar region. For laminar boundary layer flow, the local mass transfer coefficient can be estimated from the relation

$$Sh_x = \frac{k_x x}{D} = 0.332 Re_x^{0.8} Sc^{0.33}$$

or

$$\frac{k_x(0.1)}{4.2 \times 10^{-5}} = 0.332(6667)^{0.8}(0.357)^{0.33} = 270.8$$

or

$$k_x = \frac{270.8 \times 4.2 \times 10^{-5}}{0.1} = 0.114 \text{ m/s}$$

The average mass transfer coefficient can be estimated from the relation

$$\bar{k} = 2k_{x=\ell}$$

$$Sh_{x=\ell} = \frac{k_\ell \ell}{D} = 0.332 Re_\ell^{0.8} Sc^{0.33}$$

or

$$\frac{k_\ell(0.5)}{4.2 \times 10^{-5}} = 0.332\left(\frac{u\ell}{\nu}\right)^{0.8}(0.357)^{0.33}$$

or

$$k_\ell = \frac{4.2 \times 10^{-5} \times 0.332}{0.5}\left(\frac{1 \times 0.5}{15 \times 10^{-6}}\right)^{0.8} (0.357)^{0.33} = 0.082 \text{ m/s}$$

Therefore

$$\bar{k} = 2 \times 0.082 = 0.164 \text{ m/s}$$

2.11 Summary

The movement of a chemical species from a region of high concentration to a region of low concentration is called diffusion, and the process of transfer of mass as a result of concentration gradient is called mass transfer.

Thermodynamics tells us that at constant temperature and pressure, the Gibbs free energy of a system reaches a minimum value in a state of equilibrium. Application of this principle leads to the criterion for chemical equilibrium as the equality of chemical potential. That is, the chemical potential acts as a driving force for diffusion or mass transfer. The phenomenological coefficient for mass transfer is related to the diffusion coefficient.

Fick's law of diffusion gives

$$J_A^* = -D_{AB}\frac{\partial c_A}{\partial y}$$

where

J_A^* = molar flux of species A, relative to local molar average velocity

D_{AB} = diffusion coefficient

$\frac{\partial c_A}{\partial y}$ = concentration gradient.

The units of the diffusion coefficient are m^2/s

Fick's law of diffusion is analogous to Newton's law of viscosity and Fourier's law of heat conduction.

The general mass diffusion equation in a stationary medium is given by

$$\frac{1}{D_{AB}}\frac{\partial c_A}{\partial t} = \frac{\partial^2 c_A}{\partial x^2} + \frac{\partial c_A}{\partial y^2} + \frac{\partial^2 c_A}{\partial z^2}$$

which is analogous to the general heat conduction equation.

The convection mass transfer coefficient k is defined by the equation

$$J_A^* = k(c_{A1} - c_{A2})$$

in a manner which is analogous to the definition of the convection heat transfer coefficient. The units of mass transfer coefficient are m/s.

For a fluid flowing over a flat plate, there will be a momentum boundary layer, a thermal boundary layer and concentration boundary layer. The thicknesses of these boundary layers will be different. For a laminar boundary layer, the governing equation for momentum, energy and mass transfer are similar and they are given by

Momentum:

$$u\frac{\partial u}{\partial x} + v\frac{\partial u}{\partial y} = \nu\frac{\partial^2 u}{\partial y^2}$$

Energy:

$$u\frac{\partial T}{\partial x} + v\frac{\partial T}{\partial y} = \alpha\frac{\partial^2 T}{\partial y^2}$$

Continuity of species A:

$$u\frac{\partial c_A}{\partial x} + v\frac{\partial c_a}{\partial y} = D\frac{\partial^2 c_A}{\partial y^2}$$

If $\nu = \alpha$ or $Pr = \frac{\nu}{\alpha} = 1$, the velocity and temperature profiles in the boundary layer are similar. If $\nu = D$ or $SC = \nu/D = 1$, the velocity and concentration profiles in the boundary layer will be similar. If $\alpha = D$ or the Lewis number $Le = \alpha/D = 1 = Sc/Pr$, then the temperature and concentration profiles in the boundary layer will be similar. When $Pr = Sc = 1$, the velocity, temperature and concentration profiles in the boundary layer will be similar.

In convection mass transfer, the local mass transfer coefficient for flow over a flat plate is expressed by the functional relationship

$$\text{Sherwood Number } Sh_x = \frac{kx}{D} = f(Re_x, Sc)$$

The useful correlations for flow over a flat plate are

$$Sh_x = 0.332 Re_x^{0.8} Sc^{0.33} \quad \text{for laminar flow}$$

$$Sh_x = 0.0288 Re_x^{0.8} Sc^{0.33} \quad \text{for turbulent flow}$$

and the average value $\bar{k}$ of the mass transfer coefficient is

$$\bar{k} = 2k_{x=\ell}$$

For turbulent flow through tubes, the Gilliland and Sherwood correlation is

$$Sh = 0.023 Re^{0.83} Sc^{0.44}$$

which is valid for $2000 < Re < 35,000$ and $0.6 < Sc < 2.5$. The Reynolds analogy for predicting the mass transfer coefficient in terms of the friction factor is given by

$$\frac{k}{u} Sc^{2/3} = f/8$$

Review Questions

1. What is meant by diffusion?
2. At a microscopic level, which property or characteristic of the molecules of a species leads to the diffusion of that species?
3. Explain the fact that the diffusivity D_{AB} increases with increasing temperature.
4. What is the driving force for mass transfer?
5. Consider the presence of liquid and vapor phases, each phase containing alcohol and water, in equilibrium at a specified temperature and pressure. Will the concentration of alcohol in both the phases be identical? Is there any net diffusion of alcohol from one phase to the other?
6. What is the thermodynamic criterion for chemical equilibrium between two phases?
7. Define local mass average velocity and local molar average velocity.
8. Define mass flux and molar flux.
9. State Fick's law of diffusion.
10. What are the SI units of diffusivity?
11. What is the phenomenological coefficient for mass transfer and how is it related to diffusivity?
12. Define diffusional resistance.
13. Define the mass transfer coefficient.
14. What are the SI units of mass transfer coefficient?
15. What is meant by concentration boundary layer for flow of a fluid over a soluble flat plate? Specify the boundary conditions.
16. What is the continuity equation for a species which diffuses into the fluid from the surface of a flat plate over which a fluid flows?
17. Define Schmidt number and state its physical significance.
18. Define Lewis number and state its physical significance.
19. Define Sherwood number.

20. For flow of a fluid over a flat plate, what conclusions can be drawn if the thermal diffusivity α is equal to the mass diffusivity D ?
21. Consider the flow of fluid over a flat plate. What conclusions can be drawn if the viscosity and mass diffusivity are equal?

Problems

12.1 The diffusivity of CCl_4 in air is determined by observing the steady state evaporation of CCl_4 in a tube of 1 cm diameter exposed to air. The CCl_4 liquid level is 10 cm below the top level of the tube. The system is held at 25°C and 1 bar pressure. The saturation pressure of CCl_4 at 25°C is 14.76 kPa. If it is observed that the rate of evaporation of CCl_4 is 0.1 g per hour, determine the diffusivity of CCl_4 into air.

12.2 Methanol at 30°C and 1 bar pressure is held in a glass tube of 2 cm diameter. The liquid level is 10 cm below the top of the tube and the tube is exposed to air. If the saturation pressure of methanol at 30°C is 21.86 kPa and the diffusivity of methanol in air is 0.6×10^{-4} m²/s, calculate the rate of evaporation of methanol in g/hr.

12.3 In Section 12.8, the rate of evaporation [Eq. (12.59)] was obtained by differentiating the concentration profile given by Eq. (12.57). Show that the same result can be obtained by directly integrating Eq. (12.47) between the limits y_1 and y_2.

12.4 Liquid methanol is placed in a long glass tube of 1 cm diameter and air is blown past the top of the tube. The tube is maintained at 1 bar pressure. At steady-state, it is found that at the liquid air interface, the partial pressure of methanol is 20 kPa and the top of the tube is 10 cm above the liquid level. Estimate the partial pressure of methanol at 5 cm above the liquid level.

12.5 Consider steady-state diffusion through a plane membrane. Suppose the diffusivity D_{AB} varies with concentration as $D_{AB} = D_o(1+kc_A)$, where D_o and k are constants. Solve the differential equation (12.23) with the boundary conditions given by (12.44) and show the profiles for positive and negative values of k.

12.6 A long porous cylinder of radius r_i is concentrically placed inside another cylinder of radius r_o. The annulus is filled with a liquid, and a salt solution is placed inside the porous cylinder. The salt diffuses through the porous cylinder into the liquid in the annulus. At steady-state, the concentration of salt at r_i is c_i and at r_o is c_o. Derive an expression to determine the concentration profile in the annulus.

12.7 A natural gas helium mixture contained in a semi-permeable tube of radius 10 cm is concentrically placed inside another tube of radius 30 cm, and the annulus is filled with argon. The walls of the semi-permeable tube allow only helium to pass through but do not allow argon to flow through. The diffusivity of helium through argon is 0.8×10^{-4} m²/s. At steady, state, if the concentration of helium at r = 10 cm is 0.01 mol/m^3 and at r = 30 cm is 0.004 mol/m^3, calculate the rate at which helium diffuses through argon, per meter length of tube.

12.8 A flat plate of 1 m length is covered with a thin layer of methanol. Dry air at 20°C (ρ = 1.2 kg/m^3; ν = 15 × 10^{-6} m^2/s; D = 0.16 × 10^{-4} m^2/s) flows over the plate at a velocity of 1 m/s. Estimate the local mass transfer coefficient at a distance of 20 cm from the leading edge.

12.9 Estimate the average mass transfer coefficient for Problem 12.8.

12.10 The floor of a hall of size 10 m × 10 m is covered with a layer of water at 20°C. If dry air at 20°C (ρ = 1.2 kg/m^3; ν = 15 × 10^{-6} m^2/s and D = 4.2× 10^{-5} m^2/s) is blown over the floor at a velocity of 0.4 m/s, determine the average mass transfer coefficient and the mass of water evaporated from the floor in one hour. The saturation pressure of water at 20°C is 2.336 kPa.

12.11 The hardening of low carbon steel is usually done at high temperature by the process of carburization. In this process, the steel is maintained at a high temperature while the top surface is exposed to a constant concentration of carbon. The carbon diffuses through the steel. In a particular process the top surface of a steel slab which is at 1000°C is exposed to a carbon concentration of 10 mole per cent. The initial carbon concentration in the steel is 0.2 per cent. If the diffusivity of carbon in steel is 6 × 10^{-10} m^2/s, estimate the time required to raise the carbon concentration to 1.5 per cent at a depth of 2 mm.

Appendices

Table A1 *Properties of some selected metals at 20°C*

Metal	ρ (kg/m^3)	c_p (kJ/kg K)	k (W/m K)	$\alpha \times 10^5$ (m^2/s)
Aluminium	2707	0.896	204	8.418
Duralumin	2787	0.883	164	6.676
Silumin	2659	0.867	137	5.933
Lead	11373	0.130	35	2.343
Iron	7897	0.452	73	2.034
Wrought-iron	7849	0.46	59	1.626
Carbon-steel	7833	0.465	54	1.474
Nickel-steel	7933	0.46	19	0.526
Invar	8137	0.46	10.7	0.286
Chrome-steel	7833	0.46	40	1.110
Chrome-Nickel	7865	0.46	19	0.527
Tungsten-steel	8073	0.435	54	1.525
Copper	8954	0.3831	386	11.234
Bronze	8666	0.343	26	0.859
Brass	8522	0.385	111	3.412
German silver	8618	0.394	24.9	0.733
Constantan	8922	0.410	22.7	0.612
Magnesium	1746	1.013	171	9.708
Molybdenum	10220	0.251	123	4.790
Nickel	8906	0.446	90	2.266
Silver	10525	0.234	407	16.563
Tin	7304	0.2265	64	3.884
Tungsten	19350	0.1344	163	6.271
Zinc	7144	0.3843	112.2	4.106

Table A2 Properties of a few selected insulating and building materials

Material	ρ (kg/m^3)	c_p (kJ/kg K)	k (W/m K)	$\alpha \times 10^7$ (m^2/s)
Asphalt	2110	2.093	0.6978	
Asbestos	470	0.816	0.1105	
Brick, Masonry	1600	0.84	0.69	5.2
Brick, Fireclay	2000	0.96	1.04	5.4
Cement, Portland	1900	1.13	0.3024	
Concrete	1900–2300	0.88	1.37	8.2–6.8
Cork	45–120	1.88	0.045	2–5.3
Earth	320		0.061	
Glass	2500	0.67	0.7443	
Glass wool	24	0.7	0.038	22.6
Plaster, gypsum	1440	0.84	0.48	
Mica	290	8.792	0.5815	
Refractory clay	1845	1.089	1.035	
Sand	1500	0.796	0.3256	
Stone				
Granite	2640	0.82	1.73–3.98	8–18
Marble	2500–2700	0.8	2.07–2.94	10–13.6
Sand stone	2160–2300	0.71	1.83	11.2–11.9
Wood: Oak	540	2.4	0.166	1.28
Leather			0.16	
Saw dust			0.0698	

Table A3 *Properties of saturated liquid water*

T (°C)	ρ (kg/m^3)	c_p (kJ/kg K)	k (W/m K)	$\alpha \times 10^6$ (m^2/s)	$\nu \times 10^6$ (N s/m^2)	Pr	h_{fg} (kJ/kg)
0	999.9	4.226	0.558	0.131	1.789	13.7	2501
5	1000	4.206	0.568	0.135	1.535	11.4	2490
10	999.7	4.195	0.577	0.137	1.300	9.5	2477
15	999.1	4.187	0.585	0.141	1.146	8.1	2466
20	998.2	4.182	0.597	0.143	1.006	7.0	2453
25	997.1	4.178	0.606	0.146	0.884	6.1	2442
30	995.7	4.176	0.615	0.149	0.805	5.4	2430
35	994.1	4.175	0.624	0.150	0.725	4.8	2419
40	992.2	4.175	0.633	0.151	0.658	4.3	2406
45	990.2	4.176	0.640	0.155	0.611	3.9	2395
50	98.1	4.178	0.647	0.157	0.556	3.55	2382
75	974.9	4.190	0.671	0.164	0.366	2.23	2321
100	958.4	4.211	0.682	0.169	0.294	1.75	2257
120	943.5	4.232	0.685	0.171	0.244	1.43	2202
140	926.4	4.257	0.684	0.172	0.212	1.23	2144
160	907.6	4.285	0.680	0.173	0.191	1.10	2082
180	886.6	4.396	0.673	0.172	0.173	1.01	2014
200	862.8	4.501	0.665	0.170	0.160	0.95	1939
220	837.0	4.605	0.652	0.167	0.149	0.90	1856
240	809.0	4.731	0.634	0.162	0.141	0.86	1764
260	779.0	4.982	0.613	0.156	0.135	0.86	1660
280	750.0	5.234	0.588	0.147	0.131	0.89	1541
300	712.5	5.694	0.564	0.132	0.128	0.98	1403

Table A4 *Properties of saturated liquid Dichlorodifluoromethane (Freon-12),* CCl_2F_2

T (°C)	ρ (kg/m^3)	c_p (kJ/kg K)	$\nu \times 10^6$ (m^2/s)	k (W/m K)	$\alpha \times 10^7$ (m^2/s)	Pr
−50	1546.75	0.8750	0.310	0.067	0.501	6.2
−40	1518.71	0.8847	0.279	0.069	0.514	5.4
−30	1489.56	0.8956	0.253	0.069	0.526	4.8
−20	1460.57	0.9073	0.235	0.071	0.539	4.4
−10	1429.49	0.9203	0.221	0.073	0.550	4.0
0	1397.5	0.9345	0.214	0.073	0.557	3.8
10	1364.30	0.9496	0.203	0.073	0.560	3.6
20	1330.18	0.9659	0.198	0.073	0.560	3.5
30	1295.10	0.9835	0.194	0.071	0.560	3.5
40	1257.13	1.0019	0.191	0.069	0.555	3.5
50	1215.96	1.0216	0.190	0.067	0.545	3.5

Table A5 *Properties of saturated liquid glycerin,* $C_3H_5(OH)_3$

T (°C)	ρ (kg/m^3)	c_p (kJ/kg K)	$\nu \times 10^6$ (m^2/s)	k (W/m K)	$\alpha \times 10^7$ (m^2/s)	Pr
0	1276.03	2.261	0.00831	0.282	0.983	84.7
10	1270.11	2.311	0.00300	0.284	0.965	31.0
20	1264.02	2.386	0.00118	0.286	0.947	12.5
30	1258.09	2.445	0.00050	0.286	0.929	5.38
40	1252.01	2.512	0.00022	0.286	0.914	2.45
50	1244.96	2.583	0.00015	0.287	0.893	1.63

Table A6 Properties of air at atmospheric pressure

T K	ρ (kg/m^3)	c_p (kJ/kg K)	$\mu \times 10^5$ (kg/m s)	$\nu \times 10^6$ (m^2/s)	k (W/m K)	$\alpha \times 10^4$ (m^2/s)	Pr
100	3.6010	1.0266	0.6924	1.923	0.009246	0.02501	0.770
150	2.3676	1.0099	1.0283	4.343	0.013735	0.05745	0.753
200	1.7684	1.0061	1.3289	7.490	0.01890	0.10165	0.739
250	1.4128	1.0053	1.5990	11.31	0.02227	0.15675	0.722
300	1.1774	1.0057	1.8462	15.69	0.02624	0.22160	0.708
350	0.9980	1.0090	2.075	20.76	0.03003	0.2983	0.697
400	0.8826	1.0140	2.286	25.90	0.03365	0.3760	0.689
450	0.7833	1.0207	2.484	31.71	0.03707	0.4222	0.683
500	0.7048	1.0295	2.671	37.90	0.04038	0.5554	0.680
550	0.6423	1.0392	2.848	44.34	0.04360	0.6532	0.680
600	0.5879	1.0551	3.018	51.34	0.04659	0.7512	0.680
650	0.5430	1.0635	3.177	58.15	0.04953	0.8578	0.682
700	0.5030	1.0752	3.332	66.25	0.05230	0.9672	0.684
750	0.4709	1.0856	3.481	73.91	0.05509	1.0774	0.686
800	0.4405	1.0978	3.625	82.29	0.05779	1.1951	0.689
850	0.4149	1.1095	3.765	90.75	0.06028	1.3097	0.692
900	0.3925	1.1212	3.899	99.3	0.06279	1.4271	0.696
950	0.3716	1.1321	4.023	108.2	0.06525	1.5510	0.699
1000	0.3524	1.1417	4.152	117.8	0.06752	1.6779	0.702

The values of μ, k, c_p and Pr are not strongly pressure dependent for air and hence they may be used over a wide range of pressures.

Table A7 Properties of Helium at atmospheric pressure

T K	ρ (kg/m^3)	c_p (kJ/kg K)	$\mu \times 10^7$ (kg/m s)	$\nu \times 10^6$ (m^2/s)	k (W/m K)	$\alpha \times 10^4$ (m^2/s)	Pr
144	0.3379	5.200	125.5	37.11	0.0928	0.5275	0.70
200	0.2435	5.200	156.6	64.38	0.1177	0.9288	0.694
255	0.1906	5.200	181.7	95.50	0.1357	1.3675	0.70
366	0.13280	5.200	230.5	173.6	0.1691	2.449	0.71
477	0.10204	5.200	275.0	269.3	0.197	3.716	0.72
589	0.08282	5.200	311.3	375.8	0.225	5.215	0.72
700	0.07032	5.200	347.5	494.2	0.251	6.661	0.72
800	0.06023	5.200	381.7	634.1	0.275	8.774	0.72

The values of μ, k, c_p and Pr are not strongly pressure dependent for Helium and hence they may be used over a wide range of pressures.

ANSWERS TO PROBLEMS

Chapter 1

1.1 69 W/m^2
1.2 228.3 W/m^2
1.3 25.4°C
1.4 0.26 m
1.5 3.02 kg
1.6 0.992 kW/m^2; 6.889 kW/m^2; 22.765 kW/m^2; 56.24 kW/m^2
1.7 264.1 kW/m^2
1.8 632.6°C
1.9 550 W
1.10 8.654 kW/m^2 K
1.11 62.978 MW/m^2
1.12 $\dfrac{T_1 - T_2}{\frac{1}{h_1 A} + \frac{\ell}{kA} + \frac{1}{h_2 A}}$
1.13 83.08 W/m^2 K; 5.4 kW/m^2
1.14 50.14°C; 15.33 W/m^{2}K
1.15 0.2 m

Chapter 2

2.1 0.032 m
2.2 $k = 39.60\ (1 - 0.0133\ t)$, where k is in W/m K
2.3 2.867 kW
2.4 1.891 kW
2.6 132.9 W
2.7 157.2 W
2.8 82.5 W/m K
2.9 4 W/m K
2.10 40
2.11 118.2°C; 36.7 W/m^2 K
2.12 3.752 kW; 111.3°C; 88.7°C; 72.6°C; 60.2°C
2.13 119°C; 118.8°C; 178.7 W
2.14 0.079 m
2.15 0.015 m
2.16 Glass wool
2.17 (a) 825.8°C; (b) 364.3°C; (c) 332.5°C
2.18 Heat loss increases by 23.18%

2.19 Heat loss increases by 21.73%
2.20 5.88 kW
2.21 $$\frac{T_i - T}{T_i - T_o} = \frac{r_o}{r}\left(\frac{r - r_i}{r_o - r_i}\right)$$
2.22 41.5 W
2.23 0.014 m
2.24 907.9°C
2.25 957.9°C
2.26 370°C; 371.2°C
2.27 0.027 m
2.28 $$T = T_o + \frac{\dot{q}r_o^2}{6k}\left[1 - \left(\frac{r}{r_o}\right)^2\right]$$
2.29 277.4°C
2.30 66.8°C

Chapter 3

3.1 $T - T_\infty = (T_s - T_\infty)e^{-mx}$
3.2 198.4 W; 456.1°C; 358.2°C; 267.4°C
3.3 (a) $T = 25 + 375e^{-3.835x}$ (b) $T = 25 + 375e^{-5.774x}$ (c) $T = 25 + 375e^{-12.566x}$
3.4 (a) 246 W (b) 77.8 W (c) 38.9 W
3.5 (a) 82.6°C (b) 44.3°C (c) 18.63 W
3.6 278 W/m K
3.7 21.84 W/m^2 K
3.8 33.25 W/m^2 K
3.9 (a) 118.1°C (b) 62.83 W
3.10 37.37
3.12 0.111 m
3.13 (a) 128.4°C (b) 20 W
3.14 0.2838
3.15 (a) 0.7824 (b) 0.7899
3.16 236.3°C
3.17 471.2°C
3.18 329.9°C
3.19 (a) $$T - T_\infty = \frac{(T_B - T_\infty)\sinh(mx) + (T_A - T_\infty)\sinh[m(\ell - x)]}{\sinh(m\ell)}$$

(b) $$q = \frac{\sqrt{hpkA}}{\sinh(m\ell)}(T_A + T_B - 2T_\infty)[\cosh(m\ell) - 1]$$

3.21 35.95 W

Chapter 4

4.1 $$\frac{\partial^2 T}{\partial r^2}+\frac{1}{r}\frac{\partial T}{\partial r}+\frac{1}{r^2}\frac{\partial^2 T}{\partial \theta^2}+\frac{\partial^2 T}{\partial z^2}+\frac{\dot{q}}{k}=\frac{1}{\alpha}\frac{\partial T}{\partial t}$$

4.2 $$T=T_1+(T_2-T_1)\frac{x}{\ell}$$

4.3 $$T=T_i+\frac{(T_o-T_i)\ln(r/r_i)}{\ln(r_o/r_i)}$$

4.4 $$T=T_o+\frac{\dot{q}}{2k}(\ell^2-x^2)$$

4.5 $$T=T_o+\frac{\dot{q}}{4k}(R^2-r^2)$$

4.6 (a) $T=(C_1+C_2x)(C_3+C_4y)$

Does not satisfy the boundary condition $T=T_1+T_m\sin(\pi x/\ell)$ at $y=b$

(b) $T=(C_1e^{-\lambda x}+C_2e^{\lambda x})(C_3\cos\lambda y+\sin\lambda y)$

Does not satisfy the boundary condition $T=T_1+T_m\sin(\pi x/\ell)$ at $y=b$

4.7 174.5 W/m

4.8 (a) $T=300+\dfrac{100\sin\left(\frac{\pi y}{b}\right)\sin h\left(\frac{\pi x}{b}\right)}{\sin h\left(\frac{\pi a}{b}\right)}+\dfrac{200\sin\left(\frac{\pi x}{a}\right)\sin h\left(\frac{\pi y}{a}\right)}{\sin h\left(\frac{\pi b}{a}\right)}$ (b) 340°C

4.12 $$T_2+T_4-2T_3\left[\frac{2h(\Delta x)^2}{kD}+L\right]+T_\infty\left[\frac{4h(\Delta x)^2}{kD}\right]=0$$

4.13 $$T_5-T_6\left[\frac{2h(\Delta x)^2}{kD}+1+\frac{h(\Delta x)}{k}\right]+T_\infty\left[\frac{2h(\Delta x)^2}{kD}+\frac{h(\Delta x)}{k}\right]=0$$

4.14 $$\frac{T_1}{a(a+1)}+\frac{T_2}{b+1}+\frac{T_3}{a+1}+\frac{T_4}{b(b+1)}-\left(\frac{1}{a}+\frac{1}{b}\right)T_0=0$$

4.15

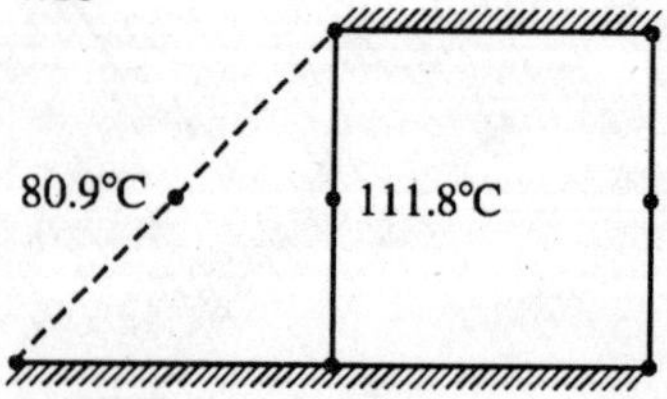

4.16

Node	1	2	3	4
Temperature (°C)	625	550	475	550

4.17

Node	1	2	3	4	5	6
Temperature (°C)	300	272.5	248.9	228.8	211.9	207.9

Chapter 5

5.1 4201 s
5.2 8361 s
5.3 (a) 489.2°C (b) 425°C (c) 369.9°C
5.4 (a) 597.6°C (b) 250.9°C
5.5 351 s
5.6 181 s
5.7 (a) 2587.4 J (b) 2482 J
5.8 2859 s
5.9 87.38 W/m^2 K
5.10 6744 s
5.11 21.2 hours
5.12 (a) 206°C; (b) 55°C
5.13 1594 s
5.14 220°C
5.15 182°C; 19.1 MJ
5.16 (a) 910 s (b) 366°C (c) 393°C
5.17 64.25 MJ
5.18 500 W/m^2 K
5.19 909 s
5.20 36 MJ/m
5.21 600 s
5.22 (a) 300°C (b) 246°C (c) 280°C (d) 21.63 MJ/m
5.23 50 W/m^2 K
5.24 619 s
5.25 192°C
5.26 200°C; 169.3°C; 153.1°C
5.27 51.9°C
5.28 18.7 s
5.29 132.9°C
5.30 305°C
5.31 48.08 MJ
5.32 25,000 s
5.33 715.2°C
5.34 321.6°C
5.35 1900 s
5.36 7000 s
5.37 57.3°C
5.38 2.658 MJ
5.39 25,000 s
5.40 622 s
5.41 150°C
5.42 23 hours
5.43 0.019 m

5.44 757.5°C
5.45 25 minutes
5.46 8 hours

Chapter 6

6.1

Distance (cm)	0	5	10	15	20	25	30
Temperature (°C)	200	158 17	123.45	88.78	70.5	52.26	52.26

6.2 0.111 m

6.3

Distance (cm)	0	3	6	9	12	15
Temperature (°C)	800	559.38	378.91	198.44	138.29	78.13
Distance (cm)	18	21	24	27	30	
Temperature (°C)	138.29	198.44	378.91	559.38	800	

6.4 $T_{m,n}^{p-1} = -(Fo)_\Delta[T_{m+1,n}^p + T_{m-1,n}^p + T_{m,n+1}^p + T_{m,n-1}^p] + [1 + 4(Fo)_\Delta]T_{m,n}^p$

where $(Fo)_\Delta = \dfrac{\alpha\Delta t}{(\Delta x)^2}$

6.5 $$T_{m,n}^{p+1} = \frac{\alpha\Delta t}{(\Delta x)^2}\left[\frac{2h\Delta x}{k}T_\infty + 2T_{m-1,n}^p + T_{m,n+1}^p + T_{m,n-1}^p + \left(\frac{(\Delta x)^2}{\alpha\Delta t} - \frac{2h\Delta x}{k} - 4\right)T_{m,n}^p\right] \quad \frac{(\Delta x)^2}{\alpha\Delta t} \geq 2\left(\frac{h\Delta x}{k} + 2\right)$$

6.6 $$T_{m,n}^{p+1} = 2(Fo)_\Delta\left[T_{m-1,n}^p + T_{m,n-1}^p + 2(Bi)T_\infty\right] + [1 - 4(Fo)_\Delta - 4(Fo)_\Delta(Bi)]\,T_{m,n}^p$$

$(Fo)_\Delta(1 + Bi) \leq 1/4$

6.7 340 s

6.8 $$T_{m,n}^{P+1} = 2(Fo)_\Delta(Bi)T_\infty - (Fo)_\Delta\left(2T_{m,n-1}^p + T_{m-1,n}^p + T_{m+1,n}^p\right) + [1 + 4(Fo)_\Delta - 2(Fo)_\Delta(Bi)]\,T_{m,n}^p$$

6.9 $$T_m^{P+1} = \frac{\alpha\Delta t}{(\Delta x)^2}\left[2T_{m-1}^p + \frac{2h\Delta x}{k}T_\infty + \left(\frac{(\Delta x)^2}{\alpha\Delta t} - \frac{2h\Delta x}{k} - 2\right)T_m^p\right]$$

$Fo \leq \dfrac{1}{2(1 + Bi)}$

6.10

Distance from surface (cm)	Temperature (°C)
0	227.24
3	163.31
6	99.38
9	73.12
12	46.86
15	46.86
18	46.86
21	73.12
24	99.38
27	163.31
30	227.24

6.11

Distance (cm)	0	2	4	6	8	10
Temperature (°C)	84.24	73.8	63.36	57.45	51.54	51.54

6.12

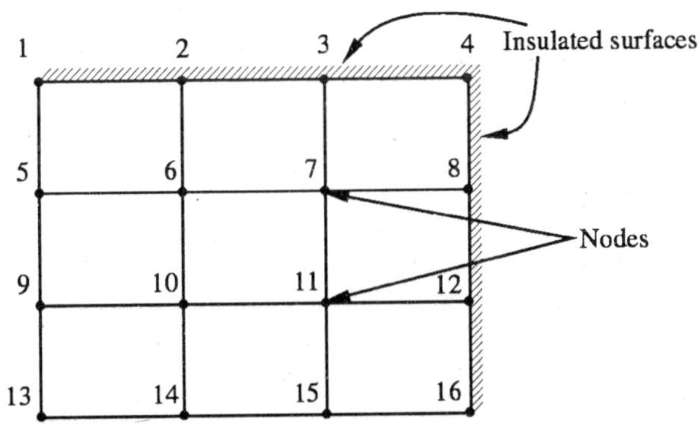

Node	1	2	3	4	5	6	7	8
Temperature (°C)	350	291.9	260.7	233.8	350	305.8	277.4	260.7
Node	9	10	11	12	13	14	15	16
Temperature (°C)	350	320.8	305.2	291.9	350	350	350	350

6.13

Node	1	2	3	4
Temperature (°C)	153.91	196.1	208.6	241.41

6.14

Node	1	2	3	4	5	6	7	8
Temperature (°C)	120.5	87.4	69.3	30	89.15	58.8	42.8	30
Node	9	10	11	12	13	14	15	16
Temperature (°C)	82.7	47	35.5	30	77.3	46	33	30

6.15

Distance (cm) from base	0	5	10	15
Temperature (°C)	350	295.9	268.8	262.1

6.16 14.85 W
6.17 600 s

Chapter 7

7.1 459.2 W/m^2; 9.659 μm; 56.69 kW/m^2; 2.898 μm; 907.04 kW/m^2; 1.449 μm
7.3 96.6%
7.4 55.4%
7.5 0.05; 0.3125
7.6 103.2 W
7.7 (a) 0.19 (b) 0.2
7.8 18.4%
7.9 (a) 0.12 (b) 0.07
7.10 1; $\left(\dfrac{r_1}{r_2}\right)^2$
7.11 $\dfrac{r_1}{r_2}; 1-\dfrac{r_1}{r_2}$
7.13 10.07 kW; 3.96 kW; 14.03 kW
7.14 $$\frac{1}{2}\sqrt{\left(\frac{d}{\ell_1}\right)^2+\left(1+\frac{\ell_2}{\ell_1}\right)^2}+\frac{d}{2\ell_1}-\frac{1}{2}\sqrt{1+\left(\frac{d}{\ell_1}\right)^2}-\frac{1}{2}\sqrt{\left(\frac{\ell_2}{\ell_1}\right)^2+\left(\frac{d}{\ell_1}\right)^2}$$
7.15 3.536 kW; 1.418 kW; 4.954 kW
7.17 418 W
7.19 3.426 kW/m^2
7.20 0.23 kg
7.21 20%
7.22 90.7
7.23 $1/(n+1)$

Chapter 8

8.1 $\dfrac{T - T_\infty}{T_i - T_\infty} = f(Bi, Fo)$

8.2 $\dfrac{T - T_\infty}{T_i - T_\infty} = f(Bi, Fo, x/\ell)$

8.3 $\dfrac{T - T_\infty}{T_i - T_\infty} = f\left(\dfrac{y}{\sqrt{\alpha t}}\right)$

8.4 9.2 mm; 14.55 mm
8.5 14.55 mm; 3.865×10^{-3}; 5.466×10^{-3}
8.6 12.98 mm; 2.7 W/m^2 K
8.7 3.81 W/m^2 K; 539.4 W/m^2

8.8 $\delta = \dfrac{3.4641x}{\sqrt{Re_x}}$

8.9 $\dfrac{u}{u_\infty} = \dfrac{3}{28}y - \dfrac{1}{2}\left(\dfrac{y}{\delta}\right)^3$

8.11 $\delta = 5.477\sqrt{\dfrac{\nu x}{u_\infty}}$

8.12 155.7 W/m^2; 168.8 W/m^2
8.13 5.014 kW/m^2 K
8.14 903.9 W/m^2 K

Chapter 9

9.1 2.63 W/m^2 K
9.2

y (cm)	0	0.25	0.42	0.5	0.75	1.0	1.254
u/u_∞	0	0.1278	0.14815	0.1442	0.0966	0.0327	0

9.3

y (cm)	0	0.25	0.5	0.75	1	1.25	1.491
u/u_∞	0	0.11616	0.14814	0.12424	0.0727	0.0219	0

9.4 0.034 m
9.5 1.407 kW
9.6 2.018 kW
9.7 3.77 W/m^2 K; 3.17 W/m^2 K; 2.66 W/m^2 K; 3.55 W/m^2 K
9.10 6.848 kW/m
9.11 3.9 W/m
9.12 129 W/m

9.13 18.17 m
9.14 442 W/m
9.15 19.35 W
9.16 662.8 W/m^2 K
9.17 44.4 W/m^2 K
9.18 902 W
9.19 1.66 m
9.20 0.19 m
9.21 12.686 kW/m^2 K
9.22 144 W/m^{2}K

Chapter 10

10.1 6.25×10^{-5} m
10.2 1.05×10^{-4} m
10.3 10.895 kW/m^2 K
10.4 7.6 kW/m^2 K; 33.56
10.5 7.5 kW/m^2 K
10.6 8.17 kW/m^2 K; 0.004 kg/s
10.7 203; Laminar
10.8 8.17 kW/m^2 K; 3.6×10^{-3} kg/s
10.9 8.91 kW/m^2 K; 3.9×10^{-3} kg/s
10.10 5.62 kW/m^2 K; 4.8×10^{-3} kg/s
10.11 109.44 kg
10.12 125.64 kg
10.13 590 kg
10.14 11.16 kg
10.15 5.85 MW
10.16 29.81 kW/m^2
10.17 107.3°C
10.18 103.4°C
10.19 0.81 kW
10.20 9.53 kW
10.21 1.11 MW/m^2
10.22 1.11 MW/m^2

Chapter 11

11.1 30.83°C
11.2 98.65°C
11.3 3.4 kW; 0.074 m^2
11.4 (a) 0.0985 m^2 (b) 0.1277 m^2
11.5 (a) 53.6°C (b) 53.6°C (c) 53.6°C
11.6 1.0

11.7 Heat exchanger 1
11.8 1.98 m^2
11.9 h_i = 485.6 W/m^2 K; h_0 = 22.95 kW/m^2 K; U_0 = 318.3 W/m^2 K
11.10 75.75°C; 60.5°C
11.11 40.55°C; 41.65°C; 745.5 kW

Chapter 12

12.1 3.565 × 10^{-6} m^2/s
12.2 0.567 g/hr
12.4 10.56 kPa

12.5 $$J_A^* = D_o \frac{(C_{A1} - C_{A2})}{\ell}\left[1 + k\frac{(C_{A1} + C_{A2})}{2}\right]$$

12.6 $$C_r = C_i - \frac{(C_i - C_o)}{\ln(r_o/r_i)}\ln(r/r_i)$$

12.7 2.745 × 10^{-6} mol/s
12.8 0.052 m/s
12.9 0.075 m/s
12.10 0.0436 m/s; 270.93 kg
12.11 1472 s

Bibliography

Arpaci, V.S., *Conduction Heat Transfer*, Addison-Wesley, Mass., USA, 1966.

Bayley, F.J., Owen, J.M. and Turner, A.B., *Heat Transfer*, Thomas Nelson & Sons, London, 1972.

Becker, M., *Heat Transfer — A Modern Approach*, Plenum Press, New York, 1986.

Bennett, C.O. and Myers, J.E., *Momentum, Heat and Mass Transfer*, McGraw-Hill, New York, 1974.

Bird, R.B., Stewart, W.E., and Lightfoot, E.N., *Transport Phenomena*, John Wiley, New York, 1960.

Carslaw, H.S. and Jaeger, J.C., *Conduction of Heat in Solids*, Oxford University Press, London, 1959.

Chapman, A.J., *Heat Transfer*, Macmillan, New York, 1974.

Eckert, E.R.G. and Drake, R.M., *Analysis of Heat and Mass Transfer*, McGraw-Hill, Kogakusha, 1972.

Holman, J.P., *Heat Transfer*, McGraw-Hill, New York, 1997.

Jakob, M., *Heat Transfer*, John Wiley, New York, 1949.

Kay, J.M., *An Introduction to Fluid Mechanics and Heat Transfer*, Cambridge University Press, New York, 1963.

Kays, W.M. and London, A.L., *Compact Heat Exchangers*, McGraw-Hill, New York, 1964.

Kern, D.Q., *Process Heat Transfer*, McGraw-Hill, New York, 1965.

Knudsen, J.D. and Katz, D.L., *Fluid Dynamics and Heat Transfer*, McGraw-Hill, New York, 1958.

Kreith, F. and Bohn, M.S., *Principles of Heat Transfer*, PWS Publishing Company, Boston, 1997.

McAdams, W.H., *Heat Transmission*, McGraw-Hill, New York, 1954.

Rohsenow, W.M. and Choi, H.Y., *Heat, Mass and Momentum Transfer*, Prentice Hall Inc., Englewood Cliffs, New Jersey, 1961.

INDEX

E

F

G

H